THE METALS RED BOOK™

Nonferrous Metals

Volume 2

The Metals Data Book Series™

Published By:

CASTI **Publishing Inc.**
14820 - 29 Street
Edmonton, Alberta, T5Y 2B1, Canada
Tel: (403) 478-1208 Fax: (403) 473-3359
E-Mail: castipub@compusmart.ab.ca
URL: http://www.casti-publishing.com

ISBN 0-9696428-1-4
Printed in Canada

Important Notice

The material presented herein has been prepared for the general information of the reader and should not be used or relied upon for specific applications without first securing competent technical advice. Nor should it be used as a replacement for current complete engineering standards. In fact, it is highly recommended that current engineering standards be reviewed in detail prior to any decision-making. See the list of standards organizations, technical societies and associations in Chapter 35, many of which prepare engineering standards, to acquire the appropriate metal standards or specifications.

While the material in this book was compiled with great effort and is believed to be technically correct, *CASTI* Publishing Inc. and its staff do not represent or warrant its suitability for any general or specific use and assume no liability or responsibility of any kind in connection with the information herein.

Nothing in this book shall be construed as a defense against any alleged infringement of letters of patents, copyright, or trademark, or as defense against liability for such infringement.

First printing, November 1993
Second printing, November 1995
ISBN 0-9696428-1-4

Dedication

The Metals Red Book™ is dedicated to Michael, Jo-Ann, Mary-Ann and Angela-Marie, my brother and sisters, for their love and support.

John E. Bringas, P.Eng.
Edmonton, Alberta

Acknowledgments

CASTI Publishing Inc. has been greatly assisted by the technical reviewers of *The Metals Red Book*™ - Nonferrous Metals, namely: The Aluminum Association, Copper Development Association, Nickel Development Institute, Titanium Development Association, INCO Ltd., and Haynes International Ltd. Grammatical editing was performed by Jade DeLang Hart. These acknowledgments cannot, however, adequately express the publisher's appreciation and gratitude for their valued assistance, patience, and advice.

Authors

The metallurgy sections of each chapter were written by Dr. Michael L. Wayman, Ph.D., P.Eng., Professor of Metallurgy, University of Alberta, Edmonton, Alberta, Canada.

The metals data in the book was researched, compiled and edited by John E. Bringas, P.Eng., Publisher and Metallurgical Engineer, *CASTI* Publishing Inc.

We Would Like To Hear From You

Our mission at CASTI Publishing Inc. is to provide industry and educational institutions with practical technical books at low cost. To do so, the book must have a valuable topic and be current with today's technology. *The Metals Red Book*™ - Nonferrous Metals, is the second volume in *The Metals Data Book Series*™, containing almost 500 pages with more than 150,000 pieces of practical metals data. Since accurate data entry of more than 150,000 numbers is contingent on normal human error, we extend our apologies for any errors that may have occurred. However, should you find errors, we encourage you to inform us so that we may keep our commitment to the continuing quality of *The Metals Data Book Series*™.

If you have any comments or suggestions we would like to hear from you:

CASTI Publishing Inc., 14820 - 29 Street
Edmonton, Alberta, T5Y 2B1, Canada
tel: (403) 478-1208, fax: (403) 473-3359
e-mail: castipub@compusmart.ab.ca

Browse Through Our Books On The Internet

You can browse through our electronic bookstore and read the table of contents and selected pages from each of our books. You can find our home page at http://www.casti-publishing.com.

Tradenames & Trademarks

To make this book as useful as possible, tradenames and trademarks have been included where feasible. Many of these names are widely known, some are more obscure. A list of tradenames and trademarks with their associated companies is catalogued in Appendix 8.

CONTENTS

CONTENTS (Continued)

Chapter 1

ALUMINUM AND ALUMINUM ALLOYS

Although aluminum compounds such as alum have been known since antiquity, aluminum metal was first isolated in minute quantities only during the first few decades of the 19th century. By mid-century aluminum had become more widely available but it was an expensive material. In 1884, when the six-pound ornamental pyramid of aluminum was placed on the top of the Washington Monument in Washington D.C., its cost was close to that of silver. However, in 1886 the simultaneous discovery by Hall in the U.S. and Héroult in France of an electrolytic smelting process for aluminum revolutionized its role in industry by making it economically attractive. The Hall-Héroult process, somewhat modified, remains to this day the only commercial aluminum smelting process.

Aluminum is the most abundant metal in the earth's crust, but its most common ore is bauxite, a hydrated oxide of aluminum, iron, and silicon. Alumina (aluminum oxide, Al_2O_3) is extracted from this ore by the Bayer process in which the bauxite is treated with hot sodium hydroxide. The resultant alumina is then dissolved in a bath of molten cryolite (sodium aluminum fluoride) and electrolyzed in electrolytic cells with carbon anodes and cathodes. The product is molten aluminum, which is degassed, alloyed if desired, and cast into ingots which are suitable for remelting or for mechanical processing, typically by rolling or extrusion. Aluminum is amenable to recycling, and the use of scrap in both primary and secondary production and in fabrication is an increasing trend.

Bauxite is mined in diverse parts of the world, but aluminum smelting is carried out in localities where electricity costs are low, as energy is more important than transportation in the economics of aluminum production. This is because aluminum is such a reactive metal. In alumina, aluminum and oxygen are so strongly bonded that a great deal of energy is required to separate them. When an atomically clean aluminum surface is exposed to air, it reacts immediately forming a thin oxide layer within seconds. Fortunately, this thin layer is normally adherent and protective, preventing further oxidation of the metal and endowing it with excellent corrosion resistance in most environments. Furthermore, the surface oxide layer can be controlled and even coloured by the anodization process, adding to its esthetic versatility. A less positive consequence of the reactivity of aluminum is that successful welding procedures require effective shielding from the air.

Aluminum and its alloys have face-centered cubic (fcc) crystal structures, and therefore have excellent formability properties, except when these are inhibited by the degree and type of alloying. The strength obtainable by heat treating some alloys can be greater than in many structural steels. The metal has an attractive surface appearance with good reflectivity, good electrical and thermal conductivity, and low density. This latter property, combined with the high strengths available, gives high strength-to-weight ratios which are responsible for these alloys being the primary structural materials in the aerospace industry. The stiffness (elastic modulus) of aluminum is only about one-third as high as that of steel, and this factor must be considered in component design in order to prevent unduly large deflections in service.

The transportation industry and the container and packaging industry are each responsible for more than 20% of the tonnage of aluminum produced, while the building industry provides more than 15% of the market. Consumer durables, electrical products, and the machinery and equipment industry each provide between 5% and 10% of the market.

Unalloyed (commercial purity) aluminum and some 300 commonly recognized alloys are available, in both cast and wrought forms. In most of these, the total alloy content is less than 15%. Some of the alloys are solid solutions, strengthenable by cold work, while others, both cast and wrought, are heat-treatable (i.e., strengthened by a precipitation hardening heat treatment). The alloy classification system developed by The Aluminum Association makes use of a 4 digit system for wrought aluminum and its alloys, while cast alloys are designated by 3 digits plus a period and a supplementary digit. These nomenclatures are followed by an alphanumeric temper designation which provides information on the thermal and mechanical processing. The system is described in Tables 1 and 2.

The major impurities in aluminum are iron and silicon, with common minor contaminants including zinc, gallium, titanium, and vanadium. These impurities are responsible for the presence, in all aluminum-based materials, of nonmetallic inclusions which are intermetallic compounds containing iron and silicon as well as other elements where available.

Coarse intermetallic constituents, such as particles of $FeAl_3$ (or its precursor $FeAl_6$) and $Al_{12}(Fe,Mn)_3Si$ form during solidification and, although they may be broken up by subsequent fabrication, they remain in the size range of 1-10 micrometers. Other intermetallic phases form as smaller (submicrometer) particles during the initial stages of processing; some of these (e.g. $Al_{18}Mg_3Cr_2$ and $Al_{20}Cu_2Mn_3$) can have the positive effect of retarding recrystallization and grain growth, thus contributing to grain refinement. Coarse, brittle intermetallic compounds are detrimental to the strength and ductility of the alloys, but especially to the fracture toughness and the appearance after anodizing. In casting alloys, they can also be detrimental to castability, since large insoluble particles in the liquid can dramatically reduce its fluidity. To alleviate these problems, some alloys are available in special grades with restricted levels of impurities. Because of the importance of these impurities, grades of unalloyed aluminum which are intended primarily for remelting are designated by a system using the letter P followed by the maximum contents of silicon and iron respectively in hundredths of a percent. Thus P1020 contains no more than 0.10%Si and 0.20%Fe. A letter may follow this term to designate different elemental limits for impurities other than iron and silicon.

Fabrication of wrought aluminum and its alloys can be carried out by virtually all metalworking processes, notably extrusion, forging, drawing, stamping, and machining. About 50% of wrought products are flat rolled (sheet , plate, foil) with more than 15% as extruded shapes and tubes. Combinations of rolling and extrusion are used to produce bar, rod, and wire. Most wrought alloys are available in many of these product forms. The severe deformations associated with the production of beverage cans provides a vivid illustration of the impressive fabrication properties of some aluminum-based materials. An increasing trend is to the continuous production of shapes starting from the liquid state, for example slab, strip, or bar casting with in-line hot rolling.

The Alclad alloys are composite materials in which thin layers of a cladding alloy are bonded to the surfaces of a high strength alloy to provide corrosion protection. The composition of the cladding alloy is chosen to make it corrosion resistant and also anodic to the core alloy. For example commercial purity cladding is used on aluminum-copper (2xxx) high strength alloys, while aluminum-zinc or aluminum-zinc-magnesium cladding is used on high strength aluminum-zinc-magnesium-copper alloys in the 7xxx series.

Castings are produced by sand mold, permanent mold, pressure-die, investment, and plaster casting. A variety of powder metallurgy processes are also applicable to aluminum and its alloys. In addition, aluminum is a popular matrix material for metal matrix composites.

High Temperature Properties

Pure aluminum melts at 660°C (1221°F), and alloying elements generally lower the range of melting temperatures, hence aluminum and its alloys are not truly high temperature materials in the usual sense. Furthermore, the strength of the precipitation hardened alloys declines rapidly if they are exposed to temperatures much above 100-200°C (212-392°F) as a result of the coarsening of the strengthening precipitates. Some solution strengthened alloys are capable of maintaining useful strengths at temperatures somewhat above 200°C (392°F), but 300°C (572°F) is beyond the useful range for most aluminum alloy applications. Thus when aluminum alloys are described below as useful for "elevated" temperature applications, this refers to temperatures much lower than would be the case in some other alloy systems.

At sub-zero temperatures aluminum alloys are particularly useful since, unlike ferritic steels, they do not undergo a low temperature embrittlement. They are thus used for structural applications at temperatures as low as -270°C (-454°F).

Aluminum and its alloys are discussed below, beginning with the wrought non-heat-treatable alloys, namely unalloyed aluminum, the aluminum-manganese (3xxx) alloys and the aluminum-magnesium (5xxx) alloys. This is followed by considerations of the wrought heat-treatable alloys and the cast alloys.

Wrought Non-Heat-Treatable Alloys

Unalloyed (Commercial Purity) Aluminum (1xxx Series)

These materials, which contain in excess of 99%Al, have microstructures consisting mainly of the relatively pure aluminum fcc matrix, the only other phases present being nonmetallic inclusions which result from the presence of the impurities iron and silicon. Unalloyed aluminum is not heat-treatable but can be strengthened by cold working. The most common grade is 1100 (UNS A91100), which has 99.0%Al with Si, Fe, and Cu the principal impurities. This material, with its attractive combination of workability and weldability properties finds major applications as sheet metal work and heat exchangers and in the food and chemical storage and handling industries. Other useful 1xxx alloys include 1050 (UNS A91050) and 1060 (UNS A91060), which have higher purity (99.5% and 99.6%Al, respectively), giving them lower strength but excellent corrosion resistance and formability. These alloys find applications as tubing and containers for the food, chemical and brewing industries. Also in the 1xxx category are 1145 (UNS A91145), which is used for foil, and 1199 (UNS A91199), a super-purity unalloyed grade having a minimum of 99.99%Al which makes it useful for electrolytic capacitor foil and vapour deposited coatings for optical reflectors. Electrical conductor grade aluminum, alloy 1350 (UNS A91350), sometimes referred to as EC grade, is similar in composition to 1050 but it has stricter limits on a number of impurity elements. It is used for wire, stranded and bus conductors, and transformer strip.

Aluminum-Manganese Alloys (3xxx Series)

The 3xxx series alloys are similar to unalloyed aluminum with additions of manganese or manganese plus magnesium totaling between 1% and 2.4%. For example alloy 3003 (UNS A93003) contains 1.2%Mn, while 3004 (UNS A93004) contains 1.2%Mn and 1.0%Mg, and 3105 (UNS A91305) contains 0.5%Mn and 0.5%Mg. The manganese and magnesium serve as solid solution strengtheners but the manganese also becomes involved in the formation of manganese-rich precipitates such as $(Mn,Fe)Al_6$ and $Al_{12}(Fe,Mn)_3Si$ which inhibit recrystallization and therefore contribute to grain refinement, for example during hot work. These alloys are used in large tonnages for beverage containers as well as for cooking utensils, chemical equipment, sheet metal work, building products and pressure vessels. Alloys 3003 and 3004 are also available as Alclad 3003 and Alclad 3004, the cladding layer being alloy 7072, an aluminum-zinc alloy containing 1%Zn. These clad alloys find applications as building products such as siding and roofing.

Aluminum-Magnesium Alloys (5xxx Series)

Dissolved in aluminum, magnesium provides good solid solution strengthening and high work hardening; it does not cause appreciable precipitation hardening at concentrations below 7%. This is the basis for the 5xxx series of non-heat-treatable aluminum alloys. There are numerous alloys in this series, with only a few being Al-Mg binary alloys, while most contain a third element such as manganese or chromium for additional strength through solution strengthening and grain refinement. These alloys have the usual combination of good formability and corrosion resistance, but the higher strength alloys in the series have particularly good weldability when argon-shielded welding procedures are used.

These alloys are solid solutions up to about 3.5%Mg, above which there exists the possibility of the formation of magnesium-aluminum intermetallic phase precipitates such as Mg_5Al_3 or Mg_5Al_8. These can create susceptibility to stress corrosion cracking if they precipitate at grain boundaries, as can happen at temperatures in the 120-180°C (248-356°F) range. Other inclusion phases which form in commercial alloys include Mg_2Si and particles containing Mn and/or Cr as well as Fe.

Common binary alloys include 5005 (UNS A95005), 5050 (UNS A95050) and 5252 (UNS A95252) which contain 0.8%, 1.4%, and 2.5%Mg respectively. These are used for appliances, conductors, pipe, hardware and trim for appliances and automobiles. Common ternary alloys include 5052 (UNS A95052) and 5154 (UNS A95154) both of which contain 0.25%Cr along with magnesium at the level of 2.5% and 3.5% respectively. These find applications in marine service and as tanks and pressure vessels. Alloys 5652 (UNS A95652) and 5254 (UNS A95254) are similar but have lower limits on the iron, silicon, copper, and manganese impurity levels, making them useful for hydrogen peroxide and chemical storage vessels. Alloys containing magnesium, manganese, and chromium include 5056 (UNS A95056), 5083 (UNS A95083), and 5086 (UNS A95086) which have 4.0-5.1%Mg, 0.1-0.7%Mn, and 0.12-0.15%Cr; applications of these materials include zipper stock, nails, construction, marine, automotive and aircraft components, and cryogenics. Other aluminum-magnesium-manganese-chromium alloys such as 5454 (UNS A95454) and 5456 (UNS A95456) find uses in welded structures, tanks, and pressure vessels.

A number of these alloys have particularly attractive appearance when anodized (5005, 5252, 5457, 5657). Alloy 5056 is also available as Alclad 5056 (the cladding layer being Alloy 5253, an aluminum-magnesium-silicon-zinc alloy) which is used in wire form as a screening material. Alclad 5052, 5086, and 5154 are also available.

Wrought Heat-Treatable Alloys

In heat-treatable aluminum alloys, the heat treatment provides strengthening by the mechanism of precipitation hardening. Precipitation hardening requires that the alloying element exhibit a strongly decreasing solubility in aluminum with decreasing temperature. For example, in the aluminum-copper system the equilibrium solubility of copper in aluminum falls dramatically from about 5.6% at 548°C (1018°F) to a negligible solubility at room temperature. This permits the development of a finely dispersed distribution of precipitate particles if an alloy containing less than 5.6%Cu is given a two-step heat treatment as follows:

1. The alloy is subjected to a solution treatment in the single phase region, for example 525°C (977°F) to allow the copper and aluminum to

form a solid solution, then quenched rapidly to room temperature in water. This gives a supersaturated solid solution of copper in aluminum.

2. The supersaturated solid solution is unstable and, given an aging treatment at a relatively low temperature, a second phase will precipitate out as a controlled distribution of fine particles. The aging may occur at room temperature (natural aging) or during heating to a low temperature, e.g. 130-190°C (266-374°F), for a number of hours. Normally it is not the equilibrium phase (e.g. $CuAl_2$, also called *theta*-phase, in the Al-Cu system) but rather a series of non-equilibrium transition phases which form progressively during the aging treatment. The presence of a fine dispersion of particles of these transition phases provides the strength to the alloy; however, if the dispersion becomes too coarse (i.e., a smaller number of larger particles) then the strength drops. This phenomenon, known as overaging, occurs if the aging temperature is too high and/or the aging time is too long. For each system there exists an optimum aging time and temperature which gives the maximum strength. In this peak strength condition, the precipitate particles are too fine to be seen in the optical microscope, but are clearly revealed by transmission electron microscopy. In some alloys precipitation can be stimulated and strength raised further by cold work between solution treatment and aging.

In general the formability and corrosion resistance are degraded by the precipitation hardening process, but they remain satisfactory for many purposes. Fabrication in many cases is carried out before aging, and corrosion resistance can be improved by using Alclad versions of the alloys. Some but not all of these alloys are weldable and in those that are it may be necessary to heat treat after welding.

Among the heat-treatable aluminum alloy systems are those whose principal alloying elements are: copper, alone or with magnesium (the 2xxx system); silicon, alone or with magnesium (the 4xxx and 6xxx systems); and zinc, alone or with magnesium and copper (the 7xxx system).

Aluminum-Copper and Aluminum-Copper-Magnesium Alloys (2xxx Series)

Copper is one of the most important alloying elements in aluminum. It contributes solution strengthening, but also with appropriate heat treatment provides the possibility of precipitation hardening as described above. Several binary aluminum-copper alloys have commercial importance, although most commercial aluminum-copper alloys contain other elements as well, notably magnesium.

The important binary aluminum-copper alloys are 2011 (UNS A92011) and 2219 (UNS A92219). All of these heat-treatable alloys can be given a wide range of strengths depending on the exact heat treatment conditions. For example, alloy 2219 (6.3%Cu, 0.3%Mn) in the T87 condition has a yield strength which is nearly six times higher than the same alloy in the O temper condition. The presence of small amounts of Mn, Zr, Ti and V in this alloy raise its recrystallization temperature so that it maintains useful strength at elevated temperatures, i.e. above 300°C (572°F). This is a readily weldable alloy which is used for supersonic aircraft skin and structural components. It is also available as Alclad 2219 with the cladding material being the aluminum-zinc alloy 7072. Alloy 2011 (UNS A92011) contains 6.5%Cu as well as small amounts of bismuth and lead which form chip-breakers during machining making this a common aluminum alloy for screw-machine products.

The presence of as little as 0.5% magnesium in aluminum-copper alloys has a strongly intensifying effect on the precipitation hardening. The exact mechanisms for this are not fully understood despite these alloys having been known since 1911. Among the important aluminum-copper-magnesium alloys are 2014 (UNS A92014), 2024 (UNS A92024), 2124 (UNS A92124), and 2218 (UNS A92218), all of which contain 4-4.5%Cu and 0.5-1.5%Mg with some also containing smaller quantities of manganese, silicon, and nickel. Alloy 2014 is used for heavy-duty forgings, plate and extrusions for structural components on aircraft and motor vehicles. Alloy 2024, one of the most common aluminum alloys, is a higher strength alloy which also finds applications in the aircraft industry for structural components and fasteners, as well as being used for truck wheels and other miscellaneous structural applications. Alclad 2014 and Alclad 2024 are also available, the cladding material on 2014 being alloy 6006 while 2024 is clad with alloy 1230. Alloy 2124 is similar to 2024 but contains lower levels of impurity elements in order to obtain improved fracture toughness; it is used for thick plate for aircraft structures. Alloy 2218, which contains 2% nickel, is used for aircraft engine components such as pistons, compressor rings, and cylinder heads.

Other aluminum-copper-magnesium alloys include 2048 (UNS A92048), which has a lower copper content than the alloys discussed above and is used as sheet and plate structural components for aerospace applications, and 2618 (UNS A92618), which contains small additions of iron, silicon, nickel, and titanium for use as forgings, tire molds, and aircraft engine components.

Aluminum-Silicon Alloys (4xxx Series)

This series includes alloy 4032 (UNS A94032), an alloy containing 12%Si, along with about 1% each of magnesium, copper, and nickel. This alloy is used for pistons and other components for service at "elevated" temperatures. The alloy 4043 (UNS A94043) which contains 0.8%Si is used in rod or wire form as a filler metal for the welding of all aluminum alloys other than those rich in magnesium.

Aluminum-Magnesium-Silicon Alloys (6xxx Series)

This heat-treatable alloy series contains 0.6-1.2%Mg and 0.4-1.3%Si with manganese, copper, and chromium often present as well to provide solid solution strengthening and control over grain size. These are referred to as "balanced" alloys when the amounts of magnesium and silicon are in the appropriate ratio to form the useful intermetallic compound Mg_2Si. With its precursor transition phases, Mg_2Si provides precipitation hardening when the alloy is solution treated and aged. These alloys are typically solution treated at 520-540°C (968-1004°F), and aged at 160-175°C (320-347°F). In some cases, the composition is specified in terms of %Mg_2Si, rather than referring to the individual elements. Unbalanced alloys generally contain excess silicon.

Alloy 6061 (UNS A96061) is a balanced alloy, containing 1.5%Mg_2Si, which is extensively used for intermediate strength, general purpose applications in the transportation and building industries (e.g., for trucks, towers, pipes and rail cars). It is also available as Alclad 6061, the cladding material being 7072 aluminum-zinc alloy. Higher strength modifications of 6061 include 6066 (UNS A96066) and 6070 (UNS A96070). In order to obtain improved extrusion properties, a lower strength alloy such as alloy 6063 (UNS A96063) with 1%Mg_2Si is employed for applications such as structural shapes, pipe, railings, etc. A modification of 6063 is 6463 (UNS A96463), which has a lower iron impurity content in order to give better appearance after anodizing.

Alloy 6005 (UNS A96005), an unbalanced alloy which has good extrusion properties but higher strength, has applications which include ladders and TV antennae. Alloys 6009 (UNS A96009) and 6011 (UNS A96011) are formable sheet alloys used for automobile body sheet. Here the aging reaction which creates the precipitation hardening occurs during the paint baking heat treatment. Alloys 6101 (UNS A96101) and 6201 (UNS A96201) have applications as high strength bus bars and electrical conductors, while 6151 (UNS A96151) is a forging alloy used in automotive and machine applications. Other useful extrusion alloys include 6070 (UNS A96070), for heavy duty welded structures, 6205

(UNS A96205) which also exhibits high impact strength, and 6351 (UNS A96351).

Aluminum-Zinc-Magnesium and Aluminum-Zinc-Magnesium-Copper Alloys (7xxx Series)

This series of wrought heat-treatable alloys is based on aluminum containing 4-8%Zn and 1-3%Mg. Zinc and magnesium are both very effective in developing strong precipitation hardening characteristics, and the addition of 1-2% copper provides further strengthening. The result is that this series includes the strongest of the precipitation hardening aluminum alloys, which are widely used for structural applications, including many in the aircraft industry. The strengthening precipitate here is $MgZn_2$ and its transition phase precursors. Magnesium in excess of the $MgZn_2$ balance further increases strength. Double aging treatments are common in developing the desired dispersion of precipitates. In general these alloys are more susceptible to corrosion than many of the other aluminum alloys, in particular to stress corrosion cracking and exfoliation corrosion. Various strategies are used to improve the resistance to stress corrosion, including overaging and careful control over thermal treatment.

The original high strength alloy in the series is 7075 (UNS A97075), which contains 5.6%Zn, 2.5%Mg, 1.6%Cu, and 0.23%Cr, the latter for improved resistance to stress corrosion cracking. Modified versions of this alloy include 7178 (UNS A97178) and 7001 (UNS A97001), with progressively higher zinc, magnesium, and copper contents and progressively higher strengths. Alloys 7175 (UNS A97175) and 7475 (UNS A97475) are comparable to 7075, but with lower limits on impurities to give improved fracture toughness. Both 7075 and 7178 are produced as Alclad alloys, with the cladding being 7072 (aluminum-zinc) or 7011 (aluminum-zinc-magnesium).

Other high strength aluminum-zinc-magnesium-copper alloys include 7049 (UNS A97049), which is used for aircraft forgings and extrusions, 7050 (UNS A97050), which is designed for improved resistance to exfoliation corrosion and stress corrosion cracking as well as for high fracture toughness and fatigue resistance, and 7076 (UNS A97076) which is a forging alloy used for aircraft propellers.

The aluminum-zinc-magnesium alloys without copper addition include 7005 (UNS A97005), which was developed specifically to be weldable. It is used in situations where moderate strength and weldability are required, such as in large heat exchangers and structural members for trucks and trailers. It is also used in sporting equipment.

The alloy 7072 (UNS A97072), a binary aluminum-zinc alloy containing 1%Zn, has a high corrosion potential and finds extensive use as cladding on a wide range of Alclad alloys; it is also used for sacrificial anodes.

Cast Aluminum Alloys

Cast aluminum alloys have compositions which have been developed not only for strength, ductility, and corrosion resistance but also for castability properties such as fluidity in the liquid state. This has resulted in their having different compositions than the wrought alloys, in particular higher silicon contents. Silicon forms a simple eutectic system with aluminum, thus at a late stage of solidification there will be a highly fluid liquid available to feed the normal solidification shrinkage cavities. Thus the presence of silicon in the alloy, at levels ranging from 4% to 12%, permits production of high quality intricate castings with large variations in section thickness while keeping scrap loss minimal by reducing cracking and shrinkage porosity. Aluminum-silicon-based alloys are the most common of the aluminum casting alloys.

Casting alloys (Table 1) are available in a number of alloy systems and in both non-heat-treatable and heat-treatable grades, although here the distinctions are not as clear as in wrought alloys.

Aluminum-Copper Alloys

Aluminum-copper alloys are heat-treatable and capable of high strength, but are relatively difficult to cast, have poor weldability, and can be susceptible to corrosion. Alloy 242.0 (UNS A02420), which contains 4%Cu, 2%Ni, and 2.5%Mg, finds applications in pistons and air-cooled cylinder heads for motorcycle, diesel, and aircraft engines.

Aluminum-Copper-Silicon Alloys

Aluminum-copper-silicon alloys are extensively used, with the copper contributing to strength and the silicon to castability. A wide range of compositions is produced, ranging from high copper to high silicon, thus these alloys fall into both the 2xx.x and the 3xx.x series. These alloys are heat-treatable when the copper content is above 3-4%, especially when magnesium is present as well. Alloys in this category which have 4-5%Cu and 1-3%Si, include 208.0 (UNS A02080) which is used for pressure-tight castings such as manifolds and valve bodies, as well as 295.0 (UNS A02950) and 296.0 (UNS A02960) which are used for wheels, fittings, and housings where strength and machinability are required. Alloys which contain more silicon than copper include 308.0 (UNS A03080) and 319.0 (UNS A03190), which contain 5-6%Si and 3-5%Cu;

these are used for general purpose castings and in automotive applications. Alloys with 8-12%Si and 2-4%Cu include 380.0 (UNS A03800), 383.0 (UNS A03830), and 384.0 (UNS A03840) which are die casting alloys. The 380.0 alloy in particular is very widely used in the automotive and consumer electrical industries, although it has poor weldability and brazeability. Some aluminum-silicon-copper alloys also contain small amounts of magnesium for strengthening purposes; examples include 354.0 (UNS A03540) and 355.0 (UNS A03550). Hypereutectic alloys, i.e. those with more than 12%Si, contain primary silicon particles in their microstructures which impart wear resistance, giving these alloys applications as automotive pistons and engine blocks. A popular example is 390.0 (UNS A03900), which contains 17%Si, 4.5%Cu, and 0.6%Mg.

Aluminum-Silicon Alloys

When a combination of castability and corrosion resistance is deemed more important than strength, aluminum-silicon alloys without copper are used. Examples of this type of alloy include 413.0 (UNS A04130) which has 12%Si and is used for intricate die castings which may require pressure tightness, and 443.0 (UNS A04430) with 5.2%Si which is used mainly for sand and permanent mold castings including cooking utensils and food handling equipment. These non-heat-treatable alloys have good weldability and are among the most important commercial casting alloys.

Aluminum-Silicon-Magnesium Alloys

Magnesium additions to the aluminum-silicon alloys render them heat-treatable, and hence provide increased strength. Alloys 356.0 (UNS A03560), which is very widely used for sand and permanent mold castings, and its stronger variants 357.0 (UNS A03570) and 359.0 (UNS A03590) fall into this category with 7-10%Si plus 0.3-0.6%Mg. So does the alloy 360.0 (UNS A03600), a die casting alloy for use where the corrosion resistance of 380.0 is inadequate. In some alloys, sodium or calcium additions are made to modify the Al-Si eutectic so as to provide a further increment of strength.

Aluminum-Magnesium Alloys

Aluminum-magnesium alloys are not heat-treatable, being mainly single phase solid solutions. They are used where high corrosion resistance is desired, especially in marine environments and food handling. Low impurity content is important here as well. These alloys have poorer castability than the silicon-bearing alloys but have good machinability and attractive appearance when anodized. Examples include 514.0 (UNS

A05140) with 4%Mg, 518.0 (UNS A05180) with 8%Mg, 520.0 (UNS A05200) with 10%Mg, and 535.0 (UNS A05350) which contains 7%Mg.

Aluminum-Zinc-Magnesium Alloys

Aluminum-zinc-magnesium casting alloys have poor castability and poor temperature resistance but can be heat treated to reasonable strength levels and are amenable to brazing and machining. Alloy 712.0 (UNS A07120), 713.0 (UNS A07130), and 771.0 (UNS A07710) contain 6-8%Zn and 0.3-0.9%Mg with minor additions of other elements. One application of 713.0 is cast aluminum furniture.

Aluminum-Tin Alloys

Aluminum-tin alloys, such as 850.0 (UNS A08500), with about 6%Sn and minor amounts of copper and nickel, find applications as cast bearing alloys, for example as connecting rod and crankcase bearings in diesel engines.

Many of these casting alloys are also available with lower specified impurity levels, for example 380.0 (UNS A13800) and 384.0 (UNS A13840). The 443.0 alloy comes additionally as A443.0 (UNS A14430), B443.0 (UNS A24430), and C443.0 (UNS A34430) with different impurity levels specified in each case.

Weldability

Welding of aluminum and its alloys is complicated by the strong tendency of aluminum to react with oxygen in the air to form a thin surface oxide layer. Thus shielding by gas atmosphere or by flux is critical in the welding process. GTAW (gas tungsten arc welding) and GMAW (gas metal arc welding) techniques are standard and most alloys, wrought and cast, are readily weldable. There are, however, a number which require special welding techniques and some for which welding is not recommended.

Generally, the non-heat-treatable alloys are readily weldable as are the heat-treatable 6xxx alloys. Special techniques may be required for the heat-treatable 2xxx and 4xxx series. The 7xxx high strength heat-treatable alloys are generally not recommended for arc welding, with the exception of 7005 and 7039 which were developed specifically for welding. In general, the heat associated with welding has a strong local effect on the microstructure and properties of the material, for example the strength of the weld region of a strain hardened alloy may be similar to that of the alloy in the annealed condition. In the heat-treatable alloys, solution treating, annealing, and overaging can occur in the heat affected zone, so it may be necessary to re-solution treat and re-age the materials after welding to recover the strength properties; however, the ductility may not be fully recovered.

In addition to arc welding, aluminum and its alloys can in general be resistance welded and can be brazed with an appropriate flux.

Table 1 Aluminum Alloy Groups

(a) Wrought aluminum and its alloys grouped by major alloying element:

1xxx	unalloyed aluminum, 99%Al minimum
2xxx	copper
3xxx	manganese
4xxx	silicon
5xxx	magnesium
6xxx	magnesium and silicon
7xxx	zinc
8xxx	other elements

In the 1xxx series, where the aluminum content is greater than 99.00%, the last two digits give the minimum aluminum percentage. They are the same as the two digits to the right of the decimal point in the minimum aluminum percentage in the specification (e.g. 1050 contains ≥ 99.50%Al). The second digit is 0 for unalloyed aluminum with natural impurity limits; 1 through 9 for special limits on one or more individual elements.

In the other series, 2xxx to 7xxx, the last two digits have no special significance. The second digit indicates alloy modifications, i.e., it is assigned 0 in the original alloy developed, with 1 through 9 assigned consecutively indicating alloying modifications.

(b) Cast aluminum and its alloys grouped by major alloying element:

1xx.x	unalloyed aluminum, 99%Al minimum
2xx.x	copper
3xx.x	silicon, with added copper and/or magnesium
4xx.x	silicon
5xx.x	magnesium
7xx.x	zinc
8xx.x	tin
9xx.x	other

In the 1xx.x series, where the aluminum content is greater than 99.00%, the second and third digits give the minimum aluminum percentage. They are the same as the two digits to the right of the decimal point in the minimum aluminum percentage. Thus 150.0 contains ≥99.50%Al. The last digit indicates the product form: 1xx.0 for castings and 1xx.1 for ingot.

In the other series, 2xx.x to 7xx.x, the second and third digits have no special significance. In these series, the final digit defines the product form: xxx.0 for castings, xxx.1 and xxx.2 for ingots suitable for foundry use, where the final digits 1 and 2 refer to composition limits on iron, magnesium, and zinc in the ingots. Prefix letters are used to denote differences in impurity limits (especially iron) in castings (e.g. B443.0).

Table 2 Basic Temper Designations

F - As fabricated, with no special control over thermal or mechanical treatment and no mechanical property limits for wrought products
O - Annealed to obtain lowest strength temper (wrought products) and to improve ductility and dimensional stability (cast products). May be followed by a digit other than zero.
H - Strain hardened (wrought products only). The H is always followed by two or more digits (see below).
W - Solution heat treated. Applicable to alloys which age at room temperature over a period of months or years.
T - Heat treated to produce stable tempers other than F, O, or H. Applicable to alloys that are thermally treated, with or without supplementary strain hardening, to produce stable tempers. The T is always followed by one or more digits (see next page).

Strain Hardened Products

The first digit following the H indicates the specific sequence of basic operations as described below. The second digit indicates the degree of strain hardening on a scale from 1 through 9, where 0 corresponds to full annealing and 8 corresponds to 75% cold work following full annealing. Thus when the second digit is 2 (e.g., alloy 5005-H12) the material is said to be quarter hard, 4 - half hard, 8 - full hard. The numeral 9 is used when the UTS exceeds that of 8 by more than 10 MPa. A third digit, when used, refers to a specific variation of a 2-digit temper. There are special 3 digit designations for patterned or embossed sheet.

H1 - Strain hardened only, with no supplementary thermal treatment.

H2 - Strain hardened and partially annealed, i.e. strain hardened more than the desired final amount, then annealed to reduce the strength. The second digit refers to the strength level after the partial anneal.
H3 - Strain hardened and stabilized by a low temperature thermal treatment, for those alloys which gradually age-soften at room temperature. The second digit refers to the strength level after stabilization.

Heat Treated Products

The T is followed by a number from 1 to 10, indicating a specific sequence of basic treatments as described below. This is followed by one or more additional digits, the first of which cannot be 0, to identify a variation of one of the 10 basic tempers (e.g. alloy 2024-T3, alloy 7075-T651). In this context "naturally aged" means aged at room temperature, and "artificially aged" means aged above room temperature. Specific designations have been assigned for the cases of stress relieving by stretching, compressing, or a combination of the two. These include TX51, TX510, TX511, while TX52 refers to stress relief by compressing and TX54 by stretching and compressing.

T1 - Cooled from an elevated temperature shaping process and naturally aged to a substantially stable condition (not cold worked).
T2 - Cooled from an elevated temperature shaping process, cold worked, and naturally aged to a substantially stable condition.
T3 - Solution treated, cold worked, and naturally aged to a substantially stable condition.
T4 - Solution treated and naturally aged to a substantially stable condition.
T5 - Cooled from an elevated temperature shaping process and artificially aged.
T6 - Solution heat treated and artificially aged.
T7 - Solution heat treated and overaged or stabilized. This applies to wrought products which have been precipitation hardened past the point of maximum strength and to cast products which are artificially aged to provide dimensional and strength stability.
T8 - Solution heat treated, cold worked, and artificially aged.
T9 - Solution heat treated, artificially aged, and cold worked.
T10 - Cooled from an elevated temperature shaping process, cold worked, and artificially aged.

SAE/AMS SPECIFICATIONS - ALUMINUM & ALUMINUM ALLOYS	
AMS	**Title**
2201	Tolerances Aluminum and Aluminum Alloy Bar, Rod, Wire, and Forging Stock Rolled or Cold Finished
2202	Tolerances, Aluminum Alloy and Magnesium Alloy Sheet and Plate
2203	Tolerances, Aluminum Alloy Drawn Tubing
2204	Tolerances Aluminum Alloy Standard Structural Shapes
2205	Tolerances, Aluminum Alloy and Magnesium Alloy Extrusions
2450	Sprayed Metal Finish Aluminum
2470	Anodic Treatment of Aluminum Alloys, Chromic Acid Process
2471	Anodic Treatment of Aluminum Alloys, Sulfuric Acid Process, Undyed Coating
2472	Anodic Treatment of Aluminum Alloys, Sulfuric Acid Process, Dyed Coating
2473	Chemical Treatment for Aluminum Alloys, General Purpose Coating
2474	Chemical Treatment for Aluminum Alloys Low Electrical Resistance Coating
2482	Teflon-Impregnated or Codeposited Hard Coating Treatment of Aluminum Alloys
2672	Brazing, Aluminum
2673	Aluminum Molten Flux (Dip) Brazing
2811	Identification Aluminum and Magnesium Alloy Wrought Products
3412	Flux, Aluminum Brazing
3414	Flux, Aluminum Welding
3415	Flux, Aluminum Dip Brazing 1030°F (554°C) or Lower Liquidus
4000	Aluminum Sheet and Plate, (1060) NONCURRENT January 1987, UNS A91060
4001	Aluminum Sheet and Plate 0.12Cu, (1100-0) Annealed, UNS A91100
4003	Aluminum Sheet and Plate 0.12Cu, (1100-H14) Strained Hardened, UNS A91100
4004	Aluminum Alloy Foil, 2.5Mg - 0.25Cr, (5052-H191) Strain Hardened, UNS A95052
4005	Aluminum Alloy Foil 5.0Mg - 0.12Mn - 0.12Cr, (5056-H191) Strain Hardened, UNS A95056
4006	Aluminum Alloy Sheet and Plate 1.25Mn - 0.12Cu - (3003-0) Annealed, UNS A93003
4007	Aluminum Alloy Foil 4.4Cu - 1.5Mg - 0.6Mn (2024-0), UNS A92024
4008	Aluminum Alloy Sheet and Plate 1.25Mn - 0.12Cu (3003-H14) Strain Hardened, UNS A93003
4009	Aluminum Alloy Foil 1.0Mg - 0.6Si - 0.30Cu - 0.20Cr (6061-0) Annealed, UNS A96061

SAE/AMS SPECIFICATIONS - ALUMINUM & ALUMINUM ALLOYS (Continued)

AMS	Title
4010	Aluminum Alloy Foil 1.2Mn - 0.12Cu (3003-H18), UNS A93003
4011	Aluminum Foil 99.45Al (1145-0) Annealed, UNS A91145
4013	Aluminum Sheet, Laminated, Surface Bonded
4014	Aluminum Alloy Plate 4.5Cu - 0.85Si - 0.80Mn - 0.50Mg (2014-T651), Superseded by AMS 4029
4015	Aluminum Alloy Sheet and Plate 2.5Mg - 0.25Cr (5052-0) Annealed, UNS A95052
4016	Aluminum Alloy Sheet and Plate 2.5Mg - 0.25Cr (5052-H32) Strain Hardened, Quarter-Hard, and Stabilized, UNS A95052
4017	Aluminum Alloy Sheet and Plate, 2.5Mg - 0.25Cr - (5052-H34) Strained Hardened, Half-Hard, and Stabilized, UNS A95052
4021	Aluminum Alloy Sheet and Plate, Alclad 1.0Mg - 0.60Si - 0.28Cu - 0.20Cr (Alclad 6061-0) Annealed, UNS A86061
4025	Aluminum Alloy Sheet and Plate 1.0Mg - 0.60Si - 0.28Cu - 0.20Cr (6061-0) Annealed, UNS A96061
4026	Aluminum Alloy Sheet and Plate 1.0Mg - 0.60Si - 0.28Cu - 0.20Cr (6061; -T4 Sheet, -T451 Plate) Solution Heat Treated and Naturally Aged, UNS A96061
4027	Aluminum Alloy Sheet and Plate 1.0Mg - 0.60Si - 0.28Cu - 0.20Cr (6061; -T6 Sheet, -T651 Plate) Solution and Precipitation Heat Treated, UNS A96061
4028	Aluminum Alloy Sheet and Plate 4.4Cu - 0.85Si - 0.80Mn - 0.50Mg (2014-0) Annealed, UNS A92014
4029	Aluminum Alloy Sheet and Plate 4.5Cu - 0.85Si - 0.80Mn - 0.50Mg (2014; -T6 Sheet, -T651 Plate) Solution and Precipitation Heat Treated, UNS A92014
4031	Aluminum Alloy Sheet and Plate 6.3Cu - 0.30Mn - 0.18Zr - 0.10V - 0.06Ti (2219-0) Annealed, UNS A92219
4035	Aluminum Alloy Sheet and Plate 4.4Cu - 1.5Mg - 0.60Mn (2024-0) Annealed, UNS A92024
4036	Aluminum Alloy Sheet and Plate, Alclad One Side, 4.4Cu - 1.5Mg - 0.60Mn (Alclad One Side 2024, -T3 Sheet; -T351 Plate), UNS A82024
4037	Aluminum Alloy Sheet and Plate 4.4Cu - 1.5Mg - 0.60Mn (2024; -T3 Flat Sheet, - T351 Plate) Solution Heat Treated, UNS A92024
4040	Aluminum Alloy Sheet and Plate, Alclad 4.4Cu - 1.5Mg - 0.60Mn (Alclad 2024 and 1½% Alclad 2024-0) Annealed, UNS A82024
4041	Aluminum Alloy Sheet and Plate, Alclad 4.4 Cu - 1.5Mg - 0.60Mn (Alclad 2024 and 1½% Alclad 2024, -T3 Flat Sheet; 1½% Alclad 2024-T351 Plate, UNS A92024
4044	Aluminum Alloy Sheet and Plate 5.6Zn - 2.5Mg - 1.6Cu - 0.23Cr (7075-0) Annealed, UNS A97075
4045	Aluminum Alloy Sheet and Plate 5.6Zn - 2.5Mg - 1.6Cu - 0.23Cr (7075; -T6 Sheet, -T651 Plate) Solution and Precipitation Heat Treated, UNS A97075
4046	Aluminum Alloy Sheet and Plate, Alclad one side 5.6Zn - 2.5Mg - 1.6Cu - 0.23Cr (Alclad One Side 7075; -T6 Sheet, -T651 Plate) Solution and Precipitation Heat Treated, UNS A87075

SAE/AMS SPECIFICATIONS - ALUMINUM & ALUMINUM ALLOYS (Continued)

AMS	Title
4048	Aluminum Alloy Sheet and Plate, Alclad 5.6Zn - 2.5Mg - 1.6Cu - 0.23Cr (Alclad 7075-0) Annealed, UNS A87075
4049	Aluminum Alloy Sheet and Plate, Alclad 5.6Zn - 2.5Mg - 1.6Cu - 0.23Cr (Alclad 7075; -T6 Sheet - T651 Plate) Solution and Precipitation Heat Treated, UNS A97075
4050	Aluminum Alloy Plate 6.2Zn - 2.3Cu - 2.2Mg - 0.12Zr (7050-T7451) Solution Heat Treated, Stress Relieved, and Overaged, UNS A97050
4056	Aluminum Alloy Sheet and Plate 4.4Mg - 0.70Mn - 0.15Cr (5083-0) Annealed, UNS A95083
4062	Aluminum Tubing, Seamless, Drawn, Round (1100-H14) Strain Hardened, UNS A91100
4063	Aluminum Alloy Sheet, Clad One Side 1.25Mn - 0.12Cu (No. 11-0 Brazing Sheet) Annealed, UNS A83003
4064	Aluminum Alloy Sheet, Clad Two Sides 1.25Mn - 0.12Cu (No. 12-0 Brazing Sheet) Annealed, UNS 83003
4065	Aluminum Alloy Tubing, Seamless, Drawn 1.2Mn - 0.12Cu (3003-0) Annealed, UNS A93003
4066	Aluminum Alloy Tubing, Seamless, Drawn 6.3Cu - 0.30Mn - 0.18Zr - 0.10V - 0.06Ti (2219-T8511) (See 8.2) Solution Heat Treated, Stress Relieved by Stretching, and Precipitation Heat Treated, UNS A92219
4067	Aluminum Alloy Tubing, Seamless, Drawn, Round 1.25Mn - 0.12Cu (3003 - H14) Strain Hardened, UNS A93003
4068	Aluminum Alloy Tubing, Seamless, Drawn 6.3Cu - 0.30Mn - 0.18Zr - 0.10V - 0.06Ti (2219-T3511) Solution Heat Treated and Stress Relieved by Stretching, UNS A92219
4069	Aluminum Alloy Tubing, Seamless, Drawn, Round Close Tolerance 2.5Mg - 0.25Cr (5052-0) Annealed, UNS A95052
4070	Aluminum Alloy Tubing, Seamless, Drawn, Round 2.5Mg - 0.25Cr (5052-0), UNS A95052
4071	Aluminum Alloy Tubing, Hydraulic, Seamless, Drawn, Round 2.5Mg - 0.25Cr (5052-0) Annealed, UNS A95052
4077	Aluminum Alloy Sheet and Plate, Alclad One Side 4.4Cu - 1.5Mg - 0.60Mn (Clad One Side 2024-0) Annealed, UNS A82024
4078	Aluminum Alloy Plate 5.6Zn - 2.5Mg - 1.6Cu - 0.23Cr (7075-T7351) Solution Heat Treated, Stress Relieved, and Overaged, UNS A97075
4079	Aluminum Alloy Tubing, Seamless, Drawn, Round, Close Tolerance 1.0Mg - 0.60Si - 0.28Cu - 0.20Cr (6061-0) Annealed, UNS A96061
4080	Aluminum Alloy Tubing, Seamless, Drawn, 1.0Mg - 0.60Si - 0.28Cu - 0.20Cr (6061-0), UNS A96061
4081	Aluminum Alloy Tubing, Hydraulic, Seamless, Drawn, Round 1.0Mg - 0.60Si - 0.28Cu - 0.20Cr (6061-T4) Solution Heat Treated and Naturally Aged, UNS A96061
4082	Aluminum Alloy Tubing, Seamless, Drawn 1.0Mg - 0.60Si - 0.28Cu - 0.20Cr (6061-T6) Solution and Precipitation Heat Treated, UNS A96061
4083	Aluminum Alloy Tubing, Hydraulic, Seamless, Drawn, Round 1.0Mg - 0.60Si - 0.28Cu - 0.20Cr (6061-T6) Solution and Precipitation Heat Treated, UNS A96061
4084	Aluminum Alloy Sheet 5.7Zn - 2.2Mg - 1.6Cu - 0.22Cr (7475T-61) Solution and Precipitation Heat Treated, UNS A97475

SAE/AMS SPECIFICATIONS - ALUMINUM & ALUMINUM ALLOYS (Continued)	
AMS	**Title**
4085	Aluminum Alloy Sheet 5.7Zn - 2.2Mg - 1.6Cu - 0.22Cr (7475-T761) Solution Heat Treated and Overaged, UNS A97475
4086	Aluminum Alloy Tubing, Hydraulic, Seamless, Drawn 4.4Cu - 1.5Mg - 0.60Mn (2024-T3) Solution Heat Treated, Cold Worked, and Naturally Aged, UNS A92024
4087	Aluminum Alloy Tubing, Seamless, Drawn 4.4Cu - 1.5Mg - 0.60Mn (2024-0) Annealed, UNS A92024
4088	Aluminum Alloy Tubing, Seamless, Drawn 4.4Cu - 1.5Mg - 0.60Mn (2024-T3) Solution Heat Treated and Cold Worked, UNS A92024
4089	Aluminum Alloy Plate 5.7Zn - 2.2Mg - 1.6Cu - 0.22Cr (7475-T7651) Solution Heat Treated, Stress Relieved by Stretching, and Precipitation Heat Treated, UNS A97475
4090	Aluminum Alloy Plate 5.7Zn - 2.2Mg - 1.6Cu - 0.22Cr (7475-T651) Solution Heat Treated, Stress Relieved, and Precipitation Heat Treated, UNS A97475
4094	Aluminum Alloy Sheet and Plate, Alclad 6.3Cu - 0.30Mn - 0.18Zr - 0.10V - 0.06Ti Alclad 2219-T81 Sheet Solution Heat Treated, Cold Worked, and Precipitation Heat Treated Alclad 2219-T851 Plate Solution Heat Treated, Stress Relieved and Precipitation Heat Treated), UNS A82219
4095	Aluminum Alloy Sheet and Plate, Alclad 6.3Cu - 0.30Mn - 0.18Zr - 0.10V - 0.06Ti Alclad 2219-T31; Sheet, Solution Heat Treated and Cold Worked Alclad 2219-T351; Plate, Solution Heat Treated and Stress Relieved, UNS A82219
4096	Aluminum Alloy Sheet and Plate, Alclad 6.3Cu - 0.30Mn - 0.18Zr - 0.10V - 0.06Ti (Alclad 2219-0) Annealed, UNS A92219
4100	Aluminum Alloy Sheet, Alclad 5.7Zn - 2.2Mg - 1.6Cu - 0.22Cr (7475- T761) Solution and Precipitation Heat Treated, UNS A97475
4101	Aluminum Alloy Plate 4.4Cu - 1.5Mg - 0.60Mn (2124-T851) Solution Heat Treated, Stretched, and Precipitation Heat Treated, UNS A92124
4102	Aluminum Alloy Bars, Rods, and Wire, Rolled or Cold-Finished 99.0Al (1100-F) as Fabricated, UNS A9100
4107	Aluminum Alloy Die Forgings 6.2Zn - 2.3Cu - 2.2Mg - 0.12Zr (7050-T74) Solution Heat Treated and Overaged, UNS A97050
4108	Aluminum Alloy Hand Forgings 6.2Zn - 2.3Cu - 2.2Mg - 0.12Zr (7050-T7452) (formerly -T73652) Solution Heat Treated, Compression Stress-Relieved, and Overaged, UNS A97050
4111	Aluminum Alloy Forgings 7.7Zn - 2.5Mg - 1.5Cu - 0.16Cr (7049-T73) Solution and Precipitation Heat Treated, UNS A97049
4113	Aluminum Alloy Extruded Structural Shapes 1.0Mg - 0.60Si - 0.28Cu - 0.20Cr (6061- T6) Solution and Precipitation Heat Treated, UNS A96061
4114	Aluminum Alloy Bars and Rods, Rolled or Cold-Finished 2.5Mg - 0.25Cr (5052-F) as Fabricated, UNS A95052
4115	Aluminum Alloy Bars, Rods, and Wire, Rolled or Cold-Finished, and Rings 1.0Mg - 0.60Si - 0.28Cu - 0.20Cr (6061-0) Annealed, UNS A96061
4116	Aluminum Alloy Bars, Rods, and Wire, Cold-Finished, 1.0Mg - 0.60Si - 0.30Cu - 0.20Cr (6061-T4), UNS A96061
4117	Aluminum Alloy Bars, Rods, And Wire, Rolled or Cold Finished and Flash Welded Rings 1.0Mg - 0.60Si - 0.28Cu - 0.20Cr (6061; -T6, -T651) Solution and Precipitation Heat Treated, UNS A96061

SAE/AMS SPECIFICATIONS - ALUMINUM & ALUMINUM ALLOYS (Continued)

AMS	Title
4118	Aluminum Alloy Bars, Rods, and Wire, Rolled or Cold Finished 4.0Cu - 0.70Mn - 0.60Mg - 0.50Si (2017; -T4, -T451) Solution Heat Treated, UNS A92017
4120	Aluminum Alloy Bars, Rods, and Wire, Rolled or Cold Finished 4.4Cu - 1.5Mg - 0.60Mn (2024) Solution Heat Treated and Naturally Aged (T4) Solution Heat Treated, Cold Worked, and Naturally Aged (T351), UNS A92024
4121	Aluminum Alloy Bars, Rods, and Wire, Rolled or Cold Finished 4.5Cu - 0.85Si - 0.80Mn - 0.50Mg (2014-T6) Solution and Precipitation Heat Treated, UNS A92014
4122	Aluminum Alloy Bars, Rods, and Wire Rolled or Cold Finished, and Rings 5.6Zn - 2.5Mg - 1.6Cu - 0.23Cr (7075-T6) Solution and Precipitation Heat Treated, UNS A97075
4123	Aluminum Alloy Bars and Rods, Rolled or Cold Finished 5.6Zn - 2.5Mg - 1.6Cu - 0.23Cr (7075-T651) Solution and Precipitation Heat Treated, UNS A97075
4124	Aluminum Alloy Bars Rods and Wire, Rolled, Drawn or Cold Finished 5.6Zn - 2.5Mg - 1.6Cu - 0.23Cr (7075-T7351), UNS A97075
4125	Aluminum Alloy Die Forgings, and Rolled or Forged Rings 0.90Si - 0.62Mg - 0.25Cr (6151-T6) Solution and Precipitation Heat Treated, UNS A96151
4126	Aluminum Alloy Forgings 5.6Zn - 2.5Mg - 1.6Cu - 0.23Cr (7075-T6) Solution and Precipitation Heat Treated, UNS A97075
4127	Aluminum Alloy Forgings and Rolled or Forged Rings 1.0Mg - 0.6Si - 0.28Cu - 0.20Cr (6061- T6) Solution and Precipitation Heat Treated, UNS A96061
4128	Aluminum Alloy Bars, Rolled or Cold Finished 1.0Mg - 0.60Si - 0.30Cu - 0.20Cr (6061- T451) Solution Heat Treated and Stress Relieved by Stretching, UNS A96061
4130	Aluminum Alloy Die Forgings 4.4Cu - 0.85Si - 0.80Mn (2025-T6) Solution and Precipitation Heat Treated, UNS A92025
4131	Aluminum Alloy Forgings 5.6Zn - 2.5Mg - 1.6Cu - 0.23Cr (7075-T74) Solution and Precipitation Heat Treated, UNS A97075
4132	Aluminum Alloy Die and Hand Forgings, Rolled Rings, and Forging Stock 2.3Cu - 1.6Mg - 1.1Fe - 1.0Ni - 0.18Si - 0.07Ti (2618-T61) Solution and Precipitation Heat Treated, UNS A92618
4133	Aluminum Alloy Forgings and Rolled Rings 4.4Cu - 0.85Si - 0.80Mn - 0.50Mg (2014-T6) Solution and Precipitation Heat Treated, UNS A92014
4134	Aluminum Alloy Die Forgings 4.4Cu - 0.85Si - 0.80Mn - 0.50Mg (2014-T4) Solution Heat Treated, UNS A92014
4140	Aluminum Alloy Die Forgings 4.0Cu - 2.0Ni - 0.68Mg (2018-T61) Solution Heat Treated, UNS A92018
4141	Aluminum Alloy Die Forgings 5.6Zn - 2.5Mg - 1.6Cu - 0.23Cr (7075-T73) Solution and Precipitation Heat Treated, UNS A97075
4143	Aluminum Alloy Forgings and Rolled or Forged Rings 6.3Cu - 0.30Mn - 0.18Zr - 0.10V - 0.06Ti (2219-T6) Solution and Precipitation Heat Treated, UNS A92219

SAE/AMS SPECIFICATIONS - ALUMINUM & ALUMINUM ALLOYS (Continued)

AMS	Title
4144	Aluminum Alloy Hand Forgings and Rolled Rings 6.3Cu - 0.30Mn - 0.18Zr - 0.10V - 0.06Ti (2219-T852) Solution Heat Treated, Stress Relief Compressed, and Precipitation Heat Treated, UNS A92219
4146	Aluminum Alloy Forgings and Rolled or Forged Rings 1.0Mg - 0.60Si - 0.28Cu - 0.20Cr (6061-T4) Solution Heat Treated and Naturally Aged, UNS A96061
4147	Aluminum Alloy Forgings 5.6Zn - 2.5Mg - 1.6Cu - 0.23Cr (7075-T7352) Solution Heat Treated Stress Relieved by Compression, and Overaged, UNS A97075
4148	Aluminum Alloy Die Forgings 5.6Zn - 2.5Mg - 1.6Cu - 0.23Cr (7175-T66) Solution and Precipitation Heat Treated, UNS A97175
4149	Aluminum Alloy Forgings 5.6Zn - 2.5Mg - 1.6Cu - 0.23Cr (7175-T74) (Formerly T736) Solution and Precipitation Heat Treated, UNS A97175
4150	Aluminum Alloy Extrusions and Rings 1.0Mg - 0.60Si - 0.28Cu - 0.20Cr (6061-T6) Solution and Precipitation Heat Treated, UNS A96061
4152	Aluminum Alloy Extrusions 4.4Cu - 1.5Mg - 0.60Mn (2024-T3) Solution Heat Treated, UNS A92024
4153	Aluminum Alloy Extrusions 4.5Cu - 0.85Si - 0.80Mn - 0.50Mg (2014 -T6) Solution and Precipitation Heat Treated, UNS A92014
4154	Aluminum Alloy Extrusions 5.6Zn - 2.5Mg - 1.6Cu - 0.23Cr (7075-T6) Solution and Precipitation Heat Treated, UNS A97075
4156	Aluminum Alloy Extrusions 0.68Mg - 0.40Si (6063-T6) Solution and Precipitation Heat Treated, UNS A96063
4157	Aluminum Alloy Extrusions 7.7Zn - 2.4Mg - 1.6Cu - 0.16Cr (7049-T73511) Solution Heat Treated, Stress Relieved, and Overaged, UNS A97049
4159	Aluminum Alloy Extrusions 7.7Zn - 2.4Mg - 1.6Cu - 0.16Cr (7049-T76511) Solution Heat Treated, Stress Relieved, and Overaged, UNS A97049
4160	Aluminum Alloy Extrusions 1.0Mg - 0.60Si - 0.28Cu - 0.20Cr (6061-0) Annealed, UNS A96061
4161	Aluminum Alloy Extrusions 1.0Mg - 0.60Si - 0.28Cu - 0.20Cr (6061-T4) Solution Heat Treated and Naturally Aged, UNS A96061
4162	Aluminum Alloy Extrusions 6.3Cu - 0.30Mn - 0.18Zr - 0.10V - 0.06Ti (2219-T8511) Solution Treated, Stress Relief Stretched, Precipitation Heat Treated, UNS A92219
4163	Aluminum Alloy Extrusions 6.3Cu - 0.30Mn - 0.18Zr - 0.10V - 0.06Ti (2219-T3511) Solution Heat Treated and Stress Relieved by Stretching, UNS A92219
4164	Aluminum Alloy Extrusions 4.4Cu - 1.5Mg - 0.60Mn (2024-T3510) Stress-Relief Stretched, Unstraightened, UNS A92024
4165	Aluminum Alloy Extrusions 4.4Cu - 1.5Mg - 0.60Mn (2024-T3511) Stress-Relieved Stretched, Straightened, UNS A92024
4166	Aluminum Alloy Extrusions 5.6Zn - 2.5Mg - 1.6Cu - 0.23Cr (7075-T73) Solution Treated and Overaged, UNS A97075
4167	Aluminum Alloy Extrusions 5.6Zn - 2.5Mg - 1.6Cu - 0.23Cr (7075-T73511) Solution Heat Treated, Stress Relieved by Stretching, and Precipitation Heat Treated Straightened, UNS A97075

SAE/AMS SPECIFICATIONS - ALUMINUM & ALUMINUM ALLOYS (Continued)

AMS	Title
4168	Aluminum Alloy Extrusions 5.6Zn - 2.5Mg - 1.6Cu - 0.23Cr (7075-T6510) Solution Heat Treated, Stress Relieved by Stretching, and Precipitation Heat Treated, Unstraightened, UNS A97075
4169	Aluminum Alloy Extrusions 5.6Zn - 2.5Mg - 1.6Cu - 0.23Cr (7075-T6511) Solution Heat Treated, Stress Relieved by Stretching, and Precipitation Heat Treated, Straightened, UNS A97075
4172	Aluminum Alloy Extrusions 1.0Mg - 0.60Si - 0.28Cu - 0.20Cr (6061- T4511) Solution Heat Treated and Stress Relieved by Stretching, UNS A96061
4173	Aluminum Alloy Extrusions 1.0Mg - 0.60Si - 0.30Cu - 0.20Cr (6061-T6511) Solution Heat Treated and Stress Relieved by Stretching, and Precipitation Heat Treated, UNS A96061
4174	Aluminum Alloy Flash Welded Rings 5.6Zn - 2.5Mg - 1.6Cu - 0.23Cr (7075-T73) Solution and Precipitation Heat Treated, UNS A97075
4177	Core, Flexible Honeycomb, Aluminum Alloy, Treated for Sandwich Construction 5056, 175 (347), UNS A95056
4178	Core, Flexible Honeycomb, Aluminum Alloy, Treated for Sandwich Construction 5052, 175 (347), UNS A95052
4179	Aluminum Alloy Forgings 5.6Zn - 2.5Mg - 1.6Cu - 0.23Cr (7175-T7452) Solution Heat Treated, Stress Relieved, and Precipitation Heat Treated, UNS A97175
4180	Aluminum Wire 99.0Al Minimum (1100-H18), UNS A91100
4181	Aluminum Alloy Welding Wire 7.0Si - 0.38Mg - 0.10Ti (4008), UNS A94008
4182	Aluminum Alloy Wire 5.0Mg - 0.12Mn - 0.12Cr (5056-0) Annealed, UNS A95056
4184	Filler Metal, Aluminum Brazing 10Si - 4.0Cu (4145), UNS A94145
4185	Filler Metal, Aluminum Brazing 12Si (4047), UNS A94047
4186	Aluminum Alloy Bars, Rods, and Wire, Rolled or Cold Finished 5.6Zn - 2.5Mg - 1.6Cu - 0.23Cr (7075-F) As Fabricated, UNS A97075
4187	Aluminum Alloy Bars and Rods, Rolled, Drawn, or Cold Finished 5.6Zn - 2.5Mg - 1.6Cu - 0.23Cr (7075-O) Annealed, UNS A97075
4189	Aluminum Alloy Welding Wire 4.1Si - 0.2Mg (4643), UNS A94643
4190	Aluminum Alloy Welding Wire 5.2Si (4043), UNS A94043
4191	Aluminum Alloy Welding Wire, 6.3Cu - 0.30Mn - 0.18Zr - 0.15Ti - 0.10V (2319), UNS A92319
4193	Aluminum Alloy Sheet and Plate 4.4Cu - 1.5Mg - 0.60Mn (2024-T861) Solution Heat Treated, Cold Worked, and Precipitation Heat Treated, UNS A92024
4194	Aluminum Alloy Sheet and Plate, Alclad 4.4Cu - 1.5Mg - 0.60Mn (Alclad 2024 and 1½% Alclad 2024-T361 Flat Sheet; 1½% Alclad 2024-T361 Plate) Solution Heat Treated and Cold Worked, UNS A82024

SAE/AMS SPECIFICATIONS - ALUMINUM & ALUMINUM ALLOYS (Continued)	
AMS	**Title**
4195	Aluminum Alloy Sheet and Plate, Alclad 4.4Cu - 1.5Mg - 0.60Mn (Alclad 2024 and 1½% Alclad 2024-T861 Flat Sheet; 1½% Alclad 2024-T861 Plate) Solution Heat Treated, Cold Worked, and Precipitation Heat Treated, UNS A82024
4200	Aluminum Alloy Plate 7.7Zn - 2.4Mg - 1.6Cu - 0.16Cr (7049-T7351) Solution Heat Treated, Stress Relieved, and Precipitation Heat Treated, UNS A97049
4201	Aluminum Alloy Plate 6.2Zn - 2.3Cu - 2.2Mg - 0.12Zr (7050-T7651) Solution Heat Treated, Stress Relieved, and Overaged, UNS A97050
4202	Aluminum Alloy Plate 5.7Zn - 2.2Mg - 1.6Cu - 0.22Cr (7475-T7351) Solution Heat Treated, Stress Relieved by Stretching, and Precipitation Heat Treated, UNS A97475
4203	Aluminum Alloy Plate 6.2Zn - 1.8Cu - 2.4Mg - 0.13Zr (7010-T7351) Solution Heat Treated, Stress Relieved, and Precipitation Heat Treated, UNS A97010
4204	Aluminum Alloy Plate 6.2Zn - 1.8Cu - 2.4Mg - 0.13Zr (7010-T7651) Solution Heat Treated, Stress Relieved, and Precipitation Heat Treated, UNS A97010
4205	Aluminum Alloy Plate 6.2Zn - 1.8Cu - 2.4Mg - 0.13Zr (7010-T73651) Solution Heat Treated, Stress Relieved, and Precipitation Heat Treated, UNS A97010
4207	Aluminum Alloy Sheet, Alclad 5.7Zn - 2.2Mg - 1.6Cu - 0.22Cr (7475-T61) Solution and Precipitation Heat Treated, UNS A87475
4208	Aluminum Alloy Sheet 6.0Cu - 0.40Zr (2004-F) As Rolled, UNS A92004
4209	Aluminum Alloy Sheet, Alclad 6.0Cu - 0.40Zr (2004-F) As Rolled, UNS A82004
4210	Aluminum Alloy Castings Sand 5.0Si - 1.2Cu - 0.50Mg (355.0-T51) Precipitation Heat Treated, UNS A03550
4212	Aluminum Alloy Sand Castings 5.0Si - 1.2Cu - 0.50Mg (355.0-T6) Solution and Precipitation Heat Treated, UNS A03550
4214	Aluminum Alloy Sand Castings 5.0Si - 1.2Cu - 0.50Mg (355.0-T71) Solution Heat Treated and Stabilized, UNS A03550
4215	Aluminum Alloy Castings 5.0Si - 1.2Cu - 0.50Mg (C355.0-T6P) Solution and Precipitation Heat Treated, UNS A33550
4216	Aluminum Alloy Sheet 1.0Mg - 0.8Si - 0.8Cu - 0.50Mn (6013-T6) Solution Heat Treated and Artificially Aged, UNS A96013
4217	Aluminum Alloy Sand Castings 7.0Si - 0.32Mg (356.0-T6) Solution and Precipitation Heat Treated, UNS A03560
4218	Aluminum Alloy Castings 7.0Si - 0.35Mg (A356.0-T61P) Solution and Precipitation Heat Treated, UNS A13560
4219	Aluminum Alloy Castings 7.0Si - 0.55Mg - 0.12Ti - 0.06Be (A357.0 T61) Solution and Precipitation Heat Treated, UNS A13570
4221	Aluminum Alloy Plate 4.4Cu - 1.5Mg - 0.60Mn (2124-T8151) Solution Heat Treated, Stress Relieved and Precipitation Heat Treated, UNS A92124
4223	Aluminum Alloy Castings 4.5Cu - 0.70Ag - 0.30Mn - 0.25Mg - 0.25Ti (A201.0-T4) Solution Heat Treated and Naturally Aged, UNS A12010
4224	Aluminum Alloy Sand Castings 4.0Cu - 2.1Ni - 2.0Mg - 0.30Cr - 0.30Mn - 0.13Ti - 0.13V (243.0) Stabilized, UNS A02430

SAE/AMS SPECIFICATIONS - ALUMINUM & ALUMINUM ALLOYS (Continued)

AMS	Title
4225	Aluminum Alloy Sand Castings, Heat Resistant 5.0Cu - 1.5Ni - 0.25Mn - 0.25Sb - 0.25Co - 0.20Ti - 0.20Zr (203.0P) Solution Heat Treated and Stabilized, UNS A02030
4229	Aluminum Alloy Castings, High Strength 4.5Cu - 0.70Ag - 0.30Mn - 0.25Mg - 0.25Ti (A201.0-T7) Solution Heat Treated and Overaged, UNS A02010
4232	Aluminum Alloy Extrusions 2.7Cu - 2.2Li - 0.12Zr (2090-T86) Solution Heat Treated, Cold Worked, and Precipitation Heat Treated, UNS A92090
4233	Aluminum Alloy Welding Wire 4.5Cu - 0.70Ag - 0.30Mn - 0.25Mg - 0.25Ti For Welding 201 Type Alloys
4235	Aluminum Alloy Castings 4.6Cu - 0.35Mn - 0.25Mg - 0.22Ti (A206.0-T71) Solution and Precipitation Heat Treated, UNS A12060
4236	Aluminum Alloy Castings 4.6Cu - 0.35Mn - 0.25Mg - 0.22Ti (A206.0-T4) Solution Heat Treated and Naturally Aged, UNS A12060
4237	Aluminum Alloy Sand Castings 4.6Cu - 0.35Mn - 0.25Mg - 0.22Ti (206.0-T71) Solution Heat Treated and Naturally Aged, UNS A02060
4238	Aluminum Alloy Sand Castings 6.8Mg - 0.18Ti - 0.18Mn (535.0-F) As Cast, UNS A05350
4239	Aluminum Alloy Sand Castings 6.8Mg - 0.18Ti - 0.18Mn (535.0-0) Annealed, UNS A05350
4241	Aluminum Alloy Castings 7.0Si - 0.58Mg - 0.15Ti - 0.06Be (D357.0-T6) Solution and Precipitation Heat Treated, Dendrite Arm Spacing (DAS) Controlled, UNS A43570
4242	Aluminum Alloy Castings, Sand Composite 4.7Cu - 0.60Ag - 0.35Mn - 0.25Mg - 0.25Ti (B201.0-T7) Solution Heat Treated and Overaged, Aircraft Structural Quality, UNS A02010
4243	Aluminum Alloy Sheet, Alclad 6.2Zn - 2.3Cu - 2.2Mg - 0.12Zr (Alclad 7050-T76) Solution Heat Treated and Overaged, UNS A87050
4244	Aluminum Alloy Welding Wire 4.6Cu - 0.35Mn - 0.25Mg - 0.22Ti for Welding 206 Type Alloys
4245	Aluminum Alloy Welding Wire 5.0Si - 1.2Cu - 0.50Mg For Welding 355 Type Alloys
4246	Aluminum Alloy Welding Wire 7.0Si - 0.52Mg For Welding 357 Type Alloys, UNS A03570
4247	Aluminum Alloy Hand Forgings 7.7Zn - 2.4Mg - 1.6Cu - 0.16Cr (7049-T7352) Solution Heat Treated, Stress Relieved by Compression, and Precipitation Heat Treated, UNS A97049
4248	Aluminum Alloy Hand Forgings and Rings 1.0Mg - 0.60Si - 0.28Cu - 0.20Cr (6061-T652) Solution Heat Treated, Stress Relief Compressed, and Precipitation Heat Treated, UNS A96061
4249	Aluminum Alloy Castings 7.0Si - 0.58Mg - 0.15Ti - 0.06Be (D357.0-T6) Solution and Precipitation Heat Treated, UNS A43570
4251	Aluminum Alloy Sheet 2.7Cu - 2.2Li - 0.12Zr (2090-T83) Solution Heat Treated, Cold Worked, and Precipitation Heat Treated, UNS A92090
4252	Aluminum Alloy Plate 6.4Zn - 2.4Mg - 2.2Cu - 0.12Zr (7150-T7751) Solution Heat Treated, Stress Relieved, and Overaged, UNS A97150
4253	Aluminum Alloy-Aramid Fiber Reinforced-Laminated Sheet 5.7Zn - 2.2Mg - 1.6Cu - 0.22Cr (7475-T61) 3, 5, 7, or 9 Ply
4254	Aluminum Alloy, Aramid-Fiber-Reinforced Laminated Sheet 4.4Cu - 1.5Mg - 0.6Mn (2024-T3) 3, 5, 7, or 9 Ply

SAE/AMS SPECIFICATIONS - ALUMINUM & ALUMINUM ALLOYS (Continued)	
AMS	**Title**
4255	Aluminum Alloy Sheet, Clad One Side 0.6Mg - 0.35Si - 0.28Cu (No. 21 Brazing Sheet) As Fabricated
4256	Aluminum Alloy Sheet, Clad Two Sides 0.6Mg - 0.35Si - 0.28Cu (No. 22 Brazing Sheet) as Fabricated, UNS A86951
4258	Aluminum Alloy, Aramid-Fiber-Reinforced, Laminated Sheet 5.7Zn - 2.2Mg - 1.6Cu - 0.22Cr (Alclad, One Side 7475-T761) 5 or 7 Ply
4260	Aluminum Alloy Investment Castings 7.0Si - 0.32Mg (356.0-T6) Solution and Precipitation Heat Treated, UNS A03560
4261	Aluminum Alloy Investment Castings 7.0Si - 0.32Mg (356.0-T51) Precipitation Heat Treated, UNS A03560
4280	Aluminum Alloy Permanent Mold Castings 5.0Si - 1.2Cu - 0.5Mg (355.0-T71) Solution Heat Treated and Overaged, UNS A03550
4281	Aluminum Alloy Permanent Mold Castings 5.0Si - 1.2Cu - 0.50Mg (355.0-T6) Solution and Precipitation Heat Treated, UNS A03550
4284	Aluminum Alloy Permanent Mold Castings 7.0Si - 0.30Mg (356.0-T6) Solution and Precipitation Heat Treated, UNS A03560
4285	Aluminum Alloy Centrifugal Mold Castings 7.0Si - 0.30Mg (356.0-T6) Solution and Precipitation Heat Treated, UNS A03560
4286	Aluminum Alloy Permanent Mold Castings 7.0Si - 0.32Mg (356.0-T51) Precipitation Heat Treated, UNS A03560
4290	Aluminum Alloy Die Castings 9.5Si - 0.50Mg - (360.0) as Cast, UNS A03600
4291	Aluminum Alloy Die Castings 8.5Si - 3.5Cu (A380.0) as Cast, UNS A13800
4300	Boron-Aluminum Composite Sheet 50 v/o 5.6B, 6061-0, For Diffusion Bonding
4302	Aluminum Alloy, Aramid-Fiber-Reinforced, Laminated Sheet 5.7Zn - 2.2Mg - 1.6Cu - 0.22Cr (7475-T761) 3, 5, 7, or 9 Ply
4303	Aluminum Alloy Plate 1.7Cu - 2.2Li - 0.12Zr (2092-T81) Solution Heat Treated, Cold Worked, and Aged, UNS A92090
4306	Aluminum Alloy Plate 6.4Zn - 2.4Mg - 2.2Cu - 0.12Zr (7150-T6151) Solution Heat Treated, Stress Relieved, and Aged, UNS A97150
4307	Aluminum Alloy Extrusions 6.4Zn - 2.4Mg - 2.2Cu - 0.12Zr (7150-T61511) Solution Heat Treated, Stress Relieved, and Precipitation Heat Treated, UNS A97150
4310	Aluminum Alloy Rings, Rolled or Forged 5.6Zn - 2.5Mg - 1.6Cu - 0.23Cr (7075-T651, 7075-T652) Solution Heat Treated, Mechanically Stress Relieved, and Precipitation Heat Treated, UNS A97075
4311	Aluminum Alloy Rings, Rolled or Forged 5.6Zn - 2.5Mg - 1.6Cu - 0.23Cr (7075-T7351, 7075-T7352) Solution Heat Treated, Mechanically Stress Relieved, and Precipitation Heat Treated, UNS A97075
4312	Aluminum Alloy Rings, Rolled or Forged 1.0Mg - 0.60Si - 0.28Cu - 0.20Cr (6061-T651, 6061-T652) Solution Heat Treated, Mechanically Stress Relieved, and Precipitation Heat Treated, UNS A96061
4313	Aluminum Alloy Rings, Rolled or Forged 6.3Cu - 0.30Mn - 0.18Zr - 0.10V - 0.06Ti (2219-T351, 2219-T352) Solution Heat Treated and Mechanically Stress Relieved, UNS A92219

SAE/AMS SPECIFICATIONS - ALUMINUM & ALUMINUM ALLOYS (Continued)

AMS	Title
4314	Aluminum Alloy Rings, Rolled or Forged 4.5Cu - 0.85Si - 0.80Mn - 0.50Mg (2014-T651, 2014-T652) Solution Heat Treated, Mechanically Stress Relieved, and Precipitation Heat Treated, UNS A92014
4320	Aluminum Alloy Forgings 7.7Zn - 2.5Mg - 1.5Cu - 0.16Cr (7149-T73) Solution and Precipitation Heat Treated, UNS A97149
4321	Aluminum Alloy Forgings 7.7Zn - 2.5Mg - 1.5Cu - 0.16Cr (7049-O1) High Temperature Annealed, UNS A97049
4322	Aluminum Alloy Die Forgings 4.0Mg - 1.3Li - 1.2C - 0.45O_2 (5091) Mechanically Alloyed, As Fabricated, UNS A95091
4323	Aluminum Alloy Hand Forgings 5.6Zn - 2.5Mg - 1.6Cu - 0.23Cr (7075- T7452) Solution Heat Treated, Stress Relieved, and Precipitation Heat Treated, UNS A97075
4333	Aluminum Alloy Die Forgings 6.2Zn - 2.3Cu - 2.2Mg - 0.12Zr (7050- T7452) Solution Heat Treated, Compression Stress Relieved, and Overaged, UNS A97050
4340	Aluminum Alloy Extrusions 6.2Zn - 2.3Cu - 2.2Mg - 0.12Zr (7050-T76511) Solution Heat Treated, Stress Relieved, and Overaged, UNS A97050
4341	Aluminum Alloy Extrusions 6.2Zn - 2.3Cu - 2.2Mg - 0.12Zr (7050-T73511) Solution Heat Treated, Stress Relieved, and Overaged, UNS A97050
4342	Aluminum Alloy Extrusions 6.2Zn - 2.3Cu - 2.2Mg - 0.12Zr (7050-T74511) (Formerly -T736511) Solution Heat Treated, Stress Relieved, and Overaged, UNS A97050
4343	Aluminum Alloy Extrusions 7.7Zn - 2.4Mg - 1.6Cu - 0.16Cr (7149-T73511) Solution Heat Treated, Stress Relieved, and Overaged, UNS A97149
4344	Aluminum Alloy Extrusions 5.6Zn - 2.5Mg - 1.6Cu - 0.23Cr (7175-T73511) Solution Heat Treated, Stress Relieved, and Overaged, UNS A97175
4345	Aluminum Alloy Extrusions 6.4Zn - 2.4Mg - 2.2Cu - 0.12Zr (7150-T77511) Solution Heat Treated, Stress Relieved, and Overaged, UNS A97150
4347	Aluminum Alloy Sheet 1.0Mg - 0.8Si - 0.8Cu - 0.5Mn (6013-T4) Solution Heat Treated and Naturally Aged, UNS A96013
4348	Core, Honeycomb, Aluminum Alloy, Corrosion Inhibited for Sandwich Construction 5052, 350 (177), UNS A95052
4349	Core, Honeycomb, Aluminum Alloy, Corrosion Inhibited for Sandwich Construction, 5056, 350 (177), UNS A95056

ASTM SPECIFICATIONS - ALUMINUM & ALUMINUM ALLOYS

ASTM	Title
Bars, Rods, Wire, and Shapes	
B 211	Bar, Rod, and Wire, Aluminum and Aluminum-Alloy
B 221	Extruded Bars, Rods, Wire, Shapes, and Tubes, Aluminum and Aluminum-Alloy
B 236	Aluminum Bars for Electrical Purposes (Bus Bars)

ASTM SPECIFICATIONS - ALUMINUM & ALUMINUM ALLOYS (Continued)

ASTM	Title
Bars, Rods, Wire, and Shapes (Continued)	
B 308	Standard Structural Shapes, Rolled or Extruded, Aluminum-Alloy 6061-T6
B 316	Rivet and Cold Heading Wire and Rods, Aluminum and Aluminum-Alloy
B 317	Aluminum-Alloy Extruded Bar, Rod, Pipe, and Structural Shapes for Electrical Purposes
Castings	
B 26	Sand Castings, Aluminum-Alloy
B 85	Die Castings, Aluminum-Alloy
B 108	Permanent Mold Castings, Aluminum-Alloy
B 618	Aluminum-Alloy Investment Castings
B 686	Aluminum-Alloy Castings, High-Strength
Products for Electrical Purposes	
B 230	Wire for Electrical Purposes, Aluminum 1350-H19
B 236	Bars for Electrical Purposes (Bus Bars), Aluminum
B 314	Wire for Communication Cable, Aluminum 1350
B 317	Extruded Bar, Rod, Pipe, and Structural Shapes for Electrical Purposes (Bus Conductors), Aluminum-Alloy
B 324	Wire for Electrical Purposes, Rectangular and Square Aluminum
B 373	Foil for Capacitors, Aluminum
B 396	Wire for Electrical Purposes, 5005-H19 Aluminum-Alloy
B 398	Wire for Electrical Purposes, 6201-T81 Aluminum-Alloy
Heat Treatment	
B 597	Heat Treatment of Aluminum Alloys
B 807	Extrusion Press Solution Heat Treatment for Aluminum Alloys
Fasteners	
F 467	Nonferrous Nuts for General Use
F 468	Nonferrous Bolts, Hex Cap Screws, and Studs for General Use
Forgings	
B 247	Die Forgings, Hand Forgings, and Rolled Ring Forgings, Aluminum and Aluminum-Alloy

ASTM SPECIFICATIONS - ALUMINUM & ALUMINUM ALLOYS (Continued)

ASTM	Title
Ingots	
B 179	Sand Castings, Permanent Mold Castings, and Die Castings, Aluminum Alloys in Ingot Form for
Pipe and Tubes	
B 210	Drawn Seamless Tubes, Aluminum and Aluminum-Alloy
B 221	Extruded Bars, Rod, Wire, Shape, and Tube, Aluminum-Alloy
B 234	Drawn Seamless Tubes for Condensers and Heat Exchangers, Aluminum and Aluminum-Alloy
B 241	Seamless Pipe and Seamless Extruded Tube, Aluminum and Aluminum-Alloy
B 313	Round Welded Tubes, Aluminum and Aluminum-Alloy
B 345	Seamless Extruded Tube and Seamless Pipe for Gas and Oil Transmission and Distribution Piping Systems, Aluminum and Aluminum-Alloy
B 404	Seamless Condenser and Heat-Exchanger Tubes with Integral Fins, Aluminum and Aluminum-Alloy
B 429	Extruded Structural Pipe and Tube, Aluminum-Alloy
B 483	Drawn Tubes for General Purpose Applications, Aluminum and Aluminum-Alloy
B 491	Extruded Round Tubes for General Purpose Applications, Aluminum and Aluminum-Alloy
B 547	Formed and Arc-Welded Round Tube, Aluminum and Aluminum-Alloy
B 745	Corrugated Aluminum Pipe for Sewers and Drains
B 789	Installing Corrugated Aluminum Structural Plate Pipe for Culverts and Sewers
B 790	Structural Design of Corrugated Aluminum, Pipe, Pipe Arches and Arches for Culverts, Storm Sewers and Other Buried Conduits
Sheet, Plate and Foil	
B 209	Sheet and Plate, Aluminum and Aluminum-Alloy
B 373	Foil for Capacitors, Aluminum
B 479	Foil for Flexible Barrier Applications, Annealed Aluminum and Aluminum Alloy
B 632	Tread Plate, Aluminum-Alloy Rolled
B 736	Cable Shielding Stock, Aluminum, Aluminum Alloy, and Aluminum-Clad, Steel
B 744	Sheet for Corrugated Aluminum Pipe, Aluminum Alloy for
B 746	Structural Plate for Field Bolted Pipe, Pipe Arches and Arches, Corrugated Aluminum Alloy
Welding Fittings	
B 361	Welding Fittings, Factory-Made Wrought, Aluminum and Aluminum-Alloy

AMERICAN CROSS REFERENCED SPECIFICATIONS - ALUMINUM AND ALUMINUM ALLOYS							
AA	**UNS**	**AMS**	**MIL**	**FED**	**ASTM**	**ASME**	**AWS**
1060	A91060	4000	---	---	B 209 (1060), B 210 (1060), B 211 (1060), B 221 (1060), B 234 (1060), B 241 (1060), B 345 (1060), B 361 (1060), B 404 (1060), B 483 (1060), B 548 (1060)	---	---
1100	A91100	4001, 4003, 4062, 4102, 4180, 7220	MIL-W-85, MIL-R-5674, MIL-W-6712, MIL-A-12545, MIL-E-15597 (MIL-1100), MIL-S-24149/5, MIL-C-26094, MIL-A-52174, MIL-A-52177, MIL-I-23413 (MIL-1100)	QQ-A-225/1, QQ-A-250/1, QQ-A-430, QQ-A-1876, WW-T-700/1	B 209 (1100), B 210 (1100), B 211 (1100), B 221 (1100), B 241 (1100), B 247 (1100), B 313 (1100), B 316 (1100), B 361 (1100), B 479 (1100), B 483 (1100), B 491 (1100), B 547 (1100), B 548 (1100)	SFA5.3 (E1100), SFA5.10 (ER1100)	A5.3 (E1100), A5.10 (ER1100)
1145	A91145	4011	---	QQ-A-1876	B 373 (1145, B 479 (1145)	---	---
1235	A91235	---	---	QQ-A-1876	B 373 (1235), B 479 (1235), B 491 (1235)	---	---
1350	A91350	---	---	---	B 230, B 231, B 232 (1350), B 233, B 236, B 314 (1350), B 324, B 400 (1350), B 401, B 524, B 549, B 609, B 778 (1350), B 779 (1350)	---	---
2011	A92011	---	---	QQ-A-225/3	B 210 (2011), B 211 (2011)	---	---
2014	A92014	4028, 4029, 4121, 4133, 4134, 4153, 4314	MIL-F-5509, MIL-A-12545, MIL-T-15089, MIL-F-18280, MIL-A-22771	QQ-A-200/2, QQ-A-225/4, QQ-A-367	B 209 (2014), B 210 (2014), B 211 (2014), B 221 (2014), B 241 (2014), B 247 (2014)	---	---
2017	A92017	4118	MIL-R-5674	QQ-A-225/5, QQ-A-430	B 211 (2017), B 316 (2017)	---	---

AMERICAN CROSS REFERENCED SPECIFICATIONS - ALUMINUM AND ALUMINUM ALLOYS (Continued)

AA	UNS	AMS	MIL	FED	ASTM	ASME	AWS
2018	A92018	4140	---	QQ-A-367	B 247 (2018)	---	---
2024	A92024	4007, 4035, 4037, 4086, 4087, 4088, 4112, 4120, 4152, 4164, 4165, 4192, 4193, 7223	MIL-F-5509, MIL-R-5674, MIL-B-6812, MIL-T-15089, MIL-F-18280, MIL-T-50777, MIL-A-81596,	QQ-A-200/3, QQ-A-225/6, QQ-A-250/4, QQ-A-430, WW-T-700/3	B 209 (2024), B 210 (2024), B 211 (2024), B 221 (2024), B 241 (2024), B 316 (2024), F 467, F 468	---	---
2025	A92025	4130	---	QQ-A-367	B 247	---	---
2117	A92117	7222	MIL-R-5674, MIL-R-8814	QQ-A-430	B 316 (2117)	---	---
2124	A92124	4101, 4221	---	QQ-A-250/9, QQ-A-250/29	B 209 (2124)	---	---
2218	A92218	---	---	QQ-A-367	B 247 (2218)	---	---
2219	A92219	4031, 4066, 4068, 4143, 4144, 4162, 4163, 4313	MIL-A-22771, MIL-A-46118, MIL-A-46808	QQ-A-250/30, QQ-A-367, QQ-A-430	B 209 (2219), B 211 (2219), B 221 (2219), B 241 (2219), B 247 (2219), B 316 (2219)	---	---
2319	A92319	4191	---	---	---	SFA5.10 (ER2319)	A5.10 (ER2319)
2618	A92618	4132	MIL-A-22771 (2618)	QQ-A-367	B 247 (2618)	---	---
3003	A93003	4006, 4008, 4010, 4065, 4067	MIL-S-12875, MIL-E-15597 (MIL-3003), MIL-P-25995, MIL-A-52174, MIL-A-81596	QQ-A-200/1, QQ-A-225/E, QQ-A-250/2, QQ-A-430, WW-T-700/2	B 209 (3003), B 210 (3003), B 211 (3003), B 221 (3003), B 234 (3003), B 241 (3003), B 247 (3003), B 313 (3003), B 316 (3003), B 345 (3003), B 404 (3003), B 483 (3003), B 491 (3003), B 547 (3003)	SFA5.3 (E3003)	A5.3 (E3003)

AMERICAN CROSS REFERENCED SPECIFICATIONS - ALUMINUM AND ALUMINUM ALLOYS (Continued)

AA	UNS	AMS	MIL	FED	ASTM	ASME	AWS
3004	A93004	---	---	---	B 209 (3004), B 221 (3004) B 313 (3004), B 547 (3004), B 548 (3004)		---
3005	A93005	---	---	---	B 209 (3005)		---
3105	A93105	---	---	---	B 209 (3105)		---
4032	A94032	4145	---	QQ-A-367	B 247 (4032)		---
4043	A94043	4190	MIL-W-6712, MIL-E-15597 (MIL-4043), MIL-I-23413 (MIL-4043)	---	---	SFA5.3 (E4043), SFA5.10 (ER4043)	A5.3 (E4043), A5.10 (ER4043)
4045	A94045	---	MIL-B-7883 (BAlSi-5)	---	---	SFA5.8 (BAlSi-5)	A5.8 (BAlSi-5)
4047	A94047	4185	MIL-B-7883 (BAlSi-4), MIL-B-20148	---	---	SFA5.8 (BAlSi-4), SFA5.10 (ER4047)	A5.8 (BAlSi-4), A5.10 (ER4047)
4145	A94145	4184	MIL-B-7883 (BAlSi-3)	---	---	SFA5.8 (BAlSi-3), SFA5.10 (ER4145)	A5.8 (BAlSi-3), A5.10 (ER4145)
4343	A94343	---	MIL-B-7883 (BAlSi-2), MIL-B-20148	---	---	SFA5.8 (BAlSi-2)	A5.8 (BAlSi-2)
5005	A95005	---	MIL-C-26094	QQ-A-430 (5005)	B 209 (5005), B 210 (5005) B 316 (5005), B 396 (5005), B 397 (5005), B 483 (5005), B 531 (5005)	---	---
5050	A95050	---	---	---	B 209 (5050), B 210 (5050), B 313 (5050), B 483 (5050), B 547 (5050), B 548 (5050)	---	---

AMERICAN CROSS REFERENCED SPECIFICATIONS - ALUMINUM AND ALUMINUM ALLOYS (Continued)

AA	UNS	AMS	MIL	FED	ASTM	ASME	AWS
5052	A95052	4004, 4015, 4016, 4017, 4069, 4070, 4071, 4114, 4175, 4178, 4348	MIL-S-12875, MIL-G-18014, MIL-G-18015, MIL-C-26094, MIL-A-81596	QQ-A-225/7, QQ-A-250/8, QQ-A-430, WW-T-700/4	B 209 (5052), B 210 (5052), B 211 (5052), B 221 (5052), B 234 (5052), B 241 (5052), B 313 (5052), B 316 (5052), B 404 (5052), B 483 (5052), B 547 (5052), B 548 (5052)	---	---
5056	A95056	4005, 4176, 4177, 4182, 4349	MIL-R-5674, MIL-R-8814, MIL-A-81596	QQ-A-430	B 211 (5056), B 316 (5056)	---	---
5083	A95083	4056, 4057, 4058, 4059	MIL-A-45225, MIL-A-46027, MIL-A-46083, MIL-G-S-24149/2	QQ-A-200/4, QQ-A-250/6, QQ-A-367	B 209 (5083), B 210 (5083), B 221 (5083), B 241 (5083), B 247 (5083), B 345 (5083), B 361 (5083), B 547 (5083), B 548 (5083)	---	---
5086	A95086	---	MIL-G-18014, MIL-S-24149/2, MIL-C-26094	QQ-A-200/5, QQ-A-250/7, WW-T-700/5	B 209 (5086), B 210 (5086), B 221 (5086), B 241 (5086), B 313 (5086), B 345 (5086), B 361 (5086), B 547 (5086), B 548 (5086)	---	---
5154	A95154	---	MIL-C-26094	---	B 209 (5154), B 210 (5154), B 211 (5154), B 221 (5154), B 313 (5154), B 361 (5154), B 547 (5154), B 548 (5154)	---	---
5183	A95183	---	---	---	---	SFA5.10 (ER5183)	A5.10 (ER5183)
5252	A95252	---	---	---	B 209 (5252)	---	---
5254	A95254	---	---	---	B 209 (5254), B 241 (5254), B548 (5254)	---	---
5356	A95356	---	MIL-S-24149/2	---	---	SFA5.10 (ER5356)	A5.10 (ER5356)

AMERICAN CROSS REFERENCED SPECIFICATIONS - ALUMINUM AND ALUMINUM ALLOYS (Continued)

AA	UNS	AMS	MIL	FED	ASTM	ASME	AWS
5454	A95454	---	---	QQ-A-200/6, QQ-A-250/10	B 209 (5454), B 221 (5454), B 234 (5454), B 241 (5454), B 404 (5454), B 547 (5454), B 548 (5454)	---	---
5456	A95456	---	MIL-G-18014, MIL-S-24149/2, MIL-A-45225, MIL-A-46027, MIL-A-46083	QQ-A-200/7, QQ-A-250/9	B 209 (5456), B 210 (5456), B 221 (5456), B 241 (5456), B 548 (5456)	---	---
5457	A95457	---	---	---	B 209 (5457)	---	---
5554	A95554	---	MIL-G-18014	---	---	SFA5.10 (ER5554)	A5.10 (ER5554)
5556	A95556	---	---	---	---	SFA5.10 (ER5556)	A5.10 (ER5556)
5652	A95652	---	---	---	B 209 (5652), B 241 (5652), B 548 (5652)	---	---
5654	A95654	---	---	---	---	SFA5.10 (ER5654)	A5.10 (ER5654)
5657	A95657	---	---	---	B 209 (5657)	---	---
6005, 6005A	A96005	---	---	---	B 221 (6005, 6005A)	---	---
6053	A96053	---	---	QQ-A-430	B 316 (6053)	---	---

AMERICAN CROSS REFERENCED SPECIFICATIONS - ALUMINUM AND ALUMINUM ALLOYS (Continued)

AA	UNS	AMS	MIL	FED	ASTM	ASME	AWS
6061	A96061	4009, 4025, 4026, 4027, 4079, 4080, 4081, 4082, 4083, 4113, 4115, 5116, 4117, 4127, 4128, 4146, 4150, 4160, 4161, 4172, 4173, 4248, 4312 MAM 4248	MIL-W-85, MIL-F-3922, MIL-T-7081, MIL-A-12545, MIL-G-18014, MIL-G-18015, MIL-F-18280, MIL-A-22771, MIL-W-23351, MIL-P-25995, MIL-F-39000	QQ-A-200/8, QQ-A-200/16, QQ-A-225/8, QQ-A-250/11, QQ-A-367, QQ-A-430, WW-T-700/6	B 209 (6061), B 210 (6061), B 211 (6061), B 221 (6061), B 234 (6061), B 241 (6061), B 247 (6061), B 308 (6061), B 313 (6061), B 316 (6061), B 345 (6061), B 361 (6061), B 404 (6061), B 429 (6061), B 483 (6061), B 547 (6061), B 548, B 632, F 467, F 468	---	---
6063	A96063	4156	MIL-W-85, MIL-G-18014, MIL-G-18015, MIL-P-25995	QQ-A-200/9	B 210 (6063), B 221 (6063), B 241 (6063), B 345 (6063), B 361 (6063), B 429 (6063), B 483 (6063), B 491 (6063)	---	---
6066	A96066	---	---	QQ-A-200/10, QQ-A-367	B 221 (6066), B 247 (6066)	---	---
6070	A96070	---	MIL-A-12545, MIL-A-46104	---	B 221 (6070), B 345 (6070)	---	---
6101	A96101	---	---	---	B 317 (6101)	---	---
6105	A96105	---	---	---	B 221 (6105)	---	---
6151	A96151	4125	MIL-C-10387, MIL-A-22771	QQ-A-367	B247	---	---
6162	A96162	---	---	QQ-A-200/17		---	---
6201	A96201	---	---	---	B 398 (6201), B 399 (6201), B 524 (6201), B 711 (6201)	---	---

AMERICAN CROSS REFERENCED SPECIFICATIONS - ALUMINUM AND ALUMINUM ALLOYS (Continued)							
AA	**UNS**	**AMS**	**MIL**	**FED**	**ASTM**	**ASME**	**AWS**
6262	A96262	---	---	QQ-A-225/10	B210 (6262), B 211 (6262), B 221 (6262), B 467, B 483 (6262)	---	---
6351	A96351	---	---	---	B 221 (6351), B 241 (6351), B 345 (6351)	---	---
6463	A96463	---	---	---	B 221 (6463)	---	---
7005	A97005	---	---	---	B 221 (7005)	---	---
7049	A97049	4111, 4157, 4159, 4200, 4247, 4321 MAM 4247	---	QQ-A-367	B 247 (7049)	---	---
7050	A97050	4050, 4107, 4108, 4201, 4340, 4341, 4342, 4343	---	QQ-A-430	B 247 (7050), B 316 (7050)	---	---

AMERICAN CROSS REFERENCED SPECIFICATIONS - ALUMINUM AND ALUMINUM ALLOYS (Continued)							
AA	**UNS**	**AMS**	**MIL**	**FED**	**ASTM**	**ASME**	**AWS**
7075	A97075	4044, 4045, 4078, 4122, 4123, 4124, 4126, 4131, 4141, 4147, 4154, 4166, 4167, 4168, 4169, 4174, 4186, 4187, 4310, 4311, 4049, 4139, 4323 MAM 4131, 4141, 4323	MIL-F-5509, MIL-A-12545, MIL-F-18280, MIL-A-22771	QQ-A-200/11, QQ-A-200/15, QQ-A-225/9, QQ-A-250/13, QQ-A-250/24, QQ-A-367, QQ-A-430, WW-T-700/7	B 209 (7075), B 210 (7075) B 211 (7075), B 221 (7075), B 241 (7075), B 247 (7075), B 316 (7075), B 468 (7075)	---	---
7175	A97175	4148, 4149, 4179, 4344	---	---	B 247 (7175)	---	---
7178	A97178		---	QQ-A-200/13, QQ-A-200/14, QQ-A-250/14, QQ-A-250/21, QQ-A-250/28, QQ-A-430	B 209 (7178), B 221 (7178), B 241 (7178), B 316 (7178)	---	---
7475	A97475	4084, 4085, 4089, 4090, 4202	---	---	---	---	---

a. This cross-reference table lists the basic specification or standard number, and since these standards are constantly being revised, it should be kept in mind that they are presented herein as a guide and may not reflect the latest revision.

AWS WELDING FILLER METAL SPECIFICATIONS - ALUMINUM & ALUMINUM ALLOYS	
AWS	**Title**
A5.3	Aluminum and Aluminum Alloy Electrodes for Shielded Metal Arc Welding
A5.10	Bare Aluminum and Aluminum Alloy Welding Electrodes and Rods

BRITISH BSI GENERAL SPECIFICATIONS - ALUMINUM & ALUMINUM ALLOYS	
BS	**Title**
1470	Specification for wrought aluminium and aluminium alloys for general engineering purposes: plate, sheet and strip
1471	Specification for wrought aluminium and aluminium alloys for general engineering purposes - drawn tube
1472	Specification for wrought aluminium and aluminium alloys for general engineering purposes - forging stock and forgings
1473	Specification for wrought aluminium and aluminium alloys for general engineering purposes - rivet, bolt and screw stock
1474	Specification for wrought aluminium and aluminium alloys for general engineering purposes - bars, extruded round tubes and sections
1475	Specification for wrought aluminium and aluminium alloys for general engineering purposes - wire
1490	Specification for aluminium and aluminium alloy ingots and castings for general engineering purposes
2627	Specification for wrought aluminium for electrical purposes - wire
2901	Filler rods and wires for gas-shielded arc welding
2901 Part 4	Specification for aluminium and aluminium alloys and magnesium alloys
3019 Part 1	Specification for TIG welding of aluminium, magnesium and their alloys
3571 Part 1	Specification for MIG welding of aluminium and aluminium alloys
5812 Part 1	Specification for aluminium alloy for solid bearings

BRITISH BSI AEROSPACE SERIES SPECIFICATIONS - ALUMINUM & ALUMINUM ALLOYS	
BS	**Title**
6L 16	Specification for sheet and strip of 99% aluminium (temper designation H14 or H24)
6L 17	Specification for sheet and strip of 99% aluminium (temper designation-O)
5L 34	Specification for forging stock, bars, extruded sections and forgings of 99% aluminium
4L 35	Specification for ingots and castings of 'Y' aluminium alloy (heat treated) (suitable for pistons) (Cu 4, Mg 1.5, Ni 2.1)
5L 36	Specification for wire for solid, cold-forged rivets of 99.5% aluminium (not exceeding 10 mm diameter)
7L 37	Specification for wire for solid cold-forged rivets of aluminium-copper-magnesium-silicon-manganese alloy (for use in the solution treated and naturally aged condition) (not exceeding 10 mm diameter) (Cu 4.4, Mg 0.5, Si 0.7, Mn 0.8) (2014A)
5L 44	Specification for forging stock, bars, extruded sections and forgings of aluminium - 2 ¼% magnesium alloy
3L 51	Specification for ingots and castings of aluminium-silicon-copper-iron-nickel-magnesium alloy (precipitation treated) (Si 2.5, Cu 1.2, Fe 1, Ni 1, Mg 0.1)
3L 52	Specification for ingots and castings of aluminium-copper-silicon-nickel-magnesium-iron alloy (solution treated and precipitation treated) (suitable for pistons) (Cu 2.2, Si 1.3, Ni 1.3, Mg 1.1, Fe 1.1)
4L 53	Specification for ingots and castings of aluminium-magnesium alloy (solution treated) Mg 10.2)
4L 54	Specification for tube of 99% aluminium (cold drawn: seamless: tested hydraulically) (not exceeding 12 mm wall thickness)
4L 56	Specification for tube of aluminium - 2 ¼% magnesium alloy (temper designation-O) (seamless: tested hydraulically) (not exceeding 12 mm wall thickness)
3L 58	Specification for wire for solid, cold-forged rivets of aluminium - 5% magnesium alloy (not exceeding 10 mm diameter)
4L 59	Specification for sheet and strip of aluminium-manganese alloy (temper designation H16 or H26)
4L 60	Specification for sheet and strip of aluminium-manganese alloy (temper designation H12 or H22)
4L 61	Specification for sheet and strip of aluminium-manganese alloy (temper designation-O)
3L 63	Specification for tube of aluminium-copper-magnesium-silicon-manganese alloy (solution treated and precipitation treated) (Cu 4.4, Mg 0.5, Si 0.7, Mn 08)
2L 77	Specification for forging stock and forgings of aluminium-copper-magnesium-silicon-manganese alloy (solution treated and precipitation treated) (Cu 4.4, Mg 0.5, Si 0.7, Mn 0.8)
3L 78	Specification for ingots and castings of aluminium-silicon-copper-magnesium alloy (solution treated and precipitation treated) (Si 5, Cu 1.2, Mg 0.5)
3L 80	Specification for sheet and strip of aluminium - 2 ¼% magnesium alloy (temper designation-O)
3L 81	Specification for sheet and strip of aluminium - 2 ¼% magnesium alloy (temper designation H16 or H26)

BRITISH BSI AEROSPACE SERIES SPECIFICATIONS - ALUMINUM & ALUMINUM ALLOYS (Continued)

BS	Title
2L 83	Specification for forging stock, bars, extruded sections and forgings of aluminium-copper-nickel-magnesium-iron-silicon alloy (solution treated and precipitation treated) (Cu 2, Ni 1, Mg 1, Si 0.9, Fe 0.9)
2L 84	Specification for bars and extruded sections of aluminium-copper-silicon-magnesium alloy (solution treated and aged at room temperature) (not exceeding 200 mm diameter or minor sectional dimension) (Cu 1.5, Si 1, Mg 0.8)
2L 85	Specification for forging stock, bars, extruded sections and forgings of aluminium-copper-silicon-magnesium alloy (solution treated and precipitation treated) (Cu 1.5, Si 1, Mg 0.8)
3L 86	Specification for wire for solid, cold-forged rivets of aluminium-copper-magnesium alloy (not exceeding 10 mm diameter) (Cu 2.5, Mg 0.3)
2L 87	Specification for hexagonal bars for nuts, couplings and hollow machined parts of aluminium-copper-magnesium-silicon-manganese alloy (solution treated and precipitation treated) (free from peripheral and asymmetric coarse grain) (not less than 14 mm nor more than 36 mm across flats) (Cu 4.4, Mg 0.5, Si 0.7, Mn 0.8)
2L 88	Specification for aluminium-alloy-coated sheet and strip of aluminium-zinc-magnesium-copper-chromium alloy (solution treated and precipitation treated) (Zn 5.8, Mg 2.5, Cu 1.6, Cr 0.15)
2L 91	Specification for ingots and castings of aluminium-copper alloy (165 MN/m^2 0.2 per cent proof stress) (Cu 4.5)
2L 92	Specification for ingots and castings of aluminium-copper alloy (200 MN/m^2 0.2 per cent proof stress) (Cu 4.5)
2L 93	Specification for plate of aluminium-copper-magnesium-silicon-manganese alloy (solution treated, controlled stretched and precipitation treated) (Cu 4.4, Mg 0.5, Si 0.7, Mn 0.8)
2L 95	Specification for plate of aluminium-zinc-magnesium-copper-chromium alloy (solution treated, controlled stretched and precipitation treated) (Zn 5.8, Mg 2.5, Cu 1.6, Cr 0.15)
2L 97	Specification for plate of aluminium-copper-magnesium-manganese alloy (solution treated, controlled stretched and aged at room temperature) (Cu 4.4, Mg 1.5, Mn 0.6)
2L 98	Specification for plate of aluminium-copper-magnesium-manganese alloy (solution treated and aged at room temperature: not controlled stretched) (Cu 4.4, Mg 1.5, Mn 0.6)
2L 99	Specification for ingots and castings of aluminium-silicon-magnesium alloy (solution treated and precipitation treated) (Si 7, Mg 0.3)
3L 100	Procedure for inspection and testing of wrought aluminium and aluminium alloys
4L 101	Procedure for inspection, testing and acceptance of aluminium-base and magnesium-base ingots and castings
L 102	Specification for bars and extruded sections of aluminium-copper-magnesium-silicon-manganese alloy (solution treated and aged at room temperature) (not exceeding 200 mm diameter or minor sectional dimension) (Cu 4.4, Mg 0.5, Si 0.7, Mn 0.8)

BRITISH BSI AEROSPACE SERIES SPECIFICATIONS - ALUMINUM & ALUMINUM ALLOYS (Continued)

BS	Title
L 103	Specification for forging stock and forgings of aluminium-copper-magnesium-silicon-manganese alloy (solution treated and aged at room temperature) (Cu 4.4, Mg 0.5, Si 0.7, Mm 0.8)
L 105	Specification for tube of aluminium-copper-magnesium-silicon-manganese alloy (solution treated and aged at room temperature) (not exceeding 10 mm wall thickness) (Cu 4.4, Mg 0.5, Si 0.7, Mn 0.8)
L 109	Specification for aluminium-coated sheet and strip of aluminium-copper-magnesium-manganese alloy (solution treated and aged at room temperature) (Cu 4.4, Mg 1.5, Mn 0.6)
L 110	Specification for aluminium-coated sheet and strip of aluminium-copper-magnesium-manganese alloy (supplied for solution treatment by the user) (Cu 4.4, Mg 1.5, Mn 0.6)
L 111	Specification for bars and extruded sections of aluminium-magnesium-silicon-manganese alloy (solution treated and precipitation treated) (suitable for welding) (Mg 0.8, Si 1, Mn 0.7)
L 112	Specification for forging stock and forgings of aluminium-magnesium-silicon-manganese alloy (solution treated and precipitation treated) (suitable for welding) (Mg 0.8, Si 1, Mn 0.7)
L 113	Specification for sheet and strip of aluminium-magnesium-silicon-manganese alloy (solution treated and precipitation treated) (suitable for welding) (Mg 0.8, Si 1, Mn 0.7)
L 114	Specification for tube of aluminium-magnesium-silicon-manganese alloy (solution treated and precipitation treated) (not exceeding 10 mm wall thickness) (suitable for welding) (Mg 0.8, Si 1, Mn 0.7)
L 115	Specification for plate of aluminium-magnesium-silicon-manganese alloy (solution treated, controlled stretched and precipitation treated) (not exceeding 25 mm wall thickness) (suitable for welding) (Mg 0.8, Si 1, Mn 0.7)
2L 116	Specification for tube of 99% aluminium (cold drawn: seamless, not tested hydraulically) (not exceeding 12 mm wall thickness)
L 117	Specification for tube of aluminium-magnesium-silicon-copper-chromium alloy (solution treated and artificially aged: not tested hydraulically) (not exceeding 10 mm wall thickness) (Mg 1.0, Si 0.6, Cu 0.28, Cr 0.2)
L 118	Specification for tube of aluminium-magnesium-silicon-copper-chromium alloy (solution treated and artificially aged: tested hydraulically) (not exceeding 10 mm wall thickness) (Mg 1.0, Si 0.6, Cu 0.28, Cr 0.2)
L 119	Specification for ingots and castings of aluminium-copper-nickel-manganese-titanium-zirconium-cobalt-antimony alloy (solution treated and artificially aged) (Cu 5.0, Ni 1.5, Mn 0.25, Ti 0.2, Zr 0.2, Co 0.2, Sb 0.2)
3L 122	Specification for ingots and castings of magnesium - 8% aluminium-zinc-manganese alloy (solution treated) (Al 8, Zn 0.5, Mn 0.3)
L 154	Specification for ingots and castings of aluminium-copper-silicon alloy (solution treated and aged at room temperature) (Cu 4, Si 1)
L 155	Specification for ingots and castings of aluminium-copper-silicon alloy (solution treated and artificially aged) (Cu 4, Si 1)

BRITISH BSI AEROSPACE SERIES SPECIFICATIONS - ALUMINUM & ALUMINUM ALLOYS (Continued)

BS	Title
L 156	Specification for sheet and strip of aluminium-copper-magnesium-silicon-manganese alloy (solution treated and aged at room temperature) (Cu 4.4, Mg 0.5, Si 0.8, Mn 0.8)
L 157	Specification for sheet and strip of aluminium-copper-magnesium-silicon-manganese alloy (solution treated and artificially aged) (Cu 4.4, Mg 0.5, Si 0.8, Mn 0.8)
L 158	Specification for close toleranced sheet and strip of aluminium-copper-magnesium-silicon-manganese alloy (solution treated and aged at room temperature) (Cu 4.4, Mg 0.5, Si 0.8, Mn 0.8)
L 159	Specification for close toleranced sheet and strip of aluminium-copper-magnesium-silicon-manganese alloy (solution treated and artificially aged) (Cu 4.4, Mg 0.5, Si 0.8, Mn 0.8)
L 160	Specification for bars and extruded sections of aluminium-zinc-magnesium-copper-chromium alloy (solution treated and artificially aged to an overaged condition) (Zn 5.6, Mg 2.5, Cu 1.6, Cr 0.22)
L 161	Specification for hand and die forgings of aluminium-zinc-magnesium-copper-chromium alloy (solution treated and artificially aged to an overaged condition) (Zn 5.6, Mg 2.5, Cu 1.6, Cr 0.22)
L 162	Specification for cold compressed hand forgings of aluminium-zinc-magnesium-copper-chromium alloy (solution treated and artificially aged to an overaged condition) (Zn 5.6, Mg 2.5, Cu 1.6, Cr 0.22)
L 163	Specification for sheet and strip of aluminium-coated aluminium-copper-magnesium-silicon-manganese alloy (solution treated, cold worked for flattening and aged at room temperature) (Cu 4.4, Mg 0.5, Si 0.8, Mn 0.8)
L 164	Specification for sheet and strip of aluminium-coated aluminium-copper-magnesium-silicon-manganese alloy (solution treated and aged at room temperature) (Cu 4.4, Mg 0.5, Si 0.8, Mn 0.8)
L 165	Specification for sheet and strip of aluminium-coated aluminium-copper-magnesium-silicon-manganese alloy (solution treated and artificially aged) (Cu 4.4, Mg 0.5, Si 0.8, Mn 0.8)
L 166	Specification for close toleranced sheet and strip of aluminium-coated aluminium-copper-magnesium-silicon-manganese alloy (solution treated and aged at room temperature) (Cu 4.4, Mg 0.5, Si 0.8, Mn 0.8)
L 167	Specification for close toleranced sheet and strip of aluminium-coated aluminium copper-magnesium-silicon-manganese alloy (solution treated and artificially aged) (Cu 4.4, Mg 0.5, Si 0.8, Mn 0.8)
L 168	Specification for bars and extruded sections of aluminium-copper-magnesium-silicon-manganese alloy (solution treated and artificially aged) (not exceeding 200 mm diameter or minor sectional dimension) (Cu 4.4, Mg 0.5, Si 0.7, Mn 0.8)
L 169	Specification for ingots and castings of aluminium-silicon-magnesium alloy (solution treated and artificially aged) (Si 7, Mg 0.6)

BRITISH BSI AEROSPACE SERIES SPECIFICATIONS - ALUMINUM & ALUMINUM ALLOYS (Continued)

BS	Title
L 170	Specification for extruded bars and sections of aluminium-zinc-copper-chromium alloy (solution treated, controlled stretched and artificially aged) (not exceeding 150 mm diameter or minor sectional dimension) (Zn 5.6, Mg 2.5, Cu 1.6, Cr 0.23) (7075)
L 171	Specification for forging of aluminium-zinc-magnesium-manganese-copper alloy (supplied as-forged or annealed for subsequent heat treatment) (not exceeding 150 mm diameter or minor sectional dimension) (Zn 5.7, Mg 2.7, Mn 0.5, Cu 0.5) (7014)
L 172	Specification for extruded, rolled or cast forging stock of aluminium-zinc-magnesium-manganese-copper alloy (for manufacture of forgings to BS L 171) (Zn 5.7, Mg 2.7, Mn 0.5, Cu 0.5) (7014)
L 173	Specification for castings of aluminium-silicon-magnesium alloy, chill cast (solution treated and precipitation treated to an overaged (T7) condition)
L 174	Specification for castings of aluminium-silicon-magnesium alloy, sand case (solution treated and precipitation treated to an overaged (T7) condition)

GERMAN DIN AEROSPACE SPECIFICATIONS - ALUMINUM & ALUMINUM ALLOYS

DIN	Title
EN 2094	Aerospace series; aluminium alloy AL-P-7009T74; die forgings 3 mm $\leq$ a $\leq$ 150 mm; inactive for new design
EN 2100	Aerospace series; aluminium alloy AL-P2014A-T4511; extruded bards and sections a or D $\leq$ 200 mm
EN 2101	Aerospace series; chromic acid anodizing of aluminium and wrought aluminium alloys
EN 2126	Aerospace series; aluminium alloy AL-P7075-T651; plate 6 mm < a $\leq$ 80 mm; inactive for new design
EN 2127	Aerospace series; aluminium alloy AL-P7075-T73511; extruded bars and sections a or D $\leq$ 100 mm
EN 2128	Aerospace series; aluminium Alloy AL-P7075-T3511; drawn bars 6 mm < a or D $\leq$ 75 mm
EN 2289	Aerospace series; rod bodies, flight controls, in aluminium alloys, technical specification
EN 2318	Aerospace series; aluminium alloy AL-P2024-T3511; extruded bars and sections 1.2 mm $\leq$ a or D $\leq$ 150 mm
EN 2326	Aerospace series; aluminium alloy AL-P6082-T6; extruded bars and sections a or D $\leq$ 200 mm
EN 2381	Aerospace series; aluminium alloy AL-P7009-T7452; hand forgings 40 mm $\leq$ a $\leq$ 150 mm
EN 2384	Aerospace series; aluminium alloy AL-P2014A-T6511; extruded bars and sections a or D $\leq$ 150 mm
EN 2385	Aerospace series; aluminium alloy AL-P7009-T74511; extruded bars and sections a or D $\leq$ 125 mm; inactive for new design
EN 2500 Part 2	Aerospace series; instructions for the drafting and use of metallic material standards; part 2: specific requirements for aluminium, aluminium alloys and magnesium alloys
EN 2599	Aerospace series; strip in aluminium and aluminium alloys 0.3 $\leq$ a $\leq$ 3.2 mm; dimensions

GERMAN DIN AEROSPACE SPECIFICATIONS - ALUMINUM & ALUMINUM ALLOYS (Continued)

DIN	Title
EN 2615	Aerospace series; wire to close tolerance in aluminium and aluminium alloys $1.6 \leq D \leq 9.6$ mm; dimensions
EN 2616	Aerospace series; wire for rivets in aluminium and aluminium alloys, large tolerances $D \leq 10$ mm; dimensions
EN 2637	Aerospace series; aluminium alloy 7075-T73; extruded bar and section $1.2 \leq$ (a or D) ≤ 100 mm with coarse peripheral grain control
EN 2638	Aerospace series; aluminium alloy 2024-T3; extruded bar and section $1.2 \leq$ (a or D) ≤ 150 mm with coarse peripheral grain control
EN 2639	Aerospace series; aluminium alloy 2014A-T6; extruded bar and section $1.2 \leq$ (a or D) ≤ 150 mm with coarse peripheral grain control
EN 2644	Aerospace series; rod assemblies for flight controls; technical specification
EN 2691	Aerospace series; aluminium alloy (2017A); solution treated, water quench, cold worked and naturally aged (T3); sheet and strip $0.4 \leq a \leq 6$ mm
EN 2697	Aerospace series; aluminium alloy (2214); solution treated, water quench and artificially aged (T6); extruded bar and section $1.2 \leq$ (a or D) ≤ 100 mm with coarse peripheral grain control
EN 2698	Aerospace series; aluminium alloy (7075); solution treated, water quench, controlled stretched and artificially aged (T6510); extruded bar and section $1.2 \leq$ (a or D) ≤ 100 mm
EN 2699	Aerospace series; aluminium alloy (5086); annealed and straightened (H111); drawn bar $6 \leq D \leq 50$ mm
EN 2700	Aerospace series; aluminium alloy (6061); solution treated, water quench and artificially aged (T6); drawn bar $6 \leq D \leq 75$ mm, with coarse peripheral grain control
EN 2701	Aerospace series; aluminium alloy (2024); solution treated, water quench, cold worked and naturally aged (T3); drawn tube for structures $0.6 \leq a \leq 12.5$ mm
EN 2702	Aerospace series; aluminium alloy 6061-T6; extruded bar and section $1.2 \leq$ (a or D) ≤ 150 mm
EN 2715	Aerospace series; macrographic examination of aluminium and aluminium alloy forging stock, forgings and wrought products
EN 2716	Aerospace series; test method for susceptibility to intergranular corrosion of wrought products in 2XXX series aluminium alloys
EN 2717	Aerospace series; test method for susceptibility to intergranular corrosion of wrought products in 5XXX series aluminium alloys with a magnesium content $\geq 3.5\%$
EN 2720	Aerospace series; test method for metallic materials; testing of susceptibility to exfoliation corrosion in 2XXX and 7XXX series wrought aluminium alloy products for aerospace constructions
EN 2721	Aerospace series; aluminium alloy Al-C12-T4; sand castings
EN 2722	Aerospace series; aluminium alloy Al-C12-T4; chill castings
EN 2723	Aerospace series; aluminium alloy Al-C12-T6; sand castings

GERMAN DIN AEROSPACE SPECIFICATIONS - ALUMINUM & ALUMINUM ALLOYS (Continued)

DIN	Title
EN 2724	Aerospace series; aluminium alloy Al-C12-T6; chill castings
EN 2725	Aerospace series; aluminium alloy Al-C14-T6; sand castings
EN 2726	Aerospace series; aluminium alloy A1-C26-T6; sand castings
EN 2727	Aerospace series; aluminium alloy Al-C26-T6; chill castings
EN 2728	Aerospace series; aluminium alloy Al-C27-T6; sand castings
EN 2729	Aerospace series; aluminium alloy Al-C27-T6; chill castings
EN 2802	Aerospace series; aluminium alloy 7475-T761; sheet and strip $0.8 \leq a \leq 6$ mm
EN 2803	Aerospace series; aluminium alloy 7475-T761; clad sheet and strip $0.8 \leq a \leq 6$ mm
EN 2804	Aerospace series; aluminium alloy 7075-T7651; plate $6 < a \leq 25$ mm
EN 2805	Aerospace series; aluminium alloy 7475-T7651; plate $6 < a \leq 25$ mm
EN 2806	Aerospace series; aluminium alloy 2024-T42; extruded section $1.2 \leq a \leq 100$ mm with coarse peripheral grain control
EN 2807	Aerospace series; aluminium alloy 7020-T6; extruded section $1.2 \leq a \leq 100$ mm with coarse peripheral grain control
EN 2813	Aerospace series; aluminium alloy 6061-T6; tube for hydraulics $0.6 \leq a \leq 12.5$ mm
EN 3350	Aerospace series; aluminium, aluminium alloys; temper designations
EN 6025	Aerospace series; plates; close-tolerance flatness; aluminium alloys
WL 3.1124 Part 1	Aerospace; wrought aluminium alloy with approx. 2.6Cu-0.4Mg; rivet wire and rivets
WL 3.1124 Part 100	Aerospace; wrought aluminium alloy with approx. 2.6Cu-0.4Mg; rivet wire and rivets; design and production data, conversion
WL 3.1254 Part 1	Aerospace; wrought aluminium alloy with approx. 4.5Cu-0.9Si-0.8Mn-0.5Mg; die forgings
WL 3.1254 Part 2	Aerospace; wrought aluminium alloy with approx. 4.5Cu-0.9Si-0.8Mn-0.5Mg; hand forgings
WL 3.1254 Part 100	Aerospace; wrought aluminium alloy with approx. 4.5Cu-0.9Si-0.8Mn-0.5Mg; die forgings and hand forgings; design and production data, conversion
WL 3.1324 Part 1	Aerospace; wrought aluminium alloy with approx. 4.0Cu-0.7Mg; rivet wire and rivets
WL 3.1324 Part 100	Aerospace; wrought aluminium alloy with approx. 4.0Cu-0.7Mg; rivet wire and rivets; design and production data, conversion
WL 3.1354 Part 1	Aerospace; wrought aluminium alloy with approx. 4.4Cu-1.5Mg-0.6Mn; sheet and plate
WL 3.1354 Part 4	Aerospace; wrought aluminium alloy with approx. 4.4Cu-1.5Mg-0.6Mn; structural tubes
WL 3.1354 Part 100	Aerospace; wrought aluminium alloy with approx. 4.4Cu-1.5Mg-0.6Mn; sheet, plate, bars, extruded sections, structural tubes; design and production data, conversion

GERMAN DIN AEROSPACE SPECIFICATIONS - ALUMINUM & ALUMINUM ALLOYS (Continued)

DIN	Title
WL 3.1364 Part 1	Aerospace; wrought aluminium alloy with approx. 4.4Cu-1.5Mg-0.6Mn; clad sheet, strip and plate
WL 3.1364 Part 2	Aerospace; wrought aluminium alloy with approx. 4.4Cu-1.5Mg-0.6Mn; clad sheet and strip sections
WL 3.1364 Part 3	Aerospace; wrought aluminium alloy with approx. 4.4Cu-1.5Mg-0.6Mn; clad wedges
WL.3.1364 Part 100	Aerospace: wrought aluminium alloy with approx. 4.4Cu-1.5Mg-0.6Mn; clad sheet, strip and plate, sheet and strip sections and wedges; design and production data, conversion
WL 3.1734	Aluminium cast alloy with about 4Cu-1.5Mg-2Ni; sand casting
WL 3.1734 Supp. 1	Aluminium cast alloy with about 4Cu-1.5Mg-2Ni; sand casting
WL 3.1754 Part 1	Aerospace; cast aluminium alloy with approx. 5Cu-1.5Ni-0.25Mn-0.25Sb-0.25Co-0.2Ti-0.2Zr; sand casting and investment casting
WL 3.1754 Part 100	Aerospace; cast aluminium alloy with approx. 5Cu-1.5Ni-0.25Mn-0.25Sb-0.25Co-0.2Ti-0.2Zr; sand casting and investment casting; design and production data, conversion
WL 3.1854	Aluminium cast alloy with about 4.5Cu-0.2Mg-0.2Ti; sand casting and chill casting
WL 3.1854 Supp. 1	Aluminium cast alloy with about 4.5Cu-0.2Mg-0.2Ti; sand casting and chill casting
WL 3.1924 Supp. 1	Aluminium wrought alloy with about 2.3Cu-1.5Mg-1.1Ni-1.1Fe-0.2Si; plates, rods, drop forgings and smith hammer forgings
WL 3.1924 Part 1	Aluminium wrought alloy, Al-Cu2MgNi; plates
WL 3.1924 Part 2	Aluminium wrought alloy; Al-Cu2MgNi; rods
WL 3.1924 Part 4	Aluminium wrought alloy; Al-Cu2MgNi; drop forgings
WL 3.1924 Part 5	Aluminium wrought alloy; Al-Cu2MgNi; smith hammer forgings
WL 3.2134 Supp. 1	Aluminium cast alloy with about 5Si-1.3Cu-0.5Mg; sand casting and chill casting
WL 3.2134 Part 1	Aluminium cast alloy with about 5Si-1.3Cu-0.5Mg; sand casting
WL 3.2134 Part 2	Aluminium cast alloy with about 5Si-1.3Cu-0.5Mg; chill casting
WL 3.2364 Part 1	Cast aluminium alloy (Alcoa 356)
WL 3.2364 Part 2	Cast aluminium alloy (Alcoa 356)
WL 3.2374 Part 1	Aerospace; cast aluminium alloy with approx. 7Si-0.3Mg; sand casting
WL 3.2374 Part 2	Aerospace; cast aluminium alloy with approx. 7Si-0.3Mg; investment casting
WL 3.2374 Part 3	Aerospace; cast aluminium alloy with approx. 7Si-0.3Mg; permanent mould casting
WL 3.2374 Part 100	Aerospace; cast aluminium alloy with approx. 7Si-0.3Mg; sand casting, investment casting and permanent mould casting; design and production data, conversion

GERMAN DIN AEROSPACE SPECIFICATIONS - ALUMINUM & ALUMINUM ALLOYS (Continued)

DIN	Title
WL 3.2384 Part 1	Aerospace; cast aluminium alloy with approx. 7Si-0.6Mg; sand casting
WL 3.2384 Part 2	Aerospace; cast aluminium alloy with approx. 7Si-0.6Mg; special sand casting
WL 3.2384 Part 3	Aerospace; cast aluminium alloy with approx. 7Si-0.6Mg; investment casting
WL 3.2384 Part 4	Aerospace; cast aluminium alloy with approx. 7Si-0.6Mg; special investment casting
WL 3.2384 Part 5	Aerospace; wrought aluminium alloy with approx. 7Si-0.6Mg; filler metal for welding
WL 3.2384 Part 100	Aerospace; cast aluminium alloy with approx. 7Si-0.6Mg; sand casting, special sand casting and investment casting; design and production data, conversion
WL 3.3214 Part 1	Aerospace; wrought aluminium alloy with approx. 1.0Mg-0.6Si-0.3Cu-0.20Cr; sheet and plate
WL 3.3214 Part 2	Aerospace; wrought aluminium alloy with approx. 1.0Mg-0.6Si-0.3Cu-0.20Cr; bars
WL 3.3214 Part 3	Aerospace; wrought aluminium alloy with approx. 1.0Mg-0.6Si-0.3Cu-0.20Cr; sections
WL 3.3214 Part 4	Aerospace; wrought aluminium alloy with approx. 1.0Mg-0.6Si-0.3Cu-0.20Cr; die forgings and hand forgings
WL 3.3214 Part 5	Aerospace; wrought aluminium alloy with approx. 1.0Mg-0.6Si-0.3Cu-0.20Cr; internal pressure tubes
WL 3.3214 Part 6	Aerospace; wrought aluminium alloy with approx. 1.0Mg-0.6Si-0.3Cu-0.20Cr; structural tubes
WL 3.3214 Part 100	Aerospace; wrought aluminium alloy with approx. 1.0Mg-0.6Si-0.3Cu-0.20Cr; sheet, plate, bars, sections, die forgings and hand forgings, internal pressure tubes and structural tubes; design and production data, conversion
WL 3.3354	Aluminium wrought alloy with about 5Mg; rivet wires and rivets
WL 3.3354 Supp. 1	Aluminium wrought alloy with about 5Mg; rivet wires and rivets
WL 3.3354 Part 1	Aerospace; wrought aluminium alloy with approx. 5Mg; rivet wire and rivets
WL 3.3354 Part 100	Aerospace; wrought aluminium alloy with approx. 5Mg; rivet wire and rivets; design and production data, conversion
WL 3.3524 Supp. 1	Aluminium wrought alloy with about 2.5Mg-0.25Cr; sheets, strips and conduit pipes
WL 3.3524 Part 1	Aluminium wrought alloy with about 2.5Mg-0.25Cr; sheets and strips
WL 3.3524 Part 2	Aerospace; wrought aluminium alloy with approx. 2.5Mg-0.25Cr; internal pressure tubes and tubes for fluids, seamless
WL 3.3524 Part 100	Aerospace; wrought aluminium alloy with approx. 2.5Mg-0.25Cr; sheet, strip, internal pressure tubes and tubes for fluids, seamless; design and production data, conversion
WL 3.4144 Part 1	Aerospace; wrought aluminium alloy with approx. 6.2Zn-2.3Cu-2.2Mg-0.12Zr; stretched plate
WL 3.4144 Part 2	Aerospace; wrought aluminium alloy with approx. 6.2Zn-2.3Cu-2.2Mg-0.12Zr; die forgings
WL 3.4144 Part 3	Aerospace; wrought aluminium alloy with approx. 6.2Zn-2.3Cu-2.2Mg-0.12Zr; hand forgings

GERMAN DIN AEROSPACE SPECIFICATIONS - ALUMINUM & ALUMINUM ALLOYS (Continued)	
DIN	**Title**
WL 3.4144 Part 4	Aerospace; wrought aluminium alloy with approx. 6.2Zn-2.3Cu-2.2Mg-0.12Zr; rivet wire and rivets
WL 3.4144 Part 100	Aerospace; wrought aluminium alloy with approx. 6.2Zn-2.3Cu-2.2Mg-0.12Zr; stretched plate, die and hand forgings; design and production data, conversion
WL 3.4144 Part 100	Aerospace; wrought aluminium alloy with approx. 6.2Zn-2.3Cu-2.2Mg-0.12Zr; stretched plate, die and hand forgings; rivet wire and rivets; design and production data, conversion
WL 3.4334 Part 1	Aerospace; wrought aluminium alloy with approx. 5.6Zn-2.5Mg-1.6Cu-0.23Cr; die forgings
WL 3.4334 Part 2	Aerospace; wrought aluminium alloy with approx. 5.6Zn-2.5Mg-1.6Cu-0.23Cr; hand forgings
WL 3.4334 Part 100	Aerospace; wrought aluminium alloy with approx. 5.6Zn-2.5Mg-1.6Cu-0.23Cr; die forgings and hand forgings; design and production data, conversion
WL 3.4354 Supp. 1	Aluminium wrought alloy with about 6.0Zn-2.5Mg-0.95Cu-0.32Ag-0.18Cr; rods, pressed sections, drop forgings and smith hammer forgings
WL 3.4354 Part 1	Aluminium wrought alloy with about 6.0Zn-2.5Mg-0.95Cu-0.32Ag-0.18Cr; rods
WL 3.4354 Part 1	Aerospace; wrought aluminium alloy with approx. 6.0Zn-2.5Mg-0.95Cu-0.32Ag-0.18Cr; bars
WL 3.4354 Part 4	Aluminium wrought alloy with about 6.0Zn-2.5Mg-0.95Cu-0.32Ag-0.18Cr; smith hammer forgings
WL 3.4354 Part 4	Aerospace; wrought aluminium alloy with approx. 6.0Zn-2.5Mg-0.95Cu-0.32Ag-0.18Cr; hand forgings
WL 3.4354 Part 100	Aerospace; wrought aluminium alloy with approx. 6.0Zn-2.5Mg-0.95Cu-0.32Ag-.018Cr; bars, extruded sections, die forgings and hand forgings; design and production data, conversion
WL 3.4364 Part 1	Aerospace; wrought aluminium alloy with approx. 5.6Zn-2.5Mg-1.6Cu-0.23Cr; sheet and plate
WL 3.4364 Part 4	Aluminium wrought alloy; Al-Zn6MgCu; drop forgings
WL 3.4364 Part 4	Aerospace; wrought aluminium alloy with approx. 5.6Zn-2.5Mg-1.6Cu-0.23Cr; die forgings
WL 3.4364 Part 5	Aluminium wrought alloy; Al-Zn6MgCu; smith hammer forgings
WL 3.4364 Part 5	Aerospace; wrought aluminium alloy with approx. 5.6Zn-2.5Mg-1.6Cu-0.23Cr; hand forgings
WL 3.4364 Part 100	Aerospace; wrought aluminium alloy with approx. 5.6Zn-2.5Mg-1.6Cu-0.23Cr; sheet, plate, bars, extruded sections, die forgings and hand forgings; design and production data, conversion
WL 3.4374 Part 1	Aerospace; wrought aluminium alloy with approx. 5.6Zn-2.5Mg-1.6Cu-0.23Cr; clad sheet
WL 3.4374 Part 2	Aluminium wrought alloy with about 6Zn-2.5Mg-1.5Cu; key sheets, clad

GERMAN DIN AEROSPACE SPECIFICATIONS - ALUMINUM & ALUMINUM ALLOYS (Continued)

DIN	Title
WL 3.4374 Part 100	Aerospace; wrought aluminium alloy with approx. 5.6Zn-2.5Mg-1.6Cu-0.23Cr; sheet and tapered sheet, clad; design and production data, conversion
WL 3.4377 Part 1	Aerospace; wrought aluminium alloy with approx. 5.7Zn-2.2Mg-1.6Cu-0.22Cr; sheet and strip, clad
WL 3.4377 Part 2	Aerospace; wrought aluminium alloy with approx. 5.7Zn-2.2Mg-1.6Cu-0.22Cr; sheet and strip sections, clad
WL 3.4377 Part 100	Aerospace; wrought aluminium alloy with approx. 5.7Zn-2.2Mg-1.6Cu-0.22Cr; clad sheet and strip; design and production data, conversion
WL 3.4384 Part 1	Aerospace; wrought aluminium alloy with approx. 5.7Zn-2.2Mg-1.6cu-0.22Cr; stretched plate
WL 3.4384 Part 100	Aerospace; wrought aluminium alloy with approx. 5.7Zn-2.2Mg-1.6Cu-0.22Cr; stretched plate. design and production data, conversion
WL 3.4394 Part 1	Aerospace; wrought aluminium alloy with approx. 6.2Zn-2.4Mg-1.8Cu-0.13Zr; stretched plate
WL 3.4394 Part 2	Aerospace; wrought aluminium alloy with approx. 6.2Zn-2.4Mg-1.8Cu-0.13Zr; die forgings
WL 3.4394 Part 3	Aerospace; wrought aluminium alloy with approx. 6.2Zn-2.4Mg-1.8Cu-0.13Zr; hand forgings
WL 3.4394 Part 100	Aerospace; wrought aluminium alloy with approx. 6.2Zn-2.4Mg-1.8Cu-0.13Zr; stretched plate, die and hand forgings; design and production data, conversion

JAPANESE JIS SPECIFICATIONS - ALUMINUM & ALUMINUM ALLOYS

JIS	Title
C 3107	Half-hard-drawn aluminium wires for electric purposes
C 3108	Hard drawn aluminium wires for electric purposes
C 3109	Hard drawn aluminium stranded conductors
C 3110	Aluminium conductors steel reinforced
H 0001	Temper designation for aluminium and aluminium alloys
H 0201	Glossary of terms used in the surface treatment of aluminium
H 0521	Testing method for atmospheric corrosion of aluminium and aluminium alloys
H 0522	Methods of radiographic test and classification of radiographs for aluminium castings
H 4000	Aluminium and aluminium alloy sheets and plates, strip and coiled sheets
H 4040	Aluminium and aluminium alloy rods, bars, and wires
H 4080	Aluminium and aluminium alloy seamless pipes and tubes

JAPANESE JIS SPECIFICATIONS - ALUMINUM & ALUMINUM ALLOYS (Continued)

JIS	Title
H 4090	Aluminium and aluminium alloy welded pipes and tubes
H 4100	Aluminium and aluminium alloy extruded shapes
H 4140	Aluminium and aluminium alloy forgings
H 4160	Aluminium and aluminium alloy foils
H 4170	High purity aluminium foils
H 4180	Aluminium and aluminium alloy bus conductors
H 5202	Aluminium alloy castings
H 5302	Aluminium alloys die castings
H 8301	Aluminium spraying on iron and steel
H 8642	Aluminium coatings (hot-dipped) on iron or steel
H 8672	Methods of test for aluminium coating (hot-dipped) on iron or steel
H 8680	Test methods for thickness of anodic oxidation coatings on aluminium and aluminium alloys
H 8681	Test methods for corrosion resistance of anodic oxidation coatings on aluminium and aluminium alloys
H 8682	Test methods for abrasion resistance of anodic oxidation coatings on aluminium and aluminium alloys
H 8683	Test methods for sealing quality of anodic oxidation coatings on aluminium and aluminium alloys
H 8684	Test method for resistance to cracking by deforming of anodic oxidation coatings on aluminium and aluminium alloys
H 8685	Accelerated test methods for light fastness of coloured anodic oxidation coatings on aluminium and aluminium alloys
H 8686	Test methods for image clarity of anodic oxidation coatings on aluminium and aluminium alloys
H 8771	Test methods for stress corrosion cracking on aluminium alloys
H 9126	Recommended practice for aluminium coating (hot-dipped)
H 9151	Recommended practice for aluminium alloy castings
H 9301	Recommended practice for aluminium spraying on iron and steel
H 9500	Recommended practice for anodizing on aluminium and aluminium alloys
H 9502	Recommended practice for combined coatings of anodic oxide and organic coatings on aluminium and aluminium alloys
W 1103	Heat treatment of aluminium alloys for aircraft
W 1111	Anodic coatings on aluminium and aluminium alloys for aircraft

CHEMICAL COMPOSITIONS OF WROUGHT ALUMINUM & ALUMINUM ALLOYS	
UNS	**Chemical Composition**
A82004	Core Alloy A92004 Cladding Alloy A91070
A82014	Core Alloy A92014 Cladding Alloy A96003
A82024	Core Alloy A92024 Cladding Alloy A91230
A82219	Core Alloy A92219 Cladding Alloy A97072
A83003	Core Alloy A93003 Cladding Alloy A94343
A86061	Core Alloy A96061 Cladding Alloy A97072
A87050	Core Alloy A97050 Cladding Alloy A97072
A87075	Core Alloy A97075 Cladding Alloy A97072 or A97011
A87178	Core Alloy A97178 Cladding Alloy A97072
A87475	Core Alloy A97475 Cladding Alloy A97072
A91035	Al 99.35 min Cu 0.10 max Fe 0.6 max Mg 0.05 max Mn 0.05 max Si 0.35 max Ti 0.03 max V 0.05 max Zn 0.10 max Other each 0.03 max
A91045	Al 99.45 min Cu 0.10 max Fe 0.45 max Mg 0.05 max Mn 0.05 max Si 0.30 max Ti 0.03 max V 0.05 max Zn 0.05 max Other each 0.03 max
A91050	Al 99.50 min Cu 0.05 max Fe 0.40 max Mg 0.05 max Mn 0.05 max Si 0.25 max Ti 0.03 max V 0.05 max Zn 0.05 max Other each 0.03 max
A91060	Al 99.60 min Cu 0.05 max Fe 0.35 max Mg 0.03 max Mn 0.03 max Si 0.25 max Ti 0.03 max V 0.05 max Zn 0.05 max Other each 0.03 max (Be 0.0008 max for welding electrode and filler metal only)
A91080	Al 99.80 min Cu 0.03 max Fe 0.15 max Ga 0.03 max Mg 0.02 max Mn 0.02 max Si 0.15 max Ti 0.03 max V 0.03 max Zn 0.03 max Other each 0.02 max
A91100	Al 99.00 min Cu 0.05-0.20 Mn 0.05 max Zn 0.10 max Other each 0.05 max (Be 0.0008 max for welding electrode and filler wire only), total 0.15 max, Si+Fe 0.95 max
A91145	Al 99.45 min Cu 0.05 max Mg 0.05 max Mn 0.05 max Ti 0.03 max Zn 0.05 max Other each 0.03 max, Si+Fe 0.55 max
A91188	Al 99.88 min Cu 0.005 max Fe 0.06 max Ga 0.03 max Mg 0.01 max Mn 0.01 max Si 0.06 max Ti 0.01 max V0.03 max Zn 0.03 max Other each 0.01 max
A91200	Al 99.00 min Cu 0.05 max Mn 0.05 max Zn 0.10 max Other each 0.05 max, total 0.15 max, Si+Fe 1.0 max
A91230	Al 99.30 min Cu 0.10 max Mn 0.05 max Zn 0.10 max Other each 0.05 max, plus Si+Fe 0.7 max
A91235	Al 99.35 min Cu 0.05 max Mg 0.05 max Mn 0.05 max Ti 0.03 max Zn 0.010 max Other each 0.03 max, Si+Fe 0.65 max
A91350	Al 99.50 min B 0.05 max Cr 0.01 max Cu 0.05 max Fe 0.40 max Mn 0.01 max Si 0.10 max Zn 0.05 max Other Gallium 0.03 max; Vanadium + Titanium 0.02 max; other unspecified elements each 0.03 max, total 0.10 max

CHEMICAL COMPOSITIONS OF WROUGHT ALUMINUM & ALUMINUM ALLOYS (Continued)

UNS	Chemical Composition
A91435	Al 99.35 min Cu 0.02 max Fe 0.30-0.50 Mg 0.05 max Mn 0.05 max Si 0.15 max Ti 0.03 max Zn 0.10 max Other each 0.03 max
A92004	Al rem Cu 5.5-6.5 Fe 0.20 max, Mg 0.50 max Mn 0.10 max Si 0.20 max Ti 0.05 max Zn 0.10 max Zr 0.30-0.50 Other each 0.05 max, total 0.15 max
A92011	Al rem Bi 0.20-0.6 Cu 5.0-6.0 Fe 0.7 max Pb 0.20-0.6 Si 0.40 max Zn 0.30 max Other each 0.05 max, total 0.15 max
A92014	Al rem Cr 0.10 max Cu 3.9-5.0 Fe 0.7 max Mg 0.20-0.8 Mn 0.40-1.2 Si 0.50-1.2 Ti 0.15 max Zn 0.25 max Other each 0.05 max (Be 0.0008 max for welding electrode and filler metal only), total 0.15 max, Ti+Zn 0.20 max
A92017	Al rem Cr 0.10 max Cu 3.5-4.5 Fe 0.7 max Mg 0.40-0.8 Mn 0.40-1.0 Si 0.8 max Zn 0.25 max Other each 0.05 max, total 0.15 max, Ti+Zn 0.20 max
A92018	Al rem Cr 0.10 max Cu 3.5-4.5 Fe 1.0 max Mg 0.45-0.9 Mn 0.20 max Ni 1.7-2.3 Si 0.9 max Zn 0.25 max Other each 0.05 max, total 0.15 max
A92024	Al rem Cr 0.10 max Cu 3.8-4.9 Fe 0.50 max Mg 1.2-1.8 Mn 0.30-0.9 Si 0.50 max Zn 0.25 max Other each 0.05 max,total 0.15 max
A92025	Al rem Cr 0.10 Cu 3.9-5.0 Fe 1.0 max Mg 0.05 max Mn 0.40-1.2 Si 0.50-1.2 Ti 0.15 max Zn 0.25 max Other each 0.05 max, total 0.15 max
A92090	Al rem Cr 0.05 max Cu 2.4-3.0 Fe 0.12 max Li 1.9-2.6 Mg 0.25 max Mn 0.05 max Si 0.10 max Ti 0.15 max Zn 0.10 max Zr 0.08-0.15 Other each 0.05 max, total 0.15 max
A92117	Al rem Cr 0.10 max Cu 2.2-3.0 Fe 0.7 max Mg 0.20-0.50 Mn 0.20 max Si 0.8 max Zn 0.25 max Other each 0.05 max, total 0.15 max
A92124	Al rem Cr 0.10 max Cu 3.8-4.9 Fe 0.30 max Mg 1.2-1.8 Mn 0.30-0.90 Si 0.20 max Ti 0.15 max Zn 0.25 max Other each 0.05 max, total 0.15 max plus Ti+Zn 0.20 max
A92218	Al rem Cr 0.10 max Cu 3.5-4.5 Fe 1.0 max Mg 1.2-1.8 Mn 0.20 max Ni 1.7-2.3 Si 0.9 max Zn 0.25 max Other each 0.05 max, total 0.15 max
A92219	Al rem Cu 5.8-6.8 Fe 0.30 max Mg 0.02 max Mn 0.20-0.40 Si 0.20 max Ti 0.02-0.10 V 0.05-0.15 Zn 0.10 max Zr 0.10-0.25 Other each 0.05 max, total 0.15 max
A92319	Al rem Be 0.0008 max Cu 5.8-6.8 Fe 0.30 max Mg 0.02 max Mn 0.20-0.40 Si 0.20 max Ti 0.10-0.20 V 0.05-0.15 Zn 0.10 max Zr 0.10-0.25 Other each 0.05 max (Be 0.0008 max for welding electrode and filler wire only), total 0.15 max
A92618	Al rem Cu 1.9-2.7 Fe 0.9-1.3 Mg 1.3-1.8 Ni 0.9-1.2 Si 0.25 max Ti 0.04-0.10 Other each 0.05 max, total 0.15 max
A93003	Al rem Cu 0.05-0.20 Fe 0.7 max Mn 1.0-1.5 Si 0.6 max Zn 0.10 max Other each 0.05 max (Be 0.0008 max for welding electrode and filler wire only), total 0.15 max
A93004	Al rem Cu 0.25 max Fe 0.7 max Mg 0.8-1.3 Mn 1.0-1.5 Si 0.30 max Zn 0.25 max Other each 0.05 max (Be 0.0008 max for welding electrode and filler wire only), total 0.15 max
A93005	Al rem Cr 0.10 max Cu 0.30 max Fe 0.7 max Mg 0.20-0.6 Mn 1.0-1.5 Si 0.6 max Ti 0.10 max Zn 0.25 max Other each 0.05 max, total 0.15 max
A93102	Al rem Cu 0.10 max Fe 0.7 max Mn 0.05-0.40 Si 0.40 max Ti 0.10 max Zn 0.30 max Other each 0.05 max, total 0.15 max

CHEMICAL COMPOSITIONS OF WROUGHT ALUMINUM & ALUMINUM ALLOYS (Continued)

UNS	Chemical Composition
A93105	Al rem Cr 0.20 max Cu 0.30 max Fe 0.7 max Mg 0.20-0.8 Mn 0.30-0.8 Si 0.6 max Ti 0.10 max Zn 0.40 max Other each 0.05 max, total 0.15 max
A93303	Al rem Cu 0.05-0.20 Fe 0.7 max Mn 1.0-1.5 Si 0.6 max Zn 0.30 max Other each 0.05 max, total 0.15 max
A94004	Al rem Cu 0.25 max Fe 0.8 max Mg 1.0-2.0 Mn 0.10 max Si 9.0-10.5 Zn 0.20 max Other each 0.05 max, total 0.15 max
A94008	Al rem Be 0.0008 max Cu 0.05 max Fe 0.09 max Mg 0.30-0.45 Mn 0.05 max Si 6.5-7.5 Ti 0.04-0.15 Zn 0.05 max Other each 0.05 max, total 0.15 max
A94011	Al rem Cu 0.20 max Fe 0.20 max Mg 0.45-0.7 Mn 0.10 max Si 6.5-7.5 Ti 0.20 max Zn 0.10 max Other each 0.05 max, total 0.15 max; Be 0.0008 for welding electrode and filler wire
A94032	Al rem Cr 0.10 max Cu 0.50-1.3 Fe 1.0 max Mg 0.8-1.3 Ni 0.50-1.3 Si 11.0-13.5 Zn 0.25 max Other each 0.05 max, total 0.15 max
A94043	Al rem Cu 0.30 max Fe 0.8 max Mg 0.05 max Mn 0.05 max Si 4.5-6.0 Ti 0.20 max Zn 0.10 max Other each 0.05 max (Be 0.0008 max for welding electrode and filler wire only), total 0.15 max
A94045	Al rem Cu 0.30 max Fe 0.8 max Mg 0.05 max Mn 0.05 max Si 9.0-11.0 Ti 0.20 max Zn 0.10 max Other each 0.05 max, total 0.15 max
A94047	Al rem Cu 0.30 max Fe 0.8 max Mg 0.10 max Mn 0.15 max Si 11.0-13.0 Zn 0.20 max Other each 0.05 max (Be 0.0008 max for welding electrode and filler wire only), total 0.15 max
A94104	Al rem Bi 0.02-0.20 Cu 0.25 max Fe 0.8 max Mg 1.0-2.0 Mn 0.10 max Si 9.5-10.5 Zn 0.20 max Other each 0.05 max, total 0.15 max
A94145	Al rem Cr 0.15 max Cu 3.3-4.7 Fe 0.8 max Mg 0.15 max Mn 0.15 max Si 9.3-10.7 Zn 0.20 max Other each 0.05 max (Be 0.0008 max for welding electrode and filler wire only), total 0.15 max
A94147	Al rem Be 0.0008 max Cu 0.25 max Fe 0.8 max Mg 0.10-0.50 Mn 0.10 max Si 11.0-13.0 Zn 0.20 max Other 0.05 max each, 0.15 max total
A94343	Al rem Cu 0.25 max Fe 0.8 max Mn 0.10 max Si 6.8-8.2 Zn 0.20 max Other each 0.05 max, total 0.15 max
A94643	Al rem Be 0.0008 max Cu 0.10 max Fe 0.8 max Mg 0.10-0.30 Mn 0.05 max Si 3.6-4.6 Ti 0.15 max Zn 0.10 max Other each 0.05 max, total 0.15 max
A95005	Al rem Cr 0.10 max cu 0.20 max Fe 0.7 max Mg 0.50-1.1 Mn 0.20 max Si 0.40 max Zn 0.25 max Other each 0.05 max, total 0.15 max
A95010	Al rem Cr 0.15 max Cu 0.25 max Fe 0.7 max Mg 0.20-0.6 Mn 0.10-0.30 Si 0.40 max Ti 0.10 max Zn 0.30 max Other each 0.05 max, total 0.15 max
A95050	Al rem Cr 0.10 max Cu 0.20 max Fe 0.7 max Mg 1.1-1.8 Mn 0.10 max Si 0.40 max Zn 0.25 max Other each 0.05 max (Be 0.0008 max for welding electrode and filler wire only), total 0.15 max
A95052	Al rem Cr 0.15-0.35 Cu 0.10 max Mg 2.2-2.8 Mn 0.10 max Zn 0.10 max Other each 0.05 max (Be 0.0008 max for welding electrode and filler wire only), total 0.15 max, plus Si+Fe 0.45 max
A95056	Al rem Cr 0.05-0.20 Cu 0.10 max Fe 0.40 max Mg 4.5-5.6 Mn 0.05-0.20 Si 0.30 max Zn 0.10 max Other each 0.05 max (Be 0.0008 max for welding electrode and filler wire only), total 0.15 max

CHEMICAL COMPOSITIONS OF WROUGHT ALUMINUM & ALUMINUM ALLOYS (Continued)	
UNS	**Chemical Composition**
A95083	Al rem Cr 0.05-0.25 Cu 0.10 max Fe 0.40 max Mg 4.0-4.9 Mn 0.40-1.0 Si 0.40 max Ti 0.15 max Zn 0.25 max Other each 0.05 max, total 0.15 max
A95086	Al rem Cr 0.05-0.25 Cu 0.10 max Fe 0.50 max Mg 3.5-4.5 Mn 0.20-0.7 Si 0.40 max Ti 0.15 max Zn 0.25 max Other each 0.05 max, total 0.15 max
A95154	Al rem Cr 0.15-0.35 Cu 0.10 max Fe 0.40 max Mg 3.1-3.9 Mn 0.10 max Si 0.25 max Ti 0.20 max Zn 0.20 max Other each 0.05 max (Be 0.0008 max for welding electrode and filler wire only), total 0.15 max
A95183	Al rem Be 0.0008 max Cr 0.05-0.25 Cu 0.10 max Fe 0.40 max Mg 4.3-5.2 Mn 0.50-1.0 Si 0.40 max Ti 0.15 max Zn 0.25 max Other each 0.05 max, total 0.15 max
A95252	Al rem Cu 0.10 max Fe 0.10 max Mg 2.2-2.8 Mn 0.10 max Si 0.08 max Other each 0.03 max, total 0.10 max
A95254	Al rem Cr 0.15-0.35 Cu 0.05 max Fe 0.40 max Mg 3.1-3.9 Mn 0.01 max Si 0.25 max Ti 0.05 max Zn 0.20 max Other each 0.05 max (Be 0.0008 max for welding electrode and filler wire only), total 0.15 max
A95356	Al rem Cr 0.05-0.20 Cu 0.10 max Fe 0.40 max Mg 4.5-5.5 Mn 0.05-0.20 Si 0.25 max Ti 0.06-0.20 Si 0.25 max Ti 0.06-0.20 Zn 0.10 max Other each 0.05 max (Be 0.0008 max for welding electrode and filler wire only), total 0.15 max
A95454	Al rem Cr 0.05-0.20 Cu 0.10 max Mg 2.4-3.0 Mn 0.50-1.0 Ti 0.20 max Zn 0.25 max Other each 0.05 max, total 0.15 max, Si+Fe 0.40 max
A95457	Al rem Cu 0.20 max Fe 0.10 max Mg 0.8-1.2 Mn 0.15-0.45 Si 0.08 max Zn 0.03 max Other each 0.03 max, total 0.10 max
A95554	Al rem Be 0.0008 max Cr 0.05-0.20 Cu 0.10 max Fe 0.40 max Mg 2.4-3.0 Mn 0.50-1.0 Si 0.25 max Ti 0.05-0.20 Zn 0.25 max Other each 0.05 max, total 0.15 max, Si+Fe 0.40 max
A95556	Al rem Be 0.0008 max Cr 0.05-0.20 Cu 0.10 max Fe 0.40 max Mg 4.7-5.5 Mn 0.50-1.0 Si 0.25 max Ti 0.05-0.20 Zn 0.25 max Other each 0.05 max, total 0.15 max
A95652	Al rem Cr 0.15-0.35 Cu 0.04 max Mg 2.2-2.8 Mn 0.01 max Zn 0.10 max Other each 0.05 max (Be 0.0008 max for welding electrode and filler wire only), total 0.15 max, Si+Fe 0.40 max
A95654	Al rem Be 0.0008 max Cr 0.15-0.35 Cu 0.05 max Mg 3.1-3.9 Mn 0.01 max Ti 0.05-0.15 Zn 0.20 max Other each 0.05 max, total 0.15 max, Si+Fe 0.45 max
A95657	Al rem Cu 0.10 max Fe 0.10 max Ga 0.03 max Mg 0.6-1.0 Mn 0.03 max Si 0.08 max Zn 0.03 max Other 0.02 max, total 0.05 max
A96003	Al rem Cr 0.35 max Cu 0.10 max Fe 0.06 max Mg 0.8-1.5 Mn 0.8 max Si 0.35-1.0 Ti 0.10 max Zn 0.20 max
A96005	Al rem Cr 0.10 max Cu 0.10 max Fe 0.35 max Mg 0.40-0.6 Mn 0.10 max Si 0.6-0.9 Ti 0.10 max Zn 0.10 max Other each 0.05 max, total 0.15 max
A96013	Al rem Cr 0.10 max Cu 0.6-1.1 Fe 0.50 max Mg 0.8-1.2 Mn 0.20-0.8 Si 0.6-1.0 Ti 0.10 max Zn 0.25 max Other each 0.05 max, total 0.15 max
A96053	Al rem Cr 0.15-0.35 Cu 0.10 max Fe 0.35 max Mg 1.1-1.4 Si 45.0-65.0 of Mg Zn 0.10 max Other each 0.05 max, total 0.15 max
A96060	Al rem Cr 0.05 max Cu 0.10 max Fe 0.10-0.30 Mg 0.35-0.6 Mn 0.10 max Si 0.30-0.6 Ti 0.10 max Zn 0.15 max Other each 0.05 max, total 0.15 max

CHEMICAL COMPOSITIONS OF WROUGHT ALUMINUM & ALUMINUM ALLOYS (Continued)

UNS	Chemical Composition
A96061	Al rem Cr 0.04-0.35 Cu 0.15-0.40 Fe 0.7 max Mg 0.8-1.2 Mn 0.15 max Si 0.40-0.8 Ti-0.15 max Zn 0.25 max Other each 0.05 max, total 0.15 max
A96063	Al rem Cr 0.10 max Cu 0.10 max Fe 0.35 max Mg 0.45-0.9 Mn 0.10 max Si 0.20-0.6 Ti 0.10 max Zn 0.10 max Other each 0.05 max, total 0.15 max
A96066	Al rem Cr 0.40 max Cu 0.7-1.2 Mg 0.8-1.4 Mn 0.6-1.1 Pb 0.50 max Si 0.9-1.8 Ti 0.20 max Zn 0.25 max Other each 0.05 max, total 0.15 max
A96070	Al rem Cr 0.10 max Cu 0.15-0.40 Fe 0.50 max Mg 0.50-1.2 Mn 0.40-1.0 Si 1.0-1.7 Ti 0.15 max Zn 0.25 max Other each 0.05 max, total 0.15 max
A96101	Al rem B 0.06 max Cr 0.03 max Cu 0.10 max Fe 0.50 max Mg 0.35-0.8 Mn 0.03 max Si 0.30-0.7 Zn 0.10 max Other each 0.03 max, total 0.10 max
A96105	Al rem Cr 0.10 max Cu 0.10 max Fe 0.35 max Mg 0.45-0.8 Mn 0.10 max Si 0.6-1.0 Ti 0.10 max Zn 0.10 max Other each 0.05 max, total 0.15 max
A96110	Al rem Cr 0.04-0.25 Cu 0.20-0.7 Fe 0.8 max Mg 0.50-1.1 Mn 0.20-0.7 Si 0.7-1.5 Ti 0.15 max Zn 0.30 max Other each 0.05 max, total 0.15 max
A96151	Al rem Cr 0.15-0.35 Cu 0.35 max Fe 1.0 max Mg 0.45-0.8 Mn 0.20 max Si 0.6-1.2 Ti 0.15 max Zn 0.25 max Other each 0.05 max, total 0.15 max
A96201	Al rem B 0.06 max Cr 0.03 max Cu 0.10 max Fe 0.50 max Mg 0.6-0.9 Mn 0.03 max Si 0.50-0.9 Zn 0.10 max Other each 0.03 max, total 0.10 max
A96253	Al rem Cr 0.15-0.35 Cu 0.10 max Fe 0.50 max Mg 1.0-1.5 Si 45.0-65.0 of Mg Zn 1.6-2.4 Other each 0.05 max, total 0.15 max
A96262	Al rem Bi 0.40-0.7 Cr 0.04-0.14 Cu 0.15-0.40 Fe 0.7 max Mg 0.8-1.2 Mn 0.15 max Pb 0.40-0.7 Si 0.40-0.8 Ti 0.15 max Zn 0.25 max Other each 0.05 max, total 0.15 max
A96351	Al rem Cu 0.10 max Fe 0.50 max Mg 0.40-0.8 Mn 0.40-0.8 Si 0.7-1.3 Ti - 0.20 max Zn 0.20 max Other each 0.05 max, total 0.15 max
A96463	Al rem Cu 0.20 max Fe 0.15 max Mg 0.45-0.9 Mn 0.05 max Si 0.20-0.6 Other each 0.05 max, total 0.15 max
A97001	Al rem Cr 0.18-0.35 Cu 1.6-2.6 Fe 0.40 max Mg 2.6-3.4 Mn 0.20 max Si 0.35 max Ti 0.20 max Zn 6.8-8.0 Other each 0.05 max, total 0.15 max
A97005	Al rem Cr 0.06-0.20 Cu 0.10 max Fe 0.40 max Mg 1.0-1.8 Mn 0.20-0.7 Si 0.35 max Ti 0.01-0.06 Zn 4.0-5.0 Zr 0.08-0.20 Other each 0.05 max, total 0.15 max
A97008	Al rem Cr 0.12-0.25 Cu 0.05 max Fe 0.10 max Mg 0.7-1.4 Mn 0.05 max Si 0.10 max Ti 0.05 max Zn 4.5-5.5 Other each 0.05 max, total 0.10 max
A97010	Al rem Cr 0.05 max Cu 1.5-2.0 Fe 0.15 max Mg 2.1-2.6 Mn 0.10 max Si 0.12 max Ti 0.6 max Zn 5.7-6.7 Zr 0.10-0.16 Other each 0.05 max, total 0.15 max
A97039	Al rem Cr 0.15-0.25 Cu 0.10 max Fe 0.40 max Mg 2.3-3.3 Mn 0.10-0.40 Si 0.30 max Ti 0.10 max Zn 3.5-4.5 Other each 0.05 max, total 0.15 max
A97049	Al rem Cr 0.10-0.22 Cu 1.2-1.9 Fe 0.35 max Mg 2.0-2.9 Mn 0.20 max Si 0.25 max Ti 0.10 max Zn 7.2-8.2 Other each 0.05 max, total 0.15 max
A97050	Al rem Cr 0.04 max Cu 2.0-2.6 Fe 0.15 max Mg 1.9-2.6 Mn 0.10 max Si 0.12 max Ti 0.06 max Zn 5.7-6.7 Zr 0.08-0.15 Other each 0.05 max, total 0.15 max
A97072	Al rem Cu 0.10 max Mg 0.10 max Mn 0.10 max Zn 0.8-1.3 Other each 0.05 max, total 0.15 max Si+Fe 0.7 max
A97075	Al rem Cr 0.18-0.28 Cu 1.2-2.0 Fe 0.50 max Mg 2.1-2.9 Mn 0.30 max Si 0.40 max Ti 0.20 max Zn 5.1-6.1 Other each 0.05 max, total 0.15 max
A97076	Al rem Cu 0.30-1.0 Fe 0.6 max Mg 1.2-2.0 Mn 0.30-0.8 Si 0.40 max Ti 0.20 max Zn 7.0-8.0 Other each 0.05 max, total 0.15 max

CHEMICAL COMPOSITIONS OF WROUGHT ALUMINUM & ALUMINUM ALLOYS (Continued)

UNS	Chemical Composition
A97116	Al rem Cu 0.50-1.1 Fe 0.30 max Ga 0.03 max Mg 0.8-1.4 Mn 0.05 max Si 0.15 max Ti 0.05 max V 0.05 max Zn 4.2-5.2 Other each 0.05 max, total 0.15 max
A97149	Al rem Cr 0.10-0.22 Cu 1.2-1.9 Fe 0.20 max Mg 2.0-2.9 Mn 0.20 max Si 0.15 max Ti 0.10 max Zn 7.2-8.2 Other each 0.05 max. total 0.15 max
A97150	Al rem Cr 0.04 max Cu 1.9-2.5 Fe 0.15 max Mg 2.0-2.7 Mn 0.10 max Si 0.12 max Ti 0.06 max Zn 5.9-6.9 Zr 0.08-0.15 Other each 0.05 max, total 0.15 max
A97175	Al rem Cr 0.18-0.28 Cu 1.2-2.0 Fe 0.20 max Mg 2.1-2.9 Mn 0.10 max Si 0.15 max Ti 0.10 max Zn 5.1-6.1 Other each 0.05 max, total 0.15 max
A97178	Al rem Cr 0.18-0.35 Cu 1.6-2.4 Fe 0.50 max Mg 2.4-3.1 Mn 0.30 max Si 0.40 max Ti 0.20 max Zn 6.3-7.3 Other each 0.05 max, total 0.15 max
A97277	Al rem Cr 0.18-0.35 Cu 0.8-1.7 Fe 0.7 max Mg 1.7-2.3 Si 0.50 max Ti 0.10 max Zn 3.7-4.3 Other each 0.05 max, total 0.15 max
A97475	Al rem Cr 0.18-0.25 Cu 1.2-1.9 Fe 0.12 max Mg 1.9-2.6 Mn 0.06 max Si 0.10 max Ti 0.06 max Zn 5.2-6.2 Other each 0.05 max, total 0.15 max
A98017	Al rem B 0.04 max Cu 0.10-0.20 Fe 0.55-0.8 Li 0.003 max Mg 0.01-0.05 Si 0.10 max Zn 0.05 max Other each 0.03 max, total 0.10 max
A98030	Al rem B 0.001-0.04 Cu 0.15-0.30 Fe 0.30-0.8 Mg 0.05 max Si 0.10 max Zn 0.05 max Other each 0.03 max, total 0.10 max
A98076	Al rem B 0.04 max Cu 0.04 max Fe 0.6-0.9 Mg 0.08-0.22 Si 0.10 max Zn 0.05 max Other each 0.03 max, total 0.10 max
A98079	Al rem Cu 0.05 max Fe 0.7-1.3 Si 0.05-0.30 Zn 0.10 max Other each 0.05 max, total 0.15 max
A98111	Al rem Cr 0.05 Cu 0.10 Fe 0.40-1.0 Mg 0.05 Mn 0.10 Si 0.30-1.1 Ti 0.08 Zn 0.10 Other each 0.05, total 0.15
A98130	Al rem Cu 0.05-0.15 Fe 0.40-1.0 Si 0.15 max Zn 0.10 max Other each 0.03 max, total 0.10 max
A98176	Al rem Fe 0.40-1.0 Ga 0.03 max Si 0.03-0.15 Zn 0.10 max Other each 0.05 max, total 0.15 max
A98177	Al rem B 0.04 Cu 0.04 Fe 0.25-0.45 Mg 0.04-0.12 Si 0.10 Zn 0.05 Other each 0.03, total 0.10
A98280	Al rem Cu 0.7-1.3 Fe 0.7 max Ni 0.20-0.7 Si 1.0-2.0 Sn 5.5-7.0 Ti 0.10 max Other each 0.05 max. total 0.15 max

rem - remainder; nom - nominal

CHEMICAL COMPOSITIONS OF CAST ALUMINUM & ALUMINUM ALLOYS

UNS	Title
A02010	Ag 0.40-1.2 Al rem Cu 4.0-5.2 Fe 0.15 max Mg 0.15-0.55 Mn 0.20-0.50 Si 0.10 max Ti 0.15-0.35 Other each 0.05 max,total 0.10 max
A02040	Al rem Cu 4.2-5.0 Fe 0.35 max Mg 0.15-0.35 Mn 0.10 max Ni 0.05 max Si 0.20 max Sn 0.05 max Ti 0.15-0.30 Zn 0.10 max Other each 0.05 max, total 0.15 max
A02080	Al rem Cu 3.5-4.5 Fe 1.2 max Mg 0.10 max Mn 0.50 max Ni 0.35 max Si 2.5-3.5 Ti 0.25 max Zn 1.0 max Other total 0.50 max

CHEMICAL COMPOSITIONS OF CAST ALUMINUM & ALUMINUM ALLOYS (Continued)

UNS	Title
A02220	Al rem Cu 9.2-10.7 Fe 1.5 max Mg 0.15-0.35 Mn 0.50 max Ni 0.50 max Si 2.0 max Ti 0.25 max Zn 0.8 max Other total 0.35 max
A02420	Al rem Cr 0.25 max Cu 3.5-4.5 Fe 1.0 max Mg 1.2-1.8 Mn 0.35 max Ni 1.7-2.3 Si 0.7 max Ti 0.25 max Zn 0.35 max Other each 0.05 max, total 0.15 max
A02950	Al rem Cu 4.0-5.0 Fe 1.0 max Mg 0.03 max Mn 0.35 max Si 0.7-1.5 Ti 0.25 max Zn 0.35 max Other each 0.05 max, total 0.15 max
A03190	Al rem Cu 3.0-4.0 Fe 1.0 max Mg 0.10 max Mn 0.50 max Ni 0.35 max Si 5.5-6.5 Ti 0.25 max Zn 1.0 max Other total 0.50 max
A03280	Al rem Cr 0.35 max Cu 1.0-2.0 Fe 1.0 max Mg 0.20-0.6 Mn 0.20-0.6 Ni 0.25 max Si 7.5-8.5 Ti 0.25 max Zn 1.5 max Other total 0.50 max
A03320	Al rem Cu 2.0-4.0 Fe 1.2 max Mg 0.50-1.5 Mn 0.50 max Ni 0.50 max Si 8.5-10.5 Ti 0.25 max Zn 1.0 max Other total 0.50 max
A03330	Al rem Cu 3.0-4.0 Fe 1.0 max Mg 0.05-0.50 Mn 0.50 max Ni 0.50 max Si 8.0-10.0 Ti 0.25 max Zn 1.0 max Other total 0.50 max
A03360	Al rem Cu 0.5-1.5 Fe 1.2 max Mg 0.7-1.3 Mn 0.35 max Ni 2.0-3.0 Si 11.0-13.0 Ti 0.25 max Zn 0.35 max Other each 0.05 max
A03540	Al rem Cu 1.6-2.0 Fe 0.20 max Mg 0.40-0.6 Mn 0.10 max Si 8.6-9.4 Ti 0.20 max Zn 0.10 max Other each 0.05 max, total 0.15 max
A03550	Al rem Cr 0.25 max Cu 1.0-1.5 Fe 0.6 max Mg 0.40-0.6 Mn 0.50 max Si 4.5-5.5 Ti 0.25 max Zn 0.35 max Other each 0.05 max, total 0.15 max. Note: If Fe exceeds 0.45, Mn shall not be less than 0.5 x Fe
A03560	Al rem Cu 0.25 max Fe 0.6 max Mg 0.20-0.45 Mn 0.35 max Si 6.5-7.5 Ti 0.25 max Zn 0.35 max Other each 0.05 max, total 0.15 max
A03570	Al rem Cu 0.05 max Fe 0.15 max Mg 0.45-0.6 Mn 0.03 max Si 6.5-7.5 Ti 0.20 max Zn 0.05 max Other each 0.05 max, total 0.15 max
A03590	Al rem Cu 0.20 max Fe 0.20 max Mg 0.50-0.7 Mn 0.10 max Si 8.5-9.5 Ti 0.20 max Zn 0.10 max Other each 0.05 max, total 0.15 max
A03600	Al rem Cu 0.6 max Fe 2.0 max Mg 0.40-0.6 Mn 0.35 max Ni 0.50 max Si 9.0-10.0 Sn 0.15 max Zn 0.50 max Other total 0.25 max
A04130	Al rem Cu 0.6 max Fe 2.0 max Mg 0.10 max Mn 0.35 max Ni 0.50 max Si 11.0-13.0 Sn 0.15 max Zn 0.50 max Other total 0.25 max
A04430	Al rem Cr 0.25 max Cu 0.6 max Fe 0.8 max Mg 0.05 max Mn 0.50 max Si 4.5-6.0 Ti 0.25 max Zn 0.50 max Other total 0.35 max
A05120	Al rem Cr 0.25 max Cu 0.35 max Fe 0.6 max Mg 3.5-4.5 Mn 0.8 max Si 1.4-2.2 Ti 0.25 max Zn 0.35 max Other each 0.05 max, total 0.15 max
A05130	Al rem Cu 0.10 max Fe 0.40 max Mg 3.5-4.5 Mn 0.30 max Si 0.30 max Ti 0.20 max Zn 1.4-2.2 Other each 0.05 max, total 0.15 max
A05140	Al rem Cu 0.15 max Fe 0.50 max Mg 3.5-4.5 Mn 0.35 max Si 0.35 max Ti 0.25 max Zn 0.15 max Other each 0.05 max, total 0.15 max
A05180	Al rem Cu 0.25 max Fe 1.8 max Mg 7.5-8.5 Mn 0.35 max Ni 0.15 max Si 0.35 max Sn 0.15 max Zn 0.15 max Other total 0.25 max
A05200	Al rem Cu 0.25 max Fe 0.30 max Mg 9.5-10.6 Mn 0.15 max Si 0.25 max Ti 0.25 max Zn 0.15 max Other each 0.05 max, total 0.15 max
A05350	Al rem B 0.005 max Be 0.003-0.007 Cu 0.05 max Fe 0.15 max Mg 6.2-7.5 Mn 0.10-0.25 Si 0.15 max Ti 0.10-0.25 Other each 0.05 max, total 0.15 max
A07050	Al rem Cr 0.20-0.40 Cu 0.20 max Fe 0.8 max Mg 1.4-1.8 Mn 0.40-0.6 Si 0.20 max Ti 0.25 max Zn 2.7-3.3 Other each 0.05 max, total 0.15 max
A07051	Al rem Cr 0.20-0.40 Cu 0.20 max Fe 0.6 max Mg 1.5-1.8 Mn 0.40-0.6 Si 0.20 max Ti 0.25 max Zn 2.7-3.3 Other each 0.05 max, total 0.15 max

CHEMICAL COMPOSITIONS OF CAST ALUMINUM & ALUMINUM ALLOYS (Continued)	
UNS	**Title**
A07070	Al rem Cr 0.20-0.40 Cu 0.20 max Fe 0.8 max Mg 1.8-2.4 Mn 0.40-0.6 Si 0.20 max Ti 0.25 max Zn 4.0-4.5 Other 0.05 max, total 0.15 max
A07100	Al rem Cu 0.35-0.65 Fe 0.50 max Mg 0.6-0.8 Mn 0.05 max Si 0.15 max Ti 0.25 max Zn 6.0-7.0 Other each 0.05 max, total 0.15 max
A07110	Al rem Cu 0.35-0.65 Fe 0.7-1.4 Mg 0.25-0.45 Mn 0.5 max Si 0.30 max Ti 0.20 max Zn 6.0-7.0 Other each 0.05 max, total 0.15 max
A07120	Al rem Cr 0.40-0.6 Cu 0.25 max Fe 0.50 max Mg 0.50-0.65 Mn 0.10 max Si 0.30 max Ti 0.15-0.25 Zn 5.0-6.5 Other each 0.05 max, total 0.20 max
A07130	Al rem Cr 0.35 max Cu 0.40-1.0 Fe 1.1 max Mg 0.20-0.50 Mn 0.6 max Ni 0.15 max Si 0.25 max Ti 0.25 max Zn 7.0-8.0 Other each 0.10 max, total 0.25 max
A07710	Al rem Cr 0.06-0.20 Cu 0.10 max Fe 0.15 max Mg 0.8-1.0 Mn 0.10 max Si 0.15 max Ti 0.10-0.20 Zn 6.5-7.5 Other each 0.05 max, total 0.15 max
A08500	Al rem Cu 0.7-1.3 Fe 0.7 max Mg 0.10 max Mn 0.10 max Ni 0.7-1.3 Si 0.7 max Sn 5.5-7.0 Ti 0.20 max Other total 0.30 max
A08510	Al rem Cu 0.7-1.3 Fe 0.7 max Mg 0.10 max Mn 0.10 max Ni 0.30-0.7 Si 2.0-3.0 Sn 5.5-7.0 Ti 0.20 max Other total 0.30 max
A08520	Al rem Cu 1.7-2.3 Fe 0.7 max Mg 0.6-0.9 Mn 0.10 max Ni 0.9-1.5 Si 0.40 max Sn 5.5-7.0 Ti 0.20 max Other total 0.30 max
A12010	Ag 0.40-1.0 Al rem Cu 4.0-5.0 Fe 0.10 max Mg 0.15-0.35 Mn 0.20-0.40 Si 0.05 max Ti 0.15-0.35 Other each 0.03 max, total 0.10 max
A12420	Al rem Cr 0.15-0.25 Cu 3.7-4.5 Fe 0.8 max Mg 1.2-1.7 Mn 0.10 max Ni 1.8-2.3 Si 0.6 max Ti 0.07-0.20 Zn 0.10 max Other each 0.05 max, total 0.15 max
A13560	Al rem Cu 0.20 max Fe 0.20 max Mg 0.20-0.40 Mn 0.10 max Si 6.5-7.5 Ti 0.20 max Zn 0.10 max Other each 0.05 max, total 0.15 max
A13570	Al rem Be 0.04-0.07 Cu 0.20 max Fe 0.20 max Mg 0.40-0.7 Mn 0.10 max Si 6.5-7.5 Ti 0.10-0.20 Zn 0.10 max Other each 0.05 max, total 0.15 max
A13600	Al rem Cu 0.6 max Fe 1.3 max Mg 0.40-0.6 Mn 0.35 max Ni 0.50 max Si 9.0-10.0 Sn 0.15 max Zn 0.50 max Other total 0.25 max
A13800	Al rem Cu 3.0-4.0 Fe 1.3 max Mg 0.10 max Mn 0.50 max Ni 0.50 max Si 7.5-9.5 Sn 0.35 max Zn 3.0 max Other total 0.50 max
A14130	Al rem Cu 0.6 max Fe 1.3 max Mg 0.10 max Mn 0.35 max Ni 0.50 max Si 11.0-13.0 Sn 0.15 max Zn 0.50 max Other total 0.25 max
A14440	Al rem Cu 0.10 max Fe 0.20 max Mg 0.05 max Mn 0.10 max Si 6.5-7.5 Ti 0.20 max Zn 0.10 max Other each 0.05 max, total 0.15 max
A23900	Al rem Cu 4.0-5.0 Fe 0.50 max Mg 0.45-0.65 Mn 0.50 max Ni 0.10 max Si 16.0-18.0 Ti 0.20 max Zn 1.5 max Other each 0.10 max, total 0.20 max
A24430	Al rem Cu 0.15 max Fe 0.8 max Mg 0.05 max Mn 0.35 max Si 4.5-6.0 Ti 0.25 max Zn 0.35 max Other each 0.05 max, total 0.15 max
A33550	Al rem Cu 1.0-1.5 Fe 0.20 max Mg 0.40-0.6 Mn 0.10 max Si 4.5-5.5 Ti 0.20 max Zn 0.10 max Other each 0.05 max, total 0.15 max
A34430	Al rem Cu 0.6 max Fe 2.0 max Mg 0.10 max Mn 0.35 max Ni 0.50 max Si 4.5-6.0 Sn 0.15 max Zn 0.50 max Other total 0.25 max

rem - remainder; nom - nominal

CHEMICAL COMPOSITION OF AWS ALUMINUM ELECTRODE & ROD FILLER METAL		
A5.3	**UNS**	**Chemical Composition**
E1100	A91100	Al 99.00 min Cu 0.05-0.20 Mn 0.05 max Zn 0.10 max Other each 0.05 max (Be 0.0008 max for welding electrode and filler wire only), total 0.15 max, Si+Fe 0.95 max
E3003	A93003	Al rem Cu 0.05-0.20 Fe 0.7 max Mn 1.0-1.5 Si 0.6 max Zn 0.10 max Other each 0.05 max (Be 0.0008 max for welding electrode and filler wire only), total 0.15 max
E4043	A94043	Al rem Cu 0.30 max Fe 0.8 max Mg 0.05 max Mn 0.05 max Si 4.5-6.0 Ti 0.20 max Zn 0.10 max Other each 0.05 max (Be 0.0008 max for welding electrode and filler wire only), total 0.15 max
A5.10	**UNS**	**Chemical Composition**
ER1100 R1100	A91100	Al 99.00 min Cu 0.05-0.20 Mn 0.05 max Zn 0.10 max Other each 0.05 max (Be 0.0008 max for welding electrode and filler wire only), total 0.15 max, Si+Fe 0.95 max
ER1188 R1188	A91188	Al 99.88 min Cu 0.005 max Fe 0.06 max Ga 0.03 max Mg 0.01 max Mn 0.01 max Si 0.06 max Ti 0.01 max V0.03 max Zn 0.03 max Other each 0.01 max
ER2319 R2319	A92319	Al rem Be 0.0008 max Cu 5.8-6.8 Fe 0.30 max Mg 0.02 max Mn 0.20-0.40 Si 0.20 max Ti 0.10-0.20 V 0.05-0.15 Zn 0.10 max Zr 0.10-0.25 Other each 0.05 max (Be 0.0008 max for welding electrode and filler wire only), total 0.15 max
ER4043 R4043	A94043	Al rem Cu 0.30 max Fe 0.8 max Mg 0.05 max Mn 0.05 max Si 4.5-6.0 Ti 0.20 max Zn 0.10 max Other each 0.05 max (Be 0.0008 max for welding electrode and filler wire only), total 0.15 max
ER4047 R4047	A94047	Al rem Cu 0.30 max Fe 0.8 max Mg 0.10 max Mn 0.15 max Si 11.0-13.0 Zn 0.20 max Other each 0.05 max (Be 0.0008 max for welding electrode and filler wire only), total 0.15 max
ER4145 R4145	A94145	Al rem Cr 0.15 max Cu 3.3-4.7 Fe 0.8 max Mg 0.15 max Mn 0.15 max Si 9.3-10.7 Zn 0.20 max Other each 0.05 max (Be 0.0008 max for welding electrode and filler wire only), total 0.15 max
ER4643 R4643	A94643	Al rem Be 0.0008 max Cu 0.10 max Fe 0.8 max Mg 0.10-0.30 Mn 0.05 max Si 3.6-4.6 Ti 0.15 max Zn 0.10 max Other each 0.05 max, total 0.15 max
ER5183 R5183	A95183	Al rem Be 0.0008 max Cr 0.05-0.25 Cu 0.10 max Fe 0.40 max Mg 4.3-5.2 Mn 0.50-1.0 Si 0.40 max Ti 0.15 max Zn 0.25 max Other each 0.05 max, total 0.15 max
ER5356 R5356	A95356	Al rem Cr 0.05-0.20 Cu 0.10 max Fe 0.40 max Mg 4.5-5.5 Mn 0.05-0.20 Si 0.25 max Ti 0.06-0.20 Si 0.25 max Ti 0.06-0.20 Zn 0.10 max Other each 0.05 max (Be 0.0008 max for welding electrode and filler wire only), total 0.15 max
ER5554 R5554	A95554	Al rem Be 0.0008 max Cr 0.05-0.20 Cu 0.10 max Fe 0.40 max Mg 2.4-3.0 Mn 0.50-1.0 Si 0.25 max Ti 0.05-0.20 Zn 0.25 max Other each 0.05 max, total 0.15 max, Si+Fe 0.40 max

CHEMICAL COMPOSITION OF AWS ALUMINUM ELECTRODE & ROD FILLER METAL (Continued)		
A5.10	**UNS**	**Chemical Composition**
ER5556 R5556	A95556	Al rem Be 0.0008 max Cr 0.05-0.20 Cu 0.10 max Fe 0.40 max Mg 4.7-5.5 Mn 0.50-1.0 Si 0.25 max Ti 0.05-0.20 Zn 0.25 max Other each 0.05 max, total 0.15 max
ER5654 R5654	A95654	Al rem Be 0.0008 max Cr 0.15-0.35 Cu 0.05 max Mg 3.1-3.9 Mn 0.01 max Ti 0.05-0.15 Zn 0.20 max Other each 0.05 max, total 0.15 max, Si+Fe 0.45 max
R-206.0	A02060	Al rem Cu 4.2-5.0 Fe 0.15 max Mg 0.15-0.35 Mn 0.20-0.50 Ni 0.05 max Si 0.10 max Sn 0.05 max Ti 0.15-0.30 Zn 0.10 max Other each 0.05 max, total 0.15 max
R-C355.0	A33550	Al rem Cu 1.0-1.5 Fe 0.20 max Mg 0.40-0.6 Mn 0.10 max Si 4.5-5.5 Ti 0.20 max Zn 0.10 max Other each 0.05 max, total 0.15 max
R-A356.0	A13560	Al rem Cu 0.20 max Fe 0.20 max Mg 0.20-0.40 Mn 0.10 max Si 6.5-7.5 Ti 0.20 max Zn 0.10 max Other each 0.05 max, total 0.15 max
R-357	A03570	Al rem Cu 0.05 max Fe 0.15 max Mg 0.45-0.6 Mn 0.03 max Si 6.5-7.5 Ti 0.20 max Zn 0.05 max Other each 0.05 max, total 0.15 max
R-A357.0	A13570	Al rem Be 0.04-0.07 Cu 0.20 max Fe 0.20 max Mg 0.40-0.7 Mn 0.10 max Si 6.5-7.5 Ti 0.10-0.20 Zn 0.10 max Other each 0.05 max, total 0.15 max

rem - remainder;
nom - nominal.
Single values are maximum, unless otherwise stated.

MECHANICAL PROPERTIES OF ALUMINUM & ALUMINUM NONHEAT-TREATABLE ALLOYS - SHEET & PLATE							
ASTM B 209 Alloy/Temper	**Specified Thickness, in.**	**Tensile Strength, ksi**		**Yield Strength, ksi**		**% El**	**Bend Dia. Factor N**
		min	**max**	**min**	**max**	**min**	
1060							
O	0.006-0.019	8.0	14.0	2.5	---	15	---
	0.020-0.050	8.0	14.0	2.5	---	22	---
	0.051-3.000	8.0	14.0	2.5	---	25	---
H12 or H22	0.017-0.050	11.0	16.0	9.0	---	6	---
	0.051-2.000	11.0	16.0	9.0	---	12	---
H14 or H24	0.009-0.019	12.0	17.0	10.0	---	1	---
	0.020-0.050	12.0	17.0	10.0	---	5	---
	0.051-1.000	12.0	17.0	10.0	---	10	---
H16 or H26	0.006-0.019	14.0	19.0	11.0	---	1	---
	0.020-0.050	14.0	19.0	11.0	---	4	---
	0.051-0.162	14.0	19.0	11.0	---	5	---
H18 or H28	0.006-0.019	16.0	---	12.0	---	1	---
	0.020-0.050	16.0	---	12.0	---	3	---
	0.051-0.128	16.0	---	12.0	---	4	---
H112	0.250-0.499	11.0	---	7.0	---	10	---
	0.500-1.000	10.0	---	5.0	---	20	---
	1.001-3.000	9.0	---	4.0	---	25	---
F	0.250-3.000	---	---	---	---	---	---
1100							
O	0.006-0.019	11.0	15.5	3.5	---	15	0
	0.020-0.031	11.0	15.5	3.5	---	20	0
	0.032-0.050	11.0	15.5	3.5	---	25	0
	0.051-0.249	11.0	15.5	3.5	---	30	0
	0.250-3.000	11.0	15.5	3.5	---	28	0
H12 or H22	0.017-0.019	14.0	19.0	11.0	---	3	0

MECHANICAL PROPERTIES OF ALUMINUM & ALUMINUM NONHEAT-TREATABLE ALLOYS - SHEET & PLATE (Continued)

ASTM B 209 Alloy/Temper	Specified Thickness, in.	Tensile Strength, ksi		Yield Strength, ksi		% El	Bend Dia. Factor N
		min	max	min	max	min	
1100 (Continued)							
H12 or H22	0.020-0.031	14.0	19.0	11.0	---	4	0
	0.032-0.050	14.0	19.0	11.0	---	6	0
	0.051-0.113	14.0	19.0	11.0	---	8	0
	0.114-0.499	14.0	19.0	11.0	---	9	0
	0.500-2.000	14.0	19.0	11.0	---	12	---
H14 or H24	0.009-0.012	16.0	21.0	14.0	---	1	0
	0.013-0.019	16.0	21.0	14.0	---	2	0
	0.020-0.031	16.0	21.0	14.0	---	3	0
	0.032-0.050	16.0	21.0	14.0	---	4	0
	0.051-0.113	16.0	21.0	14.0	---	5	0
	0.114-0.499	16.0	21.0	14.0	---	6	0
	0.500-1.000	16.0	21.0	14.0	---	10	---
H16 or H26	0.006-0.019	19.0	24.0	17.0	---	1	4
	0.020-0.031	19.0	24.0	17.0	---	2	4
	0.032-0.050	19.0	24.0	17.0	---	3	4
	0.051-0.162	19.0	24.0	17.0	---	4	4
H18 or H28	0.006-0.019	22.0	---	---	---	1	---
	0.020-0.031	22.0	---	---	---	2	---
	0.032-0.050	22.0	---	---	---	3	---
	0.051-0.128	22.0	---	---	---	4	---
H112	0.250-0.499	13.0	---	7.0	---	9	---
	0.500-2.000	12.0	---	5.0	---	14	---
	2001-3.000	11.5	---	4.0	---	20	---
F	0.250-3.000	---	---	---	---	---	---

MECHANICAL PROPERTIES OF ALUMINUM & ALUMINUM NONHEAT-TREATABLE ALLOYS - SHEET & PLATE (Continued)							
ASTM B 209 Alloy/Temper	**Specified Thickness, in.**	**Tensile Strength, ksi**		**Yield Strength, ksi**		**% El**	**Bend Dia.**
		min	**max**	**min**	**max**	**min**	**Factor N**
3003							
O	0.006-0.007	14.0	19.0	5.0	---	14	0
	0.008-0.012	14.0	19.0	5.0	---	18	0
	0.013-0.031	14.0	19.0	5.0	---	20	0
	0.032-0.050	14.0	19.0	5.0	---	23	0
	0.051-0.249	14.0	19.0	5.0	---	25	0
	0.250-3.000	14.0	19.0	5.0	---	23	---
H12 or H22	0.017-0.019	17.0	23.0	12.0	---	3	0
	0.020-0.031	17.0	23.0	12.0	---	4	0
	0.032-0.050	17.0	23.0	12.0	---	5	0
	0.051-0.113	17.0	23.0	12.0	---	6	0
	0.114-0.161	17.0	23.0	12.0	---	7	0
	0.162-0.249	17.0	23.0	12.0	---	8	0
	0.250-0.499	17.0	23.0	12.0	---	9	---
	0.500-2.000	17.0	23.0	12.0	---	10	---
H14 or H24	0.009-0.012	20.0	26.0	17.0	---	1	0
	0.013-0.019	20.0	26.0	17.0	---	2	0
	0.020-0.031	20.0	26.0	17.0	---	3	0
	0.032-0.050	20.0	26.0	17.0	---	4	0
	0.051-0.113	20.0	26.0	17.0	---	5	0
	0.114-0.161	20.0	26.0	17.0	---	6	2
	0.162-0.249	20.0	26.0	17.0	---	7	2
	0.250-0.499	20.0	26.0	17.0	---	8	---
	0.500-1.000	20.0	26.0	17.0	---	10	---
H16 or H26	0.006-0.019	24.0	30.0	21.0	---	1	4
	0.020-0.031	24.0	30.0	21.0	---	2	4

MECHANICAL PROPERTIES OF ALUMINUM & ALUMINUM NONHEAT-TREATABLE ALLOYS - SHEET & PLATE (Continued)

ASTM B 209	Specified Thickness,	Tensile Strength, ksi		Yield Strength, ksi		% El	Bend Dia.
Alloy/Temper	in.	min	max	min	max	min	Factor N
3003 (Continued)							
H16 or H26 (Con't)	0.032-0.050	24.0	30.0	21.0	---	3	4
	0.051-0.162	24.0	30.0	21.0	---	4	6
H18	0.006-0.019	27.0	---	24.0	---	1	---
	0.020-0.031	27.0	---	24.0	---	2	---
	0.032-0.050	27.0	---	24.0	---	3	---
	0.051-0.128	27.0	---	24.0	---	4	---
H112	0.250-0.499	17.0	---	10.0	---	8	---
	0.500-2.000	15.0	---	6.0	---	12	---
	2.001-3.000	14.5	---	6.0	---	18	---
F	0.250-0.300	---	---	---	---	---	---
Alclad Alloy 3003							
O	0.006-0.007	13.0	18.0	4.5	---	14	---
	0.008-0.012	13.0	18.0	4.5	---	18	---
	0.013-0.031	13.0	18.0	4.5	---	20	---
	0.032-0.050	13.0	18.0	4.5	---	23	---
	0.051-0.249	13.0	18.0	4.5	---	25	---
	0.250-0.499	13.0	18.0	4.5	---	23	---
	0.500-3.000	14.0	19.0	5.0	---	23	---
H12 or H22	0.017-0.031	16.0	22.0	11.0	---	4	---
	0.032-0.050	16.0	22.0	11.0	---	5	---
	0.051-0.113	16.0	22.0	11.0	---	6	---
	0.114-0.161	16.0	22.0	11.0	---	7	---
	0.162-0.249	16.0	22.0	11.0	---	8	---
	0.250-0.499	16.0	22.0	11.0	---	9	---
	0.500-2.000	17.0	23.0	12.0	---	10	---

MECHANICAL PROPERTIES OF ALUMINUM & ALUMINUM NONHEAT-TREATABLE ALLOYS - SHEET & PLATE (Continued)

ASTM B 209 Alloy/Temper	Specified Thickness, in.	Tensile Strength, ksi		Yield Strength, ksi		% El	Bend Dia. Factor N
		min	max	min	max	min	
Alclad Alloy 3003 (Continued)							
H14 or H24	0.009-0.012	19.0	25.0	16.0	---	1	---
	0.013-0.019	19.0	25.0	16.0	---	2	---
	0.020-0.031	19.0	25.0	16.0	---	3	---
	0.032-0.050	19.0	25.0	16.0	---	4	---
	0.051-0.113	19.0	25.0	16.0	---	5	---
	0.114-0.161	19.0	25.0	16.0	---	6	---
	0.162-0.249	19.0	25.0	16.0	---	7	---
	0.250-0.499	19.0	25.0	16.0	---	8	---
	0.500-1.000	20.0	26.0	17.0	---	10	---
H16 or H26	0.006-0.019	23.0	29.0	20.0	---	1	---
	0.020-0.031	23.0	29.0	20.0	---	2	---
	0.032-0.050	23.0	29.0	20.0	---	3	---
	0.051-0.162	23.0	29.0	20.0	---	4	---
H18	0.006-0.019	26.0	---	---	---	1	---
	0.020-0.031	26.0	---	---	---	2	---
	0.032-0.050	26.0	---	---	---	3	---
	0.051-0.128	26.0	---	---	---	4	---
H112	0.250-0.499	16.0	---	9.0	---	8	---
	0.500-2.000	15.0	---	6.0	---	12	---
	2.001-3.000	14.5	---	6.0	---	18	---
F	0.250-3.000	---	---	---	---	---	---
3004							
O	0.006-0.007	22.0	29.0	8.5	---	---	---
	0.008-0.019	22.0	29.0	8.5	---	10	0
	0.020-0.031	22.0	29.0	8.5	---	14	0

MECHANICAL PROPERTIES OF ALUMINUM & ALUMINUM NONHEAT-TREATABLE ALLOYS - SHEET & PLATE (Continued)							
ASTM B 209 Alloy/Temper	**Specified Thickness, in.**	**Tensile Strength, ksi**		**Yield Strength, ksi**		**% El**	**Bend Dia. Factor N**
		min	**max**	**min**	**max**	**min**	
3004 (Continued)							
O (continued)	0.032-0.050	22.0	29.0	8.5	---	16	0
	0.051-0.249	22.0	29.0	8.5	---	18	0
	0.250-3.000	22.0	29.0	8.5	---	16	---
H32 or H22	0.017-0.019	28.0	35.0	21.0	---	1	0
	0.020-0.031	28.0	35.0	21.0	---	3	1
	0.032-0.050	28.0	35.0	21.0	---	4	1
	0.051-0.113	28.0	35.0	21.0	---	5	2
	0.114-2.000	28.0	35.0	21.0	---	6	---
H34 or H24	0.009-0.019	32.0	38.0	25.0	---	1	2
	0.020-0.050	32.0	38.0	25.0	---	3	3
	0.051-0.113	32.0	38.0	25.0	---	4	4
	0.114-1.000	32.0	38.0	25.0	---	5	---
H36 or H26	0.006-0.007	35.0	41.0	28.0	---	---	---
	0.008-0.019	35.0	41.0	28.0	---	1	6
	0.020-0.031	35.0	41.0	28.0	---	2	6
	0.032-0.050	35.0	41.0	28.0	---	3	6
	0.051-0.162	35.0	41.0	28.0	---	4	8
H38 or H28	0.006-0.007	38.0	---	31.0	---	---	---
	0.008-0.019	38.0	---	31.0	---	1	---
	0.020-0.031	38.0	---	31.0	---	2	---
	0.032-0.050	38.0	---	31.0	---	3	---
	0.051-0.128	38.0	---	31.0	---	4	---
H112	0.250-3.000	23.0	---	9.0	---	7	---
F	0.250-3.000	---	---	---	---	---	---

MECHANICAL PROPERTIES OF ALUMINUM & ALUMINUM NONHEAT-TREATABLE ALLOYS - SHEET & PLATE (Continued)

ASTM B 209 Alloy/Temper	Specified Thickness, in.	Tensile Strength, ksi		Yield Strength, ksi		% El	Bend Dia. Factor N
		min	max	min	max	min	
Alclad Alloy 3004							
O	0.006-0.007	21.0	28.0	8.0	---	---	---
	0.008-0.019	21.0	28.0	8.0	---	10	---
	0.020-0.031	21.0	28.0	8.0	---	14	---
	0.032-0.050	21.0	28.0	8.0	---	16	---
	0.051-0.249	21.0	28.0	8.0	---	18	---
	0.250-0.499	21.0	28.0	8.0	---	16	---
	0.500-3.000	22.0	29.0	8.5	---	16	---
H32 or H22	0.017-0.019	27.0	34.0	20.0	---	1	---
	0.020-0.031	27.0	34.0	20.0	---	3	---
	0.032-0.050	27.0	34.0	20.0	---	4	---
	0.114-0.249	27.0	34.0	20.0	---	6	---
	0.250-0.499	27.0	34.0	20.0	---	6	---
	0.500-2.000	28.0	35.0	21.0	---	6	---
H34 or H24	0.009-0.019	31.0	37.0	24.0	---	1	---
	0.020-0.050	31.0	37.0	24.0	---	3	---
	0.051-0.113	31.0	37.0	24.0	---	4	---
	0.114-0.249	31.0	37.0	24.0	---	5	---
	0.250-0.499	31.0	37.0	24.0	---	5	---
	0.500-1.000	32.0	38.0	25.0	---	5	---
H36 or H26	0.006-0.007	34.0	40.0	27.0	---	---	---
	0.008-0.019	34.0	40.0	27.0	---	1	---
	0.020-0.031	34.0	40.0	27.0	---	2	---
	0.032-0.050	34.0	40.0	27.0	---	3	---
	0.051-0.162	34.0	40.0	27.0	---	4	---

MECHANICAL PROPERTIES OF ALUMINUM & ALUMINUM NONHEAT-TREATABLE ALLOYS - SHEET & PLATE (Continued)

ASTM B 209 Alloy/Temper	Specified Thickness, in.	Tensile Strength, ksi		Yield Strength, ksi		% El	Bend Dia. Factor N
		min	max	min	max	min	
3005							
H38	0.006-0.007	37.0	---	---	---	---	---
	0.008-0.019	37.0	---	---	---	1	---
	0.020-0.031	37.0	---	---	---	2	---
	0.032-0.050	37.0	---	---	---	3	---
	0.051-0.128	37.0	---	---	---	4	---
H112	0.250-0.499	22.0	---	8.5	---	7	---
	0.500-3.000	23.0	---	9.0	---	7	---
F	0.250-3.000	---	---	---	---	---	---
O	0.006-0.007	17.0	24.0	6.5	---	10	---
	0.008-0.012	17.0	24.0	6.5	---	12	---
	0.013-0.019	17.0	24.0	6.5	---	14	---
	0.020-0.031	17.0	24.0	6.5	---	16	---
	0.032-0.050	17.0	24.0	6.5	---	18	---
	0.051-0.249	17.0	24.0	6.5	---	20	---
H12	0.017-0.019	20.0	27.0	17.0	---	1	---
	0.020-0.050	20.0	27.0	17.0	---	2	---
	0.051-0.113	20.0	27.0	17.0	---	3	---
	0.114-0.161	20.0	27.0	17.0	---	4	---
	0.162-0.249	20.0	27.0	17.0	---	5	---
H14	0.009-0.031	24.0	31.0	21.0	---	1	---
	0.032-0.050	24.0	31.0	21.0	---	2	---
	0.051-0.113	24.0	31.0	21.0	---	3	---
	0.114-0.249	24.0	31.0	21.0	---	4	---
H16	0.006-0.031	28.0	35.0	25.0	---	1	---
	0.032-0.113	28.0	35.0	25.0	---	2	---

MECHANICAL PROPERTIES OF ALUMINUM & ALUMINUM NONHEAT-TREATABLE ALLOYS - SHEET & PLATE (Continued)

ASTM B 209 Alloy/Temper	Specified Thickness, in.	Tensile Strength, ksi		Yield Strength, ksi		% El	Bend Dia.
		min	max	min	max	min	Factor N
3005 (Continued)							
H16 (continued)	0.114-0.162	28.0	35.0	25.0	---	3	---
H18	0.006-0.031	32.0	---	29.0	---	1	---
	0.032-0.128	32.0	---	29.0	---	2	---
H19	0.006-0.012	34.0	---	---	---	---	---
	0.013-0.063	34.0	---	---	---	1	---
H25	0.016-0.019	26.0	34.0	22.0	---	1	---
	0.020-0.031	26.0	34.0	22.0	---	2	---
	0.032-0.050	26.0	34.0	22.0	---	3	---
	0.051-0.080	26.0	34.0	22.0	---	4	---
H27	0.016-0.019	29.5	37.5	25.5	---	1	---
	0.020-0.031	29.5	37.5	25.5	---	2	---
	0.032-0.050	29.5	37.5	25.5	---	3	---
	0.051-0.080	29.5	37.5	25.5	---	4	---
H28	0.016-0.019	31.0	---	27.0	---	1	---
	0.020-0.031	31.0	---	27.0	---	2	---
	0.032-0.050	31.0	---	27.0	---	3	---
	0.051-0.080	31.0	---	27.0	---	4	---
H29	0.025-0.031	33.0	---	28.0	---	1	---
	0.032-0.050	33.0	---	28.0	---	2	---
	0.051-0.071	33.0	---	28.0	---	3	---
3105							
O	0.013-0.019	14.0	21.0	5.0	---	16	---
	0.020-0.031	14.0	21.0	5.0	---	18	---
	0.032-0.080	14.0	21.0	5.0	---	20	---
H12	0.017-0.019	19.0	26.0	15.0	---	1	---

MECHANICAL PROPERTIES OF ALUMINUM & ALUMINUM NONHEAT-TREATABLE ALLOYS - SHEET & PLATE (Continued)							
ASTM B 209	**Specified Thickness,**	**Tensile Strength, ksi**		**Yield Strength, ksi**		**% El**	**Bend Dia.**
Alloy/Temper	**in.**	**min**	**max**	**min**	**max**	**min**	**Factor N**
3105 (Continued)							
H12 (continued)	0.020-0.031	19.0	26.0	15.0	---	1	---
	0.032-0.050	19.0	26.0	15.0	---	2	---
	0.051-0.080	19.0	26.0	15.0	---	3	---
H14	0.013-0.019	22.0	29.0	18.0	---	1	---
	0.020-0.031	22.0	29.0	18.0	---	1	---
	0.032-0.050	22.0	29.0	18.0	---	2	---
	0.051-0.080	22.0	29.0	18.0	---	2	---
H16	0.013-0.031	25.0	32.0	21.0	---	1	---
	0.032-0.050	25.0	32.0	21.0	---	2	---
	0.051-0.080	25.0	32.0	21.0	---	2	---
H18	0.013-0.031	28.0	---	24.0	---	1	---
	0.032-0.050	28.0	---	24.0	---	1	---
	0.051-0.080	28.0	---	24.0	---	2	---
H25	0.013-0.019	23.0	---	19.0	---	2	---
	0.020-0.031	23.0	---	19.0	---	3	---
	0.032-0.050	23.0	---	19.0	---	4	---
	0.051-0.080	23.0	---	19.0	---	6	---
5005							
O	0.006-0.007	15.0	21.0	5.0	---	12	---
	0.008-0.012	15.0	21.0	5.0	---	14	---
	0.013-0.019	15.0	21.0	5.0	---	16	---
	0.020-0.031	15.0	21.0	5.0	---	18	---
	0.032-0.050	15.0	21.0	5.0	---	20	---
	0.051-0.113	15.0	21.0	5.0	---	21	---
	0.114-0.249	15.0	21.0	5.0	---	22	---

MECHANICAL PROPERTIES OF ALUMINUM & ALUMINUM NONHEAT-TREATABLE ALLOYS - SHEET & PLATE (Continued)

ASTM B 209 Alloy/Temper	Specified Thickness, in.	Tensile Strength, ksi		Yield Strength, ksi		% El	Bend Dia. Factor N
		min	max	min	max	min	
5005 (Continued)							
O (continued)	0.250-3.000	15.0	21.0	5.0	---	22	---
H12	0.017-0.019	18.0	24.0	14.0	---	2	---
	0.020-0.031	18.0	24.0	14.0	---	3	---
	0.032-0.050	18.0	24.0	14.0	---	4	---
	0.051-0.113	18.0	24.0	14.0	---	6	---
	0.114-0.161	18.0	24.0	14.0	---	7	---
	0.162-0.249	18.0	24.0	14.0	---	8	---
	0.250-0.499	18.0	24.0	14.0	---	9	---
	0.500-2.000	18.0	24.0	14.0	---	10	---
H14	0.009-0.031	21.0	27.0	17.0	---	1	---
	0.032-0.050	21.0	27.0	17.0	---	2	---
	0.051-0.113	21.0	27.0	17.0	---	3	---
	0.114-0.161	21.0	27.0	17.0	---	5	---
	0.162-0.249	21.0	27.0	17.0	---	6	---
	0.250-0.499	21.0	27.0	17.0	---	8	---
	0.500-1.000	21.0	27.0	17.0	---	10	---
H16	0.006-0.031	24.0	30.0	20.0	---	1	---
	0.032-0.050	24.0	30.0	20.0	---	2	---
	0.051-0.162	24.0	30.0	20.0	---	3	---
H18	0.006-0.031	27.0	---	---	---	1	---
	0.051-0.128	27.0	---	---	---	3	---
H32 or H22	0.017-0.019	17.0	23.0	12.0	---	3	---
	0.020-0.031	17.0	23.0	12.0	---	4	---
	0.032-0.050	17.0	23.0	12.0	---	5	---
	0.051-0.113	17.0	23.0	12.0	---	7	---

MECHANICAL PROPERTIES OF ALUMINUM & ALUMINUM NONHEAT-TREATABLE ALLOYS - SHEET & PLATE (Continued)							
ASTM B 209 Alloy/Temper	**Specified Thickness, in.**	**Tensile Strength, ksi**		**Yield Strength, ksi**		**% El**	**Bend Dia. Factor N**
		min	**max**	**min**	**max**	**min**	
5005 (Continued)							
H32 or H22 (con't)	0.114-0.161	17.0	23.0	12.0	---	8	---
	0.162-0.249	17.0	23.0	12.0	---	9	---
	0.250-2.000	17.0	23.0	12.0	---	10	---
H34 or H24	0.009-0.012	20.0	26.0	15.0	---	2	---
	0.013-0.031	20.0	26.0	15.0	---	3	---
	0.032-0.050	20.0	26.0	15.0	---	4	---
	0.051-0.113	20.0	26.0	15.0	---	5	---
	0.114-0.161	20.0	26.0	15.0	---	6	---
	0.162-0.249	20.0	26.0	15.0	---	7	---
	0.250-0.499	20.0	26.0	15.0	---	8	---
	0.500-1.000	20.0	26.0	15.0	---	10	---
H36 or H26	0.006-0.007	23.0	29.0	18.0	---	1	---
	0.008-0.019	23.0	29.0	18.0	---	2	---
	0.020-0.031	23.0	29.0	18.0	---	3	---
	0.032-0.162	23.0	29.0	18.0	---	4	---
H38	0.006-0.012	26.0	---	---	---	1	---
	0.013-0.019	26.0	---	---	---	2	---
	0.020-0.031	26.0	---	---	---	3	---
	0.032-0.128	26.0	---	---	---	4	---
H112	0.250-0.499	17.0	---	---	---	8	---
	0.500-2.000	15.0	---	---	---	12	---
	2.001-3.000	14.5	---	---	---	18	---
F	0.250-3.000	---	---	---	---	---	---

MECHANICAL PROPERTIES OF ALUMINUM & ALUMINUM NONHEAT-TREATABLE ALLOYS - SHEET & PLATE (Continued)

ASTM B 209 Alloy/Temper	Specified Thickness, in.	Tensile Strength, ksi		Yield Strength, ksi		% El	Bend Dia. Factor N
		min	max	min	max	min	
5010							
O	0.010-0.070	15.0	21.0	5.0	---	3	---
H22	0.010-0.070	17.0	23.0	14.0	---	2	---
H24	0.010-0.070	20.0	26.0	17.0	---	1	---
H26	0.010-0.070	23.0	29.0	21.0	---	1	---
H28	0.010-0.070	26.0	---	---	---	---	---
5050							
O	0.006-0.007	18.0	24.0	6.0	---	---	0
	0.008-0.019	18.0	24.0	6.0	---	16	0
	0.020-0.031	18.0	24.0	6.0	---	18	0
	0.032-0.050	18.0	24.0	6.0	---	20	0
	0.051-0.113	18.0	24.0	6.0	---	20	0
	0.114-0.249	18.0	24.0	6.0	---	22	0
	0.250-3.000	18.0	24.0	6.0	---	20	2
H32 or H22	0.017-0.050	22.0	28.0	16.0	---	4	1
	0.051-0.249	22.0	28.0	16.0	---	6	2
H34 or H24	0.009-0.031	25.0	31.0	20.0	---	3	1
	0.032-0.050	25.0	31.0	20.0	---	4	1
	0.051-0.249	25.0	31.0	20.0	---	5	3
H36 or H26	0.006-0.019	27.0	33.0	22.0	---	2	3
	0.020-0.050	27.0	33.0	22.0	---	3	3
	0.051-0.162	27.0	33.0	22.0	---	4	4
H38	0.006-0.007	29.0	---	---	---	---	---
	0.008-0.031	29.0	---	---	---	2	---
	0.032-0.050	29.0	---	---	---	3	---
	0.051-0.128	29.0	---	---	---	4	---

MECHANICAL PROPERTIES OF ALUMINUM & ALUMINUM NONHEAT-TREATABLE ALLOYS - SHEET & PLATE (Continued)

ASTM B 209 Alloy/Temper	Specified Thickness, in.	Tensile Strength, ksi		Yield Strength, ksi		% El	Bend Dia. Factor N
		min	max	min	max	min	
5050 (Continued)							
H112	0.250-3.000	20.0	---	8.0	---	12	---
F	0.250-3.000	---	---	---	---	---	---
5052							
O	0.006-0.007	25.0	31.0	9.5	---	---	0
	0.008-0.012	25.0	31.0	9.5	---	14	0
	0.013-0.019	25.0	31.0	9.5	---	15	0
	0.020-0.031	25.0	31.0	9.5	---	16	0
	0.032-0.050	25.0	31.0	9.5	---	18	0
	0.051-0.113	25.0	31.0	9.5	---	19	0
	0.114-0.249	25.0	31.0	9.5	---	20	0
	0.250-3.000	25.0	31.0	9.5	---	18	---
H32 or H22	0.017-0.019	31.0	38.0	23.0	---	4	0
	0.020-0.050	31.0	38.0	23.0	---	5	1
	0.051-0.113	31.0	38.0	23.0	---	7	2
	0.114-0.249	31.0	38.0	23.0	---	9	3
	0.250-0.499	31.0	38.0	23.0	---	11	---
	0.500-2.000	31.0	38.0	23.0	---	12	---
H34 or H24	0.009-0.019	34.0	41.0	26.0	---	3	1
	0.020-0.050	34.0	41.0	26.0	---	4	2
	0.051-0.113	34.0	41.0	26.0	---	6	3
	0.114-0.249	34.0	41.0	26.0	---	7	4
	0.250-1.000	34.0	41.0	26.0	---	10	---
H36 or H26	0.006-0.007	37.0	44.0	29.0	---	2	---
	0.008-0.031	37.0	44.0	29.0	---	3	4
	0.032-0.162	37.0	44.0	29.0	---	4	5

MECHANICAL PROPERTIES OF ALUMINUM & ALUMINUM NONHEAT-TREATABLE ALLOYS - SHEET & PLATE (Continued)

ASTM B 209 Alloy/Temper	Specified Thickness, in.	Tensile Strength, ksi		Yield Strength, ksi		% El	Bend Dia. Factor N
		min	max	min	max	min	
5052 (Continued)							
H38 or H28	0.006-0.007	39.0	---	32.0	---	4	---
	0.008-0.031	39.0	---	32.0	---	2	---
	0.032-0.128	39.0	---	32.0	---	3	---
H112	0.250-0.499	28.0	---	16.0	---	7	---
	0.500-2.000	25.0	---	9.5	---	12	---
	2.001-3.000	25.0	---	9.5	---	16	---
F	0.250-3.000	---	---	---	---	---	---
5083							
O	0.051-1.500	40.0	51.0	18.0	29.0	16	---
	1.501-3.000	39.0	50.0	17.0	29.0	16	---
	3.001-4.000	38.0	---	16.0	---	16	---
	4.001-5.000	38.0	---	16.0	---	14	---
	5.001-7.000	37.0	---	15.0	---	14	---
	7.001-8.000	36.0	---	14.0	---	12	---
H321	0.188-1.500	44.0	56.0	31.0	43.0	12	---
	1.501-3.000	41.0	56.0	29.0	43.0	12	---
H112	0.250-1.500	40.0	---	18.0	---	12	---
	1.501-3.000	39.0	---	17.0	---	12	---
H116	0.063-0.499	44.0	---	31.0	---	10	---
	0.500-1.250	44.0	---	31.0	---	12	---
	1.251-1.500	44.0	---	31.0	---	12	---
	1.501-3.000	41.0	---	29.0	---	12	---
F	0.250-8.000	---	---	---	---	---	---

MECHANICAL PROPERTIES OF ALUMINUM & ALUMINUM NONHEAT-TREATABLE ALLOYS - SHEET & PLATE (Continued)

ASTM B 209 Alloy/Temper	Specified Thickness, in.	Tensile Strength, ksi		Yield Strength, ksi		% El	Bend Dia. Factor N
		min	max	min	max	min	
5086							
O	0.020-0.050	35.0	44.0	14.0	---	15	---
	0.051-0.249	35.0	44.0	14.0	---	18	---
	0.250-2.000	35.0	44.0	14.0	---	16	---
H32 or H22	0.020-0.050	40.0	47.0	28.0	---	6	---
	0.051-0.249	40.0	47.0	28.0	---	8	---
	0.250-2.000	40.0	47.0	28.0	---	12	---
H34 or H24	0.009-0.019	44.0	51.0	34.0	---	4	---
	0.020-0.050	44.0	51.0	34.0	---	5	---
	0.051-0.249	44.0	51.0	34.0	---	6	---
	0.250-1.000	44.0	51.0	34.0	---	10	---
H36 or H26	0.006-0.019	47.0	54.0	38.0	---	3	---
	0.020-0.050	47.0	54.0	38.0	---	4	---
	0.051-0.162	47.0	54.0	38.0	---	6	---
H38 or H28	0.006-0.020	50.0	---	41.0	---	3	---
H112	0.188-0.499	36.0	---	18.0	---	8	---
	0.500-1.000	35.0	---	16.0	---	10	---
	1.001-2.000	35.0	---	14.0	---	14	---
	2.001-3.000	34.0	---	14.0	---	14	---
H116	0.063-0.249	40.0	---	28.0	---	8	---
	0.250-0.499	40.0	---	28.0	---	10	---
	0.500-1.250	40.0	---	28.0	---	10	---
	1.251-2.000	40.0	---	28.0	---	10	---
F	0.250-3.000	---	---	---	---	---	---
5154							
O	0.020-0.031	30.0	41.0	11.0	---	12	---

MECHANICAL PROPERTIES OF ALUMINUM & ALUMINUM NONHEAT-TREATABLE ALLOYS - SHEET & PLATE (Continued)

ASTM B 209 Alloy/Temper	Specified Thickness, in.	Tensile Strength, ksi		Yield Strength, ksi		% El	Bend Dia. Factor N
		min	max	min	max	min	
5154 (Continued)							
O (continued)	0.032-0.050	30.0	41.0	11.0	---	14	---
	0.051-0.113	30.0	41.0	11.0	---	16	---
	0.114-3.000	30.0	41.0	11.0	---	18	---
H32 or H22	0.020-0.050	36.0	43.0	26.0	---	5	---
	0.051-0.249	36.0	43.0	26.0	---	8	---
	0.250-2.000	36.0	43.0	26.0	---	12	---
H34 or H24	0.009-0.050	39.0	46.0	29.0	---	4	---
	0.051-0.161	39.0	46.0	29.0	---	6	---
	0.162-0.249	39.0	46.0	29.0	---	7	---
	0.250-1.000	39.0	46.0	29.0	---	10	---
H36 or H26	0.006-0.050	42.0	49.0	32.0	---	3	---
	0.051-0.113	42.0	49.0	32.0	---	4	---
	0.114-0.162	42.0	49.0	32.0	---	5	---
H38 or H28	0.006-0.050	45.0	---	35.0	---	3	---
	0.051-0.113	45.0	---	35.0	---	4	---
	0.114-0.128	45.0	---	35.0	---	5	---
H112	0.250-0.499	32.0	---	18.0	---	8	---
	0.500-2.000	30.0	---	11.0	---	11	---
	2.001-3.000	30.0	---	11.0	---	15	---
F	0.250-3.000	---	---	---	---	---	---
5252							
H24	0.030-0.090	30.0	38.0	---	---	10	---
H25	0.030-0.090	31.0	39.0	---	---	9	---
H28	0.030-0.090	38.0	---	---	---	3	---

MECHANICAL PROPERTIES OF ALUMINUM & ALUMINUM NONHEAT-TREATABLE ALLOYS - SHEET & PLATE (Continued)							
ASTM B 209 Alloy/Temper	Specified Thickness, in.	Tensile Strength, ksi		Yield Strength, ksi		% El	Bend Dia. Factor N
		min	max	min	max	min	
5254							
O	0.051-0.113	30.0	41.0	11.0	---	16	---
	0.114-3.000	30.0	41.0	11.0	---	18	---
H32 or H22	0.051-0.249	36.0	43.0	26.0	---	8	---
	0.250-2.000	36.0	43.0	26.0	---	12	---
H34 or H24	0.051-0.161	39.0	46.0	29.0	---	6	---
	0.162-0.249	39.0	46.0	29.0	---	7	---
	0.250-1.000	39.0	46.0	29.0	---	10	---
H36 or H26	0.051-0.113	42.0	49.0	32.0	---	4	---
	0.114-0.162	42.0	49.0	32.0	---	5	---
H38 or H28	0.051-0.113	45.0	---	35.0	---	4	---
	0.114-0.128	45.0	---	35.0	---	5	---
H112	0.250-0.499	32.0	---	18.0	---	8	---
	0.500-2.000	30.0	---	11.0	---	11	---
	2.001-3.000	30.0	---	11.0	---	15	---
F	0.250-3.000	---	---	---	---	---	---
5454							
O	0.020-0.031	31.0	41.0	12.0	---	12	---
	0.032-0.050	31.0	41.0	12.0	---	14	---
	0.051-0.113	31.0	41.0	12.0	---	16	---
	0.114-3.000	31.0	41.0	12.0	---	18	---
H32 or H22	0.020-0.050	36.0	44.0	26.0	---	5	---
	0.051-0.249	36.0	44.0	26.0	---	8	---
	0.250-2.000	36.0	44.0	26.0	---	12	---
H34 or H24	0.020-0.050	39.0	47.0	29.0	---	4	---
	0.051-0.161	39.0	47.0	29.0	---	6	---

MECHANICAL PROPERTIES OF ALUMINUM & ALUMINUM NONHEAT-TREATABLE ALLOYS - SHEET & PLATE (Continued)

ASTM B 209 Alloy/Temper	Specified Thickness, in.	Tensile Strength, ksi		Yield Strength, ksi		% El	Bend Dia. Factor N
		min	max	min	max	min	
5454 (Continued)							
H34 or H24 (con't)	0.162-0.249	39.0	47.0	29.0	---	7	---
	0.250-1.000	39.0	47.0	29.0	---	10	---
H112	0.250-0.499	32.0	---	18.0	---	8	---
	0.500-2.000	31.0	---	12.0	---	11	---
	2.001-3.000	31.0	---	12.0	---	15	---
F	0.250-3.000	---	---	---	---	---	---
5456							
O	0.051-1.500	42.0	53.0	19.0	30.0	16	---
	1.501-3.000	41.0	52.0	18.0	30.0	16	---
	3.001-5.000	40.0	---	17.0	---	14	---
	5.001-7.000	39.0	---	16.0	---	14	---
	7.001-8.000	38.0	---	15.0	---	12	---
H321	0.188-0.499	46.0	59.0	33.0	46.0	12	---
	0.500-1.500	44.0	56.0	31.0	44.0	12	---
	1.501-3.000	41.0	54.0	29.0	43.0	12	---
H112	0.250-1.500	42.0	---	19.0	---	12	---
	1.501-3.000	41.0	---	18.0	---	12	---
H116	0.063-0.499	46.0	---	33.0	---	10	---
	0.500-1.250	46.0	---	33.0	---	12	---
	1.251-1.500	44.0	---	31.0	---	12	---
	1.501-3.000	41.0	---	29.0	---	12	---
	3.001-4.000	40.0	---	25.0	---	12	---
F	0.250-8.000	---	---	---	---	---	---
5457							
O	0.030-0.090	16.0	22.0	---	---	20	---

MECHANICAL PROPERTIES OF ALUMINUM & ALUMINUM NONHEAT-TREATABLE ALLOYS - SHEET & PLATE (Continued)

ASTM B 209 Alloy/Temper	Specified Thickness, in.	Tensile Strength, ksi		Yield Strength, ksi		% El	Bend Dia. Factor N
		min	max	min	max	min	
5652							
O (continued)	0.051-0.113	25.0	31.0	9.5	---	19	0
	0.114-0.249	25.0	31.0	9.5	---	20	0
	0.250-3.000	25.0	31.0	9.5	---	18	---
H32 or H22	0.051-0.113	31.0	38.0	23.0	---	7	2
	0.114-0.249	31.0	38.0	23.0	---	9	3
	0.250-0.499	31.0	38.0	23.0	---	11	---
	0.500-2.000	31.0	38.0	23.0	---	12	---
H34 or H24	0.051-0.113	34.0	41.0	26.0	---	6	3
	0.114-0.249	34.0	41.0	26.0	---	7	4
	0.250-1.000	34.0	41.0	26.0	---	10	---
H112	0.250-0.499	28.0	---	16.0	---	7	---
	0.500-2.000	25.0	---	9.5	---	12	---
	2.001-3.000	25.0	---	9.5	---	16	---
F	0.250-3.000	---	---	---	---	---	---
5657							
H241	0.030-0.090	18.0	26.0	---	---	13	---
H25	0.030-0.090	20.0	28.0	---	---	8	---
H26	0.030-0.090	22.0	30.0	---	---	7	---
H28	0.030-0.090	25.0	---	---	---	5	---

See ASTM B 209 for more testing details.

MECHANICAL PROPERTIES OF ALUMINUM & ALUMINUM HEAT-TREATABLE ALLOYS - SHEET & PLATE

ASTM B 209 Alloy/Temper	Specified Thickness, in.	Tensile Strength, ksi		Yield Strength, ksi		% El
		min	max	min	max	min
2014						
O	0.020-0.499	---	32.0	---	16.0	16
	0.500-1.000	---	32.0	---	---	10
T3	0.020-0.039	59.0	---	35.0	---	14
	0.040-0.249	59.0	---	36.0	---	14
T4	0.020-0.249	59.0	---	35.0	---	14
T42	0.020-1.000	58.0	---	34.0	---	14
T451	0.250-1.000	58.0	---	36.0	---	14
	1.001-2.000	58.0	---	36.0	---	12
	2.001-3.000	57.0	---	36.0	---	8
T6, T62	0.020-0.039	64.0	---	57.0	---	6
	0.040-0.249	66.0	---	58.0	---	7
T62, T651	0.250-0.499	67.0	---	59.0	---	7
	0.500-1.000	67.0	---	59.0	---	6
	1.001-2.000	67.0	---	59.0	---	4
	2.001-2.500	65.0	---	58.0	---	2
	2.501-3.000	63.0	---	57.0	---	2
	3.001-4.000	59.0	---	55.0	---	1
F	0.250-1.000	---	---	---	---	---
Alclad Alloy 2014						
O	0.020-0.499	---	30.0	---	14.0	16
	0.500-1.000	---	32.0	---	---	10
T3	0.020-0.024	54.0	---	33.0	---	14
	0.025-0.039	55.0	---	34.0	---	14
	0.040-0.249	57.0	---	35.0	---	15
T4	0.020-0.024	54.0	---	31.0	---	14

MECHANICAL PROPERTIES OF ALUMINUM & ALUMINUM HEAT-TREATABLE ALLOYS - SHEET & PLATE (Continued)						
ASTM B 209 Alloy/Temper	**Specified Thickness, in.**	**Tensile Strength, ksi**		**Yield Strength, ksi**		**% El**
		min	**max**	**min**	**max**	**min**
Alclad Alloy 2014 (Continued)						
T4 (continued)	0.025-0.039	55.0	---	32.0	---	14
	0.040-0.249	57.0	---	34.0	---	15
T42	0.020-0.024	54.0	---	31.0	---	14
	0.025-0.039	55.0	---	32.0	---	14
	0.040-0.499	57.0	---	34.0	---	15
	0.500-1.000	58.0	---	34.0	---	14
T451	0.250-0.499	57.0	---	36.0	---	15
	0.500-1.000	58.0	---	36.0	---	14
	1.001-2.000	58.0	---	36.0	---	12
	2.001-3.000	57.0	---	36.0	---	8
T6, T62	0.020-0.024	62.0	---	54.0	---	7
	0.025-0.039	63.0	---	55.0	---	7
	0.040-0.249	64.0	---	57.0	---	8
T62, T651	0.250-0.499	64.0	---	57.0	---	8
	0.500-1.000	67.0	---	59.0	---	6
	1.001-2.000	67.0	---	59.0	---	4
	2.001-2.500	65.0	---	58.0	---	2
	2.501-3.000	63.0	---	57.0	---	2
	3.001-4.000	59.0	---	55.0	---	1
F	0.250-1.000	---	---	---	---	---
2024						
O	0.010-0.499	---	32.0	---	14.0	12
	0.500-1.750	---	32.0	---	---	12
T3	0.008-0.009	63.0	---	42.0	---	10
	0.010-0.020	63.0	---	42.0	---	12

MECHANICAL PROPERTIES OF ALUMINUM & ALUMINUM HEAT-TREATABLE ALLOYS - SHEET & PLATE (Continued)

ASTM B 209 Alloy/Temper	Specified Thickness, in.	Tensile Strength, ksi		Yield Strength, ksi		% El
		min	max	min	max	min
2024 (Continued)						
T3 (continued)	0.021-0.128	63.0	---	42.0	---	15
	0.129-0.249	64.0	---	42.0	---	15
T351	0.250-0.499	64.0	---	42.0	---	12
	0.500-1.000	63.0	---	42.0	---	8
	1.001-1.500	62.0	---	42.0	---	7
	1.501-2.000	62.0	---	42.0	---	6
	2.001-3.000	60.0	---	42.0	---	4
	3.001-4.000	57.0	---	41.0	---	4
T361	0.020-0.062	67.0	---	50.0	---	8
	0.063-0.249	68.0	---	51.0	---	9
	0.250-0.499	66.0	---	49.0	---	9
	0.500	66.0	---	49.0	---	10
T4	0.010-0.020	62.0	---	40.0	---	12
	0.021-0.249	62.0	---	40.0	---	15
T42	0.010-0.020	62.0	---	38.0	---	12
	0.021-0.249	62.0	---	38.0	---	15
	0.250-0.499	62.0	---	38.0	---	12
	0.500-1.000	61.0	---	38.0	---	8
	1.001-1.500	60.0	---	38.0	---	7
	1.501-2.000	60.0	---	38.0	---	6
	2.001-3.000	58.0	---	38.0	---	4
T62	0.010-0.499	64.0	---	50.0	---	5
	0.500-2.000	63.0	---	50.0	---	5
T72	0.010-0.249	60.0	---	46.0	---	5
T81	0.010-0.249	67.0	---	58.0	---	5

MECHANICAL PROPERTIES OF ALUMINUM & ALUMINUM HEAT-TREATABLE ALLOYS - SHEET & PLATE (Continued)						
ASTM B 209 Alloy/Temper	**Specified Thickness, in.**	**Tensile Strength, ksi**		**Yield Strength, ksi**		**% El**
		min	**max**	**min**	**max**	**min**
2024 (Continued)						
T851	0.250-0.499	67.0	---	58.0	---	5
	0.500-1.000	66.0	---	58.0	---	5
	1.001-1.499	66.0	---	57.0	---	5
T861	0.020-0.062	70.0	---	62.0	---	3
	0.063-0.249	71.0	---	66.0	---	4
	0.250-0.499	70.0	---	64.0	---	4
	0.500	70.0	---	64.0	---	4
F	0.250-3.000	---	---	---	---	---
Alclad Alloy 2024						
O	0.008-0.009	---	30.0	---	14.0	10
	0.010-0.062	---	30.0	---	14.0	12
	0.063-0.499	---	32.0	---	14.0	12
	0.500-1.750	---	32.0	---	---	12
T3	0.008-0.009	58.0	---	39.0	---	10
	0.010-0.020	59.0	---	39.0	---	12
	0.021-0.062	59.0	---	39.0	---	15
	0.063-0.128	61.0	---	40.0	---	15
	0.129-0.249	62.0	---	40.0	---	15
T351	0.250-0.499	62.0	---	40.0	---	12
	0.500-1.000	63.0	---	42.0	---	8
	1.001-1.500	62.0	---	42.0	---	7
	1.501-2.000	62.0	---	42.0	---	6
	2.001-3.000	60.0	---	42.0	---	4
	3.001-4.000	57.0	---	41.0	---	4
T361	0.020-0.062	61.0	---	47.0	---	8

MECHANICAL PROPERTIES OF ALUMINUM & ALUMINUM HEAT-TREATABLE ALLOYS - SHEET & PLATE (Continued)						
ASTM B 209 Alloy/Temper	**Specified Thickness, in.**	**Tensile Strength, ksi**		**Yield Strength, ksi**		**% El**
		min	**max**	**min**	**max**	**min**
Alclad Alloy 2024 (Continued)						
T361	0.063-0.187	64.0	---	48.0	---	9
	0.188-0.249	64.0	---	48.0	---	9
	0.250-0.499	64.0	---	48.0	---	9
	0.500	66.0	---	49.0	---	10
T4	0.010-0.020	58.0	---	36.0	---	12
	0.021-0.062	58.0	---	36.0	---	12
	0.063-0.128	61.0	---	38.0	---	15
T42	0.008-0.009	55.0	---	34.0	---	10
	0.010-0.020	57.0	---	34.0	---	12
	0.021-0.062	57.0	---	34.0	---	15
	0.063-0.187	60.0	---	36.0	---	15
	0.188-0.249	60.0	---	36.0	---	15
	0.250-0.499	60.0	---	36.0	---	12
	0.500-1.000	61.0	---	38.0	---	8
	1.001-1.500	60.0	---	38.0	---	7
	1.501-2.000	60.0	---	38.0	---	6
	2.001-3.000	58.0	---	38.0	---	4
T62	0.010-0.062	60.0	---	47.0	---	5
	0.063-0.499	62.0	---	49.0	---	5
T72	0.010-0.062	56.0	---	43.0	---	5
	0.063-0.249	58.0	---	45.0	---	5
T81	0.010-0.062	62.0	---	54.0	---	5
	0.063-0.249	65.0	---	56.0	---	5
T851	0.250-0.499	65.0	---	56.0	---	5
	0.500-1.000	66.0	---	58.0	---	5

MECHANICAL PROPERTIES OF ALUMINUM & ALUMINUM HEAT-TREATABLE ALLOYS - SHEET & PLATE (Continued)						
ASTM B 209 Alloy/Temper	**Specified Thickness, in.**	**Tensile Strength, ksi**		**Yield Strength, ksi**		**% El**
		min	**max**	**min**	**max**	**min**
Alclad Alloy 2024 (Continued)						
T861	0.020-0.062	64.0	---	58.0	---	3
	0.063-0.187	69.0	---	64.0	---	4
	0.188-0.249	69.0	---	64.0	---	4
	0.250-0.499	68.0	---	62.0	---	4
	0.500	70.0	---	64.0	---	4
F	0.250-3.000	---	---	---	---	---
1 ½ % Alclad Alloy 2024						
O	0.188-0.499	---	32.0	---	14.0	12
	0.050-1.750	---	32.0	---	---	12
T3	0.188-0.249	63.0	---	41.0	---	15
T361	0.188-0.249	65.0	---	49.0	---	9
	0.250-0.499	65.0	---	48.0	---	9
	0.500	66.0	---	49.0	---	10
T351	0.250-0.499	63.0	---	41.0	---	12
	0.500-1.000	63.0	---	42.0	---	8
	1.001-1.500	62.0	---	42.0	---	7
	1.501-2.000	62.0	---	42.0	---	6
	2.001-3.000	60.0	---	42.0	---	4
	3.001-4.000	57.0	---	41.0	---	4
T42	0.188-0.249	61.0	---	37.0	---	15
	0.250-0.499	61.0	---	37.0	---	12
	0.500-1.000	61.0	---	38.0	---	8
	1.001-1.500	60.0	---	38.0	---	7
	1.501-2.000	60.0	---	38.0	---	6
	2.001-3.000	58.0	---	38.0	---	4

MECHANICAL PROPERTIES OF ALUMINUM & ALUMINUM HEAT-TREATABLE ALLOYS - SHEET & PLATE (Continued)						
ASTM B 209 Alloy/Temper	**Specified Thickness, in.**	**Tensile Strength, ksi**		**Yield Strength, ksi**		**% El**
		min	**max**	**min**	**max**	**min**
1 ½ % Alclad Alloy 2024 (Continued)						
T62	0.188-0.499	62.0	---	49.0	---	5
T72	0.188-0.249	59.0	---	45.0	---	5
T81	0.188-0.249	66.0	---	57.0	---	5
T851	0.250-0.499	66.0	---	57.0	---	5
	0.500-1.000	66.0	---	58.0	---	5
T861	0.188-0.249	70.0	---	65.0	---	4
	0.250-0.499	69.0	---	63.0	---	4
	0.500	70.0	---	64.0	---	4
F	0.250-3.000	---	---	---	---	---
Alclad One-Side Alloy 2024						
O	0.008-0.009	---	31.0	---	14.0	10
	0.010-0.062	---	31.0	---	14.0	12
	0.063-0.499	---	32.0	---	14.0	12
T3	0.010-0.020	61.0	---	40.0	---	12
	0.021-0.062	61.0	---	40.0	---	15
	0.063-0.128	62.0	---	41.0	---	15
	0.129-0.249	63.0	---	41.0	---	15
T351	0.250-0.499	63.0	---	41.0	---	12
T361	0.020-0.062	64.0	---	48.0	---	8
	0.063-0.249	66.0	---	49.0	---	9
	0.250-0.499	65.0	---	48.0	---	9
T42	0.010-0.020	59.0	---	35.0	---	12
	0.021-0.062	59.0	---	36.0	---	15
	0.063-0.249	61.0	---	37.0	---	15
	0.250-0.499	61.0	---	37.0	---	12

MECHANICAL PROPERTIES OF ALUMINUM & ALUMINUM HEAT-TREATABLE ALLOYS - SHEET & PLATE (Continued)

ASTM B 209 Alloy/Temper	Specified Thickness, in.	Tensile Strength, ksi		Yield Strength, ksi		% El
		min	max	min	max	min
Alclad One-Side Alloy 2024 (Continued)						
T62	0.010-0.062	62.0	---	48.0	---	5
	0.063-0.249	63.0	---	49.0	---	5
T72	0.010-0.062	58.0	---	44.0	---	5
	0.063-0.499	59.0	---	45.0	---	5
T81	0.010-0.062	64.0	---	56.0	---	5
	0.063-0.249	66.0	---	57.0	---	5
T851	0.250-0.499	66.0	---	57.0	---	5
T861	0.020-0.062	67.0	---	60.0	---	3
	0.063-0.249	70.0	---	65.0	---	4
	0.250-0.499	69.0	---	63.0	---	4
F	0.250-0.499	---	---	---	---	---
1 ½ % Alclad One-Side Alloy 2024						
O	0.188-0.499	---	32.0	---	14.0	12
T3	0.188-0.249	63.0	---	41.0	---	15
T351	0.250-0.499	63.0	---	41.0	---	12
T361	0.188-0.249	66.0	---	49.0	---	9
	0.250-0.499	65.0	---	48.0	---	9
T42	0.188-0.249	61.0	---	37.0	---	15
	0.250-0.499	61.0	---	37.0	---	12
T62	0.188-0.499	63.0	---	49.0	---	5
T72	0.188-0.249	59.0	---	45.0	---	5
T81	0.188-0.249	66.0	---	57.0	---	5
T851	0.250-0.499	66.0	---	57.0	---	5
T861	0.188-0.249	70.0	---	65.0	---	4
	0.250-0.499	69.0	---	63.0	---	4

MECHANICAL PROPERTIES OF ALUMINUM & ALUMINUM HEAT-TREATABLE ALLOYS - SHEET & PLATE (Continued)

ASTM B 209 Alloy/Temper	Specified Thickness, in.	Tensile Strength, ksi		Yield Strength, ksi		% El
		min	max	min	max	min
1 ½ % Alclad One-Side Alloy 2024 (Continued)						
F	0.250-0.499	---	---	---	---	---
2124 T851 (see page 103)						
2219						
O	0.020-2.000	---	32.0	---	16.0	12
T31 flat sheet	0.020-0.039	46.0	---	29.0	---	8
	0.040-0.249	46.0	---	28.0	---	10
T351 plate	0.250-2.000	46.0	---	28.0	---	10
	2.001-3.000	44.0	---	28.0	---	10
	3.001-4.000	42.0	---	27.0	---	9
	4.001-5.000	40.0	---	26.0	---	9
	5.001-6.000	39.0	---	25.0	---	8
T37	0.020-0.039	49.0	---	38.0	---	6
	0.040-2.500	49.0	---	37.0	---	6
	2.501-3.000	47.0	---	36.0	---	6
	3.001-4.000	45.0	---	35.0	---	5
	4.001-5.000	43.0	---	34.0	---	4
T62	0.020-0.039	54.0	---	36.0	---	6
	0.040-0.249	54.0	---	36.0	---	7
	0.250-1.000	54.0	---	36.0	---	8
	1.001-2.000	54.0	---	36.0	---	7
T81 sheet	0.020-0.039	62.0	---	46.0	---	6
	0.040-0.249	62.0	---	46.0	---	7
T851 plate	0.250-1.000	62.0	---	46.0	---	8
	1.001-2.000	62.0	---	46.0	---	7
	2.001-3.000	62.0	---	45.0	---	6

MECHANICAL PROPERTIES OF ALUMINUM & ALUMINUM HEAT-TREATABLE ALLOYS - SHEET & PLATE (Continued)						
ASTM B 209 Alloy/Temper	**Specified Thickness, in.**	**Tensile Strength, ksi**		**Yield Strength, ksi**		**% El**
		min	**max**	**min**	**max**	**min**
2219 (Continued)						
T851 plate	3.001-4.000	60.0	---	44.0	---	5
	4.001-5.000	59.0	---	43.0	---	5
	5.001-6.000	57.0	---	42.0	---	4
T87	0.020-0.039	64.0	---	52.0	---	5
	0.040-0.249	64.0	---	52.0	---	6
	0.250-1.000	64.0	---	51.0	---	7
	1.001-2.000	64.0	---	51.0	---	6
	2.001-3.000	64.0	---	51.0	---	6
	3.001-4.000	62.0	---	50.0	---	4
	4.001-5.000	61.0	---	49.0	---	3
F	0.250-2.000	---	---	---	---	---
Alclad Alloy 2219						
O	0.020-2.000	---	32.0	---	16.0	12
T31 flat sheet	0.040-0.099	42.0	---	25.0	---	10
	0.100-0.249	44.0	---	26.0	---	10
T351 plate	0.250-0.499	44.0	---	26.0	---	10
T37	0.040-0.099	45.0	---	34.0	---	6
	0.100-0.499	47.0	---	35.0	---	6
T62	0.020-0.039	44.0	---	29.0	---	6
	0.040-0.099	49.0	---	32.0	---	7
	0.100-0.249	51.0	---	34.0	---	7
	0.250-0.499	51.0	---	34.0	---	8
	0.500-1.000	54.0	---	36.0	---	8
	1.001-2.000	54.0	---	36.0	---	7
T81 flat sheet	0.020-0.039	49.0	---	37.0	---	6

MECHANICAL PROPERTIES OF ALUMINUM & ALUMINUM HEAT-TREATABLE ALLOYS - SHEET & PLATE (Continued)

ASTM B 209 Alloy/Temper	Specified Thickness, in.	Tensile Strength, ksi		Yield Strength, ksi		% El
		min	max	min	max	min
Alclad Alloy 2219 (Continued)						
T81 flat sheet (con't)	0.040-0.099	55.0	---	41.0	---	7
	0.100-0.249	58.0	---	43.0	---	7
T851 plate	0.250-0.499	58.0	---	42.0	---	8
T87	0.040-0.099	57.0	---	46.0	---	6
	0.100-0.249	60.0	---	48.0	---	6
	0.250-0.499	60.0	---	48.0	---	7
F	0.250-2.000	---	---	---	---	---
6061						
O	0.006-0.007	---	22.0	---	12.0	10
	0.008-0.009	---	22.0	---	12.0	12
	0.010-0.020	---	22.0	---	12.0	14
	0.021-0.128	---	22.0	---	12.0	16
	0.129-0.499	---	22.0	---	12.0	18
	0.500-1.000	---	22.0	---	---	18
	1.001-3.000	---	22.0	---	---	16
T4	0.006-0.007	30.0	---	16.0	---	10
	0.008-0.009	30.0	---	16.0	---	12
	0.010-0.020	30.0	---	16.0	---	14
	0.021-0.249	30.0	---	16.0	---	16
T451	0.250-1.000	30.0	---	16.0	---	18
	1.001-3.000	30.0	---	16.0	---	16
T42	0.006-0.007	30.0	---	14.0	---	10
	0.008-0.009	30.0	---	14.0	---	12
	0.010-0.020	30.0	---	14.0	---	14
	0.021-0.249	30.0	---	14.0	---	16

MECHANICAL PROPERTIES OF ALUMINUM & ALUMINUM HEAT-TREATABLE ALLOYS - SHEET & PLATE (Continued)

ASTM B 209 Alloy/Temper	Specified Thickness, in.	Tensile Strength, ksi		Yield Strength, ksi		% El
		min	max	min	max	min
6061 (Continued)						
T42 (continued)	0.250-1.000	30.0	---	14.0	---	18
	1.001-3.000	30.0	---	14.0	---	16
T6, T62	0.006-0.007	42.0	---	35.0	---	4
	0.008-0.009	42.0	---	35.0	---	6
	0.010-0.020	42.0	---	35.0	---	8
	0.021-0.249	42.0	---	35.0	---	10
T62, T651	0.250-0.499	42.0	---	35.0	---	10
	0.500-1.000	42.0	---	35.0	---	9
	1.001-2.000	42.0	---	35.0	---	8
	2.001-4.000	42.0	---	35.0	---	6
	4.001-6.000	40.0	---	35.0	---	6
F	0.250-3.000	---	---	---	---	---
Alclad Alloy 6061						
O	0.010-0.020	---	20.0	---	12.0	14
	0.021-0.128	---	20.0	---	12.0	16
	0.129-0.499	---	20.0	---	12.0	18
	0.500-1.000	---	22.0	---	---	18
	1.001-3.000	---	22.0	---	---	16
T4	0.010-0.020	27.0	---	14.0	---	14
	0.021-0.249	27.0	---	14.0	---	16
T451	0.250-0.499	27.0	---	14.0	---	18
	0.500-1.000	30.0	---	16.0	---	18
	1.001-3.000	30.0	---	16.0	---	16
T42	0.010-0.020	27.0	---	12.0	---	14
	0.021-0.249	27.0	---	12.0	---	16

MECHANICAL PROPERTIES OF ALUMINUM & ALUMINUM HEAT-TREATABLE ALLOYS - SHEET & PLATE (Continued)

ASTM B 209 Alloy/Temper	Specified Thickness, in.	Tensile Strength, ksi		Yield Strength, ksi		% El
		min	max	min	max	min
Alclad Alloy 6061 (Continued)						
T42 (continued)	0.250-0.499	27.0	---	12.0	---	18
	0.500-1.000	30.0	---	14.0	---	18
	1.001-3.000	30.0	---	14.0	---	16
T6, T62	0.010-0.020	38.0	---	32.0	---	8
	0.021-0.249	38.0	---	32.0	---	10
T62, T651	0.250-0.499	38.0	---	32.0	---	10
	0.500-1.000	42.0	---	35.0	---	9
	1.001-2.000	42.0	---	35.0	---	8
	2.001-4.000	42.0	---	35.0	---	6
	4.001-5.000	40.0	---	35.0	---	6
F	0.250-3.000	---	---	---	---	---
7075						
O	0.015-0.499	---	40.0	---	21.0	10
	0.500-2.000	---	40.0	---	---	10
T6, T62	0.008-0.011	74.0	---	63.0	---	5
	0.012-0.039	76.0	---	67.0	---	7
	0.040-0.125	78.0	---	68.0	---	8
	0.126-0.249	78.0	---	69.0	---	8
T62, T651	0.250-0.499	78.0	---	67.0	---	9
	0.500-1.000	78.0	---	68.0	---	7
	1.001-2.000	77.0	---	67.0	---	6
	2.001-2.500	76.0	---	64.0	---	5
	2.501-3.000	72.0	---	61.0	---	5
	3.001-3.500	71.0	---	58.0	---	5
	3.501-4.000	67.0	---	54.0	---	3

MECHANICAL PROPERTIES OF ALUMINUM & ALUMINUM HEAT-TREATABLE ALLOYS - SHEET & PLATE (Continued)

ASTM B 209	Specified Thickness,	Tensile Strength, ksi		Yield Strength, ksi		% El
Alloy/Temper	in.	min	max	min	max	min
7075 (Continued)						
T73 sheet	0.040-0.249	67.0	---	56.0	---	8
T7351 plate	0.250-1.000	69.0	---	57.0	---	7
	1.001-2.000	69.0	---	57.0	---	6
	2.001-2.500	66.0	---	52.0	---	6
	2.501-3.000	64.0	---	49.0	---	6
T76 sheet	0.125-0.249	73.0	---	62.0	---	8
T7651 plate	0.250-0.499	72.0	---	61.0	---	8
	0.500-1.000	71.0	---	60.0	---	6
F	0.250-4.000	---	---	---	---	---
Alclad Alloy 7075						
O	0.008-0.014	---	36.0	---	20.0	9
	0.015-0.062	---	36.0	---	20.0	10
	0.063-0.187	---	38.0	---	20.0	10
	0.188-0.499	---	39.0	---	21.0	10
	0.500-1.000	---	40.0	---	---	10
T6, T62	0.008-0.011	68.0	---	58.0	---	5
	0.012-0.039	70.0	---	60.0	---	7
	0.040-0.062	72.0	---	62.0	---	8
	0.063-0.187	73.0	---	63.0	---	8
	0.188-0.249	75.0	---	64.0	---	8
T62, T651	0.250-0.499	75.0	---	65.0	---	9
	0.500-1.000	78.0	---	68.0	---	7
	1.001-2.000	77.0	---	67.0	---	6
	2.001-2.500	76.0	---	64.0	---	5
	2.501-3.000	72.0	---	61.0	---	5

MECHANICAL PROPERTIES OF ALUMINUM & ALUMINUM HEAT-TREATABLE ALLOYS - SHEET & PLATE (Continued)						
ASTM B 209 Alloy/Temper	**Specified Thickness, in.**	**Tensile Strength, ksi**		**Yield Strength, ksi**		**% El**
		min	**max**	**min**	**max**	**min**
Alclad Alloy 7075 (Continued)						
T62, T651 (continued)	3.001-3.500	71.0	---	58.0	---	5
	3.501-4.000	67.0	---	54.0	---	3
T76 sheet	0.125-0.187	68.0	---	57.0	---	8
	0.188-0.249	70.0	---	59.0	---	8
T7651 plate	0.250-0.499	69.0	---	58.0	---	8
	0.500-1.000	71.0	---	60.0	---	6
F	0.250-4.000	---	---	---	---	---
Alclad One-Side Alloy 7075						
O	0.015-0.062	---	38.0	---	21.0	10
	0.063-0.187	---	39.0	---	21.0	10
	0.188-0.499	---	39.0	---	21.0	10
	0.500-1.000	---	40.0	---	---	10
T6, T62	0.015-0.039	73.0	---	63.0	---	7
	0.040-0.062	74.0	---	64.0	---	8
	0.063-0.187	75.0	---	65.0	---	8
	0.188-0.249	76.0	---	66.0	---	8
T62, T651	0.250-0.499	76.0	---	66.0	---	9
	0.500-1.000	78.0	---	68.0	---	7
	1.001-2.000	77.0	---	67.0	---	6
F	0.250-2.000	---	---	---	---	---
7008 Alclad Alloy 7075						
O	0.015-0.499	---	40.0	---	21.0	10
	0.500-2.000	---	40.0	---	---	10
T6, T62	0.015-0.039	73.0	---	63.0	---	7
	0.040-0.187	75.0	---	65.0	---	8

MECHANICAL PROPERTIES OF ALUMINUM & ALUMINUM HEAT-TREATABLE ALLOYS - SHEET & PLATE (Continued)						
ASTM B 209 Alloy/Temper	**Specified Thickness, in.**	**Tensile Strength, ksi**		**Yield Strength, ksi**		**% El**
		min	**max**	**min**	**max**	**min**
7008 Alclad Alloy 7075 (Continued)						
T6, T62 (continued)	0.188-0.249	76.0	---	66.0	---	8
T62, T651	0.250-0.499	76.0	---	66.0	---	9
	0.500-1.000	78.0	---	68.0	---	7
	1.001-2.000	77.0	---	67.0	---	6
	2.001-2.500	76.0	---	64.0	---	5
	2.501-3.000	72.0	---	61.0	---	5
	3.001-3.500	71.0	---	58.0	---	5
	3.501-4.000	67.0	---	54.0	---	3
T76 sheet	0.040-0.062	70.0	---	59.0	---	8
	0.063-0.187	71.0	---	60.0	---	8
	0.188-0.249	72.0	---	61.0	---	8
T7651 plate	0.250-0.499	71.0	---	60.0	---	8
	0.500-1.000	71.0	---	60.0	---	6
F	0.250-4.000	---	---	---	---	---
7178						
O	0.015-0.499	---	40.0	---	21.0	10
	0.500	---	40.0	---	---	10
T6, T62	0.015-0.044	83.0	---	72.0	---	7
	0.045-0.249	84.0	---	73.0	---	8
T62, T651	0.250-0.499	84.0	---	73.0	---	8
	0.500-1.000	84.0	---	73.0	---	6
	1.001-1.500	84.0	---	73.0	---	4
	1.501-2.000	80.0	---	70.0	---	3
T76	0.045-0.249	75.0	---	64.0	---	8
T7651	0.250-0.499	74.0	---	63.0	---	8

MECHANICAL PROPERTIES OF ALUMINUM & ALUMINUM HEAT-TREATABLE ALLOYS - SHEET & PLATE (Continued)						
ASTM B 209	**Specified Thickness,**	**Tensile Strength, ksi**		**Yield Strength, ksi**		**% El**
Alloy/Temper	**in.**	**min**	**max**	**min**	**max**	**min**
7178 (Continued)						
T7651 (continued)	0.500-1.000	73.0	---	62.0	---	6
F	0.250-2.000	---	---	---	---	---
Alclad Alloy 7178						
O	0.015-0.062	---	36.0	---	20.0	10
	0.063-0.187	---	38.0	---	20.0	10
	0.188-0.499	---	40.0	---	21.0	10
	0.500	---	40.0	---	---	10
T6, T62	0.015-0.044	76.0	---	66.0	---	7
	0.045-0.062	78.0	---	68.0	---	8
	0.063-0.187	80.0	---	70.0	---	8
	0.188-0.249	82.0	---	71.0	---	8
T62, T651	0.250-0.499	82.0	---	71.0	---	8
	0.500-1.000	84.0	---	73.0	---	6
	1.001-1.500	84.0	---	73.0	---	4
	1.501-2.000	80.0	---	70.0	---	3
T76	0.045-0.062	71.0	---	60.0	---	8
	0.063-0.187	71.0	---	60.0	---	8
	0.188-0.249	73.0	---	61.0	---	8
T7651	0.250-0.499	72.0	---	60.0	---	8
	0.500-1.000	73.0	---	62.0	---	6
F	0.250-2.000	---	---	---	---	---

See ASTM B 209 for more testing details.

MECHANICAL PROPERTIES OF ALUMINUM HEAT-TREATABLE ALLOY 2124-T851 - SHEET & PLATE

ASTM B 209 Alloy/Temper	Specified Thickness, in.	Axis of Test	Tensile Strength, ksi		Yield Strength, ksi		% El
			min	max	min	max	min
2124							
T851	1.500-2.000	L	66.0	---	57.0	---	6
		LT	66.0	---	57.0	---	5
		ST	64.0	---	55.0	---	1.5
	2.001-3.000	L	65.0	---	57.0	---	5
		LT	65.0	---	57.0	---	4
		ST	63.0	---	55.0	---	1.5
	3.001-4.000	L	65.0	---	56.0	---	5
		LT	65.0	---	56.0	---	4
		ST	62.0	---	54.0	---	1.5
	4.001-5.000	L	64.0	---	55.0	---	5
		LT	64.0	---	55.0	---	4
		ST	61.0	---	53.0	---	1.5
	5.001-6.000	L	63.0	---	54.0	---	5
		LT	63.0	---	54.0	---	4
		ST	58.0	---	51.0	---	1.5

L - Longitudinal; LT - Long Transverse; ST - Short Transverse
See ASTM B 209 for more testing details.

MECHANICAL PROPERTIES OF ALUMINUM & ALUMINUM ALLOYS - DRAWN SEAMLESS TUBES

ASTM B 210 Alloy/Temper	Specified Thickness, in.	Tensile Strength, ksi		Yield Strength ksi, min	% El, min Full-Section	% El, min Cut-Out
		min	max			
1060						
O	0.018-0.500	8.5	13.5	2.5	---	---
H12		10.0	---	4.0	---	---

MECHANICAL PROPERTIES OF ALUMINUM & ALUMINUM ALLOYS - DRAWN SEAMLESS TUBES (Continued)						
ASTM B 210 Alloy/Temper	Specified Thickness, in.	Tensile Strength, ksi		Yield Strength ksi, min	% El, min Full-Section	% El, min Cut-Out
		min	max			
1060 (Continued)						
H14	0.018-0.500	12.0	---	10.0	---	---
H18		16.0	---	13.0	---	---
H113		8.5	---	2.5	---	---
1100						
O	0.018-0.500	11.0	15.5	3.5	---	---
H12		14.0	---	11.0	---	---
H14		16.0	---	14.0	---	---
H16		19.0	---	17.0	---	---
H18		22.0	---	20.0	---	---
H113		11.0	---	3.5	---	---
2011						
T3	0.018-0.049	47.0	---	40.0	---	---
	0.050-0.500	47.0	---	40.0	10	8
T4511	0.018-0.049	44.0	---	25.0	---	---
	0.050-0.259	44.0	---	25.0	20	18
	0.260-0.500	44.0	---	25.0	20	20
2014						
O	0.018-0.500	---	32.0	16.0 max	---	---
T4, T42	0.018-0.024	54.0	---	30.0	10	---
	0.025-0.049	54.0	---	30.0	12	10
	0.050-0.259	54.0	---	30.0	14	10
	0.260-0.500	54.0	---	30.0	16	12
T6, T62	0.018-0.024	65.0	---	55.0	7	---
	0.025-0.049	65.0	---	55.0	7	6
	0.050-0.259	65.0	---	55.0	8	7

MECHANICAL PROPERTIES OF ALUMINUM & ALUMINUM ALLOYS - DRAWN SEAMLESS TUBES (Continued)

ASTM B 210 Alloy/Temper	Specified Thickness, in.	Tensile Strength, ksi		Yield Strength ksi, min	% El, min Full-Section	% El, min Cut-Out
		min	max			
2014 (Continued)						
T6,T62 (continued)	0.260-0.500	65.0	---	55.0	9	8
2024						
O	0.018-0.500	---	32.0	15.0 max	---	---
T3	0.018-0.024	64.0	---	42.0	10	---
	0.025-0.049	64.0	---	42.0	12	10
	0.050-0.259	64.0	---	42.0	14	10
	0.260-0.500	64.0	---	42.0	16	12
T42	0.018-0.024	64.0	---	40.0	10	---
	0.025-0.049	64.0	---	40.0	12	10
	0.050-0.259	64.0	---	40.0	14	10
	0.260-0.500	64.0	---	40.0	16	12
3003						
O	0.010-0.024	14.0	19.0	5.0	---	---
	0.025-0.049	14.0	19.0	5.0	30	20
	0.050-0.259	14.0	19.0	5.0	35	25
	0.260-0.500	14.0	19.0	5.0	---	30
H12	0.010-0.500	17.0	---	12.0	---	---
H14	0.010-0.024	20.0	---	17.0	3	---
	0.025-0.049	20.0	---	17.0	5	3
	0.050-0.259	20.0	---	17.0	8	4
	0.260-0.500	20.0	---	17.0	---	---
H16	0.010-0.024	24.0	---	21.0	---	---
	0.025-0.049	24.0	---	21.0	3	2
	0.050-0.259	24.0	---	21.0	5	4
	0.260-0.500	24.0	---	21.0	---	---

MECHANICAL PROPERTIES OF ALUMINUM & ALUMINUM ALLOYS - DRAWN SEAMLESS TUBES (Continued)						
ASTM B 210 Alloy/Temper	**Specified Thickness, in.**	**Tensile Strength, ksi**		**Yield Strength ksi, min**	**% El, min Full-Section**	**% El, min Cut-Out**
		min	**max**			
3003 (Continued)						
H18	0.010-0.024	27.0	---	24.0	2	---
	0.025-0.049	27.0	---	24.0	3	2
	0.050-0.259	27.0	---	24.0	5	3
	0.260-0.500	27.0	---	24.0	---	---
H113	0.010-0.500	14.0	---	5.0	---	---
Alclad 3003						
O	0.010-0.024	13.0	19.0	4.5	---	---
	0.025-0.049	13.0	19.0	4.5	30	20
	0.050-0.259	13.0	19.0	4.5	35	25
	0.260-0.500	13.0	19.0	4.5	---	30
H14	0.010-0.024	19.0	---	16.0	---	---
	0.025-0.049	19.0	---	16.0	5	---
	0.050-0.259	19.0	---	16.0	8	4
	0.260-0.500	19.0	---	16.0	---	---
H18	0.010-0.500	26.0	---	23.0	---	---
H113	0.050-0.500	13.0	---	4.5	---	---
3102						
O	0.018-0.049	11.0	17.0	3.5	30	20
	0.050-0.065	11.0	17.0	3.5	35	25
Alclad 3102						
O	0.018-0.049	10.0	17.0	3.5	30	20
	0.050-0.065	10.0	17.0	3.5	35	25
3303						
O	0.010-0.024	14.0	19.0	5.0	---	---
	0.025-0.049	14.0	19.0	5.0	30	20

MECHANICAL PROPERTIES OF ALUMINUM & ALUMINUM ALLOYS - DRAWN SEAMLESS TUBES (Continued)						
ASTM B 210 Alloy/Temper	**Specified Thickness, in.**	**Tensile Strength, ksi**		**Yield Strength ksi, min**	**% El, min Full-Section**	**% El, min Cut-Out**
		min	**max**			
3303 (Continued)						
O	0.050-0.065	14.0	19.0	5.0	35	25
Alclad 3303						
O	0.010-0.024	13.0	19.0	4.5	---	---
	0.025-0.049	13.0	19.0	4.5	30	20
	0.050-0.065	13.0	19.0	4.5	35	25
5005						
O	0.018-0.500	15.0	21.0	5.0	---	---
5050						
O	0.018-0.500	18.0	24.0	6.0	---	---
H32		22.0	---	16.0	---	---
H34		25.0	---	20.0	---	---
H36		27.0	---	22.0	---	---
H38		29.0	---	24.0	---	---
5052						
O	0.018-0.450	25.0	35.0	10.0	---	---
H32		31.0	---	23.0	---	---
H34		34.0	---	26.0	---	---
H36		37.0	---	29.0	---	---
H38		39.0	---	24.0	---	---
5083						
O	0.018-0.450	39.0	51.0	16.0	---	14
5086						
O	0.018-0.450	35.0	46.0	14.0	---	---
H32		40.0	---	28.0	---	---
H34		44.0	---	34.0	---	---

MECHANICAL PROPERTIES OF ALUMINUM & ALUMINUM ALLOYS - DRAWN SEAMLESS TUBES (Continued)						
ASTM B 210 Alloy/Temper	**Specified Thickness, in.**	**Tensile Strength, ksi**		**Yield Strength ksi, min**	**% El, min Full-Section**	**% El, min Cut-Out**
		min	**max**			
5086 (Continued)						
H36	0.018-0.450	47.0	---	38.0	---	---
5154						
O	0.010-0.450	30.0	41.0	11.0	10	10
H34		39.0	---	29.0	5	5
H38		45.0	---	34.0	---	---
5456						
O	0.018	41.0	53.0	19.0	---	14
6061						
O	0.018-0.500	---	22.0	14.0 max	15	15
T4	0.025-0.049	30.0	---	16.0	16	14
	0.050-0.259	30.0	---	16.0	18	16
	0.260-0.500	30.0	---	16.0	20	18
T42	0.025-0.049	30.0	---	14.0	16	14
	0.050-0.259	30.0	---	14.0	18	16
	0.260-0.500	30.0	---	14.0	20	18
T6,T62	0.025-0.049	42.0	---	35.0	10	8
	0.050-0.259	42.0	---	35.0	12	10
	0.260-0.500	42.0	---	35.0	14	12
6063						
O	0.018-0.500	---	19.0	---	---	---
T4, T42	0.025-0.049	22.0	---	10.0	16	14
	0.050-0.259	22.0	---	10.0	18	16
	0.260-0.500	22.0	---	10.0	20	18
T6, T62	0.025-0.049	33.0	---	28.0	12	8
	0.050-0.259	33.0	---	28.0	14	10

MECHANICAL PROPERTIES OF ALUMINUM & ALUMINUM ALLOYS - DRAWN SEAMLESS TUBES (Continued)						
ASTM B 210 Alloy/Temper	**Specified Thickness, in.**	**Tensile Strength, ksi**		**Yield Strength ksi, min**	**% El, min Full-Section**	**% El, min Cut-Out**
		min	**max**			
6063 (Continued)						
T6, T62 (continued)	0.260-0.500	33.0	---	28.0	16	12
T83	0.025-0.259	33.0	---	30.0	5	---
T831	0.025-0.259	28.0	---	25.0	5	---
T832	0.025-0.049	41.0	---	36.0	8	5
	0.050-0.259	40.0	---	35.0	8	5
6262						
T6, T62	0.025-0.049	42.0	---	35.0	10	8
	0.050-0.259	42.0	---	35.0	12	10
	0.260-0.500	42.0	---	35.0	14	12
T9	0.025-0.375	48.0	---	44.0	5	4
7075						
O	0.025-0.049	---	40.0	21.0 max	10	8
	0.050-0.500	---	40.0	21.0 max	12	10
T6, T62	0.025-0.259	77.0	---	66.0	8	7
	0.260-0.500	77.0	---	66.0	9	8
T73	0.025-0.259	66.0	---	56.0	10	8
	0.260-0.500	66.0	---	56.0	12	10

See ASTM B 210 for more testing details.

MECHANICAL PROPERTIES OF ALUMINUM ALLOY PERMANENT MOLD CASTINGS

ASTM B 108			Tensile Strength, min		Yield Strength[b], min		% El	Typical Hardness
UNS	Alloy	Temper	ksi [a]	MPa [a]	ksi [a]	MPa [a]	min	HB [c]
A02040	204.0	T4 separately cast specimens	48.0	331	29.0	200	8.0	---
A02080	208.0	T4	33.0	228	15.0	103	4.5	75
		T6	35.0	241	22.0	152	2.0	90
		T7	33.0	228	16.0	110	3.0	80
A02130	213.0	F	23.0	159	---	---	---	---
A02220	222.0	T551	30.0	207	---	---	c	115
		T65	40.0	276	---	---	c	140
A02420	242.0	T571	34.0	234	---	---	c	105
		T61	40.0	276	---	---	c	110
A03190	319.0	F	27.0	186	14.0	97	2.5	95
A03320	332.0	T5	31.0	214	---	---	c	105
A03330	333.0	F	28.0	193	---	---	c	90
		T5	30.0	207	---	---	c	100
		T6	35.0	241	---	---	c	105
		T7	31.0	214	---	---	c	90
A03360	336.0	T551	31.0	214	---	---	c	105
		T65	40.0	276	---	---	c	125
A03540	354.0	T61						
		separately cast specimens	48.0	331	37.0	255	3.0	---
		castings, designated area[d]	47.0	324	36.0	248	3.0	---
		castings, no location designated[d]	43.0	297	33.0	228	2.0	---
A03540	354.0	T62						
		separately cast specimens	52.0	359	42.0	290	2.0	---
		castings, designated area[d]	50.0	344	42.0	290	2.0	---
		castings, no location designated[d]	43.0	297	33.0	228	2.0	---

MECHANICAL PROPERTIES OF ALUMINUM ALLOY PERMANENT MOLD CASTINGS (Continued)

ASTM B 108			Tensile Strength, min		Yield Strength[b], min		% El	Typical Hardness
UNS	Alloy	Temper	ksi [a]	MPa [a]	ksi [a]	MPa [a]	min	HB [c]
A03550	355.0	T51	27.0	186	---	---	c	75
		T62	42.0	290	---	---	c	105
		T7	36.0	248	---	---	c	90
		T71	34.0	234	27.0	186	c	80
A33550	C355.0	T61						
		separately cast specimens	40.0	276	30.0	207	3.0	85-90
		castings, designated area[d]	40.0	276	30.0	207	3.0	---
		castings, no location designated[d]	37.0	255	30.0	207	1.0	85
A03560	356.0	F	21.0	145	10.0	69	3.0	---
		T6	33.0	228	22.0	152	3.0	85
		T71	25.0	172	---	---	3.0	70
A13560	A356.0	T61						
		separately cast specimens	38.0	262	26.0	179	5.0	80-90
		castings, designated area[d]	33.0	228	26.0	179	5.0	---
		castings, no location designated[d]	28.0	193	26.0	179	3.0	---
---	357.0	T6	45.0	310	---	---	3.0	---
A13570	A357.0	T61						
		separately cast specimens	45.0	310	36.0	248	3.0	100
		castings, designated area[d]	46.0	317	36.0	248	3.0	---
		castings, no location designated[d]	41.0	283	31.0	214	3.0	---
A03590	359.0	T61						
		separately cast specimens	45.0	310	34.0	234	4.0	90
		castings, designated area[d]	45.0	310	34.0	234	4.0	---
		castings, no location designated[d]	40.0	276	30.0	207	3.0	---

MECHANICAL PROPERTIES OF ALUMINUM ALLOY PERMANENT MOLD CASTINGS (Continued)

ASTM B 108			Tensile Strength, min		Yield Strength[b], min		% El	Typical Hardness
UNS	Alloy	Temper	ksi [a]	MPa [a]	ksi [a]	MPa [a]	min	HB [c]
A03590	359.0	T62						
		separately cast specimens	47.0	324	38.0	262	3.0	100
		castings, designated area[d]	47.0	324	38.0	262	3.0	---
		castings, no location designated[d]	40.0	276	30.0	207	3.0	---
A04430	443.0	F	21.0	145	7.0	49	2.0	45
A24430	B443.0	F	21.0	145	6.0	41	2.5	45
A14440	A444.0	T4-separately cast specimens	20.0	138	---	---	20	---
		T4-castings, designated area[d]	20.0	138	---	---	20	---
A05130	513.0	F	22.0	152	12.0	83	2.5	60
A05350	535.0	F	35.0	241	18.0	124	8.0	---
A07050	705.0	T1 or T5	37.0	255	17.0	117	10.0	---
A07070	707.0	T1	42.0	290	25.0	173	4.0	---
		T7	45.0	310	35.0	241	3.0	---
A07110	711.0	T1	28.0	193	18.0	124	7.0	70
A07130	713.0	T1 or T5	32.0	221	22.0	152	4.0	---
A08500	850.0	T5	18.0	124	---	---	8.0	---
A08510	851.0	T5	17.0	117	---	---	3.0	---
		T6	18.0	124	---	---	8.0	---
A08520	852.0	T5	27.0	186	---	---	3.0	---

a. SI units for information only.

b. Yield strength to be evaluated only when specified in contract or purchase order.

c. Not required.

d. These properties apply only to castings having section thicknesses not greater than 2 in. except that section thicknesses of ¾ in., maximum, shall apply to Alloy A444.0.

ASME P-No. - BASE METAL ALUMINUM & ALUMINUM ALLOYS

ASME Spec.	Size or Thickness (in)	UTS ksi (min)	AA Designation	Nominal Composition	Product Form
P No. 21					
SB-209	0.051-3.000	8	1060	99.6 min Al	Sheet, Plate
SB-209	0.006-3.000	11	1100	99.0 Min Al	Sheet, Plate
SB-209	0.051-3.000	14	3003	1.2Mn	Sheet, Plate
SB-209	0.051-0.499	13	Alclad 3003	1.2Mn	Sheet, Plate
SB-210	All	8.5	1060	99.6 min Al	Bar, Rod, Shape, Tube
SB-210	All	14	3003	1.2Mn	Bar, Rod, Shape, Pipe, Tube
SB-210	All	13	Alclad 3003	1.2Mn	Tube
SB-221	All	8.5	1060	99.6 min Al	Bar, Rod, Shape, Tube
SB-221	All	11	1100	99.0 min Al	Bar, Rod, Shape, Tube
SB-221	All	14	3003	1.2Mn	Bar, Rod, Shape, Pipe, Tube
SB-234	All	8.5	1060	99.6 min Al	Bar, Rod, Shape, Tube
SB-234	All	14	3003	1.2Mn	Bar, Rod, Shape, Pipe, Tube
SB-234	All	13	Alclad 3003	1.2Mn	Tube
SB-241	All	8.5	1060	99.6 min Al	Bar, Rod, Shape, Tube
SB-241	All	11	1100	99.0 min Al	Bar, Rod, Shape, Tube
SB-241	All	14	3003	1.2Mn	Bar, Rod, Shape, Pipe, Tube
SB-241	All	13	Alclad 3003	1.2Mn	Tube
SB-247	Up thru 4.000	14	3003	1.2Mn	Die Forging
SB-247	0.500-3.000	14	Alclad 3003	1.2Mn	Plate
P No. 22					
SB-209	0.006-3.000	22	3004	1.2Mn-1.0Mg	Sheet, Plate
SB-209	0.051-0.499	21	Alclad 3004	1.2Mn-1.0Mg	Sheet, Plate
SB-209	0.500-3.000	22	Alclad 3004	1.2Mn-1.0Mg	Plate
SB-209	0.051-3.000	25	5052	2.5Mg-0.25Cr	Sheet, Plate
SB-209	0.051-3.000	30	5254	3.5Mg-0.25Cr-Al	Sheet, Plate

ASME P-No. - BASE METAL ALUMINUM & ALUMINUM ALLOYS (Continued)

ASME Spec.	Size or Thickness (in)	UTS ksi (min)	AA Designation	Nominal Composition	Product Form
P No. 22 (Continued)					
SB-209	0.051-3.000	30	5154	3.5Mg-0.25Cr	Sheet, Plate
SB-209	0.051-3.000	31	5454	2.75Mg-0.8Mn-0.10Cr	Sheet, Plate
SB-209	0.051-3.000	25	5652	2.5Mg-0.25Cr-0.01Mn	Sheet, Plate
SB-210	0.018-0.450	25	5052	2.5Mg-0.25Cr	Seamless Tube
SB-210	All	30	5154	3.5Mg-0.25Cr	Bar, Rod, Shape, Tube
SB-221	All	30	5154	3.5Mg-0.25Cr	Bar, Rod, Shape, Tube
SB-221	All	31	5454	2.75Mg-0.8Mn-0.10Cr	Bar, Rod, Shape, Tube
SB-234	All	25	5052	2.5Mg-0.25Cr	Tube
SB-234	All	31	5454	2.75Mg-0.8Mn-0.10Cr	Bar, Rod, Shape, Tube
SB-241	All	25	5052	2.5Mg-0.25Cr	Tube
SB-241	All	31	5454	2.75Mg-0.8Mn-0.10Cr	Bar, Rod, Shape, Tube
P No. 23					
SB-209	0.051-6.000	24	6061	1.0Mg-0.6Si-0.25Cr	Sheet, Plate
SB-209	0.051-5.000	24	Alclad 6061	1.0Mg-0.6Si-0.25Cr	Sheet, Plate
SB-210	All	24	6061	1.0Mg-0.6Si-0.25Cr	Bar, Rod, Shape, Pipe, Tube
SB-210	All	17	6063	0.7Mg-0.4Si	Bar, Rod, Shape, Pipe, Tube
SB-211	All	24	6061	1.0Mg-0.6Si-0.25Cr	Bar, Rod, Shape, Pipe, Tube
SB-221	All	24	6061	1.0Mg-0.6Si-0.25Cr	Bar, Rod, Shape, Pipe, Tube
SB-221	All	17	6063	0.7Mg-0.4Si	Bar, Rod, Shape, Pipe, Tube
SB-234	All	24	6061	1.0Mg-0.6Si-0.25Cr	Bar, Rod, Shape, Pipe, Tube
SB-241	All	24	6061	1.0Mg-0.6Si-0.25Cr	Bar, Rod, Shape, Pipe, Tube
SB-241	All	17	6063	0.7Mg-0.4Si	Bar, Rod, Shape, Pipe, Tube
SB-247	Up thru 8.000	24	6061	1.0Mg-0.6Si-0.25Cr	Forging
SB-308	All	24	6061	1.0Mg-0.6Si-0.25Cr	Bar, Rod, Shape, Pipe, Tube

ASME P-No. - BASE METAL ALUMINUM & ALUMINUM ALLOYS (Continued)

ASME Spec.	Size or Thickness (in)	UTS ksi (min)	AA Designation	Nominal Composition	Product Form
P No. 25					
SB-209	0.051-1.500	40	5083	4.5Mg-0.8Mn-0.15Cr	Sheet, Plate
SB-209	1.501-3.000	39	5083	4.5Mg-0.8Mn-0.15Cr	Plate
SB-209	3.001-5.000	38	5083	4.5Mg-0.8Mn-0.15Cr	Plate
SB-209	5.001-7.000	37	5083	4.5Mg-0.8Mn-0.15Cr	Plate
SB-209	7.001-8.000	36	5083	4.5Mg-0.8Mn-0.15Cr	Plate
SB-209	0.020-2.000	35	5086	4.0Mg-0.5Mn-0.15Cr	Sheet, Plate
SB-209	0.051-1.500	42	5456	5.1Mg-0.8Mn-0.10Cr	Sheet, Plate
SB-221	Up thru 5.000	39	5083	4.5Mg-0.8Mn-0.15Cr	Bar, Rod, Shape, Tube
SB-221	Up thru 5.000	41	5456	5.1Mg-0.8Mn-0.10Cr	Bar, Rod, Shape, Tube
SB-241	Up thru 5.000	39	5083	4.5Mg-0.8Mn-0.15Cr	Bar, Rod, Shape, Tube
SB-241	Up thru 5.000	35	5086	4.0Mg-0.5Mn-0.15Cr	Tube
SB-241	1.501-3.000	41	5456	5.1Mg-0.8Mn-0.10Cr	Plate
SB-241	3.001-5.000	40	5456	5.1Mg-0.8Mn-0.10Cr	Plate
SB-241	5.001-7.000	39	5456	5.1Mg-0.8Mn-0.10Cr	Plate
SB-241	7.001-8.000	38	5456	5.1Mg-0.8Mn-0.10Cr	Plate
SB-241	Up thru 5.000	41	5456	5.1Mg-0.8Mn-0.10Cr	Bar, Rod, Shape, Tube
SB-247	Up thru 4.000	38	5083	4.5Mg-0.8Mn-0.15Cr	Forging
SB-247	2.001-3.000	34	5086	4.0Mg-0.5Mn-0.15Cr	Plate

ASME F No. - WELDING FILLER METAL ALUMINUM & ALUMINUM ALLOYS

F No.	ASME Spec. No.	AWS Classification
21	SFA-5.10	ER 1100
22	SFA-5.10	ER 5554, ER 5356, ER 5556, ER 5183, ER 5654
23	SFA-5.10	ER 4043, ER 4047, ER 4145
24	SFA-5.10	R-SC 51A, R356.0

INTERNATIONAL CROSS REFERENCES - ALUMINUM & ALUMINUM ALLOYS[a]

USA	BRITAIN	GERMANY	FRANCE	INTERNATIONAL	JAPAN	CANADA
AA	**BS**	**DIN**	**NF**	**ISO**	**JIS**	**CSA**
1050	1B	Al99.5	A5	Al 99.5	A 1050 P	9950
1100	---	---	A45	Al 99.0 Cu	A 1100 P	990C
1350	1E	E-Al99.5	A5/L	E-Al 99.5	---	---
2011	FC1	AlCuBiPb	A-U4Pb	Al Cu6BiPb	---	CB60
2014	H15	AlCuSiMn	A-U4SG	Al Cu4SiMg	A 2014 P	CS41N
2024	---	AlCuMg2	A-U4G1	Al Cu4Mg1	---	CG42
3003	N3	AlMn	A-M1	Al Mn1Cu	A 3003 P	MC10
3004	---	AlMn1Mg1	A-M1G	Al Mn1Mg1	A 3004 P	---
4043	N21	AlSi5	A-S5	Al Si5	---	S5
5005	N41	AlMg1	A-G0.6	Al Mg1(B)	A 5005 P	---
5050	3L44	---	A-G1	Al Mg1.5(C)	---	---
5052	---	AlMg2.5	A-G2.5C	Al Mg2.5	A 5052 P	GR20
5056	N6 2L.58	AlMg5	---	Al Mg5Cr	A 5056 W	GM50R
5083	N8	AlMg4.5Mn	---	Al Mg4.5Mn0.7	A 5083 P	GM41
5154	N5	---	A-G3C	Al Mg3.5	---	GR40
5454	N51	AlMg2.7Mn	A-G2.5MC	Al Mg3Mn	A 5454 P	GM31N
6061	H20	AlMg1SiCu	A-G5UC	Al Mg1SiCu	A 6061 P	GS11N
6063	H19	AlMgSi0.5	A-GS	Al Mg0.7Si	A 6063 BE	GS10
6101	91E	E-AlMgSi0.5	A-GS/L	E-Al MgSi	---	---
7075	L.95, L.96	AlZnMgCu1.5	A-Z5GU	Al Zn5.5MgCu	A 7075 P	ZG62

a. It is not practical to directly correlate the various metal designations from country to country, let alone comparing several countries and their metal designations; from the view that chemical composition may be similar, but not identical, and that manufacturing technologies may differ greatly. Consequently, the cross references made in this table are, at best, only listed as a guide to assist in finding comparable metal designations, rather than equivalent metal designations.

AA - The Aluminum Association; BS - British Standards; DIN - Deutsches Institut für Normung; NF - Normes Françaises; ISO - International Organization for Standardization; JIS - Japanese Industrial Standards; CSA - Canadian Standards Association.

Chapter 2

NICKEL & NICKEL ALLOYS

Metallic nickel was first isolated by the Swedish scientist A.F. Cronstedt in 1751, and since then it has become one of the most useful metals. Among the attractive properties of nickel and its alloys are excellent ductility and formability, good resistance to oxidation and corrosion and good high temperature strength. In addition it is ferromagnetic up to its Curie temperature of 358°C (676°F). Its relatively high cost, however, has restricted its use to specialized applications, generally involving severe service conditions.

Commercially important nickel ores include sulphides, oxides, and silicates. In all of these ores the nickel content is low, generally less than 3%, and the mineralogy and extraction processes are often complex. Many different smelting processes are used for the production of nickel including pyrometallurgical, hydrometallurgical, and vapour phase metallurgical processes. The vapour phase process involves the formation and decomposition of nickel carbonyl, and the metallic nickel produced is often referred to as carbonyl nickel.

More than half of the world's nickel production is used as an alloying addition in stainless steel, where its presence stabilizes the austenitic structure, giving improved formability and toughness, and resistance to corrosion and oxidation in particular environments. Approximately one-sixth of the nickel produced is used in nickel-based alloys. Nickel is also used as a plating material and as an alloy addition to carbon steel and copper-based alloys.

One of the many interesting aspects of nickel use is that unlike many other alloy systems, such as those based on aluminium or titanium which normally contain less than 15% total alloy content, useful nickel alloys have a wide range of nickel content. In some cases alloys with as little as 30% nickel are considered to be nickel-based alloys. Some of these alloys have been developed by experimentation with alloy additions to nickel, while others have been developed by increasing the nickel content of iron-based alloys, notably stainless steel.

Uses of commercial purity nickel are relatively restricted; however, an enormous range of nickel alloys is available, the most important being nickel-copper alloys, and nickel-chromium (and nickel-iron-chromium) alloys, many of which contain other alloying elements such as titanium, aluminum and molybdenum. The common presence of cobalt in nickel ores, and the similar behaviour of nickel and cobalt during extraction processes, has led to the specification of combined nickel plus cobalt in some alloy compositions rather than simply nickel.

Pure nickel has a face-centered cubic (fcc) crystal structure at all temperatures, and this is also true of most nickel alloys, but alloying elements have different influences on the microstructure and properties. Some produce solid solution hardening. Some form carbide or nitride particles within the nickel-based solid solution matrix. Some form brittle intermetallic phases which are detrimental to properties, and some form precipitates which provide useful precipitation hardening. Most alloying elements can contribute to several of these effects. Perhaps the most important is the precipitation of particles of the ordered fcc γ' *(gamma prime)* phase, which is generally of the form $Ni_3(Ti,Al)$. This is referred to as *gamma prime* to differentiate it from the fcc matrix phase which is conventionally termed *gamma*. Precipitation hardening by this phase contributes strength at intermediate temperatures because the γ' particles are very stable and have low energy interfaces with the matrix solid solution. Most of the nickel-based alloys which are strengthened by γ' belong to the category of "superalloys". These are alloys designed specifically for jet aircraft engine and gas turbine applications requiring high temperature strength and creep resistance as well as resistance to corrosion, oxidation, and carburization. The nickel alloy groups will be discussed in the following sections.

Commercial Purity Nickel And Low Alloy Nickels

The highest purity nickels commercially available are Nickel 270 (UNS N02270) and Nickel 290 (UNS N02290), which contain at least 99.95% Ni. This material finds applications in the electronics industry. The standard commercial purity nickels are Nickel 200 (UNS N02200) and Nickel 201 (UNS N02201); both of these contain 99.5% combined Ni + Co, but Nickel 201 has a lower maximum carbon content, 0.02%, to avoid the formation of graphite, which embrittles the material during extended use at temperatures above about 300°C (572°F). In addition to coinage and electronic industry uses, these materials find applications in chemical processing, especially for the handling of hot concentrated caustic soda and dry chlorine. A comparable material, Nickel 205 (UNS N02205), has controlled low levels of magnesium for improved electronic characteristics.

All of the above materials are solid solutions and are strengthened only by work hardening. However, two low-alloy nickels have alloy additions which provide higher strengths by precipitation hardening, making them useful for springs and other applications where both corrosion resistance and strength are required. These are Permanickel 300 (UNS N03300), which has titanium added, and Duranickel 301 (UNS N03301) which has both aluminum and titanium additions.

Nickel Copper Alloys

Nickel and copper form a complete range of solid solutions at all compositions, although some commercial nickel-copper alloys contain other alloying elements to encourage precipitation hardening. Nickel-copper alloys have excellent fabrication properties and a range of electrical properties, magnetic properties, and colour across the range of compositions. Many applications of these alloys are based on their corrosion resistance in reducing environments and in sea water, for example in the chemical and hydrocarbon processing industries and in marine engineering. Both castings and wrought products are available including pipe, shafts, tools, pumps, valves and fasteners. The most important alloys in this category are Monel 400 (UNS N04400 - 66%Ni+Co, 32%Cu), Monel R-405 (UNS N04405) which is similar but has added sulphur to make it free machining, and Monel K-500 (UNS N05500 which again has 66% Ni+Co with 30% Cu but in addition contains aluminum and titanium to provide precipitation hardening by the formation of $Ni_3(Ti,Al)$. Especially high strength is obtained when the latter alloy is cold worked before the precipitation hardening; it finds uses where a higher strength is required in a corrosion resistant application, for example for tools and springs.

Solid Solution Nickel-Chromium and Nickel-Chromium-Iron Alloys

A wide range of nickel-bearing alloys contain chromium in the approximate range of 15% to 23%, with varying amounts of iron and other elements such as molybdenum. At one extreme are the common austenitic stainless steels, having nickel contents which can be less than 10%. As the nickel content is raised and the iron content falls, the cost increases, but so does the maximum useful operating temperature, typically from around 1000°C (1832°F) to around 1200°C (2192°F), depending in part on reactions with the operating environments.

This section will consider these solid solution alloys starting with the higher nickel materials. Many of these were originally developed in response to the materials problems in jet engines. The major role of chromium in these alloys is to provide a stable surface oxide layer which protects the material against further oxidation. Of the other alloying elements, some such as molybdenum, niobium, and tungsten, provide strength by solution strengthening, while others are there for resistance to various types of corrosion, carburization, and other reactions with the environment.

Inconel alloy 600 (UNS N06600 - 76%Ni, 15%Cr, 8%Fe) is the basic alloy in this category. It is a solid solution alloy, not heat treatable, although it may exhibit precipitated chromium carbides and/or titanium carbides and nitrides. It finds applications in furnace components, such as muffles and baskets, as well as in chemical and food processing equipment.

Inconel alloy 601 (UNS N06601 - 61%Ni, 23%Cr, 14%Fe) has minor additions of aluminum and silicon, which give it improved resistance to oxidation and nitriding. The oxides of aluminum and chromium together can form a more protective surface layer as the aluminum oxide forms protrusions which prevent the chromium oxide layer from scaling. Inconel alloy 601GC is a grain-controlled modification with additional high temperature service benefits. These alloys are used as furnace parts and fixtures, radiant tubes, strand annealing tubes, furnace hearth roller tubes, and retorts for service up to 1260°C (2300°F).

Haynes alloy 214 (UNS N07214 - 75%Ni, 16%Cr, 4%Al, 3%Fe) has high temperature strength and oxidation resistance to 1260°C (2300°F). It can be precipitation hardened but is intended for use in the solution treated condition, for example in industrial heating applications.

Inconel alloy 617 (UNS N06617 - 53%Ni, 11%Cr, 12%Co, 9%Mo) combines metallurgical stability, strength, and oxidation resistance at high temperatures. This alloy is used in gas turbine applications, and in the petrochemical and thermal processing industries.

Inconel alloy 625 (UNS N06625 - 61%Ni, 22%Cr, 3%Fe, 9%Mo, 4%Nb) has improved strength compared to the 600 and 601 alloys because of the solution strengthening effects of the molybdenum and niobium additions. The molybdenum also improves its resistance to pitting corrosion in chloride-bearing environments as is the case in many chromium-containing alloys. The niobium contributes to weldability by tying up the carbon as niobium carbides thereby avoiding the formation of chromium carbides whose presence at grain boundaries can lead to intergranular corrosion as in the sensitization of stainless steel. Under some circumstances there is a possibility of the formation of the body-centered tetragonal γ'' (*gamma double prime*) precipitate in this alloy, but it is more frequently a solid solution with some carbide precipitates. Inconel alloy 625LCF (UNS N06625) is a modification designed to optimize low cycle fatigue resistance; this alloy is used extensively for bellows.

Inconel alloy 690 (UNS N06690 - 60%Ni, 30%Cr, 10%Fe) finds uses based on its resistance to nitric acid or mixtures of nitric and hydrofluoric acid, as well as at high temperatures in sulphur-bearing gases.

Haynes alloy HR-160 (UNS N12160 - 37%Ni, 28%Cr, 29%Co, 2%Fe, 3%Si) combines high temperature strength with resistance to most forms of high temperature corrosion including attack by sulphur, chloride, vanadium and other salt deposits for applications in the chemical process, waste incineration and other industries.

Hastelloy C-22 (UNS N06022 - 52%Ni, 22%Cr, 6%Fe, 13%Mo, 4%W, 3%Co) has superior corrosion resistance in oxidizing acid chlorides and wet chlorine. It also has good resistance to pitting, crevice corrosion, and stress corrosion cracking, and finds uses in the pulp and paper industry, petrochemical industry and in pollution control equipment.

Hastelloy C-276 (UNS N10276 - 54%Ni, 16%Cr, 6%Fe, 16%Mo, 4%W, 3%Co) has good resistance to seawater corrosion as well as to pitting and crevice corrosion, and is used in equipment for chemical processing, pollution control and waste treatment.

Hastelloy G-3 (UNS N06985 - 42%Ni, 22%Cr, 20%Fe, 7%Mo, 1%W, 5%Co, 2%Cu). The Mo provides resistance to pitting and crevice corrosion, and helps maintain the corrosion resistance after welding. This alloy is used for flue gas scrubbers and where resistance to phosphoric and sulphuric acids is required. Other Hastelloy G alloys include G-30 (UNS N06030 - 38%Ni, 30%Cr, 15%Fe, 6%Mo, 3%W, 5%Co, 2%Cu) which has excellent corrosion resistance in wet phosphoric acid and other mixed acids, and Hastelloy G-50 (UNS N06950 - 48%Ni, 20%Cr, 18%Fe, 9%Mo, 1%W, 3%Co) was designed specifically for sour gas service.

Hastelloy S (UNS N06635 - 62%Ni, 16%Cr, 3%Fe, 15%Mo, 1%W, 2%Co) was developed for resistance to cyclic heating.

Haynes alloy 230 (55%Ni, 22%Cr, 14%W, 5%Co, 3%Fe) has outstanding oxidation resistance to 1150°C (2102°F) and is used for gas turbine components as well as in petrochemical plants, power plants and industrial heating applications.

Hastelloy X (UNS N06002 - 49%Ni, 22%Cr, 18%Fe, 9%Mo, 1%W, 2%Co) has excellent strength and oxidation resistance to 1200°C (2192°F). Its fabrication properties are also excellent, and it is used for components of furnaces and aircraft engines and in the petrochemical industry.

Solid Solution Nickel-Molybdenum Alloys

The Hastelloy B alloys have low chromium and high (28%) molybdenum contents. Here the chromium level is too low to permit the formation of a protective chromium oxide film, so that these materials are not corrosion resistant in oxidizing environments. However the presence of molybdenum provides them with superior corrosion resistance in hydrochloric acid, and very good resistance to many nonoxidizing acids and most organic acids as well as to nonoxidizing salts. These alloys include the basic Hastelloy B (UNS N10001 - 64%Ni, 28%Mo, 1%Cr, 5%Fe, 1%Si) and its modifications Hastelloy B-2 (UNS N10665 - 68%Ni, 28%Mo, 1%Cr, 2%Fe) and Hastelloy B-3 (UNS N10675 - 68%Ni, 28%Mo, 2%Cr, 2%Fe).

Solid Solution Iron-Nickel-Chromium Alloys

These solid solution alloys are similar to the nickel-chromium-iron alloys discussed above but with higher iron and lower nickel contents, to give lower cost materials with reduced but still useful resistance to high temperatures and corrosive environments. All of these alloys exhibit good strength and resistance to oxidation and carburization at high temperatures. They include Incoloy alloy 800 (UNS N08800 - 33%Ni, 21%Cr, 46%Fe) and its modifications alloy 800H (UNS N08810) and alloy 800HT (UNS N08811) which have restricted grain size and composition limits. These are used for process piping, heat exchanger and vessel bodies. Incoloy alloy 800 is generally used for applications at temperatures below about 620°C (1148°F) and alloys 800H and 800HT at higher temperatures. Incoloy alloy 801 (UNS N08801 - 32%Ni, 21%Cr, 45%Fe) is similar but also contains titanium, which stabilizes it against sensitization and thus intergranular corrosion. Incoloy alloy 803 (35%Ni, 27%Cr, 35%Fe with minor Ti and Al) has improved resistance to carburization and oxidation and sulphidation for applications in the petrochemical industry, including pyrolysis tubing in high-severity ethylene furnaces. Incoloy alloy 825 (UNS N08825 - 42%Ni, 22%Cr, 30%Fe, 3%Mo, 2%Cu with minor Ti) contains additional nickel which improves resistance to chloride stress corrosion cracking and to polythionic acid cracking. Haynes alloy 556 (20%Ni, 22%Cr, 31%Fe, 18%Co, 3%Mo, 3%W) and Haynes alloy HR-120 (UNS R30556 - 37%Ni, 25%Cr, 33%Fe, 3%Co, 2%Mo, 2%W) both have good high temperature strength to 1095°C (2000°F) as well as resistance to oxidizing, carburizing and sulphidizing atmospheres. Other alloys in this category which were developed for corrosion resistance include Carpenter 20Cb-3 (UNS N08020 - 35%Ni, 20%Cr, 38%Fe, 2%Mo, 1%Nb, 3%Cu) which is resistant to chlorides and acid environments; Alloy 20Mo-4 (UNS N08024 - 38%Ni, 24%Cr, 32%Fe, 4%Mo, 1%Cu), designed for improved resistance to pitting and crevice corrosion and Carpenter 20Mo-6 (UNS N08026 - 35%Ni, 24%Cr, 46%Fe, 6%Mo, 3%Cu) which has particularly good resistance to hot chlorides at low pH.

Precipitation Hardening Nickel-Based Superalloys

The range of precipitation hardening nickel-based superalloys is broad, and a correspondingly wide range of properties is attainable. The individual alloying elements in these alloys play even more complex and multiple roles than they do in the solid solution alloys. These roles can include, in addition to solution strengthening of the matrix and of the γ' phase, control of the amount of γ', of the rate at which it coarsens at high temperature, and of the resistance to oxidation and corrosion.

The earliest of the precipitation hardenable nickel-based superalloys was Nimonic 80, (UNS N07080 - 77%Ni, 20%Cr with minor Ti and Al), developed in Britain in the early 1940s for jet engine component applications. Materials developments since then have resulted in a large number of nickel-based superalloys, available in cast and wrought forms, as well as nickel-iron-based and cobalt-based superalloys.

The basic microstructure of these alloys consists of a face-centered cubic nickel-based alloy matrix, solution strengthened by chromium, molybdenum, and other elements. In addition, heat treating causes the precipitation of the γ' phase in the form of particles typically 1 micrometer or less in dimension with a high volume fraction. These γ' particles are based on $Ni_3(Ti,Al)$ but can be considerably more complex, with elements such as Fe and Co substituting for some of the Ni, and other elements such as Nb substituting for some of the Ti and Al. The precipitation of this phase gives a strengthening which, unlike most strengthening mechanisms, increases with increasing temperature up to about 800°C (1472°F). The heat treatments used to strengthen these alloys are designed in part to control the dispersion and morphology of the γ' particles.

The other common microstructural constituent is alloy carbide particles. Several different types of alloy carbide particles form in these alloys depending on their composition and thermal treatment. Carbide particles can form within the grains and along grain boundaries. The carbide morphology on grain boundaries is particularly critical to the high temperature mechanical properties. Discontinuous chains of carbides along grain boundaries is the optimum condition; the total absence of boundary carbides permits grain boundary sliding at high temperatures, while a complete boundary network of carbide provides a continuous fracture path. The carbide distribution is, like the γ' dispersion, controlled by the heat treatment.

In nickel-based superalloys in which the alloy composition or heat treatment is faulty, brittle intermetallic phases can form during heat treatment or service, degrading mechanical properties. The best known of these is the sigma phase, a variant of the intermetallic compound FeCr.

In order to develop the optimum microstructure, therefore, the thermal and mechanical processing is critical. A typical heat treatment involves a solution anneal followed by a slow cool and then an aging or, more typically, a double aging treatment to develop the optimum dispersion of γ' precipitates along with the appropriate carbide distribution while avoiding the formation of brittle intermetallic phases.

Examples of the common wrought nickel-based superalloys include Inconel alloy X-750 (UNS N07750 -70%Ni, 15%Cr, 7%Fe with minor Ti, Al and Nb), Inconel alloy 718 (UNS N07718 -53%Ni, 19%Cr, 19%Fe, 5%Nb, 3%Mo with minor Ti and Al) in which the Ti and Al in the γ' phase is largely replaced by Nb, Inconel alloy 725 (UNS N07725 - 57%Ni, 21%Cr, 9%Fe, 8%Mo, 3%Nb), Waspaloy (UNS N07001 -58%Ni, 20%Cr, 14%Co, 4%Mo, 3%Ti, 1%Al), Udimet 500 (UNS N07500 -54%Ni, 18%Cr, 19%Co, 4%Mo, 3%Ti, 3%Al), and various Nimonic alloys.

Cast nickel-based superalloys include MAR-M200 (UNS N13009 -60%Ni, 9%Cr, 10%Co, 12%W, 5%Ti, 2%Al), Inconel alloy 738 (61%Ni, 16%Cr, 9%Co, 2%Mo, 3%W, 3%Ti, 3%Al, 2%Ta) and René 77 (58%Ni, 15%Cr, 15%Co, 4%Mo, 3%Ti, 4%Al). Some of the cast alloys are designed to be used in the as-cast condition, whereas others are heat treated after casting.

These alloys are used for many different jet engine and gas turbine components, including discs, blades, and fasteners. They also find applications in the chemical and petrochemical industries.

Precipitation Hardening Iron-Nickel-Based Superalloys

In comparison with the nickel-based superalloys discussed above, these materials generally have more of the nickel content replaced by iron. This lowers the material cost but also lowers the maximum service temperatures. These materials have austenitic matrices which are solution strengthened by such alloying elements as Cr and Mo, and precipitation hardened by several different phases, including the γ' phase. They are more susceptible to the precipitation of the detrimental brittle intermetallic phases such as FeCr. As in the case of the nickel-based superalloys, these alloys are strengthened by a precipitation hardening heat treatment generally consisting of solution heat treating, quenching, and double aging. Examples include Incoloy alloy 901 (UNS N09901 - 42%Ni, 13%Cr, 36%Fe, 6%Mo, 3%Ti), Incoloy alloy 925 (UNS N09925 - 44%Ni, 21%Cr, 28%Fe, 3%Mo, 2%Ti), Pyromet 860 (43%Ni, 13%Cr, 30%Fe, 6%Mo, 3%Ti, 1% Al, 4%Co) and Discaloy (UNS N66220 - 26%Ni, 14%Cr, 54%Fe, 3%Mo, 2%Ti).

The major applications for these materials are as aircraft engine and gas turbine components.

Nickel Alloys For Controlled Thermal Expansion Applications

Nickel-iron alloys containing 36%Ni have the lowest thermal expansion coefficient of any metal in the temperature range to 230°C (446°F). In nickel-iron alloys near this composition, the thermal expansion coefficients can be selected by selecting the nickel content. Low thermal expansion makes these alloys important for glass-to-metal seals, for example in the lamp and electronics industries. The commercially important alloys are Invar - Alloy 36 (UNS K93600 - 64%Fe, 36%Ni), Alloy 42 (UNS K94100 - 58%Fe-41%Ni) whose thermal expansion matches well with alumina, beryllia, and vitreous glass, and Alloy 52 (UNS K14052 - 48%Fe-51%Ni) which matches vitreous potash-soda-lead glass. These alloys find other uses in the electronics industry, for example for thermomechanical control and switchgear devices.

In addition, several nickel-iron-chromium and iron-nickel-cobalt alloys are available commercially. Alloy 902 (UNS N09902 - 50%Fe, 42%Ni, 5%Cr with minor Ti and Al) is a precipitation hardening alloy with a controllable thermoelastic coefficient for use in spring, pressure sensor and instrumentation applications. Incoloy alloy 903 (UNS N19903), Alloy 907 (UNS N19907), and Alloy 909 (UNS N19909) contain 38%Ni, 42%Fe, and 13%Co with varying amounts of the precipitation hardening elements Ti, Al, and Nb. These alloys provide a combination of high strength and low thermal expansion for applications up to 650°C (1202°F), particularly in gas turbine components where close fit is important, such as with rings, seals, shrouds, casings, and vanes.

Miscellaneous Nickel Alloys

Nickel and its alloys find further use as electrical resistance alloys, soft magnetic alloys, and shape memory alloys. Electrical resistance alloys are used for measurement and regulation of electrical characteristics, and also for resistance heating elements. The requirements for heating element alloys include high melting point, high electrical resistivity, and good oxidation resistance. Alloys employed for heating elements are nickel-chromium alloys containing minor amounts of silicon (e.g. UNS N06003 - 78%Ni, 20%Cr, 1%Si, and UNS N06008 - 68%Ni, 30%Cr, 1%Si) and nickel-chromium-iron alloys containing minor Si and Nb (e.g. UNS N06004 - 57%Ni, 16%Cr, 25%Fe, 1%Si and 37-21 alloy - 37%Ni, 21%Cr, 40%Fe, 2%Si). The resistance alloys include copper-nickel alloys with up to 45%Ni, and nickel-chromium alloys with 35 - 95%Ni and minor amounts of Al, Fe, or Si. Among the properties necessary in these alloys are uniform and stable resistivity and reproducible temperature coefficient of resistivity.

Soft magnetic alloys have been developed which have high permeability and low saturation induction making them useful in applications involving magnetic shielding. Alloys containing 79%Ni, 5%Mo, 16% Fe have the highest values of permeability and low saturation induction, while lower nickel alloys (approximately 50%Ni) are also available. Applications include rotors, armatures, transformers, inductors and memory storage devices.

Nickel-titanium alloys (approximately 50%Ni, 50%Ti) have the ability to return to a previously defined shape when subjected to appropriate thermal treatment. Typically, they can be deformed at relatively low temperature, then heated to return them to their original shape. Up to 8% memory strain is possible in nickel-titanium alloys. Applications include hydraulic couplings and force actuators as well as components which must be collapsed for compactness, then expanded for service, for example blood-clot filters inserted into veins.

Welding Of Nickel Alloys

Nickel alloys are generally weldable, but care must be taken, since many of the heat resistant alloys in particular can undergo cracking during welding or subsequent heat treatment. Welding by the GTAW (gas tungsten arc welding) process is the most common, but the GMAW (gas metal arc welding) and SMAW (shielded metal arc welding) processes are also used. Wrought nickel alloys are welded using techniques similar to those applicable to stainless steel. Cast alloys are weldable but generally require more care than their wrought counterparts. In both cases, the optimum situation is to weld with the alloys in the solution annealed condition.

Solid solution alloys are generally welded in the annealed condition without post-weld heat treatment. Precipitation hardening alloys are also normally annealed before welding, especially if they have been cold formed. Furthermore, if the welding induces residual stresses, of if the alloy is welded in the precipitation hardened condition, these alloys will require a post-weld heat treatment, consisting of solution treatment (with rapid heating to temperature) followed by aging. The SMAW process is not widely used for precipitation hardening alloys.

A problem which can arise when these alloys are welded (or improperly heat treated) is a susceptibility to intergranular corrosion known as sensitization. This phenomenon, which is also a problem in stainless steels, can develop if the alloy spends time (e.g. during slow cooling) in the critical temperature range between 425 and 760°C (797 and 1400°F). The precipitation of chromium carbide particles at the grain boundaries of the alloy removes chromium from solution in the adjacent matrix which thus becomes anodic with respect to the remainder of the grain and therefore becomes the site of localized intergranular corrosion. This problem is countered by three strategies: (1) keeping the carbon content low so that chromium carbides do not precipitate; (2) adding strong carbide-forming elements such as niobium to promote the formation of carbides other than chromium carbide; and finally (3) ensuring that a minimum amount of time is spent in the critical temperature range, by selecting appropriate thermal treatments and cooling rates. The sensitization can be removed if the alloy is heated above the critical range to redissolve the carbides, then cooled rapidly enough to avoid reprecipitation.

SAE/AMS SPECIFICATIONS - NICKEL & NICKEL ALLOYS	
AMS	**Title**
2261	Tolerances Nickel, Nickel Alloy, And Cobalt Alloy Bars, Rods, And Wire
2262	Tolerances Nickel, Nickel Alloy, And Cobalt Alloy Sheet, Strip, And Plate
2263	Tolerances Nickel, Nickel Alloy, And Cobalt Alloy Tubing
2268	Chemical Check Analysis Limits Cast Nickel And Nickel Alloys
2269	Chemical Check Analysis Limits Wrought Nickel Alloys And Cobalt Alloys
2280	Trace Element Control Nickel Alloy Castings
2316	Metallographic Evaluation Of Grain Size In Wrought Nickel And Heat Resistant Alloys
2399	Electroless Nickel-Boron Plating
2403	Nickel Plating General Purpose
2404	Electroless Nickel Plating
2405	Electroless Nickel Plating Low Phosphorus
2410	Silver Plating Nickel Strike, High Bake
2416	Nickel-Cadmium Plating Diffused
2417	Nickel-Zinc Alloy Plating
2423	Plating, Nickel Hard Deposit
2424	Nickel Plating Low- Stressed Deposit
2433	Electroless Nickel- Thallium-Boron Or Nickel-Boron Plating
2675	Nickel Alloy Brazing
4574	Nickel-Copper Alloy Tubing, Seamless, Corrosion Resistant, 67Ni - 31Cu Annealed, UNS N04400
4575	Nickel-Copper Alloy Tubing, Brazed, Corrosion Resistant, 67Ni - 31Cu Annealed, UNS N04400
4674	Bars and Forgings, Free-Machining, Corrosion Resistant, 67Ni - 30Cu - 0.04S, UNS N04405
4675	Nickel-Copper Alloy Bars & Forgings, Corrosion Resistant, 67Ni - 30Cr, UNS N04400
4676	Nickel-Copper Alloy Bars & Forgings, Corrosion Resistant 66.5Ni-3Al-0.62Ti-28Cu, Hot-Finished, Precipitation Hardenable, UNS N05500
4677	Nickel-Copper Alloy Bars And Forgings, Corrosion Resistant 66.5Ni - 2.9Al - 30Cu Annealed, UNS N05502
4730	Nickel-Copper Alloy Wire, Corrosion Resistant 67Ni - 31Cu Annealed, UNS N04400
4731	Nickel-Copper Alloy Wire And Ribbon, Corrosion Resistant, 67Ni - 31Cu, Annealed, UNS N04400
4775	Nickel Alloy Brazing Filler Metal 73Ni - 0.75C - 4.5Si - 14Cr - 3.1B - 4.5Fe 1790 to 1970°F (977 to 1077°C) Solidus-Liquidus Range, UNS N99600

SAE/AMS SPECIFICATIONS - NICKEL & NICKEL ALLOYS (Continued)

AMS	Title
4776	Nickel Alloy Brazing Filler Metal 73Ni - 4.5Si - 14Cr - 3.1B - 4.5Fe (Low Carbon) 1790 to 1970°F (977 to 1077°C) Solidus- Liquidus Range, UNS N99610
4777	Nickel Alloy Brazing Filler Metal 82Ni - 4.5Si - 7.0Cr - 3.1B - 3.0Fe 1780 to 1830°F (971 to 999°C) Solidus-Liquidus Range, UNS N99620
4778	Nickel Alloy Brazing Filler Metal 92Ni - 4.5Si - 3.1B 1800 to 1900°F Solidus-Liquidus Range, UNS N99630
4779	Nickel Alloy Brazing Filler Metal 94Ni - 3.5Si - 1.8B 1800 to 1950°F (982 to 1066°C) Solidus-Liquidus Range, UNS N99640
4782	Nickel Alloy Brazing Filler Metal 71Ni - 10Si - 19Cr 1975 to 2075°F (1080 to 1135°C) Solidus-Liquidus Range, UNS N99650
4892	Alloy Castings, Sand and Centrifugal, Corrosion and Heat Resistant 66Ni - 29Cu - 4.0Si as Cast, UNS N04019
4893	Alloy Castings, Sand and Centrifugal, Corrosion and Heat Resistant 66Ni - 29Cu - 4.0Si Solution Treated, UNS N04019
5377	Alloy Castings, Investment, Corrosion and Heat Resistant, 73Ni - 12Cr - 4.5Mo - 2.0Cb - 0.70Ti - 6.0Al - 0.010b - 0.10Zr, Vacuum Melted, As Cast, UNS N07713
5383	Alloy Castings, Investment, Corrosion and Heat Resistant, 52.5Ni - 19Cr - 3.0Mo - 5.1(Cb+Ta) - 0.90Ti - 0.60Al - 18Fe, Vacuum Melted, Homogenization and Solution Heat Treated, UNS N07718
5384	Alloy Castings, Investment, Corrosion and Heat Resistant, 53Ni - 18Cr - 18Co - 4.0Mo - 2.9Ti - 2.9Al - 0.006B, Vacuum Melted, Vacuum Cast, Solution and Precipitation Heat Treated, UNS N07500
5388	Alloy Castings, Investment, Corrosion and Heat Resistant, 53.5Ni - 16.5Cr - 17Mo - 4.5W - 5.8Fe - 0.40V, As Cast, UNS N10002
5389	Alloy Castings, Sand, Corrosion and Heat Resistant 55Ni - 16.5Cr - 17Mo - 4.5W - 5.8Fe - 0.40V Solution Heat Treated, UNS N10002
5390	Alloy Castings, Investment, Corrosion and Heat Resistant, 47.5Ni - 22Cr - 1.5Co - 9Mo - 0.60W - 18.5Fe, As Cast, UNS N06002
5391	Alloy Castings, Investment, Corrosion and Heat Resistant, 73Ni - 13Cr - 4.5Mo - 2.3(Cb+Ta) - 0.75Ti - 6.0Al - 0.010B - 0.10Zr Vacuum Melted, As Cast, UNS N07713
5396	Alloy Castings, Investment, Corrosion and Heat Resistant, 65Ni - 28Mo - 5.5Fe - 0.40V, As Cast, UNS N10001
5397	Alloy Castings, Investment, Corrosion and Heat Resistant, 50Ni - 9.5Cr - 15Co - 3.0Mo - 4.8Ti - 5.5Al - 0.015B - 0.95V - 0.06Zr, Vacuum Melted, Vacuum Cast, As Cast, UNS N13100
5399	Alloy Castings, Investment, Corrosion and Heat Resistant, 52Ni - 19Cr - 11Co - 9.8Mo - 3.2Ti - 1.6Al - 0.006B, Vacuum Melted, Vacuum Cast, Solution Heat Treated, UNS N07041
5401	Alloy Castings, Investment, Corrosion and Heat Resistant, 62Ni - 21.5Cr - 9.0Mo - 3.6(Cb+Ta), Vacuum Melted, Vacuum Cast, As Cast, UNS N06625
5402	Alloy Castings, Investment, Corrosion and Heat Resistant, 62Ni - 21.5Cr - 9.0Mo - 3.6(Cb+Ta), As Cast, UNS N06625

SAE/AMS SPECIFICATIONS - NICKEL & NICKEL ALLOYS (Continued)

AMS	Title
5403	Alloy Castings, Investment, Corrosion and Heat Resistant, 60Ni - 14Cr - 9.5Co - 4.0Mo - 4.0W - 5.0Ti - 3.0Al - 0.015B - 0.06Zr, Vacuum Melted, Vacuum Cast
5404	Alloy Castings, Investment, Corrosion and Heat Resistant, 59Ni - 12.6Cr - 9.0Co - 1.9Mo - 4.2W - 4.0Ti - 4.2Ta - 3.4Al - 0.015B - 0.10Zr - 0.90Hf, Vacuum Melted, Vacuum Cast, Solution and Precipitation Heat Treated, UNS N07013
5405	Alloy Castings, Investment, Corrosion and Heat Resistant, 64Ni - 8.0Cr - 10Co - 6.0Mo - 1.0Ti - 4.2Ta - 6.0Al - 0.15B - 0.008Zr, Vacuum Melted, Vacuum Cast, As Cast, UNS N13010
5406	Alloy Castings, Investment, Corrosion and Heat Resistant, 63Ni - 8.0Cr - 10Co - 6.0Mo - 1.0Ti - 4.2Ta - 6.0Al - 0.015B - 0.03Zr - 1.2Hf, Vacuum Melted, Vacuum Cast, As Cast, UNS N13010
5407	Alloy Castings, Investment, Corrosion and Heat Resistant, 60Ni - 9.0Cr - 10Co - 12.5W - 1.0Cb - 2.0Ti - 5.0Al - 0.015B - 0.06Zr, Vacuum Melted, Vacuum Cast, As Cast, UNS N13009
5408	Alloy Castings, Centrifugal, Corrosion and Heat Resistant 64Ni - 16Cr - 15Mo - 0.30Al - 0.05La Solution Heat Treated
5409	Alloy Castings, Investment, Corrosion and Heat Resistant, 61Ni - 16Cr - 8.5Co - 1.8Mo - 2.6W - 0.8Cb - 3.4Ti - 1.8Ta - 3.4Al - 0.010B - 0.06Zr, (0.03-0.08C), Vacuum Melted, Vacuum Cast, As Cast
5410	Alloy Castings, Investment, Corrosion and Heat Resistant, 61Ni - 16Cr - 8.5Co - 1.8Mo - 2.6W - 0.85Cb - 3.4Ti - 1.8Ta - 3.4Al - 0.010B - 0.05Zr, Vacuum Melted, Vacuum Cast, As Cast
5536	Alloy Sheet, Strip and Plate, 58Ni - 15.5Cr - 16Mo - 3.8w - 5.5fE, Solution Heat Treated, UNS N06002
5540	Alloy Sheet, Strip and Plate, 74Ni - 15.5Cr - 8.0Fe, Annealed, UNS N06600
5541	Nickel Alloy, Corrosion and Heat Resistant, Sheet and Strip 73Ni - 15.5Cr - 2.4Ti - 0.7Al - 7.0Fe Annealed, UNS N07722
5542	Alloy Sheet, Strip and Plate, 72Ni - 15.5Cr - 0.95(Cb+Ta) - 2.5Ti - 0.70Al - 7.0Fe, Annealed, UNS N07750
5544	Alloy Sheet, Strip and Plate, 57Ni - 19.5Cr - 13.5Co - 4.2Mo - 3.0Ti - 1.4Al - 0.05Zr - 0.006B, Consumable Electrode or Vacuum Induction Melted, Annealed, UNS N07001
5545	Alloy Sheet, Strip and Plate, Corrosion and Heat Resistant, 54Ni - 19Cr - 11Co - 9.8Mo - 3.2Ti - 1.5Al - 0.006B, Vacuum Induction and Consumable Electrode Melted, Solution Heat Treated, Precipitation Heat Treatable, UNS N07041
5550	Alloy Sheet, Strip, Corrosion and Heat Resistant, 80Ni - 15.5Cr - 0.62Ti - 3.2Al, Annealed, UNS N07702
5552	Alloy Sheet, Strip, and Plate, Corrosion and Heat Resistant, 46Fe - 20.5Cr - 32Ni - 1.1Ti, UNS N08801
5553	Nickel Sheet and Strip Low (0.02 Max) Carbon Annealed, UNS N02201
5555	Nickel Wire and Ribbon 99Ni, UNS N02205
5580	Alloy Tubing, Seamless, Corrosion and Heat Resistant 74Ni - 15.5Cr - 8.0Fe Annealed, UNS N06600

SAE/AMS SPECIFICATIONS - NICKEL & NICKEL ALLOYS (Continued)

AMS	Title
5581	Nickel Alloy, Corrosion and Heat Resistant, Seamless or Welded Tubing 62Ni - 21.5Cr - 9.0Mo - 3.7Cb Annealed, UNS N06625
5582	Alloy Tubing, Seamless, Corrosion and Heat Resistant 72Ni - 15.5Cr - 0.95 (Cb+Ta) - 2.5Ti - 0.70Al - 7.0Fe Annealed, UNS N07750
5583	Alloy Tubing, Seamless, Corrosion and Heat Resistant 72Ni - 15.5Cr - 0.95(Cb + Ta) - 2.6Ti - 0.70Al - 7.0Fe Solution Heat Treated, UNS N07750
5586	Alloy Tubing, Welded, Corrosion and Heat Resistant 57Ni - 19.5Cr - 13.5Co - 4.2Mo - 2.9Ti - 1.4Al - 0.08Zr - 0.006B Consumable Electrode or Vacuum Induction Melted, Annealed, UNS N07001
5587	Nickel Alloy, Corrosion and Heat Resistant, Seamless Tubing 47.5Ni - 22Cr - 1.5Co - 9.0Mo - 0.60W - 18.5Fe Solution Heat Treated, UNS N06002
5588	Alloy Tubing, Welded and Drawn, Corrosion and Heat Resistant 47.5Ni - 22Cr - 1.5Co - 9.0Mo - 0.60W - 18.5Fe Solution Heat Treated, UNS N06002
5589	Alloy Tubing, Seamless, Corrosion and Heat Resistant 52.5Ni - 19Cr - 3.0Mo - 5.1(Cb + Ta) - 0.90Ti - 0.50Al - 18Fe Consumable Electrode or Vacuum Induction Melted 1750°F (955°C) Solution Heat Treated, UNS N07718
5590	Alloy Tubing, Seamless, Corrosion and Heat Resistant 52.5Ni - 19Cr - 3.0Mo - 5.1 (Cb + Ta) - 0.90Ti - 0.50Al - 18Fe Consumable Electrode or Vacuum Induction Melted 1950°F (1065°C) Solution Heat Treated, UNS N07718
5592	Alloy Sheet, Strip, and Plate, 18.5Cr - 35Ni - 1.15Si, Solution Heat Treated, UNS N08330
5593	Alloy Sheet, Strip, Plate Corrosion and Heat Resistant, 45.5Ni - 25.5Cr - 3.2Co - 3.2Mo - 3.2W - 18.5Fe, Solution Heat Treated, UNS N06333
5596	Alloy Sheet, Strip, Plate Corrosion and Heat Resistant, 52.5Ni - 19Cr - 3.0Mo - 5.1(Cb+Ta) - 0.90Ti - 0.50Al - 18Fe, Consumable Electrode or Vacuum Induction Melted 1750°F (955°C) Solution Heat Treated, UNS N07718
5597	Alloy Sheet, Strip, Plate Corrosion and Heat Resistant, 52.5Ni - 19Cr - 3.0Mo - 5.1(Cb+Ta) - 0.90Ti - 0.50Al - 18Fe, Consumable Electrode or Vacuum Induction Melted 1950°F (1067°C) Solution Heat Treated, UNS N07718
5598	Alloy Sheet, Strip, Plate Corrosion and Heat Resistant, 72Ni - 15.5Cr - 0.95(Cb+Ta) - 2.5Ti - 0.70Al - 7.0Fe, Consumable Electrode or Vacuum Induction Melted, Solution Heat Treated, Precipitation Hardenable, UNS N07750
5599	Alloy Sheet, Strip, Plate Corrosion and Heat Resistant, 62Ni - 21.5Cr - 9.0Mo - 3.7(Cb+Ta), Annealed, UNS N06625
5605	Nickel Alloy, Corrosion and Heat Resistant, Sheet, Strip, and Plate 41.5Ni - 16Cr - 37Fe - 2.9Cb - 1.8Ti Consumable Electrode or Vacuum Induction Melted 1800°F (982°C) Solution Heat Treated, UNS N09706
5606	Sheet, Strip, Plate, 41.5Ni - 16Cr - 37Fe - 2.9Cb - 1.8Ti, Consumable Electrode or Vacuum Induction Melted 1759°F (954°C) Solution Heat Treated, UNS N09706
5607	Sheet, Strip, Plate, 73Ni - 7.0Cr - 16.5Mo, Solution Treated, UNS N10003
5633	Alloy Bars and Forgings, 38Fe - 13.2Cr - 38Ni - 5.5Mo - 0.85Cb - 2.5Ti - 1.6Al - 0.009B, Consumable Electrode Melted, Solution Heat Treated, UNS N09027

SAE/AMS SPECIFICATIONS - NICKEL & NICKEL ALLOYS (Continued)	
AMS	Title
5634	Alloy Bars and Forgings, 38Fe - 13.2Cr - 38Ni - 5.5Mo - 0.85Cb - 2.5Ti - 1.6Al - 0.009B, Consumable Electrode Melted, Solution and Precipitation Heat Treated, UNS N09027
5660	Alloy Bars, Forgings, and Rings 42.5Ni - 12.5Cr - 6.0Mo - 2.7Ti - 34Fe, Consumable Electrode or Vacuum Induction Melted Solution, Stabilization, and Precipitation Heat Treated, UNS N09901
5661	Alloy Bars, Forgings, and Rings, Corrosion and Heat Resistant 42.5Ni - 12.5Cr - 5.8Mo - 2.9Ti - 0.015B - 35Fe Consumable Electrode or Vacuum Induction Melted Solution, Stabilization, and Precipitation Heat Treated, UNS N09901
5662	Alloy Bars, Forgings, and Rings, Corrosion and Heat Resistant 52.5Ni - 19Cr - 3.0Mo - 5.1(Cb+Ta) - 0.90Ti - 0.50Al - 18Fe, UNS N07718
5663	Alloy Bars, Forgings, and Rings, Corrosion and Heat Resistant 52.5Ni - 19Cr - 3.0Mo - 5.1(Cb+Ta) - 0.90Ti - 0.50Al - 19Fe, Consumable Electrode or Vacuum Induction Melted, 1775°F (968°C) Solution and Precipitation Heat Treated, UNS N07718
5664	Alloy Bars, Forgings, and Rings, Corrosion and Heat Resistant 52.5Ni - 19Cr - 3.0Mo - 5.1(Cb+Ta) - 0.90Ti - 0.50Al - 19Fe, Consumable Electrode or Vacuum Induction Melted, 1950°F (1066°C) Solution Heat Treated, Precipitation Hardenable, UNS N07718
5665	Alloy Bars, Forgings, and Rings 74Ni - 15.5Cr - 8.0Fe, Solution Heat Treated, UNS N0660
5666	Bars, Forgings, Extrusions, and Rings 62Ni - 21.5Cr - 9.0Mo - 3.65 (Cb+Ta) Annealed, UNS N06625
5667	Alloy Bars, Forgings, and Rings 72Ni - 15.5Cr - 0.95(Cb+Ta) - 2.5Ti - 0.70Al - 7.0Fe, Equalized, Precipitation Hardenable, UNS N07750
5668	Alloy Bars, Forgings, and Rings 72Ni - 15.5Cr - 7Fe - 2.5Ti - 1(Cb+Ta) - 0.7Al, 2100°F (1095°C) Solution and Precipitation Heat Treated, UNS N07750
5669	Alloy Bars, Corrosion and Heat Resistant Nickel Base - 15.5Cr - 0.95 (Cb+Ta) - 2.5Ti - 0.70Al - 7.0Fe Consumable Electrode or Vacuum Induction Melted NONCURRENT July 1987, UNS N07750
5670	Alloy Bars, Forgings, and Rings 72Ni - 15.5Cr - 0.95(Cb+Ta) - 2.5Ti - 0.70Al - 7.0Fe, 1800°F (980°C) Solution Heat Treated, Precipitation Hardenable, UNS N07750
5671	Alloy Bars, Forgings, and Rings, Corrosion and Heat Resistant, 72Ni - 15.5Cr - 0.95(Cb+Ta) - 2.5Ti - 0.70Al - 7.0Fe, Consumable Electrode or Vacuum Induction Melted, 1800°F (982°C) Solution Heat Treated, Precipitation Hardenable, UNS N07750
5675	Alloy Welding Wire, Corrosion and Heat Resistant, 70Ni - 2.4Mn - 15.5Cr - 3.0Ti - 7.0Fe, UNS N07092
5676	Alloy Welding Wire, Corrosion and Heat Resistant, 80Ni - 20Cr, UNS N06003
5677	Alloy Covered Welding Electrodes, Corrosion and Heat Resistant, 75Ni - 19.5Cr - 1.6(Cb+Ta), UNS N06003
5679	Nickel Alloy Welding Wire, Corrosion and Heat Resistant, 73Ni - 15.5Cr - 2.2Cb - 8.0Fe, UNS N06062
5682	Coating Alloy, Corrosion and Heat Resistant, 78Ni - 20Cr, UNS N06003
5684	Alloy Welding Electrodes, Covered, Corrosion and Heat Resistant, 72Ni - 15Cr - 9.0Fe - 2.8(Cb+Ta), UNS W86132

SAE/AMS SPECIFICATIONS - NICKEL & NICKEL ALLOYS (Continued)	
AMS	**Title**
5687	Alloy Wire, Corrosion and Heat Resistant, 74Ni - 15.5Cr - 8.0Fe, Annealed, UNS N06600
5698	Alloy Wire, Corrosion and Heat Resistant, 72Ni - 15.5Cr - 0.95(Cb+Ta) - 2.5Ti - 0.70Al - 7.0Fe, No. 1 Temper, Precipitation Hardenable, UNS N07750
5699	Alloy Wire, Corrosion and Heat Resistant, 72Ni - 15.5Cr - 0.95(Cb+Ta) - 2.5Ti - 0.70Al - 7.0Fe, Spring Temper, Precipitation Hardenable, UNS N07750
5701	Alloy Bars, Forgings, and Rings, Corrosion and Heat Resistant, 41.5Ni - 16Cr - 37Fe - 2.9Cb - 1.8Ti, Consumable Electrode or Vacuum Induction Melted 1800°F (980°C) Solution Heat Treated, UNS N09706
5702	Alloy Bars, Forgings, and Rings, Corrosion and Heat Resistant, 41.5Ni - 16Cr - 37Fe - 2.9Cb - 1.8Ti, Consumable Electrode or Vacuum Induction Melted 1750°F (955°C) Solution Heat Treated, UNS N09706
5703	Alloy Bars, Forgings, and Rings, Corrosion and Heat Resistant, 41.5Ni - 16Cr - 37Fe - 2.9Cb - 1.8Ti, Consumable Electrode or Vacuum Induction Melted 1750°F (955°C) Solution, Stabilization, and Precipitation Heat Treated, UNS N09706
5704	Alloy Forgings, Corrosion and Heat Resistant, 57Ni - 19.5Cr - 13.5Co - 4.3Mo - 3.0Ti - 1.4Al - 0.05Zr - 0.006B, Consumable Electrode or Vacuum Induction Melted, 1828°-1900°F (995°-1040°C) Solution, Stabilization, and Precipitation Heat Treated, UNS N07001
5706	Alloy Bars, Forgings, and Rings, Corrosion and Heat Resistant, 57Ni - 19.5Cr - 13.5Co - 4.3Mo - 3.0Ti - 1.4Al - 0.006B - 0.05Zr, Consumable Electrode or Vacuum Induction Melted, 1828°-1900°F (995°-1040°C) Solution Heat Treated, UNS N07001
5707	Alloy Bars, Forgings, and Rings, Corrosion and Heat Resistant, 58Ni - 19.5Cr - 13.5Co - 4.3Mo - 3.0Ti - 1.4Al - 0.05Zr - 0.006B, Consumable Electrode or Vacuum Induction Melted, 1828°-1900°F (996°-1038°C) Solution, Stabilization, and Precipitation Heat Treated, UNS N07001
5708	Alloy Bars, Forgings, and Rings, Corrosion and Heat Resistant, 58Ni - 19.5Cr - 13.5Co - 4.3Mo - 3.0Ti - 1.4Al, Consumable Electrode or Vacuum Induction Melted, 1975°F (1079°C) Solution Heat Treated, UNS N07001
5709	Alloy Bars, Forgings, Corrosion and Heat Resistant, 58Ni - 19.5Cr - 13.5Co - 4.3Mo - 3.0Ti - 1.4Al - 0.05Zr - 0.006B, Consumable Electrode or Vacuum Induction Melted, 1975°F (1079°C) Solution, Stabilization, and Precipitation Heat Treated, UNS N07001
5711	Alloy Bars, Forgings, and Rings, Corrosion and Heat Resistant, 65Ni - 15.8Cr - 15.2Mo - 0.30Al - 0.05La Solution Heat Treated, UNS N06635
5712	Alloy Bars, Forgings, and Rings, Corrosion and Heat Resistant, 53Ni - 19Cr - 11Co - 9.8Mo - 3.2Ti - 1.6Al - 0.006B, Vacuum Melted, Solution Heat Treated, UNS N07041
5713	Alloy Bars, Forgings, and Rings, Corrosion and Heat Resistant, 53Ni - 19Cr - 11Co - 9.8Mo - 3.2Ti - 1.6Al - 0.006B, Vacuum Melted, Solution and Precipitation Heat Treated, UNS N07041
5714	Bars, Forgings, and Rings, Corrosion and Heat Resistant 73Ni - 15.5Cr - 2.5Ti - 0.70Al - 7.0Fe Precipitation Hardenable, UNS N07722
5715	Nickel Alloy, Corrosion and Heat Resistant, Bars, Forgings, and Rings 60.5Ni - 23Cr - 14Fe - 1.4Al Annealed, UNS N06601
5716	Alloy Bars, Wire, Forgings, and Rings, Heat Resistant, 44Fe - 18.5Cr - 35.5Ni - 1.1Si, Solution Heat Treated, UNS N08330

SAE/AMS SPECIFICATIONS - NICKEL & NICKEL ALLOYS (Continued)	
AMS	**Title**
5717	Nickel Alloy, Corrosion and Heat Resistant, Bars, Forgings, and Rings 45.5Ni - 25.5Cr - 3.2Co - 3.2Mo - 3.2W - 18.5Fe Solution Heat Treated, UNS N06333
5746	Alloy Bars, Forgings, Corrosion and Heat Resistant, 15Cr - 45Ni - 4.1Mo - 4.1W - 3.0Ti - 1.0Al - 31Fe, Consumable Electrode, Vacuum Melted, Solution and Precipitation Heat Treated, UNS N09979
5747	Nickel Alloy, Corrosion and Heat Resistant, Bars, Forgings, and Rings 72Ni - 15.5Cr - 0.95Cb - 2.5Ti - 0.70Al - 7.0Fe Solution Heat Treated, Precipitation Hardenable, UNS N07750
5750	Nickel Alloy Bars, Forgings, and Rings, Corrosion and Heat Resistant Nickel Base - 15.5Cr - 16Mo - 3.8W - 5.5Fe Solution Heat Treated, NONCURRENT July 1981, UNS N10002
5751	Alloy Bars, Forgings, and Rings, Corrosion and Heat Resistant, 54Ni - 17.5Cr - 16.5Co - 4.0Mo - 2.9Ti - 2.9Al - 0.006B, Solution, Stabilization, and Precipitation Heat Treated, Consumable Electrode or Vacuum Induction Melted, UNS N07500
5753	Alloy Bars, and Forgings, Corrosion and Heat Resistant, 54Ni - 17.5Cr - 16.5Co - 4.0Mo - 2.9Ti - 2.9Al - 0.006B, Vacuum Melted, Solution Heat Treated, UNS N07500
5754	Alloy Bars, Forgings, and Rings, Corrosion and Heat Resistant, 47.5Ni - 22Cr - 1.5Co - 9.0Mo - 0.60W - 18.5Fe, Solution Heat Treated, UNS N06002
5755	Alloy Bars, Forgings, and Rings, Corrosion and Heat Resistant, 64Ni - 5.0Cr - 24.5Mo - 5.5Fe, Solution Heat Treated, UNS N10004
5766	Alloy Bars, and Forgings, Corrosion and Heat Resistant, 21Cr - 32.5Ni - 0.38Al - 45Fe, Solution Heat Treated, UNS N08800
5771	Alloy Bars, Forgings, and Rings, 74Ni - 7.0Cr - 16.5Mo, Solution Heat Treated, UNS N10003
5778	Alloy Welding Wire, Corrosion and Heat Resistant 72Ni - 15.5Cr - 1.0(Cb + Ta) - 2.4Ti - 0.70Al - 7.0Fe, UNS N07069
5779	Alloy Welding Electrodes, Covered, Corrosion and Heat Resistant 75Ni - 15Cr - 1.5 (Cb + Ta) - 1.9Ti - 0.55Al - 5.5Fe, UNS N07750
5786	Welding Wire, 62.5Ni - 5.0Cr - 24.5Mo - 5.5Fe, UNS N10004
5787	Alloy Covered Welding Electrode, 63Ni - 5Cr - 24.5Mo - 5.5Fe, NONCURRENT January 1987, UNS W80004
5798	Welding Wire, Corrosion and Heat Resistant, 47.5Ni - 22Cr - 1.5Co - 9.0Mo - 0.60W - 18.5Fe, UNS N06002
5799	Alloy Covered Welding Electrodes, 48Ni - 22Cr - 1.5Co - 9.0Mo - 0.60W - 18.5Fe, UNS N06002
5800	Welding Wire, 54Ni - 19Cr - 11Mo - 3.2Ti - 1.5Al - 0.006B, UNS N07041
5802	Alloy Welding Wire, 41Fe - 37.5Ni - 14Co - 4.8(Cb+Ta) - 1.5Ti, Vacuum Melted, Low Expansion, UNS N19907
5806	Welding Wire, Low Expansion Alloy, 42Fe - 38Ni - 15Co - 3.0(Cb+Ta) - 1.4Ti - 0.92Al, Vacuum Melted, UNS N19903
5828	Alloy Welding Wire, 57Ni - 19.5Cr - 13.5Co - 4.2Mo - 3.1Ti - 1.4Al - 0.006B, Vacuum Induction Melted, Solution Heat Treated, UNS N07001
5829	Alloy Welding Wire, 56Ni - 19.5Cr - 18Co - 2.5Ti - 1.5Al, Vacuum Induction Melted, UNS N07090

SAE/AMS SPECIFICATIONS - NICKEL & NICKEL ALLOYS (Continued)	
AMS	**Title**
5830	Welding Wire, Corrosion and Heat Resistant Alloy 12.5Cr - 42.5Ni - 6.0Mo - 2.7Ti - 0.015B - 35Fe, UNS N09901
5832	Welding Wire, 52.5Ni - 19Cr - 5.1(Cb+Ta) - 0.90Ti - 0.50Al - 18Fe, Consumable Electrode or Vacuum Induction Melted, UNS N07718
5836	Alloy Welding Wire, 72Ni - 20Cr - 2.5Mn - 2.5Cb, UNS N06082
5837	Welding Wire, 62Ni - 21.5Cr - 9.0Mo - 3.7(Cb+Ta), UNS N06625
5838	Alloy Welding Wire, 65Ni - 16Cr - 15Mo - 0.30Al - 0.06La, UNS N06635
5846	Alloys Bars and Forgings, 53Ni - 15Cr - 18.5Co - 5.0Mo - 3.2Ti - 4.2Al - 0.03B, Double Vacuum Melted, Solution Stabilization, and Precipitation Heat Treated, UNS N13020
5851	Alloy Powder55Ni - 15Cr - 17Co - 5.0Mo - 3.5Ti - 4.0Al - 0.025B, as Fabricated, UNS N13017
5852	Alloy Billets and Preforms, 55Ni - 15Cr - 17Co - 5.0Mo - 3.5Ti - 4.0Al - 0.025B, Powder Metallurgy Product, Hot Isostatically Pressed, UNS N13017
5854	Alloy Bars, Forgings, and Rings, Corrosion and Heat Resistant 61Ni - 20.5Cr - 8.5Mo - 3.4Cb - 1.3Ti - 5.0Fe Consumable Electrode or Vacuum Induction Melted Solution Heat Treated, UNS N07716
5856	Alloy Billets and Preforms, 59.5Ni - 12Cr - 10Co - 3.0Mo - 6.0W - 3.0Ti - 1.5Ta - 4.5Al - 0.015B - 0.10Zr (0.30-0.35C), Powder Metallurgy Product, UNS N07012
5865	Nickel Sheet and Strip, Corrosion and Heat Resistant Thoria Dispersion Strengthened 2.2ThO_2,, UNS N03260
5869	Alloy Sheet, Strip, and Plate, Corrosion and Heat Resistant 62Ni - 21.5Cr - 9.0Mo - 3.7(Cb+Ta) Solution Heat Treated, UNS N06625
5870	Alloy Sheet, Strip, and Plate, Corrosion and Heat Resistant, 60.5Ni - 23Cr - 14Fe - 0.35Ti - 1.4Al, Solution Heat Treated, UNS N06601
5871	Alloy Sheet, Strip, and Plate, Corrosion and Heat Resistant, 21Cr - 32.5Ni - 0.38Ti - 0.38Al - 45Fe, Solution Heat Treated, UNS N08800
5872	Sheet, Strip, and Plate, Corrosion and Heat Resistant, 48Ni - 20Cr - 20Co - 5.9Mo - 2.2Ti - 0.45Al, Solution Heat Treated, UNS N07263
5873	Alloy Sheet, Strip, Plate, Corrosion and Heat Resistant, 65Ni - 15.8Cr - 15.2Mo - 0.30Al - 0.05La, Consumable Electrode Melted, Solution Heat Treated, UNS N06635
5878	Alloy Sheet, Strip, and Plate, Corrosion and Heat Resistant, 59Ni - 22Cr - 2Mo - 14W - 0.35Al - 0.03La, Solution Heat Treated, UNS N06230
5879	Sheet, Strip, and Foil, Corrosion and Heat Resistant Alloy, 62Ni - 21.5Cr - 9.0Mo - 3.7(Cb+Ta) Cold Rolled and Annealed, UNS N06625
5882	Alloy Forgings, Corrosion and Heat Resistant, 55Ni - 15Cr - 17Co - 5.0Mo - 3.5Ti - 4.0Al - 0.025B, Solution, Stabilization, and Precipitation Heat Treated, Powder Metallurgy Product, UNS N13017
5884	Iron-Nickel Alloy, Bars, Forgings, and Rings 42Fe - 37.5Ni - 14Co - 4.8(Cb + Ta) - 1.6Ti Solution Heat Treated, Precipitation Hardenable Multiple Melted, High Temperature, Low Expansion, UNS N19909

SAE/AMS SPECIFICATIONS - NICKEL & NICKEL ALLOYS (Continued)	
AMS	**Title**
5886	Alloy Bars, Forgings, and Rings, Corrosion and Heat Resistant, 50Ni - 20Cr - 20Co - 5.8Mo - 2.2Ti - 0.45Al, Consumable Electrode or Vacuum Induction Melted 2100°F (1149°C) Solution Heat Treated, UNS N07263
5887	Alloy Bars, Forgings, and Rings, Corrosion and Heat Resistant 54Ni - 22Cr - 12.5Co - 9.0Mo - 1.2Al Consumable Electrode or Vacuum Induction Melted Annealed, UNS N06617
5888	Alloy Plate, Corrosion and Heat Resistant 54Ni - 22Cr - 12.5Co - 9.0Mo - 1.2Al Annealed, UNS N06617
5889	Alloy Sheet and Strip, Corrosion and Heat Resistant 54Ni - 22Cr - 12.5Co - 9.0Mo - 1.2Al Consumable Electrode or Vacuum Induction Melted Annealed, UNS N06617
5890	Nickel Bars, Forgings, and Extrusions, Corrosion and Heat Resistant Thoria Dispersion Strengthened 2.2ThO2 Stress-Relieved, UNS N03260
5891	Nickel Alloy, Corrosion and Heat Resistant, Bars, Forgings, and Rings 60Ni - 22Cr - 2.0Mo - 14W - 0.35Al - 0.03La Annealed, UNS N06230
5892	Low Expansion, High Temperature, Alloy Sheet and Strip, 43Fe - 37Ni - 14Co - 4.5(Cb+Ta) - 1.5Ti, Multiple Melted, Solution Heat Treated, Precipitation Hardened, UNS N19909
5893	Low Expansion, High Temperature, Alloy Bars, Forgings, and Rings 43Fe - 37Ni - 14Co - 4.5(Cb+Ta) - 1.5Ti Multiple Melted, Solution Heat Treated, Short-Time Precipitation Hardenable, UNS N19909
7232	Alloy Rivets, Corrosion Resistant, 74Ni - 15.5Cr - 8.0Fe, UNS N06600
7233	Solid Rivets, Alloy-Corrosion Resistant, 67Ni 31Cu, Superseded by AS 7233
7234	Blind Rivets, Alloy-Corrosion Resistant, 67Ni - 31Cu, UNS N04405
7237	Alloy Rivets, Corrosion and Heat Resistant 47.5Ni - 22Cr - 1.5Co - 9.0Mo - 0.60W - 18.5Fe Solution Heat Treated, UNS N06002
7246	Thread Form Inserts, Corrosion and Heat Resistant Alloy, 72Ni - 15.5Cr - 0.95(Cb+Ta) - 2.5Ti - 0.70Al - 7.0Fe, UNS N07750
7490	Flash Welded Rings, Corrosion and Heat Resistant Austenitic Steels and Austenitic-Type Alloys, or Precipitation Hardenable Alloys, UNS N06625

SAE MISCELLANEOUS SPECIFICATIONS - NICKEL & NICKEL ALLOYS	
SAE	**Title**
AS 3074	Gasket-37 Degree Flared Tube Fitting, Nickel, AMS 5553
AS 3075	Gasket - 37 Degree Flared Tube Fitting, Nickel, AMS 5553
AS 3282	Gasket - 37 Degree Flared Tube Fitting, Nickel, AMS 5553
AS 3484	Gasket - 37 Degree Flared Tube Fitting, Nickel, AMS-5553

SAE MISCELLANEOUS SPECIFICATIONS - NICKEL & NICKEL ALLOYS (Continued)

SAE	Title
AS 7253	Nuts, Self-Locking, Nickel Alloy, UNS N07001 180 000 psi, All Metal 1400°F Use, UNJ Thread Form
AS 7466	Bolts and Screws, Nickel Alloy, Corrosion and Heat Resistant Forged Head, Roll Threaded, Fatigue Rated
AS 7467	Bolts and Screws, Nickel Alloy, Corrosion and Heat Resistant Forged Head, Roll Threaded, Stress-Rupture Rated
AS 7468	Bolts and Screws, Cobalt- Chromium-Nickel Alloy, Corrosion Resistant Forged Head, Roll Threaded After Aging, Fatigue Rated
AS 7471	Bolts and Screws, Nickel Alloy, Corrosion and Heat Resistant Forged Head, Heat Treated, Roll Threaded 1950°F Solution Heat Treatment
AS 8033	Nickel Cadmium Vented Rechargeable Aircraft Batteries (Non-Sealed, Maintainable Type)
J 207	Electroplating of Nickel and Chromium on Metal Parts - Automotive Ornamentation and Hardware, Standard
J 470C	Wrought Nickel and Nickel-Related Alloys, Information Report
MAM 2261	Tolerances, Metric Nickel, Nickel Alloys, and Cobalt Alloy Bars, Rods, and Wire
MAM 2262	Tolerances, Metric Nickel, Nickel Alloy, and Cobalt Alloy Sheet, Strip, and Plate
MAM 2263	Tolerances, Metric Nickel, Nickel Alloy, and Cobalt Alloy Tubing
MAM 4778	Nickel Alloy Brazing Filler Metal 92Ni - 4.5Si - 3.1B 980 to 1040°C Solidus-Liquidus Range
MA 2538	Gasket, Metal C-Ring Seal Nickel Alloy, Corrosion and Heat Resistant, Metric Procurement Specification for

AS - Aerospace Standard; MAM - Metric Aerospace Material; J - SAE (Automotive) Standard.

ASTM SPECIFICATIONS - NICKEL & NICKEL ALLOYS

Commercially Pure Nickel	
ASTM	**Title**
B 39	Nickel
Nickel and Nickel Alloy Castings	
A 494	Castings, Nickel and Nickel Alloy
Nickel and Nickel Alloy Fittings	
B 366	Factory-Made Wrought Nickel and Nickel Alloy Welding Fittings
B 462	Forged or Rolled UNS N08020, UNS N08024, UNS N08026 and UNS N08367 Alloy Pipe Flanges, Forged Fittings, and Valves and Parts for Corrosive High-Temperature Service

ASTM SPECIFICATIONS - NICKEL & NICKEL ALLOYS (Continued)	
ASTM	**Title**
Nickel Alloy Forgings	
B 564	Nickel Alloy Forgings
Nickel and Nickel Alloy Pipes and Tubes	
B 161	Nickel Seamless Pipe and Tube
B 163	Seamless Nickel and Nickel Alloy Condenser and Heat-Exchanger Tubes
B 165	Nickel-Copper Alloy (UNS N04400) Seamless Pipe and Tube
B 167	Nickel-Chromium-Iron Alloys (UNS N06600, N06601, and N06690) Seamless Pipe and Tube
B 407	Nickel-Iron-Chromium Alloy Seamless Pipe and Tube
B 423	Nickel-Iron-Chromium-Molybdenum-Copper Alloys (UNS N08825 and N08221) Seamless Pipe and Tube
B 445	Nickel-Chromium-Iron-Columbium-Molybdenum-Tungsten Alloy (UNS N06102) Seamless Pipe and Tube
B 444	Nickel-Chromium-Molybdenum-Columbium Alloys (UNS N06625) Pipe and Tube
B 464	Welded UNS N08020, UNS N08024, and UNS N08026 Alloy Pipe
B 468	Welded UNS N08020, UNS N08024, and UNS N08026 Alloy Tubes
B 474	Electric Fusion Welded UNS N08020, UNS N08026, and UNS N08024 Nickel Alloy Pipe
B 514	Welded Nickel-Iron-Chromium Alloy Pipe
B 515	Welded UNS N08800 and UNS N088 10 Alloy Tubes
B 516	Welded Nickel-Chromium-Iron Alloy (UNS N06600) Tubes
B 517	Welded Nickel-Chromium-Iron-Alloy (UNS N06600) Pipe
B 535	Nickel-Iron-Chromium-Silicon Alloys (UNS N08330 and UNS N08332) Seamless Pipe
B 546	Electric Fusion-Welded Ni-Fe-Cr-Si Alloys (UNS N08330 and UNS N08332) Pipe
B 619	Welded Nickel and Nickel-Cobalt Alloy Pipe
B 622	Seamless Nickel and Nickel-Cobalt Alloy Pipe and Tube
B 626	Welded Nickel and Nickel-Cobalt Alloy Tube
B 668	UNS N08028 Seamless Tubes
B 673	UNS N08904, UNS N08925, and UNS N08926 Welded Pipe
B 674	UNS N08904, UNS N08925, and UNS N08926 Welded Tube
B 675	UNS N08366 and UNS N08367 Welded Pipe

ASTM SPECIFICATIONS - NICKEL & NICKEL ALLOYS (Continued)

ASTM	Title
Nickel and Nickel Alloy Pipes and Tubes (Continued)	
B 676	UNS N08366 and UNS N08367 Welded Tube
B 677	UNS N08904, UNS N08925, and UNS N08926 Seamless Pipe and Tube
B 690	Iron-Nickel-Chromium-Molybdenum Alloys (UNS N08366 and UNS N08367) Seamless Pipe and Tube
B 704	Welded UNS N06625 and N08825 Alloy Tubes
B 705	Nickel-Alloy (UNS N06625 and N08825) Welded Pipe
B 710	Nickel-Iron-Chromium-Silicon Alloy Welded Pipe
B 720	UNS N08310 Seamless Tube
B 722	Nickel-Chromium-Molybdenum-Cobalt-Tungsten-Iron-Silicon Alloy (UNS N06333) Seamless Pipe and Tube
B 723	Nickel-Chromium-Molybdenum-Cobalt-Tungsten-Iron-Silicon Alloy (UNS N06333) Welded Pipe
B 725	Welded Nickel (UNS N02200/UNS N02201) and Nickel Copper Alloy (UNS N04400) Pipe
B 726	Nickel-Chromium-Molybdenum-Cobalt-Tungsten-Iron-Silicon Alloy (UNS N06333) Welded Tube
B 729	Seamless UNS N08020, UNS N08026, and UNS N08024 Nickel-Alloy Pipe and Tube
B 730	Welded Nickel Tube
B 739	Nickel-Iron-Chromium-Silicon Alloy Welded Tube
B 751	Nickel and Nickel Alloy Seamless and Welded Tube, General Requirements for
B 757	Nickel-Chromium-Molybdenum-Tungsten Alloys (UNS N06110) Welded Pipe
B 758	Nickel-Chromium-Molybdenum-Tungsten Alloys (UNS N06110) Welded Tube
B 759	Nickel-Chromium-Molybdenum-Tungsten Alloys (UNS N06110) Pipe and Tube
B 775	Nickel and Nickel Alloy Seamless and Welded Pipe, General Requirements for
B 804	UNS N08367 Welded Pipe
B 829	General Requirements for Nickel and Nickel Alloys Seamless Pipe and Tube
B 834	Pressure Consolidated Powder Metallurgy Iron-Nickel-Chromium-Molybdenum (UNS N08367) and Nickel-Chromium-Molybdenum-Columbium (Nb) (UNS N06625) Alloy Pipe Flanges, Fittings, Valves, and Parts
Nickel and Nickel Alloy Plates, Sheets, and Strips	
B 127	Nickel-Copper Alloy (UNS N04400) Plate, Sheet, and Strip
B 162	Nickel Plate, Sheet, and Strip

ASTM SPECIFICATIONS - NICKEL & NICKEL ALLOYS (Continued)

ASTM	Title
Nickel and Nickel Alloy Plates, Sheets, and Strips (Continued)	
B 333	Nickel-Molybdenum Alloy Plate, Sheet, and Strip
B 409	Nickel-Iron-Chromium Alloy Plate, Sheet, and Strip
B 424	Ni-Fe-Cr-Mo-Cu Alloy (UNS N08825 and N08221) Plate, Sheet, and Strip
B 434	Nickel-Molybdenum-Chromium-Iron Alloy (UNS N10003) Plate, Sheet, and Strip
B 435	UNS N06002, UNS N06230, and UNS R30556 Plate, Sheet, and Strip
B 443	Nickel-Chromium-Molybdenum-Columbium Alloy (UNS N06625) Plate, Sheet, and Strip
B 168	Nickel-Chromium-Iron Alloys (UNS N06600, N06601, and N06690) and Nickel-Chromium-Cobalt-Molybdenum Alloy (UNS N06617) Plate, Sheet, and Strip
B 463	UNS N08620, UNS N08026, and UNS N08024 Alloy Plate, Sheet, and Strip
B 519	Nickel-Chromium-Iron-Columbium-Molybdenum-Tungsten Alloy (UNS N06102) Plate, Sheet, and Strip
B 536	Nickel-Iron-Chromium-Silicon Alloys (UNS N08330 and N08332) Plate, Sheet, and Strip
B 575	Low-Carbon Nickel-Molybdenum-Chromium and Low-Carbon Nickel-Chromium Molybdenum Alloy Plate, Sheet, and Strip
B 582	Nickel-Chromium-Iron-Molybdenum-Copper Alloy Plate, Sheet, and Strip
B 599	Nickel-Iron-Chromium-Molybdenum-Columbium Stabilized Alloy (UNS N08700) Plate, Sheet, and Strip
B 620	Nickel-Iron-Chromium-Molybdenum Alloy (UNS N08320) Plate, Sheet, and Strip
B 625	UNS N08904, UNS N08925, UNS N08031, UNS N08932, and UNS N08926 Plate, Sheet, and Strip
B 670	Precipitation-Hardening Nickel Alloy (UNS N07718) Plate, Sheet, and Strip for High-Temperature Service (formerly A 670)
B 688	Chromium-Nickel-Molybdenum Iron (UNS N08366 and UNS N08367) Plate, Sheet, and Strip
B 709	Iron-Nickel-Chromium-Molybdenum Alloy (UNS N08028) Plate, Sheet, and Strip
B 718	Nickel-Chromium-Molybdenum-Cobalt-Tungsten-Iron-Silicon Alloy (UNS N06333) Plate, Sheet, and Strip
B 755	Nickel-Chromium-Molybdenum-Tungsten Alloys (UNS N06110) Plate, Sheet, and Strip
B 814	Nickel-Chromium-Iron-Molybdenum-Tungsten Alloy (UNS N06920) Plate, Sheet, and Strip
B 818	Cobalt-Chromium-Nickel-Molybdenum-Tungsten Alloy (UNS R31233) Plate, Sheet, and Strip
Nickel and Nickel Alloy Rods, Bars, and Wires	
B 160	Nickel Rod and Bar
B 164	Nickel-Copper Alloy Rod. Bar, and Wire

ASTM SPECIFICATIONS - NICKEL & NICKEL ALLOYS (Continued)	
ASTM	**Title**
Nickel and Nickel Alloy Rods, Bars, and Wires (Continued)	
B 166	Nickel-Chromium-Iron Alloys (UNS N06600, N06601, and N06690) and Nickel-Chromium-Cobalt-Molybdenum Alloy (UNS N06617) Rod, Bar, and Wire
B 335	Nickel-Molybdenum Alloy Rod
B 408	Nickel-Iron-Chromium Alloy Rod and Bar
B 425	Ni-Fe-Cr-Mo-Cu Alloy (UNS N08825 and N08221) Rod and Bar
B 446	Nickel-Chromium-Molybdenum-Columbium Alloy (UNS N06625) Rod and Bar
B 471	UNS N08020, UNS N08026, and UNS N08024 Nickel Alloy Spring Wire
B 472	UNS N08020, UNS N08026, UNS N08024, UNS N08926, and UNS N08367 Nickel Alloy Billets and Bars for Reforging
B 473	UNS N08020, UNS N08026, and UNS N08024 Nickel Alloy Bar and Wire
B 475	UNS N08020, UNS N08026, and UNS N08024 Nickel Alloy Round Weaving Wire
B 511	Nickel-Iron-Chromium-Silicon Alloy Bars and Shapes
B 512	Nickel-Chromium-Silicon Alloy (UNS N08330) Billets and Bars
B 518	Nickel-Chromium-Iron-Columbium-Molybdenum-Tungsten Alloy (UNS N06102) Rod and Bar
B 572	UNS N06002, UNS N06230, and UNS R30556 Rod
B 573	Nickel-Molybdenum-Chromium-Iron Alloy (UNS N10003) Rod
B 574	Low-Carbon Nickel-Molybdenum-Chromium and Low-Carbon Nickel-Chromium Molybdenum Alloy Rod
B 581	Nickel-Chromium-Iron-Molybdenum-Copper Alloy Rod
B 649	Ni-Fe-Cr-Mo-Cu Low-Carbon Alloy (UNS N08904) and Ni-Fe-Cr-Mo-Cu-N Low Carbon Alloys (UNS N08915, UNS N08030, and UNS N08926) Bar and Wire
B 672	Nickel-Iron-Chromium-Molybdenum-Columbium Stabilized Alloy (UNS N08700) Bar and Wire
B 691	Iron-Nickel-Chromium-Molybdenum Alloys (UNS N08366 and UNS N08367) Rod, Bar, and Wire
B 719	Nickel-Chromium-Molybdenum-Cobalt-Tungsten-Iron-Silicon Alloy (UNS N06333) Bar
B 756	Nickel-Chromium-Molybdenum-Tungsten Alloy (UNS N06110) Rod and Bar
B 621	Nickel-Iron-Chromium-Molybdenum Alloy (UNS N08320) Rod
B 637	Precipitation-Hardening Nickel Alloy Bars, Forgings, and Forging Stock for High-Temperature Service (formerly A 637)

ASTM SPECIFICATIONS - NICKEL & NICKEL ALLOYS (Continued)

ASTM	Title
Nickel and Nickel Alloy Rods, Bars, and Wires (Continued)	
B 805	Precipitation Hardening Nickel Alloys UNS N07716 and N07725, Bar and Wire
B 815	Cobalt-Chromium-Nickel-Molybdenum-Tungsten Alloy (UNS R31233) Rod

AWS WELDING FILLER METAL SPECIFICATIONS - NICKEL & NICKEL ALLOYS

AWS	Title
A5.11	Nickel and Nickel Alloy Bare Welding Electrodes for Shielded Metal Arc Welding
A5.14	Nickel and Nickel Alloy Bare Welding Electrodes and Rods

AMERICAN CROSS REFERENCED SPECIFICATIONS - NICKEL & NICKEL ALLOYS

UNS	SAE/AMS	MILITARY	ASTM	ASME	AWS
N02016	---	---	---	---	---
N02061	---	E-21562 (EN61, RN61), I-23413 (MIL-61)	---	SFA-5.14 (ERNi-1), SFA-5.30 (IN61)	A5.14 (ERNi-1), A5.30 (IN61)
N02100	---	---	A 494 (CZ-100)	---	---
N02200	---	---	B 160, B161, B 162, B 163, B366, B 725, B 730	SB-160, SB-161, SB-162, SB-163	---
N02201	5553	---	B 160, B 161, B 162, B 163, B 366, B 725, B 730	SB-160, SB-161, SB-162, SB-163	---
N02205	5555	N-46025	F 1, F 3	---	---
N02211	---	---	F 290	---	---
N02215	---	---	---	SFA-5.15 (ERNi-CI)	A5.15 (ERNi-CI)
N02216	---	---	---	SFA-5.15 (ERNiFeMn-CI)	A 5.15 (ERNiFeMn-CI)
N02220	---	---	F 239	---	---
N02225	---	---	F 239	---	---

AMERICAN CROSS REFERENCED SPECIFICATIONS - NICKEL & NICKEL ALLOYS (Continued)

UNS	SAE/AMS	MILITARY	ASTM	ASME	AWS
N02230	---	---	F 239	---	---
N02233	---	---	F 1, F 3, F 4	---	---
N02270	---	---	F 1, F 3, F 239	---	---
N02290	---	---	---	---	---
N03220	---	---	---	---	---
N03260	5865, 5890	---	---	---	---
N03300	---	---	F 290	---	---
N03301	---	---	---	---	---
N03360	---	---	---	---	---
N04019	4892, 4893	---	---	---	---
N04020	---	---	A 494 (M-35-2)	---	---
N04060	---	E-21562 (EN60, RN60), I-23413 (MIL-60)	---	SFA-5.14 (ERNiCu-7), SFA-5.30 (IN60)	A5.14 (ERNiCu-7), A5.30 (IN60)
N04400	4544, 4574, 4575, 4675, 4730, 4731, 7233	T-1368, T-23520, N-24106	B 127, B 163, B 164, B 165, B 366, B 564, F96, F 467 (400), F 468 (400)	SB-127, SB-163, SB-164, SB-165, SB-564	---
N04401	---	---	---	---	---
N04402	---	---	---	---	---
N04404	---	---	F 96	---	---
N04405	4674, 7234	---	B 164, F 467 (405), F 468 (405)	SB-164	---
N05500	4676	N-24549	F 467 (500), F 468 (500)	---	---
N05502	4677	---	---	---	---
N05504	---	E-21562 (EN64, RN64)	---	---	---
N06002	5390, 5536, 5587, 5588, 5754, 5798, 7237	---	B 366, B 435, B 572, B 619, B 622, B 626	SB-366, SB-435, SB-572, SB-619, SB-622, SB-626, SFA-5.14 (ERNiCrMo-2)	A5.14 (ERNiCrMo-2)
N06003	5676, 5677, 5682	---	B 344	---	---

AMERICAN CROSS REFERENCED SPECIFICATIONS - NICKEL & NICKEL ALLOYS (Continued)

UNS	SAE/AMS	MILITARY	ASTM	ASME	AWS
N06004	---	---	B 344	---	---
N06005	SAE J775 (VF 3)	---	---	---	---
N06006	---	---	A 297 (HX)	---	---
N06007	---	---	B 366, B 581, B 582, B 619, B 622, B 626	SB-366, SB-619, SFA-5.14 (ERNiCrMo-1)	A5.14 (ERNiCrMo-1)
N06008	---	---	---	---	---
N06009	---	---	---	---	---
N06010	---	---	---	---	---
N06013	SAE J755 (VF 9)	---	---	---	---
N06015	SAE J775 (VF 10)	---	---	---	---
N06022	---	---	B 366, B 564, B 574, B 575, B 619, B 622, B 626	SB-366, SB-564, SB-574,SB-575, SB-619, SB-622, SB-626, SFA-5.14 (ERNiCrMo-10)	A5.14 (ERNiCrMo-10)
N06030	---	---	B 366, B 581, B 582, B 619, B 622, B 626	SB-366, SB-581, SB-582, SB-619, SB-622, SB-626, SFA-5.14 (ERNiCrMo11)	A5.14 (ERNiCrMo-11)
N06040	---	---	A 494	---	---
N06050	---	---	A 608 (HX-50)	---	---
N06052	---	---	---	---	---
N06059	---	---	B 564, B 574, B 575, B 619, B 622, B 626	---	---
N06060[a]	---	---	B 622	---	---
N06062	5679	E-21562 (EN62, RN62), I-23413 (MIL-62)	---	SFA-5.14(ERNiCrFe-5), SFA-5.30 (IN62)	A5.14(ERNiCrFe-5), A5.30 (IN62)
N06075	---	---	---	---	---
N06076	---	E-21562 (EN6N, RN6N)	---	---	---

AMERICAN CROSS REFERENCED SPECIFICATIONS - NICKEL & NICKEL ALLOYS (Continued)

UNS	SAE/AMS	MILITARY	ASTM	ASME	AWS
N06082	---	E-21562(EN82, EN82H, RN82, RN82H), I-23413 (MIL-82)	---	SFA-5.14 (ERNiCr-3), SFA-5.30 (IN82)	A5.14 (ERNiCr-3), A5.30 (IN82)
N06102	---	---	B 445, B 518, B 519	---	---
N06110	---	---	B 564, B 755, B 756, B 757, B 758, B 759	---	---
N06132	---	---	---	---	---
N06230	5891	---	B 435, B 572, B 619, B 622, B 626	---	---
N06231	---	---	---	SFA-5.14 (ERNiCrWMo-1)	A5.14 (ERNiCrWMo-1)
N06250[a]	---	---	B 622	---	---
N06255[a]	---	---	B 622	---	---
N06333	5593, 5717	---	B 718, B 719, B 722, B 723, B 726	---	---
N06455	---	---	B 366, B 574, B 575, B 619, B 622, B 626	SB-366, SB-574, SB-575, SB-619, SB-622, SB-626, SFA-5.14 (ERNiCrMo-7)	A5.14 (ERNiCrMo-7)
N06600	5540, 5580, 5665, 5687, 7232	T-23227, N-23228, N-23229	B 163, B 166, B 167, B 168, B 516, B 517, B 564	SB-163, SB-166, SB-167, SB-168, SB-564	---
N06601	5715, 5870	---	B 166, B 167, B 168	---	---
N06602	---	---	---	---	---
N06617	---	---	---	SFA-5.14(ERNiCrCoMo-1)	A5.14(ERNiCrCoMo-1)
N06621[b]	---	---	---	---	---
N06625	5401, 5402, 5581, 5599, 5666, 5837, 7490	E-21562 (EN625, RN625)	B 366, B 443, B 444, B 446, B 704, B 705	SB-443, SB-444, SB-446, SFA-5.14 (ERNiCrMo-3)	A5.14 (ERNiCrMo-3)
N06635	5711, 5838, 5873	---	---	---	---

AMERICAN CROSS REFERENCED SPECIFICATIONS - NICKEL & NICKEL ALLOYS (Continued)					
UNS	SAE/AMS	MILITARY	ASTM	ASME	AWS
N06690	---	---	B 163, B 166, B 167, B 168	SB-163, SB-166, SB-167, SB-168	---
N06782	SAE J775 (VF 4)	---	---	---	---
N06804	---	---	---	---	---
N06920	---	---	B 814	---	---
N06975	---	---	B 581, B 582, B 619, B 622, B 626	SB-366, SB-581, SB-582,SB-619, SFA-5.14 (ERNiCrMo-8)	A5.14 (ERNiCrMo-8)
N06985	---	---	B 581, B 582, B 619, B 622, B 626	SB-581, SB-582, SB-619,SB-622, SB-626, SFA-5.14 (ERNiCrMo-9)	A5.14 (ERNiCrMo-9)
N07001	5544, 5586, 5704, 5706, 5707, 5708, 5709, 5828, SAE J775 (XEV 4)	---	B 637	---	---
N07002	SAE J775 (HEV 2)	---	---	---	---
N07012	5855, 5856, 5881	---	---	---	---
N07013	5404	---	---	---	---
N07031	SAE J775 (HEV 3)	---	---	---	---
N07032	SAE J775 (HEV 8)	---	---	---	---
N07041	5399, 5545, 5712, 5713, 5800, 7469	---	---	---	---
N07048[a]	---	---	---	---	---
N07069	---	E-21562 (RN69)	---	---	---
N07080	SAE J775 (HEV 5)	---	B 637	---	---
N07090	5829, SAE J775 (HEV 6)	---	---	---	---
N07092	5675	E-21562 (EN6A, RN6A), I-23413 (MIL-6A)	---	SFA-5.14 (ERNiCrFe-6), SFA-5.30 (IN6A)	A5.14 (ERNiCrFe-6), A5.30 (IN6A)
N07214	---	---	---	---	---

AMERICAN CROSS REFERENCED SPECIFICATIONS - NICKEL & NICKEL ALLOYS (Continued)

UNS	SAE/AMS	MILITARY	ASTM	ASME	AWS
N07252	---	---	B 637	---	---
N07263	5872, 5886	---	---	---	---
N07500	5384, 5751, 5753	---	B 637	---	---
N07626[a]	---	---	---	---	---
N07702	5550	---	---	---	---
N07716[a]	---	---	---	---	---
N07718	5383, 5589, 5590, 5596, 5597, 5662, 5663, 5664, 5832, SAE J775 (XEV 1)	---	B 637, B 670,	SFA-5.14 (ERNiFeCr-2)	A5.14 (ERNiFeCr-2)
N07721	---	---	---	---	---
N07722	5541, 5714	---	---	---	---
N07725	---	---	---	---	---
N07750	5542, 5582, 5583, 5598, 5667, 5668, 5669, 5670, 5671, 5698, 5699, 5747, 5779, 7246	N-7786, S23192, N-24114	B 637	SA-637	---
N08031	---	---	B 564, B 619, B 626	---	---
N08032[a]	---	---	---	---	---
N08042[a]	---	---	---	---	---
N08050	---	---	A 608 (HT-50)	---	---
N08065	---	E-21562,E-21562 (RN65)	---	SFA-5.14 (ERNiFeCr-1)	A5.14 (ERNiFeCr-1)
N08135[a]	---	---	B 622	---	---
N08151	---	---	A 351	SA-351	---
N08221	---	---	B 423, B 424, B 425	---	---
N08310	---	---	B 720	---	---

AMERICAN CROSS REFERENCED SPECIFICATIONS - NICKEL & NICKEL ALLOYS (Continued)

UNS	SAE/AMS	MILITARY	ASTM	ASME	AWS
N08320	---	---	B 619, B 620, B 621, B 622, B 626	SB-366, SB-620, SB-621, SB-622	---
N08321	---	---	---	---	---
N08330	5592, 5716, SAE J705 (30330), SAE J412 (30330)	---	B 366, B 511, B 512, B 535, B 536, B 546, B 710, B 739	SB-366, SB-511, SB-710	---
N08331	---	---	---	SFA-5.9 (ER330)	A5.9 (ER330)
N08332	---	---	B 511, B 535, B 536, B 546, B 710, B 739	SB-366, SB-511, SB-535, SB-536	---
N08334	---	---	---	---	---
N08366	---	---	B 675, B 676, B 688, B 690, B 691	SB-675	---
N08367	---	---	B 462, B 472, B 564, B 675, B 676, B 688, B 690, B 691, B 804	---	---
N08421	---	---	---	---	---
N08535[c]	---	---	B 622	---	---
N08603	---	---	A 351 (HT-30)	---	---
N08604	---	---	A 297 (HL)	---	---
N08605	---	---	A 297 (HT)	---	---
N08613	---	---	A 608 (HL-30)	---	---
N08614	---	---	A 608 (HL-30)	---	---
N08700	---	---	B 599, B 672	SB-599, SB-672	---
N08705	---	---	A 297 (HP)	---	---
N08800	5766, 5871	---	B 163, B 366, B 407, B 408, B 409, B 514, B 515, B 564	SB-163, SB-407, SB-408, SB-409, SB564	---
N08801	5552	---	---	---	---

AMERICAN CROSS REFERENCED SPECIFICATIONS - NICKEL & NICKEL ALLOYS (Continued)

UNS	SAE/AMS	MILITARY	ASTM	ASME	AWS
N08802	---	---	---	---	---
N08810	---	---	B 163, B 407, B 408, B 409, B 514, B 564	SB-163, SB-407, SB-408, SB-409, SB-514, SB-564	---
N08811	---	---	B 407, B 408, B 409	---	---
N08825	---	---	B 163, B 423, B 424, B 425, B 564, B 704, B 705	SB-163, SB-423, SB-424, SB-425	---
N08826	---	---	B 163, B 423, B 425	---	---
N08904	---	---	B 625, B 649, B 673, B 674, B 677	SB-625, SB-649, SB-673, SB-674, SB-677	---
N08925	---	---	B 625, B 649, B 673, B 674, B 677	SB-625, SB-649, SB-673, SB-674, SB-677	---
N08926	---	---	B 366, B 625, B 649, B 673, B 674, B 677	SB-366, SB-625, SB-649, SB-673, SB-674, SB-677	---
N08932	---	---	B 625	---	---
N09027	5633, 5634	---	---	---	---
N09706	5605, 5606, 5701, 5702, 5703	---	---	---	---
N09901	5660, 5661	---	---	---	---
N09902	5221, 5223, 5225	---	---	---	---
N09911	---	---	---	---	---
N09925	---	---	---	---	---
N09926	---	---	---	---	---
N09979	5746, SAE J467 (D-979)	---	---	---	---
N10001	5396	---	B 333, B 335, B 366, B 619, B 622, B 626, F 467 (335), F468 (335)	SB-333, SB-335, SB-366, SB-619, SFA-5.14 (ERNiMo-1)	A5.14 (ERNiMo-1)

AMERICAN CROSS REFERENCED SPECIFICATIONS - NICKEL & NICKEL ALLOYS (Continued)

UNS	SAE/AMS	MILITARY	ASTM	ASME	AWS
N10002	5388, 5389, 5530, 5750	---	---	---	---
N10003	5607, 5771	---	B 366, 434, B 573	SB-366, SB-434, SB-573, SFA-5.14 (ERNiMo-2)	A5.14 (ERNiMo-2)
N10004	5755, 5786	---	---	SFA-5.14 (ERNiMo-3)	A5.14 (ERNiMo-3)
N10276	---	---	B 366, B 564, B 574, B 575, B 619, B 622, B 626, F 467 (276), F 468 (276)	SB-366, SB-564, SB-574, SB-575, SB-619, SB-622, SB-626, SFA-5.14 (ERNiCrMo-4)	A5.14 (ERNiCrMo-4)
N10665	---	---	B 333, B 335, B 366, B 619, B 622, B 626	SB-333, SB-335, SB-366, SB-619, SB-622, SB-626, SFA-5.14 (ERNiMo-7)	A5.14 (ERNiMo-7)
N10675	---	---	B 333, B 335, B 366, B 564, B 619, B 622, B 626	---	---
N13009	5407	---	---	---	---
N13010	5405, 5406	---	---	---	---
N13017	5851, 5852, 5882	---	---	---	---
N13020	5846	---	---	---	---
N13021	---	---	---	---	---
N13100	5397	---	---	---	---
N13246	---	---	---	---	---
N14052	---	---	F 30 (52)	---	---
N19903	5806	---	---	---	---
N19907	---	---	---	---	---
N19909	---	---	---	---	---
N24025	---	---	A 494	---	---
N24030	---	---	A 494	---	---
N24130	---	---	A 494	---	---

AMERICAN CROSS REFERENCED SPECIFICATIONS - NICKEL & NICKEL ALLOYS (Continued)					
UNS	SAE/AMS	MILITARY	ASTM	ASME	AWS
N24135	---	---	A 494	---	---
N26022	---	---	A 494	---	---
N26055	---	---	A 494	---	---
N26455	---	---	A 494	---	---
N26625	---	---	A 494	---	---
N30002	---	---	A 494	---	---
N30007	---	---	A 494	---	---
N30012	---	---	A 494	---	---
N30107	---	---	A 494	---	---
N99600	4775	B-7883 (BNi-1)	---	SFA-5.8 (BNi-1)	A5.8 (BNi-1)
N99610	4776	B-7883 (BNi-1a)	---	SFA-5.8 (BNi-1a)	A5.8 (BNi-1a)
N99612	---	---	---	SFA-5.8 (BNi-9)	A5.8 (BNi-9)
N99620	4777	B-7883 (BNi-2)	---	SFA-5.8 (BNi-2)	A5.8 (BNi-2)
N99622	---	---	---	SFA-5.8 (BNi-10)	A5.8 (BNi-10)
N99624	---	---	---	SFA-5.8 (BNi-11)	A5.8 (BNi-11)
N99630	4778	B-7883 (BNi-3)	---	SFA-5.8 (BNi-3)	A5.8 (BNi-3)
N99640	4779	B-7883 (BNi-4)	---	SFA-5.8 (BNi-4)	A5.8 (BNi-4)
N99644	---	---	---	SFA-5.21 (ERNiCr-A)	A5.21 (ERNiCr-A)
N99645	---	R-17131 (MIL-RNiCr-B)	---	SFA-5.21 (ERNiCr-B)	A5.21 (ERNiCr-B)
N99646	---	R-17131 (MIL-RNiCr-C)	---	SFA-5.21 (ERNiCr-C)	A5.21 (ERNiCr-C)
N99650	4782	B-7883 (BNi-5)	---	SFA-5.8 (BNi-5)	A5.8 (BNi-5)
N99651	---	---	---	SFA-5.8 (BNi-5a)	A5.8 (BNi-5a)
N99700	---	---	---	SFA-5.8 (BNi-6)	A5.8 (BNi-6)
N99710	---	---	---	SFA-5.8 (BNi-7)	A5.8 (BNi-7)
N99800	---	---	---	SFA-5.8 (BNi-8)	A5.8 (BNi-8)

AMERICAN CROSS REFERENCED SPECIFICATIONS - NICKEL & NICKEL ALLOYS (Continued)

a. Other cross reference specification: NACE Std. MR0175.
b. N06621 combined with N06075.
c. N09911 combined with N09901.

This cross-reference table lists the basic specification or standard number, and since these standards are constantly being revised, it should be kept in mind that they are presented herein as a guide and may not reflect the latest revision.

COMMON NAMES OF NICKEL & NICKEL ALLOYS WITH UNS No.
20Cb3 Modified; Ni-Fe-Cr Alloy Solution Strengthened; N08321
20Mo-6HS; Ni-Fe-Cr-Mo-Cu Alloy; N08036
214; Ni-Cr Precipitation Hardenable Alloy; N07214
46Fe-32Ni; Alloy Casting; N08151
6Ni-Cr-Fe Alloy, Solid Solution Strengthened (Low Carbon); N06602
ACI CN-7M; Ni-Fe-Cr Alloy Solid Solution Strengthened; N08007
ACI CY-40; Ni-Cr Alloy Solid Solution Strengthened; N06040
ACI CZ-100; Nickel Alloy; N02100
ACI HL-30; Alloy Casting; N08613
ACI HL-40; Alloy Casting; N08614
ACI HL; Alloy Casting; N08604
ACI HP; Alloy Casting; N08705
ACI HT-30; Alloy Casting; N08603
ACI HT-30; Ni-Fe-Cr Alloy Solid Solution Strengthened; N08030
ACI HT-50; Ni-Fe-Cr Alloy Solid Solution Strengthened; N08050
ACI HT-50C; Ni-Fe-Cr Alloy Solid Solution Strengthened; N08008
ACI HT; Alloy Casting; N08605
ACI HT; Ni-Fe-Cr Alloy Solid Solution Strengthened; N08002
ACI HU-50; Ni-Fe-Cr Alloy Solid Solution Strengthened; N08005

COMMON NAMES OF NICKEL & NICKEL ALLOYS WITH UNS No. (Continued)
ACI HU; Ni-Fe-Cr Alloy Solid Solution Strengthened; N08004
ACI HW-50; Ni-Fe-Cr Alloy Solid Solution Strengthened; N08006
ACI HW; Ni-Fe-Cr Alloy Solid Solution Strengthened; N08001
ACI HX-50; Ni-Cr Alloy Solid Solution Strengthened; N06050
ACI HX; Ni-Cr Alloy Solid Solution Strengthened; N06006
AF2-1DA; Cr-Co-Mo-W Nickel Alloy; N07012
AL 6XN; Ni-Cr-Mo Wrought or Cast Alloy; N08367
AL-6X; Ni-Fe-Cr Alloy Solid Solution Strengthened; N08366
Allcorr; Austenitic Corrosion Resistant Nickel Alloy; N06110
Alloy 31; Ni-Fe-Cr-Mo-Cu Alloy; N08031
Alloy 59; Low Carbon Ni-Cr-Mo Alloy; N06059
Alloy No. 230; Ni-Cr-W-Mo Alloy, Solid Solution Strengthened; N06230
Alloy No. 52; Ni-Fe Alloy; N14052
Alumel; Ni-Mn-Al Thermocouple Alloy; N02016
ARMCO 20-45-5; Ni-Fe-Cr Alloy; N08245
Astroloy M; Co-Cr-Mo-Ni Alloy; N13017
Austenitic Cr-Ni-Cu-Mo Alloy Welding Filler Metal; N08021
Austenitic Cr-Ni-Cu-Mo Alloy Welding Filler Metal; N08022
BNi-1; Ni-Cr-B Brazing Filler Metal; N99600
BNi-10; Ni-W-Cr-B Brazing Filler Metal; N99622
BNi-11; Ni-W-Cr-B Brazing Filler Metal; N99624
BNi-1a; Ni-Cr-B Brazing Filler Metal; N99610
BNi-2; Ni-Cr-B Brazing Filler Metal; N99620
BNi-3; Ni-Si-B Brazing Filler Metal; N99630
BNi-4; Ni-Si-B Brazing Filler Metal; N99640
BNi-5; Ni-Cr-Si Brazing Filler Metal; N99650
BNi-6; Ni-P Brazing Filler Metal; N99700
BNi-7; Ni-Cr-P Brazing Filler Metal; N99710

COMMON NAMES OF NICKEL & NICKEL ALLOYS WITH UNS No. (Continued)
BNi-8; Ni-Mn-Si-Cu Brazing Filler Metal; N99800
BNi-9; Ni-Cr-B Brazing Filler Metal; N99612
BNi5a; Ni-Cr-Si-B Brazing Filler Metal; N99651
Carpenter 20Cb3; Ni-Fe-Cr Alloy Solid Solution Strengthened; N08020
Carpenter 20Mo6; Ni-Cr Wrought Alloy; N08026
Cast Ni-Fe-Cr-Mo-Cu Alloy; N08826
CG27; Ni-Fe-Cr Alloy Precipitation Hardenable; N09027
Chromel; Ni-Cr 10 Thermocouple Alloy; N06010
Co-Cr-Mo-Ni Alloy; N13020
Colmonoy 4, RNiCr-A; Ni-Cr-B Weld Filler Metal for Hard Surfacing; N99644
Colmonoy 5, RNiCr-B; Ni-Cr-B Weld Filler Metal for Hard Surfacing; N99645
Colmonoy 6, RNiCr-C; Ni-Cr-B Weld Filler Metal for Hard Surfacing; N99646
Combined with N06075; N06621
Combined with N09901; N09911
Commercially Pure Be-Ni Alloy Precipitation Hardenable; N03360
Commercially Pure Nickel; N02290
Cr-Co-Mo Nickel Alloy Castings; N07013
Cr-Ni-Fe-Mo-Cu Alloy Columbium Stabilized; N08024
Creusot UR SB 8; Ni-Fe-Cr-Mo-Cu, Low Carbon Alloy; N08932
D979; Ni-Fe-Cr Alloy Precipitation Hardenable; N09979
Duranickel 301; Nickel Alloy, Precipitation Hardenable; N03301
Eatonite 3; Ni-Cr-Mo-Fe Hard Facing Alloy; N06013
Eatonite 5; Ni-Cr-Mo-Fe Hard Facing Alloy; N06015
Eatonite; Ni-Cr Alloy Solid Solution Strengthened; N06005
ER330; Ni-Fe-Cr Welding Filler Metal; N08331
ERNi-Cl; Ni-Cr-Fe Weld Filler Metal for Cast Iron; N02215
ERNiCrMo-9; Ni-Cr Alloy replaced by N06985; N06017
ERNiFeMn-Cl; Ni-Fe-Mn Weld Filler Metal for Cast Iron; N02216

COMMON NAMES OF NICKEL & NICKEL ALLOYS WITH UNS No. (Continued)
Fe-Ni-Co Metal-to-Ceramic Sealing Alloy; N94620
Fe-Ni-Co Metal-to-Ceramic Sealing Alloy; N94630
Hastelloy B-2; Ni-Mo Alloy; N10665
Hastelloy B-3; Ni-Mo-Cr-Fe Alloy; N10675
Hastelloy B; Ni-Mo Alloy Solid Solution Strengthened; N10001
Hastelloy C; Ni-Mo Alloy Solid Solution Strengthened; N10002
Hastelloy C-4; Ni-Cr Alloy; N06455
Hastelloy C-22;Ni-Cr-Mo Alloy; N06022
Hastelloy C-276; Ni-Mo Alloy Solid Solution Strengthened; N10276
Hastelloy F; Ni-Cr Alloy Solid Solution Strengthened; N06001
Hastelloy G-2; Ni-Cr Alloy; N06975
Hastelloy G-3; Ni-Cr Alloy ERNiCrMo-9; N06985
Hastelloy G-30- Wrought Ni-Cr-Fe-Mo-Cu Alloy Solid Solution Strengthened; N06030
Hastelloy G; Ni-Cr Alloy; N06007
Hastelloy G-50; Ni-Cr-Fe-Mo Alloy, Solid Solution Strengthened; N06950
Hastelloy H-9M; Ni-Cr-Fe-Mo-W Alloy Solid Solution Strengthened; N06920
Hastelloy N; Ni-Mo Alloy Solid Solution Strengthened; N10003
Hastelloy S; Ni-Cr-Mo Alloy; N06635
Hastelloy W; Ni-Mo Alloy Solid Solution Strengthened; N10004
Hastelloy X; Ni-Cr Alloy Solid Solution Strengthened; N06002
Haynes No. 20 Mod; Ni-Fe-Cr Alloy; N08320
Haynes R-41; Ni-Cr Alloy Precipitation Hardenable: N07041
Haynes 230; Ni-Cr-W-Mo Alloy, Solid Solution Strengthened; N06230
Haynes 230W; Ni-Cr-W-Mo Welding Filler Metal; N06231
Haynes 263; Ni-Cr Alloy Precipitation Hardenable; N07263
IN-100; Ni-Co Alloy Precipitation Hardenable; N13100
IN-102; Ni-Cr Alloy Solid Solution Strengthened; N06102
IN-713; Ni-Cr Alloy Precipitation Hardenable; N07713

COMMON NAMES OF NICKEL & NICKEL ALLOYS WITH UNS No. (Continued)
INCO Alloy 032; Fe-Ni-Cr-Mo Alloy; N08032
Incoloy 800; Fe-Ni-Cr Alloy Solid Solution Strengthened; N08800
Incoloy 800H; Fe-Ni-Cr Alloy; N08810
Incoloy 800HT; Wrought Ni-Fe-Cr Alloy; N08811
Incoloy 801; Fe-Ni-Cr Alloy Solid Solution Strengthened; N08801
Incoloy 802; Ni-Fe-Cr Alloy Solid Solution Strengthened; N08802
Incoloy 804; Ni-Cr Alloy Solid Solution Strengthened; N06804
Incoloy 825; Ni-Fe-Cr Alloy Solid Solution Strengthened; N08825
Incoloy 901; Ni-Fe-Cr Alloy Precipitation Hardenable; N09901
Incoloy 903; Ni-Fe-Co Alloy Precipitation Hardenable; N19903
Incoloy 925; Ni-Fe-Cr Alloy Precipitation Hardenable; N09925
Incoloy 926; Ni-Fe-Cr-Cu Alloy Precipitation Hardenable; N09926
Incoloy FM65; Ni-Fe-Cr Welding Filler Metal; N08065
Inconel 600; Ni-Cr Alloy Solid Solution Strengthened; N06600
Inconel 601; Ni-Cr Alloy Solid Solution Strengthened; N06601
Inconel 617; Ni-Cr Alloy Solid Solution Strengthened; N06617
Inconel 625; Ni-Cr Alloy Solid Solution Strengthened; N06625
Inconel 690; Ni-Cr Alloy Solid Solution Strengthened; N06690
Inconel 702; Ni-Cr Alloy Precipitation Hardenable; N07702
Inconel 706; Ni-Fe-Cr Alloy Precipitation Hardenable; N09706
Inconel 718; Ni-Cr Alloy Precipitation Hardenable; N07718
Inconel 721; Ni-Cr Alloy Precipitation Hardenable; N07721
Inconel 722; Ni-Cr Alloy Precipitation Hardenable; N07722
Inconel 751; Ni-Cr Alloy Precipitation Hardenable; N07751
Inconel Alloy 725; Ni-Cr-Mo Precipitation Hardenable Alloy; N07725
Inconel FM52; Ni-Cr Welding Filler Metal; N06052
Inconel FM62; Ni-Cr Alloy Solid Solution Strengthened; N06062
Inconel FM69; Ni-Cr Welding Filler Metal; N07069

COMMON NAMES OF NICKEL & NICKEL ALLOYS WITH UNS No. (Continued)
Inconel FM82; Ni-Cr Welding Filler Metal; N06082
Inconel FM92; Ni-Cr Alloy Precipitation Hardenable; N07092
Inconel Filler Metal 72; Ni-Cr Alloy Filler Metal; N06072
Inconel MA754; Ni-Cr Alloy Y Dispersion Strengthened; N07754
Inconel WE 132; Ni-Cr Alloy Solid Solution Strengthened replaced by W86132; N06132
Inconel X750; Ni-Cr Alloy Precipitation Hardenable; N07750
JS 700; Fe-Ni-Cr Alloy; N08700
M220C; Beryllium Nickel, Precipitation Hardenable Castings; N03220
M252; Ni-Cr Alloy Precipitation Hardenable; N07252
MAR-M-Alloy, Hf modified; Ni-W-Co-Cr Precipitation Hardenable; N13246
Monel 400; Ni-Cu Alloy Solid Solution Strengthened; N04400
Monel 401; Ni-Cu Alloy, Solid Solution Strengthened; N04401
Monel 404; Ni-Cu Alloy Solid Solution Strengthened; N04404
Monel 502; Ni-Cu Alloy Precipitation Hardenable; N05502
Monel FM60; Ni-Cu Welding Filler Metal; N04060
Monel K500; Ni-Cu Alloy Precipitation Hardenable; N05500
Monel R405; Ni-Cu Alloy Solid Solution Strengthened; N04405
Ni-20 Cr Columbium Stabilized; N06009
Ni-Co-Cr Alloy, Precipitation Hardenable; N13021
Ni-Co-Cr-Mo Alloy Castings; N13010
Ni-Cr 30 Alloy Solid Solution Strengthened; N06008
Ni-Cr Precipitation Hardenable Alloy; N07626
Ni-Cr Precipitation Hardenable Alloy; N07716
Ni-Cr Welding Filler Metal; N06076
Ni-Cr-Bi-Sn Casting Alloy; N26055
Ni-Cr-Fe-Mo-Cu Alloy, Age Hardenable; N07048
Ni-Cr-Mo Alloy; N08904
Ni-Cr-Mo Casting Alloy; N26022

COMMON NAMES OF NICKEL & NICKEL ALLOYS WITH UNS No. (Continued)
Ni-Cr-Mo Casting Alloy; N26455
Ni-Cr-Mo Casting Alloy; N26625
Ni-Cr-Mo-Fe-Cu Alloy; N08535
Ni-Cu Alloy Solid Solution Strengthened (Low Carbon); N04402
Ni-Cu Casting Alloy; N24025
Ni-Cu Casting Alloy; N24030
Ni-Cu Casting Alloy; N24130
Ni-Cu Casting Alloy; N24135
Ni-Cu Casting Alloy (Weldable Grade); N04020
Ni-Cu Welding Filler Metal; N05504
Ni-Fe-Co Alloy Precipitation Hardenable; N19907
Ni-Fe-Co Alloy Precipitation Hardenable; N19909
Ni-Fe-Cr-Cu Alloy; N08221
Ni-Fe-Cr-Mo Alloy; N08310
Ni-Fe-Cr-Mo-Cu Alloy, Low Carbon; N08925
Ni-Fe-Cr-Mo-Cu Alloy; N08421
Ni-Fe-Cr-Mo-Cu-N Alloy, Low Carbon, Nitrogen Modified; N08926
Ni-Mo Casting Alloy; N30007
Ni-Mo Casting Alloy; N30012
Ni-Mo-Cr Casting Alloy; N30002
Ni-Mo-Cr Casting Alloy; N30107
Ni-Span-C 902; Ni-Fe-Cr Alloy Precipitation Hardenable; N09902
Ni-W-Co-Cr Alloy Castings; N13009
NiC 42M; Ni-Fe-Cr-Mo-Cu Alloy; N08042
NiC 52; Ni-Cr Alloy; N06952
Nichrome V; Ni-Cr Alloy Solid Solution Strengthened; N06003
Nichrome; Ni-Cr Alloy Solid Solution Strengthened; N06004
Nickel 200; Commercially Pure Ni Alloy; N02200

COMMON NAMES OF NICKEL & NICKEL ALLOYS WITH UNS No. (Continued)
Nickel 201; Commercially Pure Ni Alloy; N02201
Nickel 205; Commercially Pure Ni Alloy; N02205
Nickel 211; Nickel Alloy, Solution Strengthened; N02211
Nickel 220; Commercially Pure Ni Alloy; N02220
Nickel 225; Commercially Pure Ni Alloy; N02225
Nickel 230; Commercially Pure Ni Alloy; N02230
Nickel 233; Commercially Pure Ni Alloy; N02233
Nickel 270; Commercially Pure Ni Alloy; N02270
Nickel Base Castings; N04019
Nickel FM61; Ni Welding Filler Metal; N02061
Nickel Thoria Dispersion Strengthened; N03260
Nimonic 75; Ni-Cr Alloy Solid Solution Strengthened; N06075
Nimonic 80A; Ni-Cr Alloy Precipitation Hardenable; N07080
Nimonic 90; Ni-Cr Alloy Precipitation Hardenable; N07090
Nimonic Alloy 263; Ni-Cr Alloy Precipitation Hardenable; N07263
Permanickel 300; Nickel Alloy, Precipitation Hardenable; N03300
Pyromet 31; Ni-Cr Alloy Precipitation Hardenable; N07031
Pyromet 31V; Ni-Cr Alloy, Precipitation Hardenable; N07032
RA 330-04; Ni-Fe-Cr-Mn Ahoy, Solid Solution Strengthened; N08334
RA 330-04; Ni-Fe-Cr-Si Alloy; N08332
RA-330; Ni-Fe-Cr Alloy Solid Solution Strengthened; N08330
RA333; Ni-Cr Alloy Solid Solution Strengthened; N06333
Rene 41; Ni-Cr Alloy Precipitation Hardenable; N07041
Sanicro 28; Ni-Fe-Cr Alloy Solid Solution Strengthened; N08028
SM 2035; Ni-Fe-Cr-Mo-W Alloy; N08135
SM2050; Ni-Cr-Fe-Mo Alloy; N06250
SM2060Mo; Ni-Cr-Mo-Fe Alloy; N06060
SM2550; Ni-Cr-Fe-Mo-W Alloy; N06255

COMMON NAMES OF NICKEL & NICKEL ALLOYS WITH UNS No. (Continued)
TPM; Ni-Cr Alloy Precipitation Hardenable; N07002
Udimet 500; Ni-Cr Alloy Precipitation Hardenable; N07500
Waspaloy; Ni-Cr Alloy Precipitation Hardenable; N07001
X-782; Ni-Cr Alloy Solid Solution Strengthened; N06782

BRITISH BSI GENERAL SERIES SPECIFICATIONS - NICKEL & NICKEL ALLOYS

BSI	Title
375	Refined nickel
558, 564	Nickel anodes, anode nickel and nickel salts for electroplating
1224	Electroplated coatings of nickel and chromium
2857	Nickel-iron transformer and choke laminations
2901: Part 5	Filler rods and wires for gas-shielded arc welding part 5: specification for nickel and nickel alloys
3071	Nickel-copper alloy castings
3072	Nickel and nickel alloys: sheet and plate
3073	Nickel and nickel alloys: strip
3074	Nickel and nickel alloys: seamless tube
3075	Nickel and nickel alloys: wire
3076	Nickel and nickel alloys: bar
3146: Part 2	Corrosion and heat resisting steels, nickel and cobalt base alloys
3382: Parts 3, 4	Nickel or nickel plus chromium on steel components. nickel or nickel plus chromium on copper and copper alloy (including brass) components.
3504	Magnesium-activated nickel for cathodes for electronic tubes and valves
3727	Methods for the analysis of nickel for use in electronic tubes and valves (Parts 1 through 22)
4572 Part 2	Compression joints in nickel iron and plated copper conductors
4601	Electroplated coatings of nickel plus chromium on plastics materials
4758	Method for specifying electroplated coatings of nickel for engineering purposes
4937: Part 4	Nickel-chromium/nickel-aluminum thermocouples. Type K

BRITISH BSI GENERAL SERIES SPECIFICATIONS - NICKEL & NICKEL ALLOYS (Continued)

BSI	Title
4937: Part 6	Nickel-chromium/copper-nickel thermocouples. Type E
4937: Part 8	Nickel-chromium-silicon/nickel-silicon thermocouples including composition Type N
5383	Material identification of steel, nickel alloy and titanium alloy tubes by continuous character marking and colour coding of steel tubes
5411: Part 9	Measurement of coating thickness of electrodeposited nickel coatings on magnetic and non-magnetic substrates: magnetic method
5932	Sealed nickel-cadmium cylindrical rechargeable single cells
6115	Sealed nickel-cadmium prismatic rechargeable single cells
6260	Open nickel-cadmium prismatic rechargeable single cells
6404: Part 6	Methods of measurement of the magnetic properties of isotropic nickel-iron soft magnetic alloys, types E1, E3 and E4
6783: Part 1	Method for determination of silver, bismuth, cadmium, cobalt, copper, iron, manganese, lead and zinc in nickel by flame atomic absorption spectrometry
6783: Part 2	Method for determination of nickel in ferronickel (dimethylglyoxime gravimetric method)
6783: Part 3	Method for determination of cobalt in ferronickel by flame atomic absorption spectrometry
6783: Part 4	Method for determination of silver, arsenic, bismuth, cadmium, lead, antimony, selenium, tin, tellurium and thallium in nickel by electrothermal atomic absorption spectrometry
6783: Part 5	Method for determining of carbon in nickel ferronickel and nickel alloys by infra-red absorption after induction furnace combustion
6783: Part 6	Method for determining of sulphur in nickel by methylene blue molecular absorption spectrometry after generation of hydrogen sulphide
6783: Part 7	Method for determining of sulphur in nickel, ferronickel and nickel alloy by infra-red absorption after induction furnace combustion
6783: Part 8	Method for determination of sulphur in nickel, ferronickel and nickel alloys by iodimetric after induction furnace combustion
6783: Part 9	Method for determination of silicon in ferronickel (gravimetric method)
6783: Part 10	Method for determination of iron in nickel alloys (titrimetric method using potassium dichromate)
6783: Part 11	Method for the determination of cobalt in nickel alloys (potentiometric titration method using potassium hexacyanoferrate (III))
6783: Part 12	Method for the determination of phosphorus in nickel alloys by molybdenum blue molecular absorption spectrometry
6783: Part 13	Method for the determination of phosphorus in nickel, ferronickel and nickel alloys by phosphovanadomolybdate molecular absorption spectrometry

BRITISH BSI AEROSPACE SERIES SPECIFICATIONS - HEAT RESISTING ALLOY CASTINGS	
BSI	**Title**
HC 202	Precipitation hardening nickel base chromium-niobium-molybdenum-iron-tungsten alloy castings (Cr 20.0, Nb 6.6, Mo 6.0, Fe 3.0, W 2.5)
2HC 203	Nickel-chromium-aluminum-molybdenum-niobium alloy castings (Nickel base Cr 13.0, Al 6.0, Mo 4.5, Nb 2.3)
2HC 204	Nickel-cobalt-chromium-aluminum-titanium-molybdenum alloy castings (Nickel base Co 15.0, Cr 10.0, Al 5.5, Ti 4.8, Mo 3.0)
HC 205	Precipitation hardening nickel base chromium-cobalt-molybdenum-titanium alloy castings (Cr 20, Co 20, Mo 6, Ti 2)
HC 206	Precipitation hardening nickel base chromium-cobalt-molybdenum-titanium alloy precision castings (Cr 20, Co 20, Mo 6, Ti 2)
2HC 207	Nickel-cobalt-tungsten-chromium-aluminium-tantalum-hafnium-titanium alloy castings, (Nickel base Co 10.0, W 10.0, Cr 9.0, Al 5.5, Ta 2.5, Hf 1.5, Ti 1.5)
2HC 208	Nickel-cobalt-tungsten-chromium-aluminium-molybdenum-titanium-tantalum alloy castings (Nickel base Co 10.0, W 10.0, Cr 9.0, Al 5.5, Mo 2.5, Ti 1.5, Ta 1.5)
2HC 209	Low carbon, precipitation hardening nickel-chromium-aluminium-molybdenum-niobium alloy castings (Nickel base Cr 11.5, Al 6.0, Mo 4.0, Nb 2.0)
2HC 210	Nickel-tungsten-aluminum-chromium-molybdenum alloy castings
2HC 211	Nickel-chromium-cobalt-molybdenum-aluminum-titanium alloy castings (Nickel base Cr 15.5, Co 10.0, Mo 8.3, Al 4.2, Ti 3.6)
2HR 1	Nickel-chromium-titanium-aluminum heat-resisting alloy billets, bars, forgings and parts (Nickel base, Cr 19.5, Ti 2.2, Al 1.4)
2HR 2	Nickel-chromium-cobalt-titanium-aluminum heat-resisting alloy billets, bars, forgings and parts (Nickel base, Cr 19.5, Co 18.0, Ti 2.5, Al 1.5)
HR 3	Nickel-cobalt-chromium-molybdenum-aluminum-titanium heat-resisting alloy billets, bars, forgings and parts (Nickel base Co 20, Cr 14.8, Mo 5, Al 4.7, Ti 1.2)
HR 4	Nickel-chromium-cobalt-aluminum-molybdenum-titanium heat-resisting alloy billets, bars, forgings and parts (Nickel base, Cr 15, Co 14.2, Al 5, Mo 4, Ti 4)
HR 5	Nickel-chromium-titanium heat-resisting alloy billets, bars forgings and parts (Nickel base Cr 19.5, Ti 0.4)
HR 6	Nickel-chromium-iron-molybdenum-cobalt-tungsten heat-resisting alloy billets, bars, forgings and parts (Nickel base, Cr 21.7, Fe 18.5, Mo 9, Co 1.5, W 0.6)
HR 10	Nickel-cobalt-chromium-molybdenum-titanium-aluminum heat- resisting alloy billets, bars, forgings and parts (Nickel base, Co 20, Cr 20, Mo 5.9, Ti 2.1, Al 0.5)
2HR 201	Nickel-chromium-titanium-aluminum heat-resisting alloy plate, sheet, and strip (Nickel base, Cr 19.5, Ti 2.2, Al 1.4)
2HR 202	Nickel-chromium-cobalt-titanium-aluminum heat-resisting alloy sheet and strip (Nickel base, Cr 19.5, Co 18.0, Ti 2.5, Al 1.5)
HR 203	Nickel-chromium-titanium heat-resisting alloy plate, sheet and strip (Nickel base, Cr 19.5, Ti 0.4)
HR 204	Nickel-chromium-iron-molybdenum-cobalt-tungsten heat-resisting alloy plate, sheet and strip (Nickel base, Cr 22, Fe 18.5, Mo 9, Co 1.5, W 0.6)

BRITISH BSI AEROSPACE SERIES SPECIFICATIONS - HEAT RESISTING ALLOY CASTINGS (Continued)

BSI	Title
HR 205	Nickel-copper alloy sheet and strip: annealed (Nickel base, Cu 31)
HR 206	Nickel-cobalt-chromium- molybdenum-titanium-aluminum heat-resisting alloy plate, sheet and strip (Nickel base, Co 20, Cr 20, Mo 5.9, Ti 2.1, Al 0.5)
HR 207	Nickel-iron-chromium-molybdenum-aluminum-titanium heat-resisting alloy plate, sheet, strip (Ni/Co 43.5, Cr16.5, Mo 3.3, Al 1.2, Ti 1.2, Fe remainder)
HR 208	Nickel-chromium-iron alloy sheet, strip and plate, annealed (Nickel base, Cr 15.5, Fe 8)
HR 209	Nickel-chromium-cobalt-molybdenum-titanium-aluminum heat-resisting alloy sheet, strip and plate: softened (Nickel base, Cr 18, Co 14, Mo 7, Ti 2.2, Al 2.1)
2HR 401	Nickel-chromium-titanium-aluminum heat-resisting alloy cold worked and softened seamless tubes (Nickel base Cr 19.5, Ti 2.2, Al 1.4)
HR 402	Nickel-chromium-cobalt-titanium-aluminum heat-resisting alloy cold worked and softened seamless tubes (Nickel base Cr 19.5, Co 18.0, Ti 2.5, Al 1.5)
HR 403	Nickel-chromium-titanium heat-resisting alloy cold worked and softened seamless tubes (Nickel base Cr 19.5, Ti 0.4)
HR 404	Nickel-cobalt-chromium-molybdenum-titanium-aluminum heat-resisting alloy cold worked solution treated seamless tubes (Nickel base, Co 20, Cr 20, Mo 5.9, Ti 2.1, Al 0.5)
2HR 501	Nickel-chromium-cobalt-titanium-aluminum heat-resisting alloy cold drawn wire for springs and springs (Nickel base Cr 19.5, Co 18.0, Ti 2.5, Al 1.5)
2HR 502	Nickel-chromium-cobalt-titanium-aluminum heat-resisting alloy cold drawn and solution heat treated wire for springs and springs (Nickel base Cr 19.5, Co 18.0, Ti 2.5, Al 1.5)
2HR 503	Nickel-chromium-cobalt-titanium-aluminum heat-resisting alloy wire for thread inserts (Nickel base, Cr 19.5, Co 18.0, Ti 2.5, Al 1.5)
2HR 504	Nickel-chromium-titanium heat-resisting alloy bar, and wire for rivets, and rivets (Nickel base, Cr 19.5, Ti 0.4)
HR 505	Nickel-chromium-iron-titanium-niobium-aluminum heat-resisting alloy round wire for thread inserts (max. dia. 2 mm) (Nickel base Cr 15.5, Fe 7.0, Ti 2.5, Nb/Ta 0.95, Al 0.70)
HR 506	Nickel-chromium-cobalt-titanium-aluminum heat-resisting alloy wire: softened (Nickel base, Cr 19.5, Co 1.8, Ti 2.5, Al 1.5)
3HR 601	Nickel -chromium-titanium-aluminum heat- resisting alloy bar wire manufacture of fasteners (max. dia. or minor sectional dimension 30 mm) (Nickel base Cr 19.5-Ti 2.2-Al 1.4)

GERMAN DIN SPECIFICATIONS - NICKEL & NICKEL ALLOYS	
DIN	**Title**
EN 2952	Aerospace series; heat resisting nickel base alloy (Ni-P100 HT); solution treated and cold worked; bar for hot upset forging for fasteners $3 \le D \le 30$ mm
EN 2959	Aerospace series; heat resisting nickel base alloy (Ni-P101 HT); solution treated and cold worked; bar for hot upset forging for fasteners $3 \le D \le 30$ mm
EN 2960	Aerospace series; heat resisting nickel base alloy (Ni-P101 HT); solution treated and cold worked; bar for machining for fasteners $3 \le D \le 50$ mm
EN 2961	Aerospace series; heat resisting nickel base alloy (Ni-P100 HT); solution treated and cold worked; bar for machining for fasteners $3 \le D \le 50$ mm
EN 3219	Aerospace series; heat resisting nickel base alloy (Ni-P100 HT); cold worked and softened; bar and wire for continuous forging or extrusion for fasteners $3 \le D \le 30$ mm
EN 3220	Aerospace series; heat resisting nickel base alloy (Ni-P101 HT); cold worked and softened; bar and wire for continuous forging or extrusion for fasteners $3 \le D \le 30$ mm
1702	Nickel anodes
17471	Resistance alloys
17730	Nickel and nickel-copper alloys; castings
17740	Wrought nickel; chemical composition
17741	Wrought nickel alloys; low alloyed; chemical composition
17742	Wrought nickel alloys with chromium; chemical composition
17743	Wrought nickel alloys with copper; chemical composition
17744	Wrought nickel alloys with molybdenum and chromium; chemical composition
17745	Wrought alloys of nickel and iron, composition
17750	Plates, strips and sheets of wrought nickel and nickel alloys; properties
17751	Tubes of wrought nickel and nickel alloys; properties
17752	Rods and bars of wrought nickel alloys; properties
17753	Wires of wrought nickel and nickel alloys; properties
17754	Forgings and drop forgings of wrought nickel and nickel alloys; properties
50970	Electroplated coatings; nickel salts for nickel baths; requirements and testing

GERMAN DIN SPECIFICATIONS - NICKEL & NICKEL ALLOYS (Continued)

DIN	Title
59740	Sheets of wrought nickel and nickel alloys; hot rolled; dimensions; tolerances
59741	Strips of wrought nickel and nickel alloys; hot rolled; dimensions and tolerances
59742	Circular blanks and rings of wrought nickel and nickel alloys; hot rolled; dimensions; tolerances
59745	Sheet of wrought nickel and nickel alloys; cold rolled; dimensions; tolerances
59746	Strips of wrought nickel and nickel alloys; cold rolled; dimensions; tolerances
59755	Tubes of wrought nickel and nickel alloys; cold worked; dimensions; tolerances
59760	Round rods of wrought nickel and nickel alloys; hot rolled; dimensions; tolerances
59761	Square rods of wrought nickel and nickel alloys; hot rolled; dimensions; tolerances
59762	Hexagonal bars of wrought nickel and nickel alloys; hot rolled; dimensions; tolerances
59763	Rectangular bars of wrought nickel and nickel alloys; hot rolled; dimensions; tolerances; static values
59765	Round rods of wrought nickel and nickel alloys; drawn; dimensions; tolerances
59766	Square rods of wrought nickel and nickel alloys; drawn; dimensions; tolerances
59767	Hexagonal rods of wrought nickel and nickel alloys; drawn; dimensions; tolerances
59768	Rectangular bars of wrought nickel and nickel alloys; drawn with sharp edges; dimensions; tolerances; static values
59780	Round wire of wrought nickel and nickel alloys; hot rolled; dimensions; tolerances
59781	Round wire of wrought nickel and nickel alloys; drawn; dimensions; tolerances
59782	Square wire of wrought nickel and nickel alloys; drawn; dimensions; tolerances
59783	Hexagonal wire of wrought nickel and nickel alloys; drawn; dimensions; tolerances
65596	Aerospace; chemical check analysis limits of titanium and titanium alloys
LN 9286	Aerospace; sheet from nickel and cobalt alloys; cold rolled; dimensions; masses
LN 29516	Hot rolled square bars from nickel and cobalt alloys; dimensions; weights
WL 2.1504	Wrought alloy - nickel bronze; NiAlBz
WL 2.4360 Part 1	Nickel-copper wrought alloy; NiCu30Fe; creep-resistant (Monel)
WL 2.4360 Part 2	Nickel-copper wrought alloy; NiCu30Fe; creep-resistant (Monel)
WL 2.4360 Part 2	Aerospace; wrought nickel-copper alloy with approximately 67Ni-30Cu-1Fe; lockwire; rivet wire; rivets
WL 2.4360 Part 100	Aerospace; wrought nickel-copper alloy with approximately 67Ni-30Cu-1Fe; internal pressure tubes, lockwire, rivet wire, rivets; design and production data, conversion

GERMAN DIN SPECIFICATIONS - NICKEL & NICKEL ALLOYS (Continued)

DIN	Title
WL 2.4370	Nickel-copper wrought alloy; NiCu30Ti; creep-resistant; filler metal (Monel 60)
WL 2.4374 Part 1	Nickel-copper wrought alloy; NiCu30Al; creep-resistant (K-Monel)
WL 2.4374 Part 2	Nickel-copper wrought alloy; NiCu30Al; creep-resistant; filler metal (K-Monel)
WL 2.4630 Supp. 1	High temperature nickel alloy with about 0.1C-20Cr-0.4Ti; sheets; strips; rods; forgings; pipes; wire; rivet
WL 2.4630 Part 1	High temperature nickel alloy with about 0.1C-20Cr-0.4Ti; sheets and strips
WL 2.4630 Part 1	Aerospace; high-temperature nickel alloy with approximately 0.1C-20Cr-0.4Ti; sheet and strip
WL 2.4630 Part 2	High temperature nickel alloy with about 0.1C-20Cr-0.4Ti; rods and forgings
WL 2.4630 Part 2	Aerospace; high temperature nickel alloy with approximately 0.1C-20Cr-0.4Ti; bar and forging
WL 2.4630 Part 3	High temperature nickel alloy with about 0.1C-20Cr-0.4Ti; pipes
WL 2.4630 Part 3	Aerospace; high temperature nickel alloy with approximately 0.1C-20Cr-0.4Ti; tubes
WL 2.4630 Part 4	High temperature nickel alloy with about 0.1C-20Cr-0.4Ti; wire; rivet
WL 2.4630 Part 4	Aerospace; high temperature nickel alloy with approximately 0.1C-20Cr-0.4Ti; wire; rivets
WL 2.4630 Part 100	Aerospace; high temperature nickel alloy with approximately 0.1C-20Cr-0.4Ti; sheet; bar; forging; tubing; wire and rivets; notes for design and production; cross-reference
WL 2.4631 Supp. 1	High temperature nickel alloy with about 0.1C-20Cr-2.3Ti-1.4Al; sheets; strips; rods and forgings
WL 2.4631 Part 1	High temperature nickel alloy with about 0.1C-20Cr-2.3Ti-1.4Al; sheets and strips
WL 2.4631 Part 1	Aerospace; high temperature nickel alloy with approximately 0.1C-20Cr-2.3Ti-1.4Al; sheet and strip
WL 2.4631 Part 2	High temperature nickel alloy with about 0.1C-20Cr-2.3Ti-1.4Al; rods and forgings
WL 2.4631 Part 2	Aerospace; high temperature nickel alloy with approximately 0.1C-20Cr-2.3Ti-1.4Al; bar and forging
WL 2.4631 Part 100	Aerospace; high temperature nickel alloy with approximately 0.1C-20Cr-2.3Ti-1.4Al; sheet; strip; bar and forging; notes for design and production; cross-reference
WL 2.4632 Supp. 1	High temperature nickel alloy with about 0.1C-20Cr-18Co-2.5Ti-1.5Al; sheets; strips; rods; forgings; pipes and filler metal
WL 2.4632 Part 1	High temperature nickel alloy with about 0.1C-20Cr-18Co-2.5Ti-1.5Al; sheets and strips
WL 2.4632 Part 1	Aerospace; high temperature nickel alloy with approximately 0.1C-20Cr-18Co-2.5Ti-1.5Al; sheets and strips
WL 2.4632 Part 2	High temperature nickel alloy with about 0.1C-20Cr-18Co-2.5Ti-1.5Al; rods and forgings
WL 2.4632 Part 2	Aerospace; high temperature nickel alloy with approximately 0.1C-20Cr-18Co-2.5Ti-1.5Al; bar and forging
WL 2.4632 Part 3	High temperature nickel alloy with about 0.1C-20Cr-18Co-2.5Ti-1.5Al; pipes

GERMAN DIN SPECIFICATIONS - NICKEL & NICKEL ALLOYS (Continued)	
DIN	Title
WL 2.4632 Part 3	Aerospace; high temperature nickel alloy with approximately 0.1C-20Cr-18Co-2.5Ti-1.5Al; tubes
WL 2.4632 Part 4	High temperature nickel alloy with about 0.1C-20Cr-18Co-2.5Ti-1.5Al; filler metal
WL 2.4632 Part 4	Aerospace; high temperature nickel alloy with approximately 0.1C-20Cr-18Co-2.5Ti-1.5Al; welding filler metal
WL 2.4632 Part 100	Aerospace; high temperature nickel alloy with approximately 0.1C-20Cr-18Co-2.5Ti-1.5Al; sheet; strip; bar; forging; tubing and welding filler metal; notes for design and production; cross-reference
WL 2.4634	High temperature nickel alloy with about 0.15C-20Co-15Cr-5Mo-4.7Al-1.2Ti; rods and forgings
WL 2.4634 Supp. 1	High temperature nickel alloy with about 0.15C-20Co-15Cr-5Mo-4.7Al-1.2Ti; rods and forgings
WL 2.4634 Part 1	Aerospace; high temperature nickel alloy with approximately 0.15C-20Co-15Cr-5Mo-4.7Al-1.2Ti; bar and forging
WL 2.4634 Part 100	Aerospace; high temperature nickel alloy with approximately 0.15C-20Co-15Cr-5Mo-4.7Al-1.2Ti; bar and forging; notes for design and production; cross-reference
WL 2.4639	High temperature nickel alloy with about 0.2C-20Cr; filler metal
WL 2.4639 Supp. 1	High temperature nickel alloy with about 0.2C-20Cr; filler metal
WL 2.4639 Part 1	Aerospace; high temperature nickel alloy with approximately 0.2C-20Cr; welding filler metal
WL 2.4639 Part 100	Aerospace; high temperature nickel alloy with about 0.2C-20Cr; welding filler metal; notes for design and production, cross-reference
WL 2.4650 Supp. 1	High temperature nickel alloy with about 0.06C-20Co-20Cr-6Mo-2Ti-0.5Al; sheets; strips; rods and forgings
WL 2.4650 Part 1	High temperature nickel alloy with about 0.06C-20Co-20Cr-6Mo-2Ti-0.5Al; sheets and strips
WL 2.4650 Part 2	High temperature nickel alloy with about 0.06C-20Co-20Cr-6Mo-2Ti-0.5Al; rods and forgings
WL 2.4654	High temperature nickel alloy with about 0.06C-19Cr-14Co-4Mo-3Ti-1.4Al; rods and forgings
WL 2.4654 Supp. 1	High temperature nickel alloy with about 0.06C-19Cr-14Co-4Mo-3Ti-1.4Al; rods and forgings
WL 2.4662	High temperature nickel alloy with about 0.04C-35Fe-13Cr-6Mo-3Ti; rods and forgings
WL 2.4662 Supp. 1	High temperature nickel alloy with about 0.04C-35Fe-13Cr-6Mo-3Ti; rods and forgings
WL 2.4665 Supp. 1	High temperature nickel alloy with about 0.1C-22Cr-18Fe-9Mo; sheets; strips; forgings and filler wire
WL 2.4665 Part 1	High temperature nickel alloy with about 0.1C-22Cr-18Fe-9Mo; sheets and strips
WL 2.4665 Part 2	High temperature nickel alloy with about 0.1C-22Cr-18Fe-9Mo; rods and forgings
WL 2.4665 Part 3	High temperature nickel alloy with about 0.1C-22Cr-18Fe-9Mo; filler wire
WL 2.4668 Supp. 1	High temperature nickel alloy with about 0.05C-19Cr-18Fe-5Nb-3Mo; sheets; strips; rods; forgings and filler wire
WL 2.4668 Part 1	High temperature nickel alloy with about 0.05C-19Cr-18Fe-5Nb-3Mo; sheets and strips

GERMAN DIN SPECIFICATIONS - NICKEL & NICKEL ALLOYS (Continued)

DIN	Title
WL 2.4668 Part 2	High temperature nickel alloy with about 0.05C-19Cr-18Fe-5Nb-3Mo; rods and forgings
WL 2.4668 Part 3	High temperature nickel alloy with about 0.05C-19Cr-18Fe-5Nb-3Mo; filler wire
WL 2.4668 Part 4	Aerospace; heat-resisting nickel alloy with approximately 0.05C-19Cr-18Fe-5Nb-3Mo; consumable electrode remelted; bars and wire for screws and bolts
WL 2.4668 Part 100	Aerospace; heat-resisting nickel alloy with approximately 0.05C-19Cr-18Fe-5Nb-3Mo; consumable electrode remelted; sheets; strips; bars; forgings; filler wire for welding and bars and wire for screws and bolts; design and production date; conversion
WL 2.4670	High temperature nickel cast alloy with about 0.05C-12Cr-6Al-4.5Mo-2Nb; investment casting
WL 2.4670 Supp. 1	High temperature nickel cast alloy with about 0.05C-12Cr-6Al-4.5Mo-2Nb; investment casting
WL 2.4671	High temperature nickel cast alloy with about 0.15C-13Cr-6Al-4.5Mo-2.3Nb; investment casting
WL 2.4671 Supp. 1	High temperature nickel cast alloy with about 0.15C-13Cr-6Al-4.5Mo-2.3Nb; investment casting
WL 2.4672	High temperature nickel cast alloy with about 0.06C-20Co-20Cr-6Mo-2Ti-0.5Al; investment casting
WL 2.4672 Supp. 1	High temperature nickel cast alloy with about 0.06C-20Co-20Cr-6Mo-2Ti-0.5Al; investment casting
WL 2.4674	High temperature nickel cast alloy with about 0.2C-15Co-10Cr-6Al-5Ti-3Mo-1V; investment casting
WL 2.4674 Supp. 1	High temperature nickel cast alloy with about 0.2C-15Co-10Cr-6Al-5Ti-3Mo-1V; investment casting
WL 2.4676	High temperature nickel cast alloy with about 0.15C-10Co-10W-9Cr-5.5Al-2.5Mo; investment casting
WL 2.4676 Supp.	High temperature nickel cast alloy with about 0.15C-10Co-10W-9Cr-5.5Al-2.5Mo; investment casting
WL 2.4676 Part 1	Aerospace; high temperature nickel cast alloy with approximately 0.15C-10Co-10W-9Cr-5.5Al-2.5Mo; precision casting
WL 2.4676 Part 100	Aerospace; high temperature nickel cast alloy with approximately 0.15C-10Co-10W-9Cr-5.5Al-2.5Mo; precision casting; notes for design and production; cross-reference
WB 263	Nickel-copper alloy NiCu30Fe
WB 263 Supp.	Nickel-copper alloy NiCu30Fe
WB 432/1	Highly corrosion-proof nickel base alloy NiCr21Mo; material no. 2.4858
WB 432/1 Supp.	Highly corrosion-proof nickel base alloy NiCr21Mo; material no. 2.4858
WB 432/2	Highly corrosion-proof nickel base alloy NiCr21Mo; material no. 2.4858
WB 432/2 Supp.	Highly corrosion-proof nickel base alloy NiCr21Mo; material no. 2.4858
WB 432/3	Highly corrosion-proof nickel base alloy NiCr21Mo; material no. 2.4858
WB 432/3	Highly corrosion-proof nickel base alloy NiCr21Mo; material no. 2.4858

GERMAN DIN SPECIFICATIONS - NICKEL & NICKEL ALLOYS (Continued)

DIN	Title
WB 479	Highly corrosion-proof alloy NiCr21Mo14W; material no. 2.4602
WB 485	High temperature nickel base alloy NiCr23Co12Mo; material no 2.4663
WB 485 Supp.	High temperature nickel base alloy NiCr23Co12Mo; material no 2.4663

JAPANESE JIS SPECIFICATIONS - NICKEL & NICKEL ALLOYS

JIS	Title
C 8705	Sealed Nickel-Cadmium Cylindrical Rechargeable Single Cells
C 8706	Stationary Nickel Cadmium Alkaline Batteries
G 3602	Nickel and Nickel Alloy Clad Steels
H 3261	Nickel Copper Alloy Sheet and Plates
H 4501	Nickel Sheets and Strips for Electronic Tube
H 4502	Nickel Sheets and Strips for Cathode of Electronic Tube
H 4511	Nickel Bars and Wires for Electronic Tube
H 4522	Seamless Nickel Tube for Cathode of Vacuum Tube
H 4551	Nickel and Nickel Alloy Plate, Sheet and Strip
H 4552	Nickel and Nickel Alloy Seamless Pipes and Tubes
H 4553	Nickel and Nickel Alloy Rod and Bar
H 4554	Nickel Alloy Wire
H 5701	Nickel and Nickel Alloy Castings
H 8617	Electroplated Coatings of Nickel and Chromium
H 8645	Autocatalytic Nickel - Phosphorus Coatings on Metals
K 0013	Nickel Standard Solution
K 9062	Nickel
T 6101	Nickel-Chromium Alloy Wires for Dental Use
T 6102	Nickel-Chromium Alloy Plates for Dental Use
Z 3224	Nickel and Nickel-Alloy Covered Electrodes

JAPANESE JIS SPECIFICATIONS - NICKEL & NICKEL ALLOYS (Continued)	
JIS	**Title**
Z 3265	Nickel Brazing Filler Metals
Z 3334	Nickel and Nickel Alloy Filler Rods and Wires for Arc Welding

CHEMICAL COMPOSITION OF NICKEL & NICKEL ALLOYS	
UNS	**Chemical Composition**
N02016	Ni rem Al 1.75-2.25 C 0.15 max Fe 0.50 max Mn 2.0-3.0 Si 1.6 max
N02061	Ni 93.0 min Al 1.5 max C 0.15 max Cu 0.25 max Fe 1.0 max Mn 1.0 max P 0.03 max S 0.015 max Si 0.75 max Ti 2.0-3.5
N02100	Ni rem C 1.00 max Cu 1.25 max Fe 3.00 max Mn 1.50 max Si 2.00 max
N02200	Ni 99.0 min C 0.15 max Cu 0.25 max Fe 0.40 max Mn 0.35 max S 0.010 max Si 0.35 max
N02201	Ni 99.0 min C 0.02 max Cu 0.25 max Fe 0.40 max Mn 0.35 max S 0.010 max Si 0.35 max
N02205	Ni 99.0 min C 0.15 max Cu 0.15 max Fe 0.20 max Mg 0.01-0.08 Mn 0.35 max S 0.008 max Si 0.15 max Ti 0.01-0.05
N02211	Ni 93.7 min C 0.20 max Cu 0.25 max Fe 0.75 max Mn 4.25-5.25 S 0.015 max Si 0.15 max
N02215	Ni rem C 1.0 max Cu 4.0 max Fe 4.0 max Mn 2.5 max S 0.03 max Si 0.75 max Other 1.0 max total
N02216	Ni 35.0-45.0 Al 1.0 max C 0.50 max Cu 2.5 max Fe rem Mn 10.0-14.0 S 0.03 max Si 1.0 max Other 1.0 max total
N02220	Ni 99.00 min C 0.15 max Cu 0.10 max Fe 0.10 max Mg 0.01-0.08 Mn 0.20 max S 0.008 max Si 0.01-0.05 Ti 0.01-0-05
N02225	Ni 99.00 min C 0.15 max Cu 0.10 max Fe 0.10 max Mg 0.01-0.08 Mn 0.20 max S 0.008 max Si 0.15-0.25 Ti 0.01-0-05
N02230	Ni 99.00 min C 0.15 max Cu 0.10 max Fe 0.10 max Mg 0.04-0.08 Mn 0.15 max S 0.008 max Si 0.010-0.035 Ti 0.005 max
N02233	Ni 99.00 min C 0.15 max Cu 0.10 max Fe 0.10 max Mg 0.01-0.10 Mn 0.30 max S 0.008 max Si 0.10 max Ti 0.005 max
N02270	Ni 99.97 min C 0.02 max Co 0.001 max Cr 0.001 max Cu 0.001 max Fe 0.005 max Mg 0.001 max Mn 0.001 max S 0.001 max Si 0.001 max Ti 0.001 max
N02290	Ni rem Al 0.001 max C 0.006 max Cr 0.001 max Cu 0.02 max Fe 0.015 max Mg 0.001 max Mn 0.001 max N 0.001 max O 0.025 max S 0.0008 max Si 0.001 max
N03220	Ni rem Be 1.80-2-30 C 0.30-0.50
N03260	Ni rem C 0.02 max Co 0.20 max Cr 0.05 max Cu 0.15 max Fe 0.05 max S 0.0025 max Ti 0.05 max Other ThO_2 1.80-2.60
N03300	Ni 97.0 min C 0.40 max Cu 0.25 max Fe 0.60 max Mg 0.20-0.50 Mn 0.50 max S 0.01 max Si 0.35 max Ti 0.20-0.60
N03301	Ni 93.0 min Al 4.00-4.75 C 0.30 max Cu 0.25 max Fe 0.60 max Mn 0.50 max S 0.01 max Si 1.00 max Ti 0.25-1.00

CHEMICAL COMPOSITION OF NICKEL & NICKEL ALLOYS (Continued)	
UNS	**Chemical Composition**
N03360	Ni rem Be 1.85-2.05 Ti 0.4-0.6
N04019	Ni 60.0 min C 0.25 max Cu 27.0-31.0 Fe 2.50 max Mn 1.50 max S 0.015 max Si 3.50-4.50 max
N04020	Ni rem Al 0.50 max C 0.35 max Cu 26.0-33.0 Fe 2.50 max Mn 1.5 max Si 2.0 max
N04060	Ni 62.0-69.0 Al 1.25 max C 0.15 max Cu rem Fe 2.5 max Mn 4.0 max P 0.02 max S 0.015 max Si 1.25 max Ti 1.5-3.0
N04400	Ni 63.00-70.00 C 0.3 max Cu rem Fe 2.50 max Mn 2.00 max S 0.024 max Si 0.50 max
N04401	Ni 40.0-45.0 C 0.10 max Co 0.25 max Cu rem Fe 0.75 max Mn 2.25 max S 0.015 max Si 0.25 max
N04402	Ni 63.0 min C 0.04 max Cu 28.0-34.0 Fe 2.5 max Mn 2.0 max S 0.025 max Si 0.5 max
N04404	Ni 52.0-57.0 Al 0.05 max C 0.15 max Cu rem Fe 0.50 max Mn 0.10 max S 0.024 max Si 0.10 max
N04405	Ni 63.0-70.0 C 0.30 max Cu rem Fe 2.5 max Mn 2.0 max S 0.025-0.060 Si 0.50 max
N05500	Ni 63.0-70.0 Al 2.30-3.15 C 0.25 max Cu rem Fe 2.00 max Mn 1.50 max S 0.01 max Si 0.50 max Ti 0.35-0.85
N05502	Ni 63.0-70.0 Al 2.50-3.50 C 0.1 0 max Cu rem Fe 2.00 max Mn 1.50 max S 0.010 max Si 0.5 max Ti 0.50 max
N05504	Ni 63.0-70.0 Al 2.0-4.0 C 0.25 max Cu rem Fe 2.00 max Mn 1.50 max P 0.030 max Pb 0.010 max S 0.015 max Si 1.00 max Ti 0.25-1.00
N06002	Ni rem C 0.05-0.15 Co 0.5-2.5 Cr 20.5-23.0 Fe 17.0-20.0 Mn 1.00 max Mo 8.0-10.0 P 0.040 max S 0.030 max Si 1.00 max W 0.20-1.0
N06003	Ni rem C 0.15 max Cr 19-21 Fe 1.0 max Mn 2.5 max S 0.01 max Si 0.75-1.6
N06004	Ni 57 min C 0.15 max Cr 14-18 Fe rem Mn 1.0 max S 0.01 max Si 0.75-1.6
N06005	Ni 39.00 nom C 2.40 nom Co 10.00 nom Cr 29.00 nom Fe 6.5 max Si 0.70 nom W 15.00 nom
N06006	Ni 64.0-68.0 C 0.35-0.75 Cr 15.0-19.0 Fe rem Mn 2.00 max Mo 0.50 max P 0.04 max S 0.04 max Si 2.50 max
N06007	Ni rem C 0.05 max Cb 1.75-2.5 Co 2.5 max Cr 21.0-23.5 Cu 1.5-2.5 Fe 18.0-21.0 Mn 1.0-2.0 Mo 5.5-7.5 P 0.04 max S 0.03 max Si 1.0 max W 1.0 max
N06008	Ni rem Al 0.20 max C 0.15 max Cr 29.0-31.0 Fe 1.0 max Mn 0.10 max P 0.030 max S 0.010 max Si 0.75-1.60
N06009	Ni rem C 0.15 max Cb 0.75-1.50 Cr 19.0-21.0 Fe 1.00 max Mn 2.5 max S 0.010 max Si 0.75-1.60
N06010	Ni rem Al 0.20 max C 0.15 max Cr 9.0-11.0 Fe 0.50 max Mn 0.10 max Si 1.60 max
N06013	Ni rem C 1.80-2.20 Cr 28.0-30.0 Fe 1.0-8.0 Mn 1.0 max P 0.030 max S 0.030 max Si 0.80-1.20
N06015	Ni rem C 1.80-2.20 Cr 28.0-30.0 Fe 1.0-8.0 Mn 1.0 max Mo 7.0-9.0 P 0.030 max S 0.030 max Si 0.80-1.20
N06017	Replaced by N06985
N06022	Ni rem C 0.015 max Co 2.5 max Cr 20.0-22.5 Fe 2.0-6.0 Mn 0.50 max Mo 12.5-14.5 P 0.02 max S 0.02 max Si 0.08 max V 0.35 max W 2.5-3.5

CHEMICAL COMPOSITION OF NICKEL & NICKEL ALLOYS (Continued)	
UNS	**Chemical Composition**
N06030	Ni rem C 0.03 max Cb 0.30-1.50 Co 5.0 max Cr 28.0-31.5 Cu 1.0-2.4 Fe 13.0-17.0 Mn 1.5 max Mo 4.0-6.0 P 0.04 max S 0.02 max Si 0.8 max W 1.5-4.0
N06040	Ni rem C 0.40 max Cr 14.0-17.0 Fe 11.0 max Mn 1.50 max Si 3.00 max
N06050	Ni 64.0-68.0 C 0.40-0.60 Cr 15.0-19.0 Fe rem Mn 1.50 max Mo 0.50 max P 0.04 max S 0.04 max Si 0.50-2.00
N06052	Ni rem Al 1.10 max C 0.04 max Cr 28.0-31.5 Cb 0.10 max Cu 0.30 max Fe 7.0-11.0 Mn 1.0 max Mo 0.50 max P 0.020 max S 0.015 max Si 0.50 max Ti 1.0 max Other 0.50 max total, Al+Ti 1.50 max
N06059	Ni rem Al 0.1-0.4 C 0.010 max Co 0.3 max Cr 22.0-24.0 Fe 1.5 max Mn 0.5 max Mo 15.0-16.5 P 0.015 max S 0.005 max Si 0.10 max
N06060	Ni 54.0-60.0 C 0.03 max Cr 19.0-22.0 Cb 1.25 max Cu 1.00 max Fe rem Mn 1.50 max Mo 12.0-14.0 P 0.030 max Si 0.50 max S 0.005 max W 1.25 max
N06062	Ni 70.00 min C 0.08 max Cb 1.50-3.00 Cr 14.00-17.00 Cu 0.50 max Fe 6.00-10-00 Mn 1.00 max P 0.030 max S 0.015 max Si 0.35 max
N06072	Ni rem C 0.01-0.1 0 Cr 42.0-46.0 Cu 0.50 max Fe 0.50 max Mn 0.20 max P 0.020 max S 0.015 max Si 0.20 max Ti 0.30-1.00 Other 0.50 max total
N06075	Ni rem C 0.08-0.15 Cr 18.0-21.0 Cu 0.50 max Fe 5.00 max Mn 1.00 max Si 1.00 max Ti 0.20-0.60
N06076	Ni 75.0 min Al 0.40 max C 0.08-0.15 Cr 19.0-21.0 Cu 0.50 max Fe 2.00 max Mn 1.00 max P 0.030 max Pb 0.010 max S 0.015 max Si 0.30 max Ti 0.15-0.50
N06082	Ni 67.0 min C 0.10 max Cb 2.0-3.0 includes Ta, Cr 18.0-22.0 Cu 0.50 max Fe 3.0 max Mn 2.5-3.5 P 0.03 max S 0.015 max Si 0.50 max Ti 0.75 max
N06102	Ni rem Al 0.30-0.60 B 0.003-0-008 C 0.08 max Cb 2.75-3.25 Cr 14.0-16.0 Fe 5.0-9.0 Mg 0.01 -0.05 Mn 0.75 max Mo 2.75-3.25 P 0.010 max S 0.010 max Si 0.40 max Ti 0.40-0.70 W 2.75-3.25 Zr 0.01-0-05
N06110	Ni rem Al 1.50 max C 0.15 max Cb 2.00 max Co 12.0 max Cr 27.0-33.0 Mo 8.00-12.0 Ti 1.50 max W 4.00 max
N06132	Replaced by W86132
N06230	Ni rem Al 0.20-0.50 B 0.015 max C 0.05-0-15 Cr 20.0-24.0 Co 5.0 max Fe 3.0 max La 0.005-0-05 Mn 0.30-1.00 Mo 1.0-3.0 P 0.03 max S 0.015 max Si 0.25-0.75 W 13.0-15.0
N06231	Ni rem Al 0.2-0.5 B 0.003 max C 0.05-0-15 Cr 20.0-24.0 Co 3.0 max Cu 0.5 max Fe 3.0 max La 0.050 max Mn 0.3-1.0 Mo 1.0-3.0 P 0.030 max S 0.015 max Si 0.25-0.75 W 13.0-15.0
N06250	Ni 50.0-53.0 C 0.020 max Cr 20.0-23.0 Cu 1.00 max Fe rem Mn 1.0 max Mo 10.1-12.0 P 0.030 max S 0.005 max Si 0.09 max W 1.00 max
N06255	Ni 47.0-52.0 C 0.03 max Cr 23.0-26.0 Cu 1.20 max Fe rem Mn 1.00 max Mo 6.0-9.0 P 0.03 max S 0.03 max Si 1.0 max Ti 0.69 max W 3.0 max
N06333	Ni 44.00-47.00 C 0.08 max Co 2.50-4.00 Cr 24.00-27.00 Cu 0.50 max Fe rem Mn 2.00 max Mo 2.50-4.00 P 0.030 max Pb 0.025 max S 0.030 max Si 0.75-1.50 Sn 0.025 max W 2.50-4.00

CHEMICAL COMPOSITION OF NICKEL & NICKEL ALLOYS (Continued)

UNS	Chemical Composition
N06455	Ni rem C 0.015 max Co 2.0 max Cr 14.0-18.0 Fe 3.0 max Mn 1.0 max Mo 14.0-17.0 P 0.04 max S 0.03 max Si 0.08 max Ti 0.70 max
N06600	Ni 72.0 min C 0.15 max Cr 14.00-17.00 Cu 0.50 max Fe 6.00-10.00 Mn 1.00 max S 0.015 max Si 0.50 max
N06601	Ni 58.0-63.0 Al 1.0-1.7 C 0.1 max Cr 21.0-25.0 Cu 1.0 max Fe rem Mn 1.0 max S 0.015 max Si 0.50 max
N06602	Ni 72.0 min C 0.02 max Cr 14.0-17.0 Cu 0.5 max Fe 6.0-10.0 Mn 1.0 max S 0.015 max Si 0.5 max
N06617	Ni 44.5 min Al 0.80-1.50 B 0.006 max C 0.05-0.15 Co 10.0-15.0 Cr 20.0-24.0 Cu 0.50 max Fe 3.00 max Mn 1.00 max Mo 8.00-10.0 S 0.015 max Si 1.00 max Ti 0.60 max
N06621	Combined with N06075
N06625	Ni rem Al 0.40 max C 0.10 max Cb 3.15-4.15 Cr 20.0-23.0 Fe 5.0 max Mn 0.50 max Mo 8.0-10.0 P 0.015 max S 0.015 max Si 0.50 max Ti 0.40 max
N06635	Ni rem Al 0.10-0.50 B 0.015 max C 0.02 max Co 2.00 max Cr 14.5-17.0 Cu 0.35 max Fe 3.00 max Mn 0.30-1.00 Mo 14.0-16.5 P 0.020 max S 0.015 max Si 0.20-0.75 W 1.00 max Other La 0.01-0.10
N06690	Ni 58.0 min C 0.05 max Cr 27.0-31.0 Cu 0.50 max Fe 7.0-11.0 Mn 0.50 max S 0.015 max Si 0.50 max
N06782	Ni rem C 2.00 nom Co 0.50 nom Cr 26.00 nom Fe 4.0 max Mn 0.30 nom Si 0.30 nom W 8.75 nom
N06804	Ni 39.0-43.0 Al 0.60 max C 0.10 max Cr 28.0-31.0 Cu 0.50 max Fe rem Mn 1.50 max S 0.015 max Si 0.75 max Ti 1.20 max
N06920	Ni rem C 0.03 max Co 5.0 max Cr 20.5-23.0 Fe 17.0-20.0 Mn 1.0 max Mo 8.0-10.0 P 0.040 max S 0.030 max Si 1.0 max W 1.0-2.0
N06950	Ni 50.0 min C 0.015 max Cb 0.5 max Co 2.5 max Cr 19.0-21.0 Cu 0.5 max Fe15.0-20.0 Mn 1.0 max Mo 8.0-10.0 P 0.04 max S 0.015 max Si 1.0 max W 1.0 max
N06952	Ni 48.0-56.0 C 0.03 max Cr 23.0-27.0 Cu 0.5-1.0 Fe rem Mn 1.0 max Mo 6.0-8.0 P 0.03 max S 0.003 max Ti 0.6-1.5
N06975	Ni 47.0-52.0 C 0.03 max Cr 23.0-26.0 Cu 0.70-1.20 Fe rem Mn 1.0 max Mo 5.0-7.0 P 0.03 max S 0.03 max Si 1.0 max Ti 0.70-1.50
N06985	Ni rem C 0.015 max Co 5.0 max Cr 21.0-23.5 Cu 1.5-2.5 Fe 18.0-21.0 Mn 1.0 max Mo 6.0-8.0 P 0.04 max S 0.03 max Si 1.0 max W 1.5 max Other Cb+Ta 0.50 max
N07001	Ni rem Al 1.20-1.60 B 0.003-0.01 C 0.03-0.10 Co 12.00-15.00 Cr 18.00-21.00 Cu 0.50 max Fe 2.00 max Mn 1.00 max Mo 3.50-5.00 P 0.030 max S 0.030 max Si 0.75 max Ti 2.75-3.25 Zr 0.02-0.12
N07002	Ni rem Al 0.05 nom C 0.05 nom Co 0.50 nom Cr 16.00 nom Mn 2.30 nom Si 0.05 nom Ti 3.10 nom
N07012	Ni rem Al 4.20-4.80 B 0.01-0.02 Bi 0.00005 max C 0.30-0.35 Co 9.50-10.50 Cr 11.5-12.5 Fe 1.00 max Mn 0.10 max Mo 2.50-3.50 N 0.005 max O 0.010 max P 0.015 max Pb 0.002 max S 0.015 max Si 0.10 max Ta 1.00-2.00 Ti 2.75-3.25 W 5.50-6.50 Zr 0.05-0.15
N07013	Ni rem Al 3.20-3.60 B 0.010-0.020 C 0.07-0.20 Cb 0.10 max Co 8.50-9.50 Cr 12.2-13.0 Fe 0.50 max Hf 0.75-1.05 Mn 0.10 max Mo 1.70-2.10 P 0.015 max S 0.015 max Si 0.10 max Ta 3.85-4.50 Ti 3.85-4.15 W 3.85-4.50 Zr 0.05-0.14 Other Al+Ti 7.30-7.70

CHEMICAL COMPOSITION OF NICKEL & NICKEL ALLOYS (Continued)	
UNS	**Chemical Composition**
N07031	Ni 55.0-58.0 Al 1.00-1.70 B 0.003-0.007 C 0.03-0.06 Cr 22.0-23.0 Cu 0.60-1.20 Fe rem Mn 0.20 max Mo 1.70-2.30 P 0.015 max S 0.015 max Si 0.20 max Ti 2.10-2.60
N07032	Ni 55.0-58.0 Al 1.15-1.40 C 0.03-0.06 Cb 0.75-0.95 Co 1.00 max Cr 22.3-22.9 Fe rem Mo 1.70-2.30 P 0.015 max S 0.015 max Ti 2.10-2.40
N07041	Ni rem Al 1.40-1.80 B 0.0030-0.010 C 0.12 max Co 10.00-12.00 Cr 18.00-20.00 Fe 5.00 max Mn 0.10 max Mo 9.00-10.50 S 0.015 max Si 0.50 max Ti 3.00-3.30
N07048	Ni rem Al 0.4-0.9 C 0.015 max Cb 0.5 max Co 2.0 max Cr 20.0-23.5 Cu 1.0-2.2 Fe 18.0-21.0 Mn 0.8 max Mo 5.0-7.0 P 0.02 max S 0.01 max Si 0.1 max Ti 1.5-2.1
N07069	Ni 70.0 min Al 0.40-1.00 C 0.08 Cb 0.70-1.20 includes Ta, Cr 14.0-17.0 Cu 0.50 max Fe 5.0-9.0 Mn 1.0 max P 0.03 max S 0.015 max Si 0.50 max Ti 2.00-2.75
N07080	Ni rem Al 1.0-1.8 B 0.008 max C 0.10 max Co 2.0 max Cr 18.0-21.0 Cu 0.2 max Fe 3.0 max Mn 1.0 max S 0.015 max Si 1.00 max Ti 1.8-2.7 P 0.045 max
N07090	Ni rem Al 0.8-2.0 C 0.13 max Co 15.0-21.0 Cr 18.0-21.0 Fe 3.0 max Mn 1.0 max Si 1.5 max Ti 1.8-3.0
N07092	Ni 67.00 min C 0.08 max Cr 14.00-17.00 Cu 0.50 max Fe 8.0 max Mn 2.00-2.75 P 0.030 max S 0.015 max Si 0.35 max Ti 2.50-3.50
N07214	Ni rem Al 4.0-5.0 B 0.006 max C 0.05 max Co 2.0 max Cr 15.0-17.0 Fe 2.0-4.0 Mn 0.5 max Mo 0.5 max P 0.015 max S 0.015 max Si 0.2 max Ti 0.5 max W 0.5 max Y 0.002-0.040 Zr 0.05 max
N07252	Ni rem Al 0.75-1.25 B 0.003-0.01 C 0.1 0-0.20 Co 9.00-11.00 Cr 18.00-20.00 Fe 5.00 max Mn 0.50 max Mo 9.00-10.50 P 0.015 max S 0.015 max Si 0.50 max Ti 2.25-2.75
N07263	Ni rem Al 0.3-0.6 C 0.04-0.08 Co 19.0-21.0 Cr 19.0-21.0 Cu 0.20 max Fe 0.7 max Mn 0.60 max Mo 5.6-6.1 P 0.015 max S 0.007 max Si 0.40 max Ti 1.9-2.4 Other Al+Ti 2.4-2.8
N07500	Ni rem Al 2.50-3.25 B 0.003-0.01 C 0.15 max Co 13.00-20.00 Cr 15.00-20.00 Cu 0.15 max Fe 4.00 max Mn 0.75 max Mo 3.00-5.00 P 0.015 max S 0.015 max Si 0.75 max Ti 2.50-3.25
N07626	Ni rem Al 0.40-0.80 C 0.05 max Cr 20.0-23.0 Co 1.00 max Cb 4.50-5.50 Cu 0.50 max Fe 6.00 max Mn 0.50 max Mo 8.0-10.0 N 0.05 max P 0.020 max S 0.015 max Si 0.50 max Ti 0.60 max
N07702	Ni rem Al 2.75-3.75 C 0.10 max Cr 14.0-17.0 Cu 0.5 max Fe 2.0 max Mn 1.0 max S 0.01 max Si 0.7 max Ti 0.25-1.00
N07713	Ni rem Al 5.5-6.5 B 0.005-0.015 C 0.08-0-20 Cb 1.8-2.8 Cr 12.00-14.00 Fe 2.50 max Mn 0.25 max Mo 3.8-5.2 Si 0.50 max Ti 0.5-1.0 Zr 0.05-0.15
N07716	Ni 57.0-63.0 Al 0.35 max C 0.03 max Cr 19.0-22.0 Cb 2.75-4.00 Fe rem Mn 0.20 max Mo 7.00-9.50 P 0.015 max S 0.010 max Si 0.20 max Ti 1.00-1.60

CHEMICAL COMPOSITION OF NICKEL & NICKEL ALLOYS (Continued)

UNS	Chemical Composition
N07718	Ni 50.0-55.0 Al 0.20-0.80 B 0.006 max C 0.08 max Cb 4.75-5-50 Co 1.00 max Cr 17.0-21.0 Cu 0.30 max Fe rem Mn 0.35 max Mo 2.80-3.30 P 0.015 max S 0.015 max Si 0.35 max Ti 0.65-1.15
N07721	Ni rem Al 0.10 max C 0.07 max Cr 15.0-17.0 Cu 0.20 max Fe 8.00 max Mn 2.00-2.50 S 0.01 max Si 0.15 max Ti 2.75-3.35
N07722	Ni 70.0 min Al 0.4-1.0 C 0.08 max Cr 14.0-17.0 Cu 0.5 max Fe 5.0-9.0 Mn 1.0 max S 0.01 max Si 0.07 max Ti 2.00-2.75
N07725	Ni 55.0-59.0 Al 0.35 max C 0.03 max Cr 19.0-22.5 Cb 2.75-4.00 Fe rem Mn 0.35 max Mo 7.00-9.50 P 0.015 max S 0.010 max Si 0.20 max Ti 1.00-1.70
N07750	Ni 70.0 min Al 0.40-1.0 C 0.08 max Cb 0.70-1.20 Cr 14.0-17.0 Cu 0.5 max Fe 5.0-9.0 Mn 1.0 max S 0.01 max Si 0.50 max Ti 2.25-2.75
N07751	Ni 70.0 min Al 0.90-1.50 C 0.10 max Cb 0.70-1.20 Cr 14.0-17.0 Cu 0.50 max Fe 5.00-9.00 Mn 1.00 max S 0.01 max Si 0.50 max Ti 2.00-2.60
N07754	Ni rem Al 0.2-0.5 C 0.05 max Cr 19.0-23.0 Fe 2.5 max Ti 0.3-0.6 Other Y_2O_3 0.5-0.7
N08001	Ni 58.0-62.0 C 0.35-0.75 Cr 10.0-14.0 Fe rem Mn 2.00 max Mo 0.50 max P 0.04 max S 0.04 max Si 2.50 max
N08002	Ni 33.0-37.0 C 0.35-0.75 Cr 13.0-17.0 Fe rem Mn 2.00 max Mo 0.50 max P 0.04 max S 0.04 max Si 2.50 max
N08004	Ni 37.0-41.0 C 0.35-0.75 Cr 17.0-21.0 Fe rem Mn 2.00 max Mo 0.50 max P 0.04 max S 0.04 max Si 2.50 max
N08005	Ni 37.0-41.0 C 0.40-0.60 Cr 17.0-21.0 Fe rem Mn 1.50 max Mo 0.50 max P 0.04 max S 0.04 max Si 0.50-2.00
N08006	Ni 58.0-62.0 C 0.40-0.60 Cr 10.0-14.0 Fe rem Mn 1.50 max Mo 0.50 max P 0.04 max S 0.04 max Si 0.50-2.00
N08007	Ni 27.5-30.5 C 0.07 max Cr 19.0-22.0 Cu 3.00-4.00 Fe rem Mn 1.50 max Mo 2.00-3.00 Si 1.50 max
N08008	Ni 33.0-37.0 C 0.40-0.60 Cb 0.75-1-25 Cr 13.0-17.0 Fe rem Mo 0.50 max
N08020	Ni 32.00-38.00 C 0.07 max Cb (8xC-1.00) Cr 19.00-21.00 Cu 3.00-4.00 Fe rem Mn 2.00 max Mo 2.00-3.00 P 0.045 max S 0.035 max Si 1.00 max
N08021	Ni 32.0-36.0 C 0.07 max Cr 19.0-21.0 Cu 3.0-4.0 Mn 2.5 max Mo 2.0-3.0 P 0.03 max S 0.03 max Si 0.60 max Other Cb+Ta (8xC-1.0)
N08022	Ni 32.0-36.0 C 0.025 max Cr 19.0-21.0 Cu 3.0-4.0 Mn 1.5-2.0 Mo 2.0-3.0 P 0.015 max S 0.020 max Si 0.15 max Other Cb+Ta (8xC-0.40)
N08024	Ni 35.0-40.0 C 0.03 max Cb 0.15-0.35 Cr 22.5-25.0 Cu 0.50-1.50 Fe rem Mn 1.00 max Mo 3.50-5.00 P 0.035 max S 0.035 max Si 0.50 max
N08026	Ni 33.0-37.2 C 0.03 max Cr 22.0-26.0 Cu 2.00-4.00 Fe rem Mn 1.00 max Mo 5.00-6.70 N 0.10-0.16 P 0.03 max S 0.03 max Si 0.50 max
N08028	Ni 29.5-32.5 C 0.03 max Cr 26.0-28.0 Cu 0.6-1.4 Fe rem Mn 2.50 max Mo 3.0-4.0 P 0.030 max S 0.030 max Si 1.00 max
N08030	Ni 33.0-37.0 C 0.25-0.35 Cr 13.0-17.0 Fe rem Mn 2.00 max Mo 0.50 max P 0.040 max S 0.040 max Si 2.50 max
N08031	Ni 30.0-32.0 C 0.015 max Cr 26.0-28.0 Cu 1.0-1.4 Fe rem Mn 2.0 max Mo 6.0-7.0 N 0.15-0.25 P 0.03 max S 0.005 max Si 0.05 max
N08032	Ni 30.0-34.0 C 0.03 max Cr 20.0-23.0 Fe rem Mn 1.0 max Mo 4.0-5.0 P 0.03 max S 0.005 max Si 0.05 max
N08036	Ni 33.0-37.2 C 0.06 max Cr 22.0-26.0 Cu 1.00-3.00 Fe rem Mn 1.00 max Mo 5.00-6.70 N 0.17-0.40 P 0.030 max S 0.030 max Si 0.50 max
N08042	Ni 40.0-44.0 C 0.03 max Cr 20.0-23.0 Cu 1.5-3.0 Fe rem Mn 1.0 max Mo 5.0-7.0 P 0.03 max S 0.003 max Si 0.5 max Ti 0.6-1.2

CHEMICAL COMPOSITION OF NICKEL & NICKEL ALLOYS (Continued)

UNS	Chemical Composition
N08050	Ni 33.0-37.0 C 0.40-0.60 Cr 15.0-19.0 Fe rem Mn 1.50 max Mo 0.50 max P 0.04 max S 0.04 max Si 0.50-2.00
N08065	Ni 38.0-46.0 Al 0.20 max C 0.05 max Cr 19.5-23.5 Cu 1.50-3.0 Fe 22.0 min Mn 1.0 max Mo 2.5-3.5 P 0.03 max S 0.03 max Si 0.50 max Ti 0.60-1.2
N08135	Ni 33.0-38.0 C 0.03 max Cr 20.5-23.5 Cu 0.70 max Fe rem Mn 1.00 max Mo 4.0-5.0 P 0.03 max S 0.03 max Si 0.75 max W 0.20-0.80
N08151	Ni 31.0-34.0 C 0.05-0.15 Cr 19.0-21.0 Cb 0.50-1.50 Fe rem Mn 0.50-1.50 P 0.03 max S 0.03 max Si 0.50-1.50
N08221	Ni 36.0-46.0 Al 0.20 max C 0.025 max Cr 20.0-22.0 Cu 1.50-3.00 Fe rem Mn 1.00 max Mo 5.00-6.50 S 0.03 max Si 0.50 max Ti 0.60-1.00
N08310	Ni 18.0-22.0 C 0.020 max Cr 24.0-26.0 Fe rem Mn 2.00-4.00 Mo 2.00-4.00 N 0.20-0.40 P 0.035 max S 0.015 max Si 0.050 max
N08320	Ni 25.0-27.0 C 0.05 max Cr 21.0-23.0 Fe rem Mn 2.5 max Mo 4.0-6.0 P 0.04 max S 0.03 max Si 1.0 max Ti 4xC min
N08321	Ni 32.0-36.0 C 0.035 max Cb (8xC-0.4) max Cr 19.0-21.0 Cu 3.0-4.0 Fe rem Mn 1.5-2.5 Mo 2.0-3.0 P 0.02 max S 0.015 max Si 0.3 max
N08330	Ni 34.0-37.0 C 0.08 max Cr 17.0-20.0 Cu 1.00 max Fe rem Mn 2.00 max P 0.03 max Pb 0.005 max S 0.03 max Si 0.75-1.50 Sn 0.025 max
N08331	Ni 34.0-37.0 C 0.18-0.25 Cr 15.0-17.0 Cu 0.75 max Fe rem Mn 1.0-2.5 Mo 0.75 max P 0.03 max S 0.03 max Si 0.30-0.65
N08332	Ni 34.0-37.0 Al 0.10-0.50 C 0.05-0.10 Cr 17.0-20.0 Cu 1.00 max Fe rem Mn 2.00 max P 0.03 max Pb 0.005 max S 0.03 Si 0.75-1.50 Sn 0.025 max Ti 0.20-0.60
N08334	Ni 33.0-37.0 C 0.18-0.29 Cr 17.0-20.0 Cu 0.5 max Fe rem Mn 4.25-6.5 Mo 0.7 max P 0.025 max S 0.02 max Si 0.65-1.3
N08366	Ni 23.5-25.5 C 0.035 max Cr 20.0-22.0 Fe rem Mn 2.00 max Mo 6.00-7.00 P 0.030 max S 0.030 max Si 1.00 max
N08367	Ni 23.50-25.50 C 0.030 max Cr 20.0-22.0 Fe rem Mn 2.00 max Mo 6.00-7.00 N 0.18-0.25 P 0.040 max S 0.030 max Si 1.00 max
N08421	Ni 39.0-41.0 Al 0.2 max C 0.025 max Cr 20.0-22.0 Cu 1.5-2.0 Fe rem Mn 1.00 max Mo 5.0-6.5 S 0.03 max Si 0.5 max Ti 0.6-1.0
N08535	Ni 29.0-36.5 C 0.03 max Cr 24.0-27.0 Cu 1.50 max Fe rem Mn 1.00 max Mo 2.5-4.0 P 0.03 max S 0.03 max Si 0.50 max
N08603	Ni 33.0-37.0 C 0.25-0.35 Cr 13.0-17.0 Fe rem Mn 2.00 max Mo 0.50 max P 0.040 max S 0.040 max Si 2.50 max
N08604	Ni 16.0-22.0 C 0.20-0.60 Cr 28.0-32.0 Fe rem Mn 2.00 max Mo 0.50 max P 0.04 max S 0.04 max Si 2.00 max
N08605	Ni 33.0-37.0 C 0.35-0.75 Cr 15.0-19.0 Fe rem Mn 2.00 max P 0.04 max S 0.04 max Si 2.50 max
N08613	Ni 18.0-22.0 C 0.25-0.35 Cr 28.0-32.0 Fe rem Mn 1.50 max Mo 0.50 max P 0.04 max S 0.04 max Si 0.50-2.00
N08614	Ni 18.0-22.0 C 0.35-0.45 Cr 28.0-32.0 Fe rem Mn 1.50 max Mo 0.50 max P 0.04 max S 0.04 max Si 0.50-2.00
N08700	Ni 24.0-26.0 C 0.04 max Cb (8xC to 0.50) Cr 19.0-23.0 Cu 0.50 max Fe rem Mn 2.00 max Mo 4.3-5.0 P 0.04 max Pb 0.005 max S 0.03 max Si 1.00 max Sn 0.035 max
N08705	Ni 35.0-37.0 C 0.35-0.75 Cr 24.0-28.0 Fe rem Mn 2.00 max Mo 0.50 max P 0.04 max S 0.04 max Si 2.50 max
N08800	Ni 30.0-35.0 Al 0.15-0.60 C 0.10 max Cr 19.0-23.0 Cu 0.75 max Fe rem Mn 1.5 max S 0.015 max Si 1.0 max Ti 0.15-0.60
N08801	Ni 30.0-34.0 C 0.10 max Cr 19.0-22.0 Cu 0.5 max Fe rem Mn 1.5 max S 0.015 max Si 1.0 max Ti 0.75-1.5

CHEMICAL COMPOSITION OF NICKEL & NICKEL ALLOYS (Continued)	
UNS	**Chemical Composition**
N08802	Ni 30.0-35.0 Al 0.15-1.00 C 0.20-0.50 Cr 19.0-23.0 Cu 0.75 max Fe rem Mn 1.50 max S 0.015 max Si 0.75 max Ti 0.25-1.25
N08810	Ni 30.0-35.0 Al 0.15-0-60 C 0.05-0.10 Cr 19.0-23.0 Cu 0.75 max Fe rem Mn 1.5 max S 0.015 max Si 1.0 max Ti 0.15-0-60
N08811	Ni 30.0-35.0 Al 0.15-0.60 C 0.06-0.10 Cr 19.0-23.0 Cu 0.75 max Fe 39.5 min Mn 1.5 max S 0.015 max Si 1.0 max Ti 0.15-0.60 Other Al+Ti 0.85-1.20
N08825	Ni 38.0-46.0 Al 0.2 max C 0.05 max Cr 19.5-23.5 Cu 1.5-3.0 Fe rem Mn 1.0 max Mo 2.5-3.5 P 0.03 max S 0.03 max Si 0.5 max Ti 0.6-1.2
N08826	Ni 38.0-46.0 C 0.05 max Cb 0.6-1.2 Cr 19.5-23.5 Cu 1.5-3.0 Fe 22.0 min Mn 1.00 max Mo 2.5-3.5 P 0.03 max S 0.03 max Si 1.00 max
N08904	Ni 23.0-28.0 C 0.020 max Cr 19.0-23.0 Cu 1.00-2.00 Fe rem Mn 2.00 max Mo 4.00-5.00 P 0.045 max S 0.035 max Si 1.00 max
N08925	Ni 24.0-26.0 C 0.020 max Cr 19.0-21.0 Cu 0.50-1.50 Fe rem Mn 1.00 max Mo 6.00-7.00 N 0.10-0.20 P 0.045 max S 0.030 max Si 0.50 max
N08926	Ni 24.0-26.0 C 0.020 max Cr 19.0-21.0 Cu 0.5-1.5 Fe rem Mn 2.00 max Mo 6.0-7.0 N 0.15 max P 0.030 max S 0.010 max Si 0.50 max
N08932	Ni 24.0-26.0 C 0.020 max Cr 24.0-26.0 Cu 1.0-2.0 Fe rem Mn 2.0 max Mo 4.7-5.7 N 0.17-0.25 P 0.025 max S 0.010 max Si 0.50 max
N09027	Ni 36.50-39.50 Al 1.45-1.75 B 0.003-0-015 C 0.02-0.08 Cb 0.60- 1.10 Cr 12.50-14.00 Fe rem Mn 0.25 max Mo 5.00-6.00 P 0.015 max S 0.015 max Si 0.25 max Ti 2.30-2.70
N09706	Ni 39.0-44.0 Al 0.40 max B 0.006 max C 0.06 max Cb 2.5-3.3 Cr 14.5-17.5 Cu 0.30 max Fe rem Mn 0.35 max P 0.020 max S 0.015 max Si 0.35 max Ti 1.5-2.0
N09901	Ni 40.00-45.00 Al 0.35 max B 0.010-0.020 C 0.10 max Cr 11.00-14.00 Cu 0.50 max Fe rem Mn 1.00 max Mo 5.00-7.00 S 0.030 max Si 0.60 max Ti 2.35-3.10
N09902	Ni 41.0-43.5 Al 0.30-0.80 C 0.06 max Cr 4.90-5.75 Fe rem Mn 0.80 max P 0.04 max S 0.04 max Si 1.0 max Ti 2.20-2.75
N09911	Combined with N09901
N09925	Ni 38.0-46.0 Al 0.10-0.50 C 0.03 max Cb 0.50 max Cr 19.5-23.5 Cu 1.50-3.00 Fe 22.0 min Mn 1.00 max Mo 2.50-3.50 S 0.03 max Si 0.50 max Ti 1.90-2.40
N09926	Ni 26.0-30.0 Al 0.3 max C 0.04 max Cr 14.0-18.0 Cu 3.5-5.5 Fe 39.0 min Mn 1.5 max Mo 2.5-3.5 S 0.015 max Si 0.75 max Ti 1.5-2.3
N09979	Ni 42.00-48.00 Al 0.75-1.30 B 0.008-0.016 C 0.08 max Cr 14.00-16.00 Fe rem Mn 0.75 max Mo 3.75-4.50 P 0.015 max S 0.015 max Si 0.75 max Ti 2.70-3.30 W 3.75-4.50 Zr 0.050 max
N10001	Ni rem C 0.12 max Co 2.50 max Cr 1.00 max Fe 6.00 max Mn 1.00 max Mo 26.0-33.0 P 0.040 max S 0.030 max Si 1.00 max V 0.60 max
N10002	Ni rem C 0.08 max Co 2.5 max Cr 14.5-16.5 Fe 4.0-7.0 Mn 1.00 max Mo 15.0-17.0 P 0.040 max S 0.030 max Si 1.00 max V 0.35 max W 3.0-4.5
N10003	Ni rem Al 0.50 max B 0.010 max C 0.04-0.08 Co 0.20 max Cr 6.0-8.0 Cu C.35 max Fe 5.0 max Mn 1.00 max Mo 15.0-18.0 P 0.015 max S 0.020 max Si 1.00 max V 0.50 max W 0.50 max

CHEMICAL COMPOSITION OF NICKEL & NICKEL ALLOYS (Continued)	
UNS	**Chemical Composition**
N10004	Ni rem C 0.12 max Cr 4.00-6.00 Fe 4.00-7.00 Mn 1.00 max Mo 23.00-26.00 P 0.050 max S 0.050 max Si 1.00 max V 0.60 max
N10276	Ni rem C 0.02 max Co 2.5 max Cr 14.5-16.5 Fe 4.0-7.0 Mn 1.0 max Mo 15.0-17.0 P 0.030 max S 0.030 max Si 0.08 max V 0.35 max W 3.0-4.5
N10665	Ni rem C 0.02 max Co 1.0 max Cr 1.0 max Fe 2.0 max Mn 1.0 max Mo 26.0-30.0 P 0.04 S 0.03 max Si 0.10 max
N10675	Ni 65.0 min Al 0.50 max C 0.01 max Cb 0.20 max Co 3.0 max Cr 1.0-3.0 Cu 0.20 max Fe 1.0-3.0 Mn 3.0 max Mo 27.0-32.0 P 0.030 max S 0.010 max Si 0.10 max Ta 0.20 max Ti 0.20 max V 0.20 max W 3.0 max Zr 0.10 max Other Ni+Mo 94.0-98.0
N13009	Ni rem Al 4.75-5.25 B 0.010-0.020 Bi 0.00005 max C 0.12-0.17 Cb 0.75-1.25 Co 9.00-11.00 Cr 8.00-10.00 Cu 0.10 max Fe 1.50 max Mn 0.20 max Pb 0.0010 max S 0.015 max Si 0.20 max Ti 1.75-2.25 W 11.5-13.5 Zr 0.03-0.08
N13010	Ni rem Al 5.75-6.25 B 0.010-0.020 Bi 0.00005 max C 0.08-0.13 Cb 0.10 max Co 9.50-10.50 Cr 7.50-8.50 Fe 0.35 max Mn 0.20 max Mo 5.75-6.25 P 0.015 max Pb 0.0005 max S 0.015 max Si 0.25 max Ta 4.00-4.50 Ti 0.80-1.20 W 0.10 max Zr 0.05-0.10
N13017	Ni rem Al 3.85-4.15 B 0.020-0.030 Bi 0.00005 max C 0.02-0.06 Co 16.0-18.0 Cr 14.0-16.0 Cu 0.10 max Fe 0.50 max Mn 0.15 max Mo 4.50-5.50 N 0.0050 max O 0.010 max P 0.015 max Pb 0.0002 max S 0.015 max Si 0.20 max Ti 3.35-3.65 W 0.05 max Zr 0.06 max
N13020	Ni rem Al 3.75-4.75 B 0.025-0.035 Bi 0.00005 max C 0.03-0.10 Co 17.0-20.0 Cr 14.0-16.0 Cu 0.10 max Fe 2.00 max Mn 0.15 max Mo 4.50-5.50 Ti 2.75-3.75 Zr 0.06 max
N13021	Ni rem Ag 0.0005 max Al 4.5-4.9 B 0.003-0.010 Bi 0.0001 max C 0.12-0.17 Cr 14.0-15.7 Co 18.0-22.0 Cu 0.2 max Fe 1.0 max Mn 1.0 max Mo 4.5-5.5 Pb 0.0015 max S 0.015 max Si 1.0 max Ti 0.9-1.5
N13100	Ni rem Al 5.00-6.00 B 0.01-0.02 C 0.15-0.20 Co 13.0-17.0 Cr 8.0-11.0 Fe 1.0 max Mn 0.20 max Mo 2.0-4.0 S 0.015 max Si 0.20 max Ti 4.50-5.00 V 0.70-1.20 Zr 0.03-0.09
N13246	Ni rem Al 5.25-5.75 B 0.01-0.02 C 0.13-0.17 Cr 8.0-1 0.0 Co 9.0-10.0 Cu 0.10 max Hf 1.5-2.0 Fe 1.0 max Mn 0.20 max Mo 2.25-2.75 S 0.015 max Si 0.20 max Ta 1.25-1.75 Ti 1.25-1.75 W 9.0-11.0 Zr 0.03-0.08
N14052	Ni 50.5 nom Al 0.10 max C 0.05 max Co 0.50 max Cr 0.25 max Fe rem Mn 0.60 max P 0.025 max S 0.025 max Si 0.30 max
N19903	Ni 36.0-40.0 Al 0.30-1.15 B 0.012 max C 0.06 max Cb 2.40-3.50 Co 13.0-17.0 Cr 1.00 max Cu 0.50 max Fe rem Mn 1.00 max S 0.015 max Si 0.35 max Ti 1.00-1.25
N19907	Ni 35.0-40.0 Al 0.20 max B 0.012 max C 0.06 max Cb 4.3-5.2 Co 12.0-16.0 Cr 1.0 max Cu 0.5 max Fe rem Mn 1.0 max P 0.015 max S 0.015 max Si 0.35 max Ti 1.2-1.8
N19909	Ni 35.0-40.0 Al 0.15 max B 0.012 max C 0.06 max Cb 4.3-5.2 Co 12.0-16.0 Cr 1.0 max Cu 0.5 max Fe rem Mn 1.0 max P 0.015 max S 0.015 max Si 0.25-0.50 Ti 1.3-1.8
N24025	Ni rem C 0.25 max Cu 27.0-33.0 Fe 3.50 max Mn 1.50 max P 0.03 max S 0.03 max Si 3.5-4.5
N24030	Ni rem C 0.30 max Cu 27.0-33.0 Fe 3.50 max Mn 1.50 max P 0.03 max S 0.03 max Si 2.7-3.7

CHEMICAL COMPOSITION OF NICKEL & NICKEL ALLOYS (Continued)	
UNS	Chemical Composition
N24130	Ni rem C 0.30 max Cb 1.0-3.0 Cu 26.0-33.0 Fe 3.50 max Mn 1.50 max P 0.03 max S 0.03 max Si 1.0-2.0
N24135	Ni rem C 0.35 max Cb 0.5 max Cu 26.0-33.0 Fe 3.50 max Mn 1.50 max P 0.03 max S 0.03 max Si 1.25 max
N26022	Ni rem C 0.02 max Cr 20.0-22.5 Fe 2.0-6.0 Mn 1.00 max Mo 12.5-14.5 P 0.025 max S 0.025 max Si 0.80 max V 0.35 max W 2.5-3.5
N26055	Ni rem Bi 3.0-5.0 C 0.05 max Cr 11.0-14.0 Fe 2.0 max Mn 1.5 max Mo 2.0-3.5 P 0.03 max S 0.03 max Si 0.5 max Sn 3.0-5.0
N26455	Ni rem C 0.02 max Cr 15.0-17.5 Fe 2.0 max Mn 1.00 max Mo 15.0-17.5 P 0.03 max S 0.03 max Si 0.80 max W 1.0 max
N26625	Ni rem C 0.06 max Cb 3.15-4.50 Cr 20.0-23.0 Fe 5.0 max Mn 1.00 max Mo 8.0-10.0 P 0.015 max S 0.015 max Si 1.00 max
N30002	Ni rem C 0.12 max Cr 15.5-17.5 Fe 4.5-7.5 Mn 1.00 max Mo 16.0-18.0 P 0.040 max S 0.030 max Si 1.00 max V 0.20-0.40 W 3.75-5.25
N30007	Ni rem C 0.07 max Cr 1.0 max Fe 3.00 max Mn 1.00 max Mo 30.0-33.0 P 0.040 max S 0.030 max Si 1.00 max
N30012	Ni rem C 0.12 max Cr 1.00 max Fe 4.0-6.0 Mn 1.00 max Mo 26.0-30.0 P 0.040 max S 0.030 max Si 1.00 max V 0.20-0.60
N30107	Ni rem C 0.07 max Cr 17.0-20.0 Fe 3.0 max Mn 1.00 max Mo 17.0-20.0 P 0.040 max S 0.030 max Si 1.00 max
N94620	Ni 27 nom Al 0.01 max C 0.02 max Cr 0.03 max Co 25 nom Cu 0.20 max Mg 0.01 max Mn 0.15 max Mo 0.06 max P 0.006 max S 0.006 max Si 0.15 max Ti 0.01 max Zr 0.01 max
N94630	Ni 29 nom Al 0.01 max C 0.02 max Cr 0.03 max Co 17 nom Cu 0.20 max Mg 0.01 max Mn 0.35 max Mo 0.06 max P 0.006 max S 0.006 max Si 0.15 max Ti 0.01 max Zr 0.01 max
N99600	Ni rem Al 0.05 max B 2.75-3.50 C 0.6-0.9 Co 0.10 max Cr 13.0-15.0 Fe 4.0-5.0 P 0.02 max S 0.02 max Se 0.005 max Si 4.0-5.0 Ti 0.05 max Zr 0.05 max Other 0.50 max
N99610	Ni rem Al 0.05 max B 2.75-3.50 C 0.06 Co 0.10 max Cr 13.0-15.0 Fe 4.0-5.0 P 0.02 max S 0.02 max Se 0.005 max Si 4.0-5.0 Ti 0.05 max Zr 0.05 max Other 0.50 max
N99612	Ni rem Al 0.05 max B 3.25-4.0 C 0.06 max Co 0.10 max Cr 13.5-16.5 Fe 1.5 max P 0.02 max S 0.02 max Se 0.005 max Ti 0.05 max Zr 0.05 max Other total 0.50 max
N99620	Ni rem Al 0.05 max B 2.75-3.50 C 0.06 max Co 0.10 max Cr 6.0-8.0 Fe 2.5-3.5 P 0.02 S 0.02 max Se 0.005 max Si 4.0-5.0 Ti 0.05 max Zr 0.05 max Other 0.50 max
N99622	Ni rem Al 0.05 max B 2.0-3.0 C 0.40-0.55 Co 0.10 max Cr 10.0-13.0 Fe 2.5-4.5 P 0.02 max S 0.02 max Se 0.005 max Si 3.0-4.0 Ti 0.05 max W 15.0-17.0 Zr 0.05 max Other total 0.50 max
N99624	Ni rem Al 0.05 max B 2.2-3.1 C 0.30-0.50 Co 0.10 max Cr 9.0-11.75 Fe 2.5-4.0 P 0.02 max S 0.02 max Se 0.005 Si 3.25-4.25 Ti 0.05 max W 11.50-12.75 Zr 0.05 max Other total 0.50 max
N99630	Ni rem Al 0.05 max B 2.75-3.50 C 0.06 max Co 0.10 max Fe 0.5 max P 0.02 max S 0.02 max Se 0.005 max Si 4.0-5.0 Ti 0.05 max Zr 0.05 max Other 0.50 max

CHEMICAL COMPOSITION OF NICKEL & NICKEL ALLOYS (Continued)	
UNS	**Chemical Composition**
N99640	Ni rem Al 0.05 max B 1.5-2.2 C 0.06 max Co 0.10 max Fe 1.5 max P 0.02 max S 0.02 max Se 0.005 max Si 3.0-4.0 Ti 0.05 max Zr 0.05 max Other 0.50 max
N99644	Ni rem B 2.00-3.00 C 0.30-0.60 Co 1.50 max Cr 8.0-14.0 Fe 1.25-3.25 Se 0.005 max Si 1.25-3.25
N99645	Ni rem B 2.00-4.00 C 0.40-0.80 Co 1.25 max Cr 10.0-16.0 Fe 3.00-5.00 Se 0.005 max Si 3.00-5.00
N99646	Ni rem B 2.50-4.50 C 0.50-1.00 Co 1.00 max Cr 12.0-18.0 Fe 3.50-5.50 Se 0.005 max Si 3.50-5.50
N99650	Ni rem Al 0.05 max B 0.03 max C 0.10 Co 0.10 max Cr 18.5-19.5 P 0.02 max S 0.02 max Se 0.005 max Si 9.75-10.50 Ti 0.05 max Zr 0.05 max Other 0.50 max
N99651	Ni rem Al 0.05 max B 1.0-1.5 C 0.10 max Co 0.10 max Cr 18.5-19.5 Fe 0.5 max P 0.02 max S 0.02 max Se 0.005 max Si 7.0-7.5 Ti 0.05 max Zr 0.05 max
N99700	Ni rem Al 0.05 max C 0.01 max Co 0.10 max P 10.0- 12.0 S 0.02 max Se 0.005 max Ti 0.05 max Zr 0.05 max Other 0.50 max
N99710	Ni rem Al 0.05 max B 0.01 max C 0.08 max Co 0.10 max Cr 13.0-15.0 Fe 0.2 max Mn 0.04 max P 9.7-10.5 S 0.02 max Se 0.005 max Si 0.10 max Ti 0.05 max Zr 0.05 max Other 0.50 max
N99800	Ni rem Al 0.05 max C 0.10 max Co 0.10 max Cu 4.0-5.0 Mn 21.5-24.5 P 0.02 max S 0.02 max Se 0.005 max Si 6.0-8.0 Ti 0.05 max Zr 0.05 max Other 0.50 max

rem - remainder; nom - nominal

CHEMICAL COMPOSITION OF NICKEL & NICKEL ALLOY CASTINGS

ASTM A 494 Gr	UNS	Chemical Composition
CZ-100	N02100	Ni rem C 1.00 max Cu 1.25 max Fe 3.00 max Mn 1.50 max Si 2.00 max
M-35-1	N24135	Ni rem C 0.35 max Cb 0.5 max Cu 26.0-33.0 Fe 3.50 max Mn 1.50 max P 0.03 max S 0.03 max Si 1.25 max
M-35-2	---	Ni rem C 0.35 max Cb 0.5 max Cu 26.0-33.0 Fe 3.50 max Mn 1.50 max P 0.03 max S 0.03 max Si 2.00 max
M-30H	N24030	Ni rem C 0.30 max Cu 27.0-33.0 Fe 3.50 max Mn 1.50 max P 0.03 max S 0.03 max Si 2.7-3.7
M-25S[a]	N24025	Ni rem C 0.25 max Cu 27.0-33.0 Fe 3.50 max Mn 1.50 max P 0.03 max S 0.03 max Si 3.5-4.5
M-30C	N24130	Ni rem C 0.30 max Cb 1.0-3.0 Cu 26.0-33.0 Fe 3.50 max Mn 1.50 max P 0.03 max S 0.03 max Si 1.0-2.0
N-12MV[a]	N30012	Ni rem C 0.12 max Cr 1.00 max Fe 4.0-6.0 Mn 1.00 max Mo 26.0-30.0 P 0.040 max S 0.030 max Si 1.00 max V 0.20-0.60
N-7M[a]	N30007	Ni rem C 0.07 max Cr 1.0 max Fe 3.00 max Mn 1.00 max Mo 30.0-33.0 P 0.040 max S 0.030 max Si 1.00 max
CY-40[a]	N06040	Ni rem C 0.40 max Cr 14.0-17.0 Fe 11.0 max Mn 1.50 max Si 3.00 max
CW-12MW[a]	N30002	Ni rem C 0.12 max Cr 15.5-17.5 Fe 4.5-7.5 Mn 1.00 max Mo 16.0-18.0 P 0.040 max S 0.030 max Si 1.00 max V 0.20-0.40 W 3.75-5.25
CW-6M[a]	N30107	Ni rem C 0.07 max Cr 17.0-20.0 Fe 3.0 max Mn 1.00 max Mo 17.0-20.0 P 0.040 max S 0.030 max Si 1.00 max
CW-2M[a]	N26455	Ni rem C 0.02 max Cr 15.0-17.5 Fe 2.0 max Mn 1.00 max Mo 15.0-17.5 P 0.03 max S 0.03 max Si 0.80 max W 1.0 max
CW-6MC	N26625	Ni rem C 0.06 max Cb 3.15-4.50 Cr 20.0-23.0 Fe 5.0 max Mn 1.00 max Mo 8.0-10.0 P 0.015 max S 0.015 max Si 1.00 max
CY5SnBiM	N26055	Ni rem Bi 3.0-5.0 C 0.05 max Cr 11.0-14.0 Fe 2.0 max Mn 1.5 max Mo 2.0-3.5 P 0.03 max S 0.03 max Si 0.5 max Sn 3.0-5.0
CX2MW[a]	N26022	Ni rem C 0.02 max Cr 20.0-22.5 Fe 2.0-6.0 Mn 1.00 max Mo 12.5-14.5 P 0.025 max S 0.025 max Si 0.80 max V 0.35 max W 2.5-3.5

a. See ASTM A 494 for details regarding class designations and heat treat requirements. rem - remainder; nom - nominal

CHEMICAL COMPOSITION OF NICKEL & NICKEL ALLOY - WELD FILLER METALS

UNS	Chemical Composition
W80001	Ni rem C 0.07 max Co 2.5 max Cr 1.0 max Cu 0.50 max Fe 4.0-7.0 Mn 1.0 max Mo 26.0-30.0 P 0.04 max S 0.03 max Si 1.0 max V 0.60 max W 1.0 max
W80002	Ni rem C 0.10 max Co 2.50 max Cr 14.50-16.50 Cu 0.50 max Fe 4.0-7.0 Mn 1.0 max Mo 15.0-17.0 P 0.040 max S 0.030 max Si 1.0 max W 3.00-4.50

CHEMICAL COMPOSITION OF NICKEL & NICKEL ALLOY - WELD FILLER METALS (Continued)	
UNS	**Chemical Composition**
W80004	Ni rem C 0.12 max Co 2.5 max Cr 2.5-5.5 Cu 0.50 max Fe 4.0-7.0 Mn 1.0 max Mo 23.0-27.0 P 0.04 max S 0.03 max Si 1.0 max V 0.60 max W 1.0 max
W80276	Ni rem C 0.02 max Co 2.5 max Cr 14.5-16.5 Cu 0.50 max Fe 4.0-7.0 Mn 1.0 max Mo 15.0-17.0 P 0.04 max S 0.03 max Si 0.2 max V 0.35 max W 3.0-4.5
W80665	Ni rem C 0.02 max Co 1.0 max Cr 1.0 max Cu 0.50 max Fe 2.0 max Mn 1.75 max Mo 26.0-30.0 P 0.04 max S 0.03 max Si 0.2 max W 1.0 max
W82001	Ni 85.00 min Al 1.0 max C 2.00 max Cu 2.50 max Fe 8.0 max Mn 2.5 max S 0.03 max Si 4.00 max Other 1.0 max total
W82002	Ni 45.0-60.0 Al 1.0 max C 2.00 max Cu 2.50 max Fe rem Mn 2.5 max S 0.03 max Si 4.00 max Other 1.0 max total
W82003	Ni 85.0 min Al 1.00-3.00 C 2.00 max Cu 2.50 max Fe 8.00 max Mn 2.5 max S 0.03 max Si 4.00 max Other 1.0 max total
W82004	Ni 45.0-60.0 Al 1.00-3.00 C 2.00 max Cu 2.50 max Fe rem Mn 2.5 max S 0.03 max Si 4.0 max Other 1.0 max total
W82006	Ni 35.0-45.0 Al 1.0 max C 2.0 max Cu 2.5 max Fe rem Mn 10.0-14.0 S 0.03 max Si 1.0 max Other 1.0 max total
W82032	Ni 45.0-60.0 Al 1.0 max C 2.0 max Cu 2.5 max Fe rem Mn 3.0-5.0 S 0.03 max Si 1.0 max Other 1.0 max total
W82141	Ni 92.0 min Al 1.0 max C 0.10 max Cu 0.25 max Fe 0.75 max Mn 0.75 max P 0.03 max S 0.02 max Si 1.25 max Ti 1.0-4.0
W83002	Ni rem C 2.2-3.0 Cr 25.0-30.0 Co 10.0-15.0 Fe 20.0-25.0 Mn 1.0 max Mo 7.0-10.0 Si 0.60-1.5 W 2.0-4.0
W84001	Ni 50.0-60.0 C 0.35-0.55 Cu 35.0-45.0 Fe 3.0-6.0 Mn 2.25 max S 0.025 max Si 0.75 max Other 1.0 max total
W84002	Ni 60.0-70.0 C 0.35-0.55 Cu 25.0-35.0 Fe 3.0-6.0 Mn 2.25 max S 0.025 max Si 0.75 max Other 1.0 max total
W84190	Ni 62.0-68.0 Al 0.50 max C 0.15 max Cu rem Fe 2.5 max Mn 4.0 max P 0.02 max S 0.015 max Si 1.5 max Ti 1.0 max
W86002	Ni rem C 0.05-0.15 Co 0.50-2.50 Cr 20.5-23.0 Cu 0.50 max Fe 17.0-20.0 Mn 1.0 max Mo 8.0-10.0 P 0.04 max S 0.03 max Si 1.0 max W 0.20-1.0
W86003	Ni 71.0 min C 0.15 max Cb 1.0-2.2 includes Ta, Co 1.0 max Cr 18.0-21.0 Fe 4.0 max Mn 1.50 max S 0.015 max Si 0.75 max
W86007	Ni rem C 0.05 max Cb 1.75-2.50 Co 2.5 max Cr 21.0-23.5 Cu 1.5-2.5 Fe 18.0-21.0 Mn 1.0-2.0 Mo 5.5-7.5 P 0.04 max S 0.03 max Si 1.0 max W 1.0 max
W86022	Ni rem C 0.02 max Co 2.5 max Cr 20.0-22.5 Cu 0.50 max Fe 2.0-6.0 Mn 1.0 max Mo 12.5-14.5 P 0.03 max S 0.015 max Si 0.2 max V 0.35 W 2.5-3.5
W86030	Ni rem C 0.03 max Cb 0.30-1.50 Co 5.0 max Cr 28.0-31.5 Cu 1.0-2.4 Fe 13.0-17.0 Mn 1.5 max Mo 4.0-6.0 P 0.04 max S 0.02 max Si 1.0 max Ti 1.5-4.0
W86040	Ni rem C 0.03 max Cr 20.5-22.5 Cb 1.0-2.8 Cu 0.50 max Fe 5.0 max Mn 2.2 max Mo 8.8-10.0 P 0.03 max S 0.02 max Si 0.7 max Other 0.50 max total
W86082	Ni 67.0 min C 0.10 max Cr 18.0-22.0 Cb 2.0-3.0 Cu 0.50 max Fe 3.0 max Mn 2.5-3.5 P 0.03 max S 0.015 max Si 0.50 max Ti 0.75 max

CHEMICAL COMPOSITION OF NICKEL & NICKEL ALLOY - WELD FILLER METALS (Continued)

UNS	Chemical Composition
W86112	Ni 55.0 min C 0.10 max Cb 3.15-4.15 includes Ta, Co 0.12 max Cr 20.0-23.0 Cu 0.50 max Fe 7.0 max Mn 1.0 max Mo 8.0-10.0 P 0.03 max S 0.02 max Si 0.75 max
W86117	C 0.05-0.15 Cb+Ta 1.0 max Co 9.0-15.0 Cr 21.0-26.0 Cu 0.50 max Fe 5.0 max Mn 0.30-2.5 Mo 8.0-10.0 Ni rem P 0.03 max S 0.015 max Si 0.75 max
W86130	Ni 66.0 min Al 0.1-1.0 C 0.25 max Cb 4xSi min includes Ta, Co 1.0 max Cr 12.5-17.0 Fe 11.0 max Mn 1.0 max S 0.015 max Si 1.0 max Ti 1.0-2.75
W86132	Ni 62.0 min C 0.08 max Cb 1.5-4.0 includes Ta, Cr 13.0-17.0 Cu 0.50 max Fe 11.0 max Mn 3.5 max P 0.03 max S 0.02 max Si 0.75 max
W86133	Ni 62.0 min C 0.10 max Cb 0.5-3.0 includes Ta, Cr 13.0-17.0 Cu 0.50 max Fe 12.0 max Mn 1.0-3.5 Mo 0.50-2.50 P 0.03 max S 0.02 max Si 0.75 max
W86134	Ni 60.0 min C 0.20 max Cb 1.0-3.5 includes Ta, Cr 13.0-17.0 Cu 0.50 max Fe 12.0 max Mn 1.0-3.5 Mo 1.0-3.5 P 0.03 max S 0.02 max Si 1.0 max
W86182	Ni 59.0 min C 0.10 max Cb 1.0-2.5 includes Ta, Cr 13.0-17.0 Cu 0.50 Fe 10.0 Mn 5.0-9.5 P 0.03 max S 0.015 Si 1.0 max Ti 1.0 max
W86333	Ni 44.0-47.0 C 0.08 max Co 2.50-4.00 Cr 24.0-27.0 Fe rem Mn 2.00 max Mo 2.50-4.00 P 0.030 max S 0.030 max Si 0.75-1.50 W 2.50-4.00
W86455	Ni rem C 0.015 max Co 2.0 max Cr 14.0-18.0 Cu 0.50 max Fe 3.0 max Mn 1.5 max Mo 14.0-17.0 P 0.04 max S 0.03 max Si 0.2 max Ti 0.7 max W 0.5 max
W86620	Ni 55.0 min C 0.10 max Cb 0.5-2.0 includes Ta, Cr 12.0-17.0 Cu 0.50 max Fe 10.0 max Mn 2.0-4.0 Mo 5.0-9.0 P 0.03 max S 0.02 max Si 1.0 max W 1.0-2.0
W86985	Ni rem C 0.02 max Cb 0.5 max includes Ta, Co 5.0 max Cr 21.0-23.5 Cu 1.5-2.5 Fe 18.0-21.0 Mn 1.0 max Mo 6.0-8.0 P 0.04 max S 0.03 max Si 1.0 max W 1.5 max
W87718	Ni 50.0-55.0 Al 0.20-0.80 B 0.006 max C 0.08 max Cr 17.0-21.0 Co 1.0 max Cb 4.75-5.5 Cu 0.30 max Fe rem Mn 0.50 max Mo 2.8-3.3 P 0.015 max S 0.015 max Si 0.75 max Ti 0.65-1.15
W88021	Ni 32.0-36.0 C 0.07 max Cb 8xC-1.00 includes Ta, Cr 19.0-21.0 Cu 3.0-4.0 Mn 0.5-2.5 Mo 2.0-3.0 P 0.04 S 0.03 Si 0.60
W88022	Ni 32.0-36.0 C 0.03 max Cb 8xC-0.40 includes Ta, Cr 19.0-21.0 Cu 3.0-4.0 Mn 1.50-2.50 Mo 2.0-3.0 P 0.020 max S 0.015 max Si 0.30 max
W88028	Ni 30.0-33.0 C 0.03 max Cr 26.5-29.0 Cu 0.60-1.50 Mn 0.50-2.50 Mo 3.20-4.20 P 0.02 max S 0.02 max Si 0.90 max
W88331	Ni 33.0-37.0 C 0.18-0.25 Cr 14.0-17.0 Cu 0.75 max Mn 1.0-2.5 Mo 0.75 max P 0.04 max S 0.03 max Si 0.90 max
W88334	Ni 33.0-37.0 C 0.18-0.29 Cr 17.0-20.0 Cu 0.5 max Fe rem Mn 4.25-6.5 Mo 0.7 max P 0.03 max S 0.02 max Si 0.7-1.3
W88335	Ni 33.0-37.0 C 0.35-0.45 Cr 14.0-17.0 Cu 0.75 max Mn 1.0-2.5 Mo 0.50 max P 0.04 max S 0.03 max Si 0.90 max
W88338	Ni 33.0-37.0 C 0.75-0.95 Cr 17.0-20.0 Fe rem Mn 2.50 max P 0.03 max S 0.03 max Si 0.65-1.3 max
W88904	Ni 24.0-26.0 C 0.03 Cr 19.0-21.5 Cu 1.2-2.0 Mn 1.0-2.5 Mo 4.0-5.2 P 0.03 S 0.02 Si 0.90

CHEMICAL COMPOSITION OF NICKEL & NICKEL ALLOY - WELD FILLER METALS (Continued)

UNS	Chemical Composition
W89604	Ni rem B 2.00-3.00 C 0.30-0.60 Co 1.50 max Cr 8.0-14.0 Fe 1.25-3.25 Si 1.25-3.25
W89605	Ni rem B 2.00-4.00 C 0.40-0.80 Co 1.25 max Cr 10.0-16.0 Fe 3.00-5.00 Si 3.00-5.00
W89606	Ni rem B 2.50-4.50 C 0.50-1.00 Co 1.00 max Cr 12.0-18.0 Fe 3.50-5.50 Si 3.50-5.50

See Common Names & Cross Referenced Specifications - Nickel & Nickel Alloys - Weld Filler Metals (page 264) for UNS cross references to AWS and ASME specifications.

rem - remainder; nom - nominal

MECHANICAL PROPERTIES OF NICKEL

ASTM Spec.	UNS	Condition	Size OD	Tensile Strength		Yield Strength		% El
				ksi	MPa	Ksi	MPa	
B 161	N02200	Annealed	≤5 in (127 mm)	55	380	15	105	35
Pipe		Annealed	>5 in (127 mm)	55	380	12	80	40
		Stress Relieved	all sizes	65	450	40	275	15
	N02201	Annealed	≤5 in (127 mm)	50	345	12	80	35
		Annealed	>5 in (127 mm)	50	345	10	70	40
		Stress Relieved	all sizes	60	415	30	205	15
ASTM Spec.	**UNS**	**Condition**	**Thickness[b]**	**ksi**	**MPa**	**Ksi**	**MPa**	**% El**
B 162	N02200	Annealed	over 0.109	55	380	15	100	40
Plate, Sheet,		Annealed	0.010-0.049	55	380	15	100	30
Strip[a]		Annealed	0.050-0.109	55	380	15	100	35
	N02201	Annealed	over 0.109	50	345	12	80	40
		Annealed	0.010-0.049	50	345	12	80	30
		Annealed	0.050-0.109	50	345	12	80	35

a. Plate - hot-rolled; sheet - hot and cold-rolled; strip - cold-rolled

b. Applies to sheet and strip only, no thickness limitations on plate.

Single values are minimum specified.

MECHANICAL PROPERTIES OF NICKEL & NICKEL ALLOYS - TUBING & PIPING

ASTM				Tensile Strength		Yield Strength			Hardness
Spec.	Form	UNS	Condition[a]	ksi	MPa	Ksi	MPa	% El	HRB max
B 163	ST	N02200	Annealed	55	379	15	103	40	---
		N02200	Stress relieved	65	448	40	276	15	65
		N02201	Annealed	50	345	12	83	40	---
		N02201	Stress relieved	60	414	30	207	15	62
		N04400	Annealed	70	483	28	193	35	---
		N04400	Stress relieved	85	586	55	379	15	75
		N06600	Annealed	80	552	35	241	30	---
		N06690	Annealed	85	586	35	241	30	---
		N08800	Annealed	75	517	30	207	30	---
		N08801	Annealed	65	448	25	172	30	---
		N08800	Annealed	83	572	47	324	30	---
		N08810	Annealed	65	448	25	172	30	---
		N08825	Annealed	85	586	35	241	30	---
B 167	SP, ST	N06601	Annealed	80	550	30	205	30	---
B 444	P, T	N06625	gr 1 Annealed	120	827	60	414	30	---
		N06625	gr 2 Solution-annealed	100	690	40	276	30	---
B 464	WP	N08020	Stabilized-annealed	80	551	35	241	30	---
		N08024	Annealed	80	551	35	241	30	---
		N08026	Solution-annealed	80	551	35	241	30	---
B 517	WP	N06600	---	80	550	35	240	30	---
B 535	SP	N08330	Annealed	70	483	30	207	30	70-90
		N08332	Annealed	67	462	27	186	30	65-88

MECHANICAL PROPERTIES OF NICKEL & NICKEL ALLOYS - TUBING & PIPING (Continued)

ASTM Spec.	Form	UNS	Condition[a]	Tensile Strength		Yield Strength		% El	Hardness HRB max
				ksi	MPa	Ksi	MPa		
B 626 (con't)	WT	N06022	---	100	690	45	310	45	---
		N06030	---	85	586	35	241	30	---
		N06059	---	100	690	45	310	45	---
		N06230	---	110	760	45	310	40	---
		N06455	---	100	690	40	276	40	---
		N06975	---	85	586	32	221	40	---
		N06985	---	90	621	35	241	45	---
		N08031	---	94	650	40	276	40	---
		N08320	---	75	517	28	193	35	---
		N10276	---	100	690	41	283	40	---
		N10665	---	110	760	51	352	40	---
B 622	SP	N06002	---	100	690	40	276	35	---
		N06007	---	90	621	35	241	35	---
		N06022	---	100	690	45	310	45	---
		N06030	---	85	586	35	241	30	---
		N06059	---	100	690	45	310	45	---
		N06060	---	90	621	35	241	40	---
		N06230	---	110	760	45	310	40	---
		N06250	---	90	621	35	241	40	---
		N06255	---	85	586	32	221	40	---
		N06455	---	100	690	40	276	40	---
		N06975	---	85	586	32	221	40	---
		N06985	---	90	621	35	241	40	---
		N08031	---	94	650	40	276	40	---
		N08135	---	73	503	31	214	40	---
		N08320	---	75	517	28	193	35	---

MECHANICAL PROPERTIES OF NICKEL & NICKEL ALLOYS - TUBING & PIPING (Continued)									
ASTM Spec.	Form	UNS	Condition[a]	Tensile Strength ksi	Tensile Strength MPa	Yield Strength Ksi	Yield Strength MPa	% El	Hardness HRB max
B 622	SP	N08535	---	73	503	31	214	40	---
(con't)		N10001	---	100	690	45	310	40	---
		N10276	---	100	690	41	283	40	---
		N10665	---	110	760	51	352	40	---

a. See specific ASTM Standard for specific details of heat treat condition.
P - pipe, T - tube, SP- seamless pipe, ST - seamless tube, WP - welded pipe, WT - welded tube.
Single values are minimum specified.

MECHANICAL PROPERTIES OF NICKEL & NICKEL ALLOYS - PLATE SHEET & STRIP									
ASTM Spec.	Form	UNS	Condition[a]	Tensile Strength ksi	Tensile Strength MPa	Yield Strength ksi	Yield Strength MPa	% El	Hardness HRB max
B 127	Pl, Sh, St	N04400	Annealed	70	485	28	195	35	---
B 162	Pl, Sh, St	N02200	Annealed	55	380	15	100	40[b]	---
		N02201	Annealed	50	345	12	80	40[b]	---
B 168	Pl, Sh, St	N06600	Annealed	80	550	35	240	30	---
		N06601	Annealed	80	550	30	205	30	---
		N06617	Annealed	95	655	35	240	30	---
		N06690	Annealed	85	586	35	240	30	---
B 333	Sh, St	N10001	---	115	795	50	345	45	100[g]
		N10665	---	110	758	51	352	40	100[g]
	Pl	N10001	---	100	690	45	310	40	100[g]
		N10665	---	110	758	51	352	40	100[g]
B 409	Pl, Sh, St	N08800	Annealed	75	520	30	205	30	---
		N08810	Annealed	65	450	25	170	30	---

MECHANICAL PROPERTIES OF NICKEL & NICKEL ALLOYS - PLATE SHEET & STRIP (Continued)									
ASTM Spec.	Form	UNS	Condition[a]	Tensile Strength		Yield Strength		% El	Hardness HRB max
				ksi	MPa	ksi	MPa		
B 409	Pl, Sh, St	N08811	Annealed	65	450	25	170	30	---
B 424	Pl, Sh, St	N08825	Annealed	85	586	35	241	30	---
		N08221	Annealed	79	544	34	235	30	---
B 434	Pl, Sh, St	N10003	---	100	690	40	280	40	---
B 435	Pl, Sh, St	N06002	---	95	655	35	240	35	---
		N06230	---	110	760	45	310	40	---
B 443	Sh, St	N06625	gr 1 Annealed - CR	120	827	60	414	30	---
	Pl, Sh	N06625	gr 1 Annealed - HR	110	758	55	379	30	---
	Pl	N06625	gr 1 Annealed - CR	110	758	55	379	30	---
	Pl, Sh, St	N06625	gr 2 Annealed	100	690	40	276	30	---
B 463	Pl, Sh, St	N08020	Stabilize-annealed	80	551	35	241	30	95 (217 HB)
		N08024	Annealed	80	551	35	241	30	95 (217 HB)
		N08026	Solution-annealed	80	551	35	241	30	95 (217 HB)
B 575	Pl, Sh, St	N10276	---	100	690	41	283	40	100[g]
		N06022	---	100	690	45	310	45	100[g]
		N06455	---	100	690	40	276	40	100[g]
		N06059	---	100	690	45	310	45	100[g]
B 582	Pl[c]	N06007	Annealed	90	621	35	241	35	100[g]
	Pl[d]	N06007	Annealed	85	586	30	207	30	100[g]
	Sh, St[e]	N06007	Annealed	90	621	35	241	40	100[g]
	Pl,[f]	N06975	Annealed	85	586	32	221	40	100[g]
	Sh, St[e]	N06975	Annealed	85	586	32	221	40	100[g]
	Pl[c]	N06985	Annealed	90	621	35	241	45	100[g]
	Pl[d]	N06985	Annealed	85	586	30	207	30	---

MECHANICAL PROPERTIES OF NICKEL & NICKEL ALLOYS - PLATE SHEET & STRIP (Continued)

ASTM				Tensile Strength		Yield Strength			Hardness
Spec.	Form	UNS	Condition[a]	ksi	MPa	ksi	MPa	% El	HRB max
B 582	Sh[e]	N06985	Annealed	90	621	35	241	45	100[g]
(con't)	Pl, Sh, St	N06030	Annealed	85	586	35	241	30	---
B 620	Pl, Sh, St	N08320	---	75	517	28	193	35	95[g]

a. See specific ASTM Standard for specific details of heat treat condition. CR - cold-rolled; HR - hot-rolled.
b. Sheet and strip 0.010 to 0.49 in (0.25 to 1.2 mm), inclusive, in thickness requires an elongation of 30% minimum; and thicknesses from 0.050 to 0.109 in (1.3 to 2.7 mm), inclusive, requires an elongation of 35% minimum.
c. Thickness from 3⁄16 to ¾ in (4.76 to 19.05 mm) inclusive.
d. Thickness over ¾ to 2 ½ in (19.05 to 63.5 mm) inclusive.
e. Thickness over 0.020 in (0.51 mm).
f. Thickness 3/16 to 2 ½ in (4.76 to 63.5 mm) inclusive.
g. For information purpose only, not to be used as a basis for rejection or acceptance.
Pl - plate, Sh - sheet, St - strip. Single values are minimum specified.

MECHANICAL PROPERTIES OF NICKEL ALLOYS - FLANGES, FORGED FITTINGS & VALVE PARTS

ASTM				Tensile Strength		Yield Strength			
Spec.	Form	UNS	Condition[a]	ksi	MPa	ksi	MPa	% El	% RA
B 462	Fl, FFt, V	N08020	Stabilized-annealed	80	551	35	214	30.0	50.0
		N08024	Annealed	80	551	35	214	30.0	50.0
		N08026	Solution-annealed	80	551	35	214	30.0	50.0
		N08367	Solution-annealed	95	655	45	310	30.0	50.0

a. See specific ASTM Standard for specific details of heat treat condition.
Fl - flange, FFt - forged fitting, V - valve.
Single values are minimum specified.

MECHANICAL PROPERTIES OF NICKEL ALLOYS - FORGINGS

ASTM Spec.	UNS	Section Thickness max	Tensile Strength		Yield Strength		% El
			ksi	MPa	ksi	MPa	
B 564	N04400	---	70	483	25	172	35
	N06022	---	100	690	45	310	45
	N06059	---	100	690	45	310	45
	N06110	≤ 4 in (102 mm)	95	655	45	310	60
	N06110	> 4 - 10 in (102 - 254 mm)[a]	90	621	40	276	50
	N06230	---	110	758	45	310	40
	N06600	---	80	552	35	241	30
	N06617	---	95	655	35	241	35
	N06625	≤ 4 in (102 mm)	120	827	60	414	30
	N06625	> 4 - 10 in (102 - 254 mm)[a]	110	758	50	345	25
	N08031	---	94	650	40	276	40
	N08367	---	95	655	45	310	30
	N08800	---	75	517	30	207	30
	N08810	---	65	448	25	172	30
	N08825	---	85	586	35	241	30
	N10276	---	100	690	41	283	40

a. Over 4 to 10 in (102 to 254 mm) diameter for parts machined from forged bar.
Single values are minimum specified.

MECHANICAL PROPERTIES OF NICKEL & NICKEL ALLOYS - RODS & BARS								
ASTM Spec.	UNS	Condition	Thickness or Diameter	Tensile Strength		Yield Strength		% El
				ksi	MPa	ksi	MPa	
B 160	N02200	Annealed	All sizes	55	380	15	105	40[a]
	N02201	Annealed	All sizes	55	345	10	70	40[a]
B 164	N04400	Annealed	All sizes	70	480	25	170	35
	N04405	Annealed	All sizes	70	480	25	170	35
B 166	N06600	Annealed	All sizes	80	550	35	240	30[a]
	N06601	Annealed	All sizes	80	550	30	205	30
	N06617	Annealed	All sizes	95	655	35	240	30
	N06690	Annealed	All sizes	85	586	35	240	30[a]
B 335	N10001	n/s	5⁄16-1 ½ in (7.94-38.1 mm) incl	115	795	46	315	35
	N10001	n/s	> 1 ½-3 ½ in (38.1-88.9 mm) incl	100	690	46	315	30
	N10665	n/s	5⁄16 -3 ½ in (7.94-88.9 mm) incl	110	758	51	352	40
B 408	N08800	Annealed	All sizes	75	515	30	205	30
	N08810	Annealed	All sizes	65	450	25	170	30
	N08811	Annealed	All sizes	65	450	25	170	30
B 425	N08825	Annealed	All sizes	85	586	35	241	30[a]
	N08221	Annealed	All sizes	79	544	34	235	30
B 446	N06625	Annealed	Grade 1, ≤ 4 in (102 mm)	120	---	60	---	30
	N06625	Annealed	Grade 1, > 4 - 10 in (102 - 254 mm)	110	---	50	---	25
	N06625	SA	Grade 2, All sizes	100	---	40	---	30
B 511	N08330	Annealed	All sizes	70	483	30	207	30
	N08332	Annealed	All sizes	67	462	27	186	30
B 573	N10003	n/s	All sizes	100	689	40	276	35
B 574	N10276	n/s	All sizes	100	690	41	283	40
	N06022	n/s	All sizes	100	690	45	310	45
	N06455	n/s	All sizes	100	690	40	276	40

MECHANICAL PROPERTIES OF NICKEL & NICKEL ALLOYS - RODS & BARS (Continued)

ASTM Spec.	UNS	Condition	Thickness or Diameter	Tensile Strength		Yield Strength		% El
				ksi	MPa	ksi	MPa	
B574	N06059	n/s	All sizes	100	690	45	310	45
B581	N06007	n/s	5/16 -3/4 in (7.94-19.05 mm) incl	90	621	35	241	35
	N06007	n/s	> 3/4-3 1/2 in (19.05-88.9 mm) incl	85	586	30	207	30
	N06975	n/s	5/16 -3 1/2 in (7.94-88.9 mm) incl	85	586	32	221	40
	N06985	n/s	5/16 -3/4 (7.9-19.05 mm) incl	90	621	35	241	45
	N06985	n/s	> 3/4-3 1/2 in (19.05-88.9 mm) incl	85	586	30	207	35
	N06030	n/s	All sizes	85	586	35	241	30
B 621	N08320	n/s	All sizes	75	517	28	193	35
B 672	N08700	Annealed	All sizes	80	551	35	241	30[b]

a. Not applicable to diameters or cross sections under 3/32 in (2.4 mm). b. Also includes reduction of area 50.0% minimum.
n/s - not specified. SA - Solution Annealed, see ASTM B 446 for heat treatment details.

MECHANICAL PROPERTIES OF NICKEL ALLOYS - WIRES[a]

ASTM Spec.	UNS	Condition	Size	Tensile Strength		Yield Strength		% El
				ksi	MPa	ksi	MPa	
B 164	N04400	Annealed	All sizes	70	483	85	586	---
	N04405	Annealed	All sizes	70	483	85	586	---
	N04400	Regular temper	< 1/2 in (12.7 mm)	110	758	140	965	---
	N04400	Regular temper	≥ 1/2 in (12.7 mm)	90	621	130	896	---
B 166	N06600	Annealed	< 0.032 in (0.81 mm)	80	552	115	793	---
	N06600	Annealed	≥ 0.032 (0.81 mm)	80	552	105	724	---
	N06690	Annealed	< 0.032 in (0.81 mm)	80	552	115	793	---
	N06690	Annealed	≥ 0.032 (0.81 mm)	80	552	105	724	---
B 672	N08700	Annealed	All sizes	80	551	35	241	30.0[b]

a. Also includes a wrapping test. b. Also includes reduction of area 50.0% minimum.

MECHANICAL PROPERTIES OF NICKEL & NICKEL ALLOYS - CASTINGS

ASTM Spec.	Grade	UNS	Condition[a]	Tensile Strength		Yield Strength		% El	Hardness HB
				ksi	MPa	ksi	MPa		
B 494	CZ-100	N02100	As cast	50	345	18	125	10.0	---
	M-35-1	N24135	As cast	65	450	25	170	25.0	---
	M-35-2	---	As cast	65	450	30	205	25.0	---
	M-30H	N24030	As cast	100	690	60	415	10.0	243-294[d]
	M-25S[a]	N24025	As cast[b]	---	---	---	---	---	300 min[e]
	M-30C	N24130	As cast	65	450	35.5	225	25.0	125-150[d]
	N-12MV[a]	N30012	Solution-annealed	76	525	40	275	6.0	---
	N-7M[a]	N30007	Solution-annealed	76	525	40	275	20.0	---
	CY-40[a]	N06040	As cast[c]	70	485	28	195	30.0	---
	CW-12MW[a]	N30002	Solution-annealed	72	495	40	275	4.0	---
	CW-6M[a]	N30107	Solution-annealed	72	495	40	275	25.0	---
	CW-2M[a]	N26455	Solution-annealed	72	495	40	275	20.0	---
	CW-6MC	N26625	Solution-annealed	70	485	40	275	25.0	---
	CY5SnBiM	N26055	As cast	---	---	---	---	---	---
	CX2MW[a]	N26022	Solution-annealed	80	550	45	280	30.0	---

a. See ASTM A 494 for details regarding class designation and heat treat requirements.
b. M-25S class 1 - as cast; class 2 and class 3 see ASTM A494 for more details.
c. CY-40 class 1 - as cast; CY-40 class 2 - solution-annealed, see ASTM A494 for more details.
d. For information only, see ASTM A 494 for more details.
e. Hardness of 300 HB minimum for the age hardened condition.

ASME P-No. - BASE METAL NICKEL & NICKEL ALLOYS						
ASME Spec.	**Condition[a]**	**Size or Thickness (in)**	**UTS ksi[c]**	**UNS**	**Nominal Composition**	**Product Form**
P No. 41						
SB-160	A	---	55	N02200	99.0Ni	Rod, Bar
	HR	---	60	N02200	99.0Ni	Rod, Bar
	CD	1 and under	80	N02200	99.0Ni	Rod, Bar
		> 1-4	75	N02200	99.0Ni	Rod, Bar
		All	65	N02200	99.0Ni	Square, etc.
	A	All	50	N02201	Low C-99.0Ni	Rod, Bar
	HR	All	50	N02201	Low C-99.0Ni	Rod, Bar
SB-161	A	---	55	N02200	99.0Ni	Pipe, Tube
	SR	---	65	N02200	99.0Ni	Pipe, Tube
	A	---	50	N02201	Low C-99.0Ni	Smls. Pipe & Tube
	SR	---	60	N02201	Low C-99.0Ni	Smls. Pipe & Tube
SB-162	A	---	55	N02200	99.0Ni	Plate, Sheet, Strip
	HR	---	55	N02200	99.0Ni	Plate, Sheet, Strip
	A	---	50	N02201	Low C-99.0Ni	Plate, Sheet, Strip
	HR	---	50	N02201	Low C-99.0Ni	Plate, Sheet, Strip
SB-163	A	---	55	N02200	99.0Ni	Smls. Tube
	SR	---	65	N02200	99.0Ni	Smls. Tube
	A	---	50	N02201	Low C-99.0Ni	Smls. Tube
	SR	---	60	N02201	Low C-99.0Ni	Smls. Tube
P No. 42						
SB-127	A	---	70	N04400	67Ni-30Cu	Plate, Sheet, Strip
	HR	---	75	N04400	67Ni-30Cu	Plate, Sheet, Strip
SB-163	A	---	70	N04400	67Ni-30Cu	Smls. Tube
	SR	---	85	N04400	67Ni-30Cu	Smls. Tube
SB-164	A	---	70	N04400	67Ni-30Cu	Rod, Bar

ASME P-No. - BASE METAL NICKEL & NICKEL ALLOYS (Continued)

ASME Spec.	Condition[a]	Size or Thickness (in)	UTS ksi[c]	UNS	Nominal Composition	Product Form
P No. 42 (Continued)						
SB-164	HF	b	80	N04400	67Ni-30Cu	Rod, Bar
		Hex. over 2 ⅛	75	N04400	67Ni-30Cu	Rod, Bar
	SR	4-12 dia., incl.	80	N04400	67Ni-30Cu	Rod, Bar
		> 12 dia.	75	N04400	67Ni-30Cu	Rod, Bar
	CD	< ½	110	N04400	67Ni-30Cu	Round
		< ½	85	N04400	67Ni-30Cu	Square, etc.
	SR	< ½	84	N04400	67Ni-30Cu	Round
		½-3 ½, incl.	87	N04400	67Ni-30Cu	Round
		> 3 ½-4, incl.	84	N04400	67Ni-30Cu	Round
		All	84	N04400	67Ni-30Cu	Square, etc.
SB-165	A	---	70	N04400	67Ni-30Cu	Smls. Pipe & Tube
	SR	---	85	N04400	67Ni-30Cu	Smls. Pipe & Tube
SB-564	A	---	70	N04400	67Ni-30Cu	Forging
P No. 43						
SB-163	A	---	80	N06600	72Ni-15Cr-8Fe	Smls. Tube
	A	---	85	N06690	58Ni-29Cr-9Fe	Smls. Tube
SB-166	A	---	80	N06600	72Ni-15Cr-8Fe	Rod, Bar
	HF	¼-3, incl.	90	N06600	72Ni-15Cr-8Fe	Rod, Bar
		> 3 and hex.	85	N06600	72Ni-15Cr-8Fe	Rod, Bar
	A	---	85	N06690	58Ni-29Cr-9Fe	Rod, Bar
	HF	¼-3, incl.	90	N06690	58Ni-29Cr-9Fe	Rod, Bar
		> 3 and hex.	85	N06690	58Ni-29Cr-9Fe	Rod, Bar
SB-167	CD/A	---	80	N06600	72Ni-15Cr-8Fe	Smls. Pipe & Tube
	HF	5 and under	80	N06600	72Ni-15Cr-8Fe	Smls. Pipe & Tube
		> 5 dia.	75	N06600	72Ni-15Cr-8Fe	Smls. Pipe & Tube

ASME P-No. - BASE METAL NICKEL & NICKEL ALLOYS (Continued)						
ASME Spec.	**Condition[a]**	**Size or Thickness (in)**	**UTS ksi[c]**	**UNS**	**Nominal Composition**	**Product Form**
P No. 43 (Continued)						
SB-167	HF/A	5 dia. and under	80	N06600	72Ni-15Cr-8Fe	Smls. Pipe & Tube
		> 5 dia.	75	N06600	72Ni-15Cr-8Fe	Smls. Pipe & Tube
	CD/A	---	85	N06690	58Ni-29Cr-9Fe	Smls. Pipe & Tube
	HF	---	85	N06690	58Ni-29Cr-9Fe	Smls. Pipe & Tube
	HF/A	---	85	N06690	58Ni-29Cr-9Fe	Smls. Pipe & Tube
SB-168	A	---	80	N06600	72Ni-15Cr-8Fe	Plate, Sheet, Strip
	HR	---	85	N06600	72Ni-15Cr-8Fe	Plate, Sheet, Strip
	A	---	85	N06690	58Ni-29Cr-9Fe	Plate, Sheet, Strip
	HR	---	85	N06690	58Ni-29Cr-9Fe	Plate, Sheet, Strip
SB-435	A	---	100	N06002	47Ni-22Cr-9Mo-18Fe	Sheet, Strip
		---	95	N06002	47Ni-22Cr-9Mo-18Fe	Plate
SB-443	A	---	120	N06625	60Ni-22Cr-9Mo-3.5Cb	Plate, Sheet, Strip
SB-444	A	---	120	N06625	60Ni-22Cr-9Mo-3.5Cb	Smls. Pipe & Tube
SB-446	A	---	120	N06625	60Ni-22Cr-9Mo-3.5Cb	Rod, Bar
SB-516	A	---	80	N06600	72Ni-15Cr-8Fe	Welded Tube
SB-517	A	---	80	N06600	72Ni-15Cr-8Fe	Welded Pipe
SB-564	A	---	80	N06600	72Ni-15Cr-8Fe	Forging
	A	Up to 4, incl.	120	N06625	60Ni-22Cr-9Mo-3.5Cb	Forging
	A	> 4-10, incl.	110	N06625	60Ni-22Cr-9Mo-3.5Cb	Forging
SB-572	A	---	95	N06002	47Ni-22Cr-9Mo-18Fe	Rod
SB-619	A	---	100	N06002	47Ni-22Cr-9Mo18Fe	Welded Pipe
SB-622	A	---	100	N06002	47Ni-22Cr-9Mo-18Fe	Smls. Pipe & Tube
SB-626	A	---	100	N06002	47Ni-22Cr-9Mo-18Fe	Welded Tube
SB-704	CW/A	---	120	N06625	60Ni-22Cr-9Mo-3.5Cb	Welded Tube
SB-705	CW/A	---	120	N06625	60Ni-22Cr-9Mo-3.5Cb	Welded Pipe

ASME P-No. - BASE METAL NICKEL & NICKEL ALLOYS (Continued)						
ASME Spec.	Condition[a]	Size or Thickness (in)	UTS ksi[c]	UNS	Nominal Composition	Product Form
P No. 44						
SB-333	A	< 3/16	115	N10001	28Mo-5Fe	Sheet, Strip
		3/1-2 ½, incl.	100	N10001	28Mo-5Fe	Plate
		< 3/16	110	N10665	28Mo	Sheet, Strip
		3/16 -2 ½, incl.	110	N10665	28Mo	Plate
SB-335	A	5/16-1 ½, incl.	115	N10001	28Mo-5Fe	Rod
		1 ½-3 ½, incl.	100	N10001	28Mo-5Fe	Rod
	A	5/16 -3 ½, incl.	110	N10665	28Mo	Rod
SB-434	A	---	100	N10003	70Ni-16Mo-7Cr-5Fe	Plate, Sheet, Strip
SB-564	---	---	100	N10276	Low C-16Mo-16Cr-5 ½Fe	Forging
SB-573	A	---	100	N10003	70Ni-16Mo-7Cr-5Fe	Rod
SB-574	A	---	100	N06455	Low C-15.5Mo-16Cr	Rod
	A	---	100	N10276	Low C-16Mo-16Cr-5.5Fe	Rod
	SA	---	100	N06022	Low C-13.5Mo-21.5Cr-4Fe-3W	Rod
SB-575	A	---	100	N06455	Low C-15.5Mo-16Cr	Plate, Sheet, Strip
	A	---	100	N10276	Low C-16Mo-16Cr-5.5Fe	Plate, Sheet, Strip
	SA	---	100	N06022	Low C-13.5Mo-21.5Cr-4Fe-3W	Plate, Sheet, Strip
SB-619	A	---	100	N06455	Low C-15.5Mo-16Cr	Welded Pipe
	A	---	100	N10001	28Mo-5Fe	Welded Pipe
	A	---	100	N10276	Low C-16Mo-16Cr-5.5 Fe	Welded Pipe
	A	---	110	N10665	28Mo	Welded Pipe
	SA	---	100	N06022	Low C-13.5Mo-21.5Cr-4Fe-3W	Welded Pipe
SB-622	A	---	100	N06455	Low C-15.5Mo-16Cr	Smls. Pipe & Tube
	A	---	100	N10001	28Mo-5Fe	Smls. Pipe & Tube
	A	---	100	N10276	Low C-16Mo-16Cr-5.5Fe	Smls. Pipe & Tube

ASME P-No. - BASE METAL NICKEL & NICKEL ALLOYS (Continued)						
ASME Spec.	**Condition[a]**	**Size or Thickness (in)**	**UTS ksi[c]**	**UNS**	**Nominal Composition**	**Product Form**
P No. 44 (Continued)						
SB-622	A	---	110	N10665	28Mo	Smls. Pipe & Tube
(continued)	SA	---	100	N06022	Low C-13.5Mo-21.5Cr-4Fe-3W	Smls. Pipe & Tube
SB-626	SA	---	100	N06455	Low C-15.5Mo-16Cr	Welded Tube
	SA	---	100	N10001	28Mo-5Fe	Welded Tube
	SA	---	100	N10276	Low C-16Mo-16Cr-5.5Fe	Welded Tube
	SA	---	110	N10665	28Mo	Welded Tube
	SA	---	100	N06022	Low C-13.5Mo-21.5Cr-4Fe-3W	Welded Tube
SB-710	A	---	70	N08330	35Ni-19Cr-1.25Si	Welded Pipe
P No 45						
SB-163	A	---	75	N08800	33Ni-21Cr	Smls. Tube
	A	---	65	N08810	33Ni-21Cr	Smls. Tube
	A	---	85	N08825	42Ni-21.5Cr-3Mo-2.3Cu	Smls. Tube
SA-351						
Gr CN7M	---	---	62	J95150	28Ni-19Cr-3Cu-2Mo	Casting
Gr CT15C	---	---	63	---	20Cr-32 ½Ni-Cb	Casting
SB-407	A	---	75	N08800	33Ni-21Cr	Smls. Pipe & Tube
	A	---	65	N08810	33Ni-21Cr	Smls. Pipe & Tube
	A	---	65	N08811	33Ni-21Cr-1Al+Ti	Smls. Pipe & Tube
SB-408	A	---	75	N08800	33Ni-21Cr	Rod, Bar
	A	---	65	N08810	33Ni-21Cr	Rod, Bar
	A	---	65	N08811	33Ni-21Cr-Al+Ti	Rod, Bar
SB-409	A	---	75	N08800	33Ni-21Cr	Plate, Sheet, Strip
	A	---	65	N08810	33Ni-21Cr	Plate, Sheet, Strip
	A	---	65	N08811	33Ni-21Cr-Al+Ti	Plate, Sheet, Strip
SB 423	CD/A	---	85	N08825	42Ni-21.5Cr-3Mo-2.3Cu	Smls. Pipe & tube

ASME P-No. - BASE METAL NICKEL & NICKEL ALLOYS (Continued)

ASME Spec.	Condition[a]	Size or Thickness (in)	UTS ksi[c]	UNS	Nominal Composition	Product Form
P No. 45 (Continued)						
SB-423	HF/A	---	75	N08825	42Ni-21.5Cr-3Mo-2.3Cu	Smls. Pipe & tube
SB-424	A	---	85	N08825	42Ni-21.5Cr-3Mo-2.3Cu	Plate, Sheet, Strip
SB-425	A	---	85	N08825	42Ni-21.5Cr-3Mo-2.3Cu	Rod, Bar
SB-435	---	---	100	R30556	22Cr-20.75Ni-18.5Co-3.25 Mo-2.75W-FeRem	Plate, Sheet, Strip
SB-462	A	---	80	N08020	35Ni-35Fe-20Cr-Cb	Forging
SB-463	A	---	80	N08020	35Ni-35Fe-20Cr-Cb	Plate, Sheet, Strip
	A	---	80	N08026	35Ni-24Cr-5Mo-2Cu	Plate, Sheet, Strip
SB-464	A	---	80	N08020	35Ni-35Fe-20Cr-Cb	Smls. & Welded Pipe
	A	---	80	N08026	35Ni-24Cr-5Mo-2Cu	Welded Pipe
SB-468	A	---	80	N08020	35Ni-35Fe-20Cr-Cb	Smls. & Welded Tube
	A	---	80	N08026	35Ni-24Cr-5Mo-2Cu	Welded Tube
SB-473	A	---	80	N08020	35Ni-35Fe-20Cr-Cb	Bar
SB-514	A	---	75	N08800	33Ni-21Cr	Welded Pipe
	A	---	65	N08810	33Ni-21Cr	Welded Pipe
SB-515	A	---	75	N08800	33Ni-21Cr	Welded Pipe
	A	---	65	N08810	33Ni-21Cr	Welded Pipe
SB-564	A	---	75	N08800	33Ni-21Cr	Forging
	A	---	65	N08810	33Ni-21Cr	Forging
SB-572	---	---	100	R30556	22Cr-20.75Ni-18.5Co-3.25 Mo-2.75W-FeRem	Rod
SB-581	A	5⁄16 -¾, incl.	90	N06007	22.3Cr-19.5Fe-6.5Mo-2Cu-2Cb	Rod
		> ¾-3 ½	85	N06007	22.3Cr-19.5Fe-6.5Mo-2Cu-2Cb	Rod
	A	5⁄16 -3 ½	85	N06975	49.5Ni-24.5Cr-6Mo-1Cu-1Ti	Rod
		5⁄16,-¾, incl.	90	N06985	22Cr-20Fe-7Mo-2Cu	Rod

ASME P-No. - BASE METAL NICKEL & NICKEL ALLOYS (Continued)

ASME Spec.	Condition[a]	Size or Thickness (in)	UTS ksi[c]	UNS	Nominal Composition	Product Form
P No. 45 (Continued)						
SB-581	A	> ¾-3 ½, incl.	85	N06985	22Cr-20Fe-7Mo-2Cu	Rod
(con't)	SA	---	85	N06030	30Cr-15Fe-5Mo-2Cu	Rod
SB-582	A	> 0.020	90	N06007	22.3Cr-19.5Fe-6.5Mo-2Cu-2Cb	Sheet & Strip
		3/16-¾, incl.	90	N06007	22.3Cr-19.5Fe-6.5Mo-2Cu-2Cb	Plate
		> ¾-2 ½, incl.	85	N06007	22.3Cr-19.5Fe-6.5Mo-2Cu-2Cb	Plate
	A	3/16,-2 ½, incl.	85	N06975	49.5Ni-24.5Cr-6Mo-1Cu-1Ti	Plate
		> 0.020	85	N06975	49.5Ni-24.5Cr-6Mo-1Cu-1Ti	Sheet, Strip
	A	5/16 -¾, incl.	90	N06985	22Cr-20Fe-7Mo-2Cu	Plate
		> ¾-2 ½, incl.	85	N06985	22Cr-20Fe-7Mo-2Cu	Plate
		> 0.020	90	N06985	22Cr-20Fe-7Mo-2Cu	Sheet, Strip
	SA	---	85	N06030	30Cr-15Fe-5Mo-2Cu	Plate, Sheet, & Strip
SB-599	A	---	80	N08700	25Ni-21Cr-4.5Mo-Cb	Plate, Sheet, & Strip
SB-619	A	---	90	N06007	22.3Cr-19.5Fe-6.5Mo-2Cu-2Cb	Welded Pipe
		---	85	N06975	49.5Ni-24.5Cr-6Mo-1Cu-1Ti	Welded Pipe
		---	90	N06985	22Cr-20Fe-7Mo-2Cu	Welded Pipe
		---	75	N08320	26Ni-22Cr-5Mo-Ti	Welded Pipe
	SA	---	85	N06030	30Cr-15Fe-5Mo-2Cu	Welded Pipe
	---	---	100	R30556	22Cr-20.75Ni-18.5Co-3.25 Mo-2.75W-FeRem	Welded Pipe
SB-620	A	---	75	N08320	26Ni-22Cr-5Mo-Ti	Plate, Sheet, & Strip
SB-621	A	---	75	N08320	26Ni-22Cr-5Mo-Ti	Rod
SB-622	A	---	90	N06007	22.3Cr-19.5Fe-6.5Mo-2Cu-2Cb	Smls. Pipe & Tube
	---	---	85	N06975	49.5Ni-24.5Cr-6Mo-1Cu-1Ti	Smls. Pipe & Tube
	---	---	90	N06985	22Cr-20Fe-7Mo	Smls. Pipe & Tube
	---	---	75	N08320	26Ni-22Cr-5Mo-Ti	Smls. Pipe & Tube

ASME P-No. - BASE METAL NICKEL & NICKEL ALLOYS (Continued)						
ASME Spec.	**Condition[a]**	**Size or Thickness (in)**	**UTS ksi[c]**	**UNS**	**Nominal Composition**	**Product Form**
P No. 45 (Continued)						
SB-622	SA	---	85	N06030	30Cr-15Fe-5Mo-2Cu	Smls. Pipe & Tube
(continued)	---	---	100	R30556	22Cr-20.75Ni-18.5Co-3.25 Mo-2.75W-FeRem	Smls. Pipe & Tube
SB-625	A	---	71	N08904	25.5Ni-21Cr-4.5Mo-1.5Cu	Plate, Sheet, & Strip
SB-626	SA	---	90	N06007	22.3Cr-19.5Fe-6.5Mo-2Cu-2Cb	Welded Tube
	SA	---	85	N06975	49.5Ni-24.5Cr-6Mo-1Cu-1Ti	Welded Tube
	SA	---	90	N06985	22Cr-20Fe-7Mo-2Cu	Welded Tube
	SA	---	75	N08320	26Ni-22Cr-5Mo-Ti	Welded Tube
	SA	---	85	N06030	30Cr-15Fe-5Mo-2Cu	Welded Tube
	---	---	100	R30556	22Cr-20.75Ni-18.5Co-3.25 Mo-2.75W-FeRem	Welded Tube
SB-668	A	---	73	N08028	Low C-31Ni-27Cr-35Mo-1Cu	Smls. Tube
SB-672	A	All	80	N08700	25Ni-21Cr-4.5Mo-Cb	Bar, Wire
SB-673	ST	All	71	N08904	25.5Ni-21Cr-4.5Mo-1.5Cu	Welded Pipe
SB-674	ST	All	71	N08904	25.5Ni-21Cr-4.5Mo-1.5Cu	Welded Tube
SB-675	ST	All	75	N08366	24.5Ni-21Cr-6.5Mo	Welded Pipe
	---	---	104	N08367	Low C-24 ½Ni-21Cr-6 ½Mo-N	Welded Pipe
SB-676	ST	All	75	N08366	24.5Ni-21Cr-6.5Mo	Welded Tube
	---	---	104	N08367	Low C-24 ½Ni-21Cr-6 ½Mo-N	Welded Tube
SB-677	ST	All	71	N08904	25.5Ni-21Cr-4.5Mo-1.5Cu	Smls. Pipe & Tube
SB-688	A	All	75	N08366	25Ni-21Cr-6Mo	Plate, Sheet, & Strip
	---	---	104	N08367	Low C-24 ½Ni-21Cr-6 ½Mo-N	Plate, Sheet, & Strip
SB-690	A	All	75	N08366	25Ni-21Cr-6Mo	Smls. Pipe & Tube
	---	---	104	N08367	Low C-24 ½Ni-21Cr-6 ½Mo-N	Smls. Pipe & Tube
SB-691	A	All	75	N08366	25Ni-21Cr-6Mo	Rod, Bar, & Wire

ASME P-No. - BASE METAL NICKEL & NICKEL ALLOYS (Continued)						
ASME Spec.	**Condition[a]**	**Size or Thickness (in)**	**UTS ksi[c]**	**UNS**	**Nominal Composition**	**Product Form**
P No. 45 (Continued)						
SB-691	---	---	104	N08367	Low C-24 ½Ni-21Cr-6 ½Mo-N	Rod & Bar
SB-704	CW/A	---	85	N08825	42Ni-21 ½Cr-3Mo-2.3Cu	Welded Tube
SB-705	CW/A	---	85	N08825	42Ni-21 ½Cr-3Mo-23Cu	Welded Pipe
P No. 46						
SB-511	A	---	70	N08330	35Ni-19Cr-1.25Si	Bar
SB-535	A	---	70	N08330	35Ni-19Cr-1.25Si	Smls. & Welded Pipe
SB-536	A	---	70	N08330	35Ni-19Cr-1.25Si	Plate, Sheet, Strip
P No. 47						
SB-435	---	---	110	N06230	22Cr-14W-2Mo-Al-La	Plate, Sheet, Strip
SB-572	---	---	110	N06230	22Cr-14W-2Mo-Al-La	Rod
SB-619	---	---	110	N06230	22Cr-14W-2Mo-Al-La	Welded Pipe
SB-622	---	---	110	N06230	22Cr-14W-2Mo-Al-La	Smls Pipe & Tube
SB-626	---	---	110	N06230	22Cr-14W-2Mo-Al-La	Welded Tube
SB-710	A	---	70	N08330	35Ni-19Cr-1.25Si	Welded Pipe

a. A- annealed; HR - hot rolled: SR - Stress Relieved; HF - hot finished; CD - Cold Drawn; SA - solution annealed; CW - cold worked; ST - solution treated; CW/A - cold worked/annealed; CD/A - cold drawn/annealed; HF/A - hot finished/annealed.

b. Except hex. over 2 ⅛.

c. Minimum value.

ASME F No. - WELDING FILLER METAL NICKEL & NICKEL ALLOYS

F No.	ASME Spec. No.	AWS Classification	F No.	ASME Spec. No.	AWS Classification
41	SFA-5.11	E Ni-1	44	SFA-5.11	E NiMo-3
41	SFA-5.14	ER Ni-1	44	SFA-5.11	E NiMo-7
41	SFA-5.30	IN 61	44	SFA-5.11	E NiCrMo-4
42	SFA-5.11	E NiCu-7	44	SFA-5.11	E NiCrMo-5
42	SFA-5.14	ER NiCu-7	44	SFA-5.11	E NiCrMo-7
42	SFA-5.30	IN 60	44	SFA-5.11	E NiCrMo-10 (C-22 alloy)
43	SFA-5.11	E NiCrFe-1	44	SFA-5.14	ER NiMo-1
43	SFA-5.11	E NiCrFe-2	44	SFA-5.14	ER NiMo-2
43	SFA-5.11	E NiCrFe-3	44	SFA-5.14	ER NiMo-7 (B-2 alloy)
43	SFA-5.11	E NiCrFe-4	44	SFA-5.14	ER NiCrMo-4
43	SFA-5.11	E NiCrMo-2	44	SFA-5.14	ER NiCrMo-5
43	SFA-5.11	E NiCrMo-3	44	SFA-5.14	ER NiCrMo-7 (C-4 alloy)
43	SFA-5.11	E NiCrMo-b	44	SFA-5.14	ER NiCrMo-10 (C-22 alloy)
43	SFA-5.14	ER NiCr-3	45	SFA-5.11	E NiCrMo-1
43	SFA-5.14	ER NiCrFe-5	45	SFA-5.11	E NiCrMo-9
43	SFA-5.14	ER NiCrFe-6	45	SFA-5.11	E NiCrMo-11 (G-30 alloy)
43	SFA-5.14	ER NiCrMo-2	45	SFA-5.14	ER NiCrMo-1
43	SFA-5.14	ER NiCrMo-3	45	SFA-5.14	ER NiFeCr-1
43	SFA-5.30	IN 82	45	SFA-5.14	ER NiCrMo-8
43	SFA-5.30	IN 62	45	SFA-5.14	ER NiCrMo-9
43	SFA-5.30	IN 6A	45	SFA-5.14	ER NiCrMo-11 (G-30 alloy)
44	SFA-5.11	E NiMo-1			

COMMON NAMES & CROSS REFERENCED SPECIFICATIONS - NICKEL & NICKEL ALLOYS - WELD FILLER METALS

Common Name	UNS	ASME	AWS	MIL-SPEC
75Ni-25Cr[a]	W86003	---	---	---
904L Electrode	W88904	SFA-5.4 (E385)	A5.4 (E385)	---
Avesta PE 12	W86040	---	A5.11 (ENiCrMo-12)	---
Carpenter 20Cb-3	W88022	SFA-5.4 (E320)	A5.4 (E320)	---
		SFA-5.9 (EC320)	A5.9 (EC320)	---
Carpenter 20Cb3L.R.	W88022	SFA-5.4 (E320LR)	A5.4 (E320LR)	---
		SFA-5.9 (EC320LR)	A5.9 (EC320LR)	---
Cast Iron Electrode ENi-CI	W82001	---	A5.15 (ENi-CI)	---
Cast Iron Electrode ENi-CI-A	W82003	SFA-5.15 (ENi-CI-A)	A5.15 (ENi-CI-A)	---
Cast Iron Electrode ENiCu-A	W84001	---	A5.15 (ENiCu-A)	---
Cast Iron Electrode ENiCu-B	W84002	---	A5.15 (ENiCu-B)	---
Cast Iron Electrode ENiFe-CI	W82002	---	A5.15 (ENiFe-CI)	---
Cast Iron Electrode ENiFe-CI-A	W82004	SFA-5.15 (ENiFe-CI-A)	A5.15 (ENiFe-CI-A)	---
Cast Iron Electrode ENiFeMn-CI-A	W82006	SFA-5.15 (ENiFeMn-CI-A)	A5.15 (ENiFeMn-CI-A)	---
Cast Iron Fux Cored Wire	W82032	SFA-5.15 (ENiFeT3-CI)	A5.15 (ENiFeT3-CI)	---
E383 Electrode	W88028	SFA-5.4 (E383)	A5.4 (E383)	---
E330 Electrode	W88331	SFA-5.4 (E330)	A5.4 (E330)	E-22200/3 (MIL-330)
E330H high carbon	W88335	SFA-5.4 (E330H)	A5.4 (E330H)	---
Hard Surfacing Electrode	W83002	SFA-5.13 (ENiCrFeCo)	A5.11 (ENiCrFeCo)	---
Hard Surfacing Electrode	W89604	---	A5.13 (ENiCr-A)	---
Hard Surfacing Electrode	W89605	---	A5.13 (ENiCr-B)	---
Hard Surfacing Electrode	W89606	---	A5.13 (ENiCr-C)	---
Hastelloy B	W80001	SFA-5.11 (ENiMo-1)	A5.11 (ENiMo-1)	E-22200/3 (MIL-3N1B)
Hastelloy B-2	W80665	SFA-5.11 (ENiMo-7)	A5.11 (ENiMo-7)	---
Hastelloy C	W80002	SFA-5.11 (ENiCrMo-5)	A5.11 (ENiCrMo-5)	E-22200/3 (MIL-3N1C)
		SFA-5.13 (ENiCrMo-5A)	A5.13 (ENiCrMo-5A)	
Hastelloy C-22	W86022	SFA-5.11 (ENiCrMo-10)	A5.11 (ENiCrMo-10)	---

COMMON NAMES & CROSS REFERENCED SPECIFICATIONS - NICKEL & NICKEL ALLOYS - WELD FILLER METALS (Continued)				
Common Name	**UNS**	**ASME**	**AWS**	**MIL-SPEC**
Hastelloy C-276	W80276	SFA-5.11 (ENiCrMo-4)	A5.11 (ENiCrMo-4)	E-22200/3 (MIL-3N1C)
Hastelloy G	W86007	SFA-5.11 (ENiCrMo-1)	A5.11 (ENiCrMo-1)	---
Hastelloy G-30	W86030	SFA-5.11 (ENiCrMo-11)	A5.11 (ENiCrMo-11)	---
Hastelloy W[b]	W80004	SFA-5.11 (ENiMo-3)	A5.11 (ENiMo-3)	E-22200/3 (MIL-4N1W)
		SFA-5.34 (ENiCrMo-4T-X)	A5.34 (ENiCrMo-4T-X)	---
Hastelloy X[c]	W86002	SFA-5.11 (ENiCrMo-2)	A5.11 (ENiCrMo-2)	---
Inconel 112 Electrode	W86112	SFA-5.11 (ENiCrMo-3)	A5.11 (ENiCrMo-3)	E-22200/3 (MIL-1N12)
Inconel 112 Electrode		SFA-5.34 (ENiCrMo-3T-X)	A5.34 (ENiCrMo-3T-X)	---
Inconel 117 Electrode	W86117	SFA-5.11 (ENiCrCoMo-1)	A5.11 (ENiCrCoMo-1)	---
Inconel 132[d]	W86132	SFA-5.11 (ENiCrFe-1)	A5.11 (ENiCrFe-1)	E-22200/3 (MIL-3N12)
		SFA-5.34 (ENiCrFe-1T-X)	A5.34 (ENiCrFe-1T-X)	E-22200/3 (MIL-3N12H)
Inconel 152 Electrode[f]	W86152	SFA-5.11 (ENiCrFe-7)	A5.11 (ENiCrFe-7)	---
Inconel 182 Electrode	W86182	SFA-5.11 (ENiCrFe-3)	A5.11 (ENiCrFe-3)	E-22200/3 (MIL-8N12)
INCO-Weld A	W86133	SFA-5.11 (ENiCrFe-2)	A5.11 (ENiCrFe-2)	E-22200/3 (MIL-4N1A)
		SFA-5.34 (ENiCrFe-2T-X)	A5.34 (ENiCrFe-2T-X)	E-22200/3 (MIL-4N1AH)
INCO-Weld B	W86134	SFA-5.11 (ENiCrFe-4)	A5.11 (ENiCrFe-4)	---
Monel 190 Electrode	W84190	SFA-5.11 (ENiCu-7)	A5.11 (ENiCu-7)	E-22200/3 (MIL-9N10)
Nickel 141 Electrode	W82141	SFA-5.11 (ENi-1)	A5.11 (ENi-1)	E-22200/3 (MIL-4N11)
Ni-Cr Electrode[e]	W86130	---	---	---
Ni-Cr Electrode[e]	W86182	SFA-5.34 (ENiCrFe-3T-X)	A5.34 (ENiCrFe-3T-X)	E-22200/3 (MIL-8N12H)
Ni-Cr Flux Cored Electrode	W86082	SFA-5.34 (ENiCr-3T-X)	A5.34 (ENiCr-3T-X)	---
RA 330-04 Electrode	W88334	---	---	---
RA 330-80	W88338	---	---	---
RA 333 Electrode	W86333	---	---	---
---	W86455	SFA-5.11 (ENiCrMo-7)	A5.11 (ENiCrMo-7)	---
---	W86620	SFA-5.11 (ENiCrMo-6)	A5.11 (ENiCrMo-6)	---

COMMON NAMES & CROSS REFERENCED SPECIFICATIONS - NICKEL & NICKEL ALLOYS - WELD FILLER METALS (Continued)				
Common Name	**UNS**	**ASME**	**AWS**	**MIL-SPEC**
---	W86985	SFA-5.11 (ENiCrMo-9)	A5.11 (ENiCrMo-9)	---
---	W87718	SFA-5.34 (ENiFeCr-2T-X)	A5.34 (ENiFeCr-2T-X)	---

a. Includes AMS specification 5677.
b. Includes AMS specification 5787.
c. Includes AMS specification 5677.
d. Includes AMS specification 5684.
e. Includes AMS specification 5779.
f. Obsolete, no longer active and are retained for reference purposes only.

INTERNATIONAL CROSS REFERENCES - NICKEL & NICKEL ALLOYS[a]							
Common Name	**USA UNS**	**Product Form**	**USA ASTM/AWS**	**BRITAIN**		**GERMANY**	
				BS	**Alloy**	**DIN**	**W. Nr.**
Nickel 200	N02200	Plate, Sheet	B 162	3072	NA11	17750 Ni99.0	2.4066
		Strip	B 162	3073	NA11	17750 Ni99.0	2.4066
		Rod	B 160	3075	NA11	17152Ni99.0	2.4066
		Bar	B 160	3076	NA11	17152 Ni99.0	2.4066
		Seamless Pipe	B 161	---	---	17751 Ni99.0	2.4066
		Seamless Tube	B 161	3074	NA11	17751 Ni99.0	2.4066
		Welded Pipe	B 725	---	---	---	---
		Welded Tube	B 730	---	---	---	---
		Fittings	B 366	---	---	---	---
		Forgings	---	---	---	17754 LCNi99.0	2.4068
Nickel 201	N02201	Plate, Sheet	B 162	3072	NA12	17750 LCNi99.0	2.4068
		Strip	B 162	3073	NA12	17750 LCNi99.0	2.4068
		Rod	B 160	3075	NA12	17152 LCNi99.0	2.4068
		Bar	B 160	3076	NA12	17152 LCNi99.0	2.4068

INTERNATIONAL CROSS REFERENCES - NICKEL & NICKEL ALLOYS[a] (Continued)

Common Name	USA UNS	Product Form	USA ASTM/AWS	BRITAIN BS	BRITAIN Alloy	GERMANY DIN	GERMANY W. Nr.
Nickel 201	N02201	Seamless Pipe	B 161	---	---	17751 LCNi99.0	2.4068
		Seamless Tube	B 161	3074	NA12	17751 LCNi99.0	2.4068
		Welded Pipe	B 725	---	---	---	---
		Welded Tube	B 730	---	---	---	---
		Fittings	B 366	---	---	---	---
		Forgings	---	---	---	17754 LCNi99.0	2.4068
Alloy 20Cb-3	N08020	Plate	B 463	---	---	NiCr20CuMo	2.4660
		Fittings	B 366	---	---	NiCr20CuMo	2.4660
		Welded Pipe	B 464	---	---	NiCr20CuMo	2.4660
		Welded Tube	B 468	---	---	NiCr20CuMo	2.4660
Alloy 400	N04400	Plate, Sheet	B 127	3072	NA13	17750 NiCu30Fe	2.4360
		Strip	B 127	3073	NA13	17750 NiCu30Fe	2.4360
		Rod	B 164	3075	NA13	17752 NiCu30Fe	2.4360
		Bar	B 164	3076	NA13	17752 NiCu30Fe	2.4360
		Seamless Tubing	B 165	3074	NA13	17751 NiCu30Fe	2.4360
		Seamless Pipe	B 165	---	---	17751 NiCu30Fe	2.4360
		Forgings	B 564	---	---	---	---
Alloy 600	N06600	Sheet, Plate	B 168	3072	NA14	17750 NiCr15Fe	2.4816
		Strip	B 168	3073	NA14	17750 NiCr15Fe	2.4816
		Welded Tube	B 516	3074	NA14	17751 NiCr15Fe	2.4816
		Bar	B 166	3076	NA14	---	2.4816
Alloy 601	N06601	Sheet, Plate, Strip	B 168	---	---	17750	2.4851
		Seamless Tube	B 167	---	---	17751	2.4851
		Seamless Pipe	B 167	---	---	17751	2.4851
		Bar	B 166	---	---	---	2.4851

INTERNATIONAL CROSS REFERENCES - NICKEL & NICKEL ALLOYS[a] (Continued)

Common Name	USA UNS	Product Form	USA ASTM/AWS	BRITAIN		GERMANY	
				BS	Alloy	DIN	W. Nr.
Alloy 617	N06617	Weld Filler Metal	AWS A5.14 ERNiCrCoMo-1	---	---	---	2.4663
Alloy 625	N06625	Sheet, Plate	B 443	3072	NA21	17750 NiCr22Mo9Nb	2.4856
		Strip	B 443	3073	NA21	17750 NiCr22Mo9Nb	2.4856
		Seamless Tube	B 444	3074	NA21	17751 NiCr22Mo9Nb	2.4856
		Bar	B 446	3076	NA21	17752 NiCr22Mo9Nb	2.4856
Alloy 690	N06690	Sheet, Plate, Strip	B 168	---	---	17750	2.4642
		Seamless Tube	B 167	---	---	17751	2.4642
		Seamless Pipe	B 167	---	---	17751	2.4642
		HE Tubes	B 163	---	---	17751	2.4642
		Bar	B 166	---	---	---	2.4642
Alloy 718	N07718	Plate, Sheet, Strip	B 670	---	---	NiCr19NbMo	2.4668
		Forgings	B 637	---	---	NiCr19NbMo	2.4668
		Weld Filler Metal	AWS A5.14 ERNiFeCr-2	---	---	---	2.4668
Alloy 800	N08800	Sheet, Plate	B 409	3072	NA15	X10NiCrAlTi	1.4876
		Strip	B 409	3073	NA15	X10NiCrAlTi	1.4876
		Welded Tube	B 515	3074	NA15	X10NiCrAlTi	1.4876
		Bar	B 408	3076	NA15	X10NiCrAlTi	1.4876
Alloy 800H	N08810	Sheet, Plate	B 409	3072	NA15H	X10NiCrAlTi	1.4876
		Strip	B 409	3073	NA15H	X10NiCrAlTi	1.4876
		Welded Tube	B 407	3074	NA15H	X10NiCrAlTi	1.4876
		Bar	B 408	3076	NA15H	X10NiCrAlTi	1.4876
Alloy 825	N08825	Sheet, Plate	B 424	3072	NA16	17750	2.4858
		Strip	B 424	3073	NA16	17750	2.4858
		Seamless Tube	B 423	3074	NA16	17751	2.4858

INTERNATIONAL CROSS REFERENCES - NICKEL & NICKEL ALLOYS[a] (Continued)

Common	USA	Product	USA	BRITAIN		GERMANY	
Name	UNS	Form	ASTM/AWS	BS	Alloy	DIN	W. Nr.
Alloy 825	N08825	Bar	B 425	3076	NA16	---	2.4858
Alloy B-2	N10665	Plate, sheet, strip	B 333	---	---	17750 NiMo28	2.4617
		Welded Pipe	B 622	---	---	17751 NiMo28	2.4617
		Welded Tube	B 622	---	---	17751 NiMo28	2.4617
Alloy C-4	N06455	Plate, Sheet, Strip	B 575	---	NA45	17750 NiMo16Cr16Ti	2.4610
		Welded Tube	B 626	---	NA45	17751 NiMo16Cr16Ti	2.4610
		Welded Pipe	B 619	---	NA45	17751 NiMo16Cr16Ti	2.4610
		Seamless Tube	B 622	---	NA45	17751 NiMo16Cr16Ti	2.4610
		Seamless Pipe	B 622	---	NA45	17751 NiMo16Cr16Ti	2.4610
Alloy C-276	N010276	Plate, Sheet, Strip	B 575	---	---	17750 NiMo16Cr5W	2.4819
		Welded Tube	B 626	---	---	17751 NiMo16Cr5W	2.4819
		Welded Pipe	B 619	---	---	17751 NiMo16Cr5W	2.4819
		Seamless Tube	B 622	---	---	17751 NiMo16Cr5W	2.4819
		Seamless Pipe	B 622	---	---	17751 NiMo16Cr5W	2.4819
Alloy G	N06007	Plate, Sheet, Strip	B 582	---	---	17750 NiCr22Mo6Cu	2.4618
		Welded Tube	B 626	---	---	17751 NiCr22Mo6Cu	2.4618
		Welded Pipe	B 619	---	---	17751 NiCr22Mo6Cu	2.4618
		Seamless Tube	B 622	---	---	17751 NiCr22Mo6Cu	2.4618
		Seamless Pipe	B 622	---	---	17751 NiCr22Mo6Cu	2.4618
Alloy G-3	N06985	Plate, Sheet, Strip	B 582	---	---	17750 NiCr22MoCu	2.4619
		Welded Tube	B 626	---	---	17751 NiCr22MoCu	2.4619
		Welded Pipe	B 619	---	---	17751 NiCr22MoCu	2.4619
Alloy G-30	N06030	Plate, Sheet, Strip	B 582	---	---	17744	2.4603
		Welded Tube	B 626	---	---	17744	2.4603
		Welded Pipe	B 619	---	---	17744	2.4603
		Seamless Tube	B 622	---	---	17744	2.4603

INTERNATIONAL CROSS REFERENCES - NICKEL & NICKEL ALLOYS[a] (Continued)

Common Name	USA UNS	Product Form	USA ASTM/AWS	BRITAIN		GERMANY	
				BS	Alloy	DIN	W. Nr.
Alloy G-30	N06030	Seamless Pipe	B 622	---	---	17744	2.4603
Alloy X-750	N07750	Forgings	B 637	---	HR505	---	2.4669
Monel 400	N04400	Plate, Sheet, Strip	B 127	---	---	---	2.4360
		HE Tubes	B 163	---	---	---	2.4360
		Rod, Bar, Wire	B 164	---	---	---	2.4360
		Seamless Tube	B 165	---	---	---	2.4360
		Seamless Pipe	B 165	---	---	---	2.4360
Monel K-500	N05500	Nuts	F 467 (500)	---	---	---	2.4375

a. It is not practical to directly correlate the various metal designations from country to country, let alone comparing several countries and their metal designations; from the view that chemical composition may be similar, but not identical, and that manufacturing technologies may differ greatly. Consequently, the cross references made in this table are, at best, only listed as a guide to assist in finding comparable metal designations, rather than equivalent metal designations.
UNS - Unified Numbering System; ASTM - American Society for Testing Materials; BS - British Standards; DIN - Deutsches Institut für Normung; W-Nr. - Werkstoffe Number.

Chapter 3

TITANIUM & TITANIUM ALLOYS

Metallic titanium was first isolated in impure form in 1887 and with higher purity in 1910; however, it was not until the 1950s that it began to come into use as a structural material. This was initially stimulated by aircraft applications. Although the aerospace industry still provides the major market, titanium and titanium alloys are finding increasingly widespread use in other industries due to their many desirable properties. Notable among these is their low densities, which fall between those of aluminum and iron and give very attractive strength to weight ratios. In addition, titanium and titanium alloys readily form stable protective surface layers which give them excellent corrosion resistance in many environments, including oxidizing acids and chlorides, and good elevated temperature properties up to about 550°C (1022°F) in some cases.

Titanium metal is abundant in the earth's crust and is extracted commercially from the ore minerals rutile (titanium dioxide) and ilmenite (iron-titanium oxide). The commercial extraction process involves treatment of the ore with chlorine gas to produce titanium tetrachloride, which is purified and reduced to a metallic titanium sponge by reaction with magnesium or sodium. The sponge, blended with alloying elements (and reclaimed scrap) as desired, is then vacuum melted. Several meltings may be necessary to achieve a homogeneous ingot which is ready for processing into useful shapes, typically by forging followed by rolling. For many applications the cost of titanium alloys can be justified on the basis of desirable properties.

Pure titanium, like iron, is allotropic. At ambient temperature it has a hexagonal close packed (hcp) crystal structure which is stable during heating up to 883°C (1621°F) where it transforms to the body centered cubic (bcc) crystal structure. It remains bcc at higher temperatures until it melts at 1668°C (3034°F). On cooling, the transformation from bcc to hcp in pure titanium cannot be suppressed by rapid cooling, the transformation occurring by a martensitic type reaction. This is not, however, the case with titanium alloys, in which the transformation can be suppressed or modified. Thus the microstructures of titanium alloys frequently contain particles of the bcc phase at ambient temperatures.

Different alloying elements have different effects. Some stabilize the hcp "alpha" structure by allowing it to exist at temperatures above 883°C (1621°F) while others stabilize the bcc "beta" structure by allowing it to remain stable below this temperature. Particular combinations of alloy composition and heat treatment can thus be used to obtain a wide range of possible microstructures, and hence a wide range of useful combinations of properties. The most important of the alpha stabilizers are aluminium, tin, and oxygen, while the important beta stabilizers include molybdenum, vanadium, manganese, chromium, and iron.

The spectrum of titanium-based materials can be divided into four classes depending their constituent phases; this in turn depends on their relative contents of alpha-stabilizing and beta-stabilizing alloying elements. The four basic classes are: (1) unalloyed or commercially pure titanium; (2) alpha and near-alpha alloys; (3) alpha-plus-beta alloys; and, (4) beta alloys. These four classes will be considered further in the sections that follow.

Unalloyed (Commercial Purity) Titanium

Unalloyed titanium typically contains between 99% and 99.5% titanium with the balance being made up of iron and the interstitial impurity elements hydrogen, nitrogen, carbon and, most importantly, oxygen. Titanium has a strong affinity for the interstitial elements, so much so that commercial purity titanium is sometimes referred to as an alpha phase titanium-oxygen alloy. The properties of titanium and its alloys are quite sensitive to the amounts of these elements present. For example, an oxygen content of only 0.1% is sufficient to raise the hardness of pure titanium by a factor of about three. At the same time the ductility and impact toughness are adversely affected, with excessive quantities causing embrittlement. Maximum contents of iron and the four interstitial impurity elements are therefore specified. For example, there are four ASTM grades of unalloyed titanium (grades 1 to 4, corresponding to UNS R50250, R50400, R50500 and R50700), differing primarily in their specified maximum oxygen contents (between about 0.1 and 0.4%). Unalloyed titanium also has a specified maximum hydrogen content, typically about 0.01%, because the presence of small amounts of hydrogen can lead to the precipitation of embrittling hydride particles, even at ambient temperature. Small amounts of iron cause the formation of microscopic particles of beta phase which act to strengthen the material. Typically a maximum of 0.2 to 0.5% is specified. Nitrogen and carbon also cause strengthening. The ASTM standards deal with the adverse effects of these residual impurities by specifying, in most grades, a maximum of 0.4% total residual content.

In addition to the four grades described above, there are several other grades of unalloyed titanium. For example there are grades (ASTM grades 7 and 11 corresponding to UNS R52400 and R52250) which contain 0.2% palladium for improved corrosion resistance especially in applications where operating conditions vary between oxidizing and mildly reducing. Another grade (ASTM grade 12, UNS R53400) of unalloyed titanium contains 0.3% molybdenum and 0.8% nickel, giving improved strength plus somewhat improved corrosion resistance. Unalloyed titanium is also produced with extra low interstitial content, as "ELI" alloys for applications requiring high ductility and/or toughness.

The microstructure of unalloyed titanium consists of grains of the alpha phase, with the possibility of small amounts of beta phase as a result of the presence of the iron impurity as mentioned above. The alpha grains in the annealed condition are either equiaxed or acicular, the former when worked alpha is recrystallized in the alpha region, and the latter when the material is cooled from above the beta transus (the temperature above which only the beta phase is present).

Unalloyed titanium is less expensive, generally more corrosion resistant, and lower in strength than the titanium alloys discussed below. Its impact toughness is comparable to that of quenched and tempered low alloy steel, and its toughness at low temperatures can be high with appropriate levels of interstitial impurity content. Although titanium has a low thermal conductivity, its heat transfer rate is high because of the characteristics of its surface oxide film and its resistance to corrosion. Unalloyed titanium is weldable and formable and finds uses in applications requiring corrosion resistance and high ductility for fabrication where strength is not of primary importance. It is not heat treatable. Typical applications include heat exchangers, chemical process piping, valves and tanks, as well as airframe sheet components.

Alpha and Near-Alpha Alloys

Titanium alloys have a fully alpha (hcp) structure only if they contain alpha stabilizers such as aluminum, tin, and oxygen. These elements also act as solid solution strengtheners; however, their total amounts must be restricted to avoid the formation of an embrittling "alpha-2" phase, which is the intermetallic compound Ti_3Al. Aluminum is particularly potent in this respect, so in some alloys aluminum is partially replaced by zirconium to add strength while avoiding this embrittlement. The oxygen lowers the ductility and thus low oxygen or extra low interstitial (ELI) alloys are available. The typical all-alpha alloy is Ti-5%Al-2.5%Sn (ASTM grade 6, UNS R54520, the ELI equivalent is UNS R54521).

In reality, small particles of beta phase can occur in the microstructures of all-alpha alloys as a result of the presence of iron impurity. In the near-alpha alloys, small amounts (1 to 2%) of beta stabilizers such as molybdenum and vanadium are added to retain some beta phase at ambient temperature. However the microstructures of these alloys are primarily alpha, and their responses to thermal and mechanical treatments are closer to those of the all-alpha alloys than to the alpha-plus-beta alloys discussed below.

Near-alpha titanium alloys include Ti-8%Al-1%Mo-1%V (UNS R54810), Ti-6%Al-2%Nb-1%Ta-0.8%Mo (UNS R56210), Ti-6%Al-2%Sn-4%Zr-2%Mo (Ti-6242 or UNS R54620), Timetal-1100 (a high temperature creep resistant alloy) and Ti-3%Al-2.5%V (UNS R56320). Several of these alloys including Ti-6-2-4-2 and Ti-3Al-2.5V are sometimes considered near-alpha alloys and sometimes alpha-beta alloys.

All-alpha alloys are not heat treatable and can be strengthened only by cold working or by cold working and annealing to control the grain size. On the other hand, some control over microstructure and properties is possible in near-alpha alloys by varying the processing conditions. This gives alpha grains with different sizes and morphologies, ranging from equiaxed to acicular (martensitic), depending on the annealing temperature and cooling rate. A common treatment for Ti-8Al-1Mo-1V is "mill annealing" which consists of annealing in a region where both alpha and beta phases are stable, e.g. 790°C (1454°F), and then furnace cooling. Improved notch sensitivity, at the expense of lower strength, can be obtained by "duplex annealing" (e.g. 1010°C (1850°F) for 15 minutes, air cool, then reheat for 15 minutes at 750°C (1382°F), then air cool). This minimizes the amount of embrittling alpha-2 phase. The alpha-2 phase is also responsible for decreased resistance to stress corrosion cracking in this alloy.

All-alpha and near-alpha alloys are slightly less corrosion resistant than unalloyed titanium, but they are higher in strength and retain high ductility. They have good toughness properties even at cryogenic temperatures, especially the ELI alloys. They are weldable and in many cases have excellent resistance to creep and oxidation and are therefore preferred for high temperature applications. All-alpha titanium alloys are employed as forgings such as gas turbine engine casings and rings. ELI alloys are useful for high pressure cryogenic applications. Typical uses of near-alpha alloys include airframe skin and structural components such as seamless tubing and honeycomb panels (foil) as well as moderately high temperature applications such as blades in the compressor sections of jet engines. The Ti-3%Al-2.5%V alloy is also used for a range of industrial applications (it is boiler code approved) as well as for golf club shafts and bicycle frames.

Alpha-Plus-Beta Alloys

This class of titanium alloy contains both alpha stabilizers and beta stabilizers, the latter in sufficient quantity to permit the development of microstructures containing appreciable amounts of beta phase, typically 10 to 50%, at ambient temperature. These alloys can thus be heat treated to develop a range of microstructures, with control over the distribution and morphology of the alpha and beta phases. There are two classes of alpha-beta alloys, as determined by the amount of beta stabilizer. The "lean" alpha-beta alloys are moderately heat treatable (i.e. have low hardenability) and are weldable, while the "rich" alpha-beta alloys have greater hardenability, and can thus be through-hardened in thicker sections by heat treatment, but are more difficult to weld. The most important of these alpha-beta alloys is Ti-6%Al-4%V (UNS R56400), a "lean" alloy which is also the most important of all the titanium alloys. Other "lean" alpha-beta alloys include Ti-6%Al-2%Sn-4%Zr-2%Mo (Ti-6-2-4-2 or UNS R54620), while the "rich" alloys include Ti-6%Al-6%V-2%Sn (UNS R56620) and Ti-6%Al-2%Sn-4%Zr-6%Mo (Ti-6-2-4-6 or UNS R56260). Ti-6Al-4V is also available in an ELI grade (UNS R56401).

The microstructures and microstructural transformations in these alloys, as developed by appropriate thermal and mechanical processing, are complex. The following examples apply to Ti-6Al-4V. Quenching from above the beta transus gives a martensitic alpha prime structure, although this does not have the extreme hardness of ferrous martensite. Slower cooling gives acicular or coarser plate-like alpha phase, with some intergranular beta phase present. If the material is annealed slightly below the beta transus and water quenched, other microstructures are developed, consisting of mixtures of primary alpha with martensitic, acicular or equiaxed alpha, in some cases with intergranular beta phase as well. High temperature processing, for example by forging, produces a range of microstructures, depending on the forging temperature and cooling rate. Alternatively, the alloy can be solution treated, quenched and aged to control the quantity and dispersion of the beta phase which precipitates from martensitic alpha formed during the quench. This latter treatment gives the highest strengths of the Ti-6Al-4V alloys, with tensile strength as high as 185,000 psi, as compared to annealed Ti-6Al-4V at 145,000 psi. In general, the wide range of microstructures possible as a result of heat treatment and thermomechanical processing gives rise to a wide range of properties.

Alpha-beta alloys are capable of excellent combinations of strength, toughness and corrosion resistance. They have acceptable forming properties and are generally weldable. Their creep resistance at elevated temperatures is poorer than that of the alpha or near alpha alloys. Typical applications include rocket motor cases, blades, and discs for aircraft turbines and compressors, structural aircraft components including fasteners and landing gear, chemical process equipment, marine components, and surgical implant materials. The ELI alloys have cryogenic applications.

Beta Alloys

These alloys contain a balance of beta stabilizers to alpha stabilizers which is sufficiently high that a fully beta phase microstructure can be retained on cooling. In this condition, the alloys are generally thermodynamically unstable, and this can be used to advantage by allowing some decomposition reactions (alpha precipitation) during an aging treatment in order to provide strengthening. Their generally high strength, high toughness, and improved formability as compared to alpha-beta alloys provides an attractiveness, although processing is sometimes critical. Furthermore, the beta alloys are more readily heat treated in thick sections.

Typical beta alloys include Ti-10%V-2%Fe-3%Al, Ti-15%V-3%Cr-3%Sn-3%Al (Ti-15-3), Ti-13%V-11%Cr-3%Al (UNS R58010), Ti-3%Al-8%V-6%Cr-4%Zr-4%Mo (Beta C or Ti-3-8-6-4-4 or UNS R58640), and Ti-15%Mo-2.7%Nb-3%Al (Timetal•21S). The beta alloys are becoming more widely used, notably Ti-15-3, Beta C and Timetal•21S.

Beta alloys are heat treatable to very high strengths, especially when they are cold worked in the solution treated and quenched condition and subsequently aged. The solution treatment plus quenching provides a metastable beta microstructure which decomposes on subsequent aging to a mixture of alpha and beta (the "near-beta" alloys) depending on the precise thermal treatment.

The cold formability of most beta alloys is generally high and their fracture toughness is generally superior to that of the alpha-beta alloys, although in the high strength condition they can have low ductility. In addition, they have relatively high density. Typical applications include high strength airframe components, fasteners, springs, pipe, and commercial consumer products.

Products

Available mill products vary by alloy but bar, billet, extrusions, strip, sheet, plate, wire, pipe and tubing are available for unalloyed titanium, while some or all of these products are available for titanium alloys. Many grades of titanium and its alloys are also available as castings and most can be forged. Some beta alloys have superior forgeability; sheet can be cold-formed in the solution-treated condition. Ti-6Al-4V is relatively difficult to cold form but is readily hot formed or even superplastically formed.

Weldability

In general, welding of titanium and its alloys can be readily performed but it is necessary to exclude reactive gases, including oxygen and nitrogen from the air, and to maintain cleanliness. Thus weld properties are heavily influenced by welding procedures, especially by the adequacy of inert gas shielding. The GTAW (gas tungsten arc welding) process is common, although GMAW (gas metal arc welding), friction welding, laser welding, resistance welding, plasma arc welding, electron beam welding, and diffusion bonding are all used in some cases. Both alloy composition and microstructure are important in determining weldability, with the presence of beta phase having a deleterious effect. Thus unalloyed titanium and alpha alloys are generally weldable and welded joints generally have acceptable strength and ductility; postweld stress-relief annealing of weldments is recommended. Some alpha-beta alloys, specifically Ti-6Al-4V, are weldable in the annealed condition as well as in the solution treated and partially aged condition (aging can be completed during the post-weld heat treatment). Strongly stabilized alpha-beta alloys can be embrittled by welding, the result of phase transformations occurring in the weld metal or the heat affected zone. Some beta alloys are weldable in the annealed or the solution treated condition.

Corrosion Resistance

Unalloyed titanium, the most corrosion-resistant of the titanium-based materials, is resistant to nitric acid and many different chlorine-bearing environments, including hot chloride solutions. It is also resistant to sulphides. The Pd-bearing unalloyed grades have improved resistance to corrosion in reducing media, so that it can be applied in hydrochloric, phosphoric, and sulphuric acid solutions. Since the corrosion resistance is based on the formation of a stable adherent protective surface oxide film, corrosion susceptibility can arise if the environment is such that the film cannot regenerate itself when damaged; such a situation can arise for example in the case of crevice corrosion, where oxygen depletion and acidic conditions can occur in confined spaces.

Although titanium and its alloys are generally resistant to stress-corrosion cracking (SCC), problems can arise under some specific service conditions, i.e. particular combinations of material (alloy composition and heat treatment), loading parameters (stress, stress concentrations, loading rate, etc.) and environment (including temperature). Although SCC can be a potential problem in other environments, the following generalizations apply to behaviour in salt solutions. Alloys containing high aluminum and oxygen levels are particularly susceptible to SCC. In near-alpha alloys, treatments which lead to the formation of alpha-2 render the alloy susceptible, for example mill annealed alloys in neutral salt solutions. In alpha-beta alloys, the alpha phase is susceptible so the above factors apply; however, the distribution of beta phase also plays a role in determining susceptibility. Some beta alloys, particularly those in which the beta phase is stabilized by manganese or chromium, are susceptible. Different concentration levels of species in the salt solutions can aggravate or ameliorate SCC susceptibility. Hot salt environments, 220 to 510°C (428 to 950°F), can also give rise to cracking problems in some highly stressed titanium alloys. Beta alloys are much more tolerant of hydrogen.

Hydrogen can have adverse effects on alpha and alpha-beta alloys in situations where brittle titanium hydride forms within the alpha phase under stress. This is a potential problem over a range of temperatures above ambient, and particularly for material in contact with high pressures of gaseous hydrogen.

Creep And Oxidation Resistance

Unalloyed titanium has good creep resistance below 315°C (599°F). Alpha alloys are generally stable for periods of 1000 hours up to 540°C (1004°F), alpha-beta alloys up to about 370°C (698°F) in the mill annealed condition and as high as 425°C (797°F) after heat treatment.

SAE/AMS SPECIFICATIONS - TITANIUM & TITANIUM ALLOYS	
AMS	**Title**
2249	Chemical check analysis limits titanium and titanium alloys
2380	Approval and control of premium-quality titanium alloys
2419	Cadmium-Titanium plating
2444	Coating, titanium nitride physical vapor deposition
2486	Conversion coating of titanium alloys fluoride-phosphate type
2488	Anodic treatment of titanium and titanium alloys
2631	Ultrasonic inspection titanium and titanium alloy bar and billet
2642	Structural examination of titanium alloys etch- anodize inspection procedure
2643	Structural examination of titanium alloys chemical etch inspection procedure
2689	Fusion welding titanium and titanium alloys
2775	Case hardening of titanium and titanium alloys
2801	Heat treatment of titanium alloy parts
2809	Identification titanium and titanium alloy wrought products
4900	Titanium sheet, strip, and plate commercially pure annealed, 55.0 ksi (379 MPa) yield strength
4901	Titanium sheet, strip, and plate commercially pure annealed, 70.0 ksi (485 MPa)
4902	Titanium sheet, strip, and plate commercially pure annealed, 40.0 ksi (276 MPa) yield strength
4905	Titanium alloy plate, damage-tolerant grade 6Al-4V, beta annealed
4906	Titanium alloy sheet and strip 6Al-4V, continuously rolled, annealed, noncurrent April 1982
4907	Titanium alloy sheet, strip, and plate 6.0Al-4.0V, extra low interstitial annealed
4908	Titanium alloy sheet and strip 8Mn annealed, 110,000 psi (760 MPa) yield strength
4909	Titanium alloy sheet, strip, and plate 5Al-2.5Sn, extra low interstitial annealed
4910	Titanium alloy sheet, strip, and plate 5Al-2.5Sn annealed
4911	Titanium alloy sheet, strip, and plate 6Al-4V annealed
4912	Titanium alloy sheet and strip 4Al-3Mo-1V solution heat treated, noncurrent July 81
4913	Titanium alloy sheet and strip 4Al-3Mo-1V solution and precipitation treated, noncurrent July 81
4914	Titanium alloy cold rolled sheet and strip 15V-3Al-3Cr-3Sn solution heat treated
4915	Titanium alloy sheet, strip, and plate 8Al-1V-1Mo single annealed

SAE/AMS SPECIFICATIONS - TITANIUM & TITANIUM ALLOYS (Continued)

AMS	Title
4916	Titanium alloy sheet, strip, and plate 8Al-1Mo-1V duplex annealed
4917	Titanium alloy sheet, strip, and plate 13.5V-11Cr-3.0Al solution heat treated
4918	Titanium alloy sheet, strip, and plate 6Al-6V-2Sn annealed
4919	Titanium alloy sheet, strip, and plate 6Al-2Sn-4Zr-2Mo-0.08Si duplex annealed
4920	Titanium alloy forgings 6Al-4V alpha-beta or beta processed, annealed
4921	Titanium bars, wire, forgings, and rings commercially pure 70.0 ksi (483 MPa) yield strength
4922	Tubing, seamless, hydraulic, titanium alloy 15V-3.0Cr-3.0Al-3.0 Sn cold worked and precipitation heat treated
4924	Titanium alloy bars, forgings, and rings 5Al-2.5Sn, extra low interstitial annealed
4926	Titanium alloy bars and rings 5Al-2.5Sn annealed, 110,000 psi (760 MPa) yield strength
4928	Titanium alloy bars, wire, forgings, and rings 6Al-4V annealed
4930	Titanium alloy bars, wire, forgings, and rings 6Al-4V, extra low interstitial annealed
4931	Titanium alloy bars, forgings, and rings 6Al-4V extra low interstitial (ELI) duplex annealed, fracture toughness
4932	Titanium alloy sheet 6Al-4V driver sheet
4933	Titanium alloy extrusions and flash welded rings 8Al-1Mo-1V solution heat treated and stabilized
4934	Titanium alloy extrusions and flash welded rings 6Al-4V solution heat treated and aged
4935	Titanium alloy extrusions and flash welded rings 6Al-4V annealed beta processed
4936	Titanium alloy extrusions and flash welded rings 6Al-6V-2Sn beta extruded plus annealed, heat treatable
4937	Titanium alloy extrusions and flash welded rings 6Al-6V-2Sn beta extruded plus annealed, heat treatable
4941	Titanium tubing, welded annealed, 40,000 psi (276 MPa) yield strength
4942	Titanium tubing, seamless annealed, 40,000 psi (275 MPa) yield strength
4943	Titanium alloy tubing, seamless, hydraulic 3.0Al-2.5V annealed
4944	Titanium alloy tubing, seamless, hydraulic 3.0Al-2.5V cold worked, stress relieved
4945	Titanium alloy tubing, seamless, hydraulic 3Al-2.5V, texture controlled 105 ksi (724 MPa) yield strength cold worked, stress relieved
4951	Titanium welding wire commercially pure environment controlled packaging
4952	Titanium alloy welding wire 6Al-2Sn-4Zr-2Mo
4953	Titanium alloy welding wire 5Al-2.5Sn
4954	Titanium alloy welding wire 6Al-4V

SAE/AMS SPECIFICATIONS - TITANIUM & TITANIUM ALLOYS (Continued)

AMS	Title
4955	Titanium alloy welding wire 8Al-1Mo-1V
4956	Titanium alloy welding wire 6Al-4V, extra low interstitial environment controlled packaging
4957	Titanium alloy bars and wire 3Al-8V-6Cr-4Mo-4Zr consumable electrode melted cold drawn
4958	Titanium alloy bars and rods 3Al-8V-6Cr-4Mo-4Zr consumable electrode melted solution heat treated and centerless ground
4959	Titanium alloy wire 13.5V-11Cr-3Al spring temper
4965	Titanium alloy bars, wire, forgings, and rings 6.0Al-4.0V solution and precipitation heat treated
4966	Titanium alloy forgings 5Al-2.5Sn annealed, 110,000 psi (758 MPa) yield strength
4967	Titanium alloy bars, forgings, and rings 6.0Al-4.0V annealed, heat treatable
4970	Titanium alloy bars, wire, and forgings 7Al-4Mo solution and precipitation heat treated
4971	Titanium alloy bars, wire, forgings, and rings 6Al-6V-2Sn annealed, heat treatable
4972	Titanium alloy bars, wire, and rings 8Al-1Mo-1V solution heat treated and stabilized
4973	Titanium alloy forgings 8Al-1Mo-1V solution heat treated and stabilized
4974	Titanium alloy bars and forgings 11Sn-5.0Zr-2.3Al-1.0Mo-0.21Si solution and precipitation heat treated
4975	Titanium alloy bars and rings 6.0Al-2.0Sn-4.0Zr-2.0Mo-0.08Si solution and precipitation heat treated
4976	Titanium alloy forgings 6.0Al-2.0Sn-4.0Zr-2.0Mo solution and precipitation heat treated
4977	Titanium alloy bars 11.5Mo-6.0Zr-4.5Sn 1275°-1350°F (690°-730°C) solution heat treated, noncurrent April 1987
4978	Titanium alloy bars, wire, forgings, and rings 6Al-6V-2Sn annealed
4979	Titanium alloy bars, wire, forgings, and rings 6Al-6V-2Sn solution and precipitation heat treated
4980	Titanium alloy bars 11.5Mo-6.0Zr-4.5Sn 1375°F (745°C) solution heat treated, noncurrent April 1987
4981	Titanium alloy bars, wire, and forgings 6.0Al-2.0Sn-4.0Zr-6.0Mo solution and precipitation heat treated
4982	Titanium alloy wire 44.5Cb
4983	Titanium alloy forgings 10V-2Fe-3Al consumable electrode melted, single solution heat treated and aged 180,000 psi (1240 MPa) tensile strength
4984	Titanium alloy forgings 10V-2Fe-3Al consumable electrode melted, solution heat treated and aged 173,000psi (1195 MPa) tensile strength
4985	Titanium alloy castings, investment or rammed graphite 6Al-4V annealed
4986	Titanium alloy forgings 10V-2Fe-3Al consumable electrode melted, single solution heat treated and overaged 160,000 psi (1105 MPa) tensile strength
4987	Titanium alloy forgings 10V-2Fe-3Al consumable electrode melted, single solution heat treated and overaged 140,000 psi (965 MPa) tensile strength
4991	Titanium alloy castings, investment 6Al-4V annealed

SAE/AMS SPECIFICATIONS - TITANIUM & TITANIUM ALLOYS (Continued)	
AMS	**Title**
4993	Titanium alloy compacts for B-Nuts 6Al-4V blended powder product
4994	Titanium alloy powdered metal products 6Al-4V hot isostatically pressed, annealed
4995	Billets and preforms 5Al-2Sn-2Zr-4Cr-4Mo-0.10O premium quality, powder-metallurgy product, noncurrent Oct 1989
4996	Billets and preforms 6Al-4V premium quality, powder- metallurgy product, noncurrent Oct 1989
4997	Titanium alloy powder 5Al-2Sn-2Zr-4Cr-4Mo-0.10O premium quality, noncurrent Jan 1992
4998	Powder 6Al-4V premium quality, noncurrent Oct 1989
7498	Rings, flash welded titanium and titanium alloys

SAE AEROSPACE STANDARDS (AS) - TITANIUM & TITANIUM ALLOYS	
AS	**Title**
1576B	Fittings, Welded, Hydraulic, Titanium and Corrosion Resistant Steel, 3000 psi Hydraulic
1580A	Ring, Tube Weld, 3000 psi Hydraulic, Titanium
1814B	Terminology for Titanium Microstructures
4076	Contractile Strain Ratio Testing of Titanium Hydraulic Tubing
4194	Sheet and Strip Surface Finish Nomenclature
4568	Hose Assembly, PTFE, Aramid Reinforced Standard Duty, 4000/3000 psi, 275°F, Titanium Fittings, Flareless, Straight to Straight
4569	Hose Assembly, PTFE, Aramid Reinforced, Standard Duty, 4000/3000 psi, 275°F, Titanium Fittings, Flareless, Straight to 45°
4570	Hose Assembly, PTFE, Aramid Reinforced, Standard Duty 4000/3000 psi, 275°F, Titanium Fittings, Flareless, Straight to 90°
4571	Hose Assembly, PTFE, Aramid Reinforced, Standard Duty, 4000/3000 psi, 275°F, Titanium Fittings, Flareless, 45° to 90°
4572	Hose Assembly, PTFE, Aramid Reinforced, Standard Duty, 4000/3000 psi, 275°F, Titanium Fittings, Flareless, 45° to 45°
4573	Hose Assembly, PTFE, Aramid Reinforced, Standard Duty, 4000/3000 Flareless, 90° to 90°
4624	Hose Assembly, Polytetrafluoroethylene, Aramid Reinforced, Heavy -Duty, 275°F, 3000 psi, Flareless, Titanium Fittings, Straight to Straight
4625	Hose Assembly, Polytetrafluoroethylene, Aramid Reinforced, Heavy -Duty, 275°F, 3000 psi, Flareless, Titanium Fittings, Straight to 45°
4626	Hose Assembly, Polytetrafluoroethylene, Aramid Reinforced, Heavy -Duty, 275°F, 3000 psi, Flareless, Titanium Fittings, Straight to 90°
4627	Hose Assembly, Polytetrafluoroethylene, Aramid Reinforced, Heavy -Duty, 275°F, 3000 psi, Flareless, Titanium Fittings, 45° to 45°
4628	Hose Assembly, Polytetrafluoroethylene, Aramid Reinforced, Heavy -Duty, 275°F, 3000 psi, Flareless, Titanium Fittings, 45° to 90°

SAE AEROSPACE STANDARDS (AS) - TITANIUM & TITANIUM ALLOYS (Continued)

AS	Title
4629	Hose Assembly, Polytetrafluoroethylene, Aramid Reinforced, Heavy -Duty, 275°F, 3000 psi, Flareless, Titanium Fittings, 90° to 90°
7460	Bolts and Screws, Titanium Alloy 6Al - 4V Procurement Specification for (Superseding AMS 7460D)
7461	Bolts and Screws, Titanium Alloy 6Al - 4V Fatigue-Rated, Procurement Specification for (Superseding AMS 7461C)

SAE AEROSPACE RECOMMENDED PRACTICES (ARP) - TITANIUM & TITANIUM ALLOYS

ARP	Title
1333	Nondestructive Testing of Electron Beam Welded Joints in Titanium-Base Alloys
1795	Stress-Corrosion of Titanium Alloys Effect of Cleaning Agents on Aircraft Engine Materials
1843	Surface Preparation for Structural Adhesive Bonding Titanium Alloy Parts
1932	Anodize Treatment of Titanium and Titanium Alloys pH 12.4 Maximum
4146	Coiled Tubing - Titanium Alloy, Hydraulic Applications
982B	Minimizing Stress- Corrosion Cracking in Wrought Titanium Alloy Products

SAE AEROSPACE TOLERANCE SPECIFICATIONS FOR TITANIUM & TITANIUM ALLOYS

AMS	Title
2241	Tolerances, Corrosion and Heat Resistant Steel, Iron Alloy, Titanium,and Titanium Alloy Bars and Wire
2242	Tolerances, Corrosion and Heat Resistant Steel, Iron Alloy, Titanium,and Titanium Alloy Sheet, Strip, and Plate
2244	Tolerances, Titanium and Titanium Alloy Tubing
2245	Tolerances, Titanium and Titanium Alloy Extruded Bars, Rods, and Shapes

ASTM SPECIFICATIONS - TITANIUM & TITANIUM ALLOYS	
ASTM	**Title**
B 265	Standard Specification for Titanium and Titanium Alloy Strip, Sheet, and Plate
B 299	Standard Specification for Titanium Sponge
B 337	Standard Specification for Seamless and Welded Titanium and Titanium Alloy Pipe
B 338	Standard Specification for Seamless and Welded Titanium and Titanium Alloy Tubes for Condensers and Heat Exchangers
B 348	Standard Specification for Titanium and Titanium Alloy Bars and Billets
B 363	Standard Specification for Seamless and Welded Unalloyed Titanium and Titanium Alloy Welding Fittings
B 367	Standard Specification for Titanium and Titanium Alloy Castings
B 381	Standard Specification for Titanium and Titanium Alloy Forgings
B 481	Standard Practice for Preparation of Titanium and Titanium Alloys for Electroplating
B 600	Standard Guide for Descaling and Cleaning Titanium and Titanium Alloy Surfaces
B 817	Standard Specification for Powder Metallurgy (P/M) Titanium Alloy Structural Components

ASTM TESTING STANDARDS - TITANIUM & TITANIUM ALLOYS	
ASTM	**Title**
E 120	Standard Test Methods for Chemical Analysis of Titanium and Titanium Alloys
E 539	Standard Test Method for X-Ray Emission Spectrometric Analysis of 6Al-4V Titanium Alloy
E 1320	Standard Reference Radiographs for Titanium Castings
E 1409	Standard Test Method for Determination of Oxygen in Titanium and Titanium Alloys by the Inert Gas Fusion Technique
E 1447	Standard Test Method for Determination of Hydrogen in Titanium Alloys by the Inert Gas Fusion Thermal Conductivity Method
F 945	Standard Test Method for Stress-Corrosion of Titanium Alloys by Aircraft Engine Cleaning Materials
G 41	Standard Practice for Determining Cracking Susceptibility of Metals Exposed Under Stress to a Hot Salt Environment

SURGICAL IMPLANT INTERNATIONAL SPECIFICATIONS - TITANIUM & TITANIUM ALLOYS	
Standard	**Title**
ASTM F 67	Standard Specification for Unalloyed Titanium for Surgical Implant Applications
ASTM F 136	Standard Specification for Wrought Titanium 6Al-4V ELI Alloy for Surgical Implant Applications
ASTM F 620	Standard Specification for Titanium 6Al-4V ELI Alloy Forgings for Surgical Implants
ASTM F 1295	Standard Specification for Wrought Titanium-6 Aluminum-7 Niobium Alloy for Surgical Implant Applications
AS 2320.2	Metals for the Manufacture of Surgical Implants - Part 2: Unalloyed Titanium
AS 2320.3	Metals for the Manufacture of Surgical Implants - Part 3: Wrought Titanium 6-Aluminium 4-Vanadium Alloy
CAN3-Z310.1 -78	Titanium Alloy (6% Aluminum and 4% Vanadium) for Surgical Implants
CAN/CSA-Z310.7	Orthopaedic Implant Standards
BS 7252: Part 2	Metallic Materials for Surgical Implants Part 2: Unalloyed Titanium
BS 7252: Part 3	Metallic Materials for Surgical Implants Part 3: Wrought Titanium 6 - Aluminium 4-Vanadium Alloy
BS 7254: Section 6.2	Orthopaedic Implants Part 6: Forgings Section 6.2: Method for Specifying Forgings
DIN 5832 Part 2	Implants for Surgery - Metallic Materials - Part II: Unalloyed Titanium
DIN 5832 Part 3	Implants for Surgery - Metallic Materials - Part 3: Wrought Titanium 6-Aluminium 4-Vanadium Alloy

ASME SPECIFICATIONS - TITANIUM & TITANIUM ALLOYS	
ASME	**Title**
SB-265	Specification for Titanium and Titanium Alloy Strip, Sheet, and Plate
SB-337	Specification for Seamless and Welded Titanium and Titanium Alloy Pipe
SB-338	Specification for Seamless and Welded Titanium and Titanium Alloy Tubes for Condensers and Heat Exchangers
SB-348	Specification for Titanium and Titanium Alloy Bars and Billets
SB-363	Specification for Seamless and Welded Unalloyed Titanium and Titanium Alloy Welding Fittings
SB-367	Specification for Titanium and Titanium Alloy Castings
SB-381	Specification for Titanium and Titanium Alloy Forgings

AWS SPECIFICATIONS - TITANIUM & TITANIUM ALLOYS	
AWS	**Title**
A5.16	Specification for Titanium and Titanium Alloy Welding Electrodes and Rods
D10.6	Recommended Practices for Gas Tungsten Arc Welding of Titanium Pipe and Tubing

AMERICAN CROSS REFERENCED SPECIFICATIONS - TITANIUM & TITANIUM ALLOYS					
UNS	**SAE/AMS**	**MILITARY**	**ASTM**	**ASME**	**AWS**
R50100	---	---	---	SFA-5.16 (ERTi-1)	A5.16 (ERTi-1)
R50120	---	---	---	SFA-5.16 (ERTi-2)	A5.16 (ERTi-2)
R50125	4951	---	---	SFA-5.16 (ERTi-3)	A5.16 (ERTi-3)
R50130	---	---	---	SFA-5.16 (ERTi-4)	A5.16 (ERTi-4)
R50250	---	T-81556, T-81915	B 265 grade 1, B 337 grade 1, B 338 grade 1, B 348 grade 1, B 381 grade F-1, F 67 grade 1, F 467 alloy Ti 1, F 468 alloy Ti 1	SB-265 grade 1,SB-337 grade 1, SB-338 grade 1,SB-348 grade 1, SB-381 grade F-1	---
R50400	4902, 4941, 4942	T-9046, T-81556	B 265 grade 2, B 337 grade 2, B 338 grade 2, B 348 grade 2, B 367 grade C-2, B 381 grade F-2, F 67 grade 2, F 467 alloy Ti 2, F 468 alloy Ti 2	SB-265 grade 2, SB-337 grade 2, SB-338 grade 2, SB-348 grade 2, SB-381 grade F-2	---
R50550	4900	T-9046, T-81556	B 265 grade 3, B 337 grade 3, B 338 grade 3, B 348 grade 3, B 367 grade C-3, B 381 grade F-3, F 67 grade 3	SB-265 grade 3, SB-337 grade 3, SB-338 grade 3, SB-348 grade 3, SB-381 grade F-3	---
R50700	4901	T-9046, T-9047, F-83142	B 265 grade 4, B 348 grade 4, B 381 grade F-4, F 67 grade 4, F 467 alloy Ti 4, F 468 alloy Ti 4	---	---

AMERICAN CROSS REFERENCED SPECIFICATIONS - TITANIUM & TITANIUM ALLOYS (Continued)					
UNS	**SAE/AMS**	**MILITARY**	**ASTM**	**ASME**	**AWS**
R52250	---	---	B 265 grade 11, B 337 grade 11, B 338 grade 11, B 348 grade 11, B 381 grade F-11	SB-265 grade 11	---
R52400	---	---	B 265 grade 7, B 337 grade 7, B 338 grade 7, B 348 grade 7, B 381 grade F-7, F 467 alloy Ti 7, F 468 alloy Ti 7	SB-265 grade 7, SB-337 grade 7, SB-338 grade 7, SB-348 grade 7, SB-381 grade F-7	---
R52401	---	---	B 265 grades 7, 11,B 337 grades 7, 11, B 338 grades 7, 11,B 348 grades 7, 11, B 367 grade Ti-Pd7B, B 367 grade Ti-Pd8A, B 381 grades 7, 11	SFA-5.16 (ERTi-7)	A5.16 (ERTi-7)
R52550	---	---	B 367 grade C-2, B 367 grade C-3	---	---
R52700	---	---	---	---	---
R53400	---	---	B 265 grade 12, B 337 grade 12, B 338 grade 12, B 348 grade 12, B 381 grade F-12	SB-265 grade 12, SB-337 grade 12, SB-338 grade 12, SB-348 grade 12, SB-381 grade F-12	---
R53401	---	---	B 265 grade 12, B 337 grade 12, B 338 grade 12, B 348 grade 12, B 381 grade 12	SFA-5.16 (ERTi-12)	A5.16 (ERTi-12)
R54520	4910, 4926, 4966	T-9046, T-9047, T-81556, T-81915, F-83142,	B 265 grade 6, B 348 grade 6, B 367 grade C-6, B 381 grade F-6	---	---
R54521	4909, 4924	T-9046, T-9047, T-81556, F-83142	---	---	---
R54522	4953	---	---	SFA-5.16 (ERTi-6)	A5.16 (ERTi-6)
R54523	---	---	---	SFA-5.16 (ERTi-6ELI)	A5.16 (ERTi-6ELI)
R54550	---	F-83142	---	---	---

AMERICAN CROSS REFERENCED SPECIFICATIONS - TITANIUM & TITANIUM ALLOYS (Continued)

UNS	SAE/AMS	MILITARY	ASTM	ASME	AWS
R54560	---	T-9046, T-9047	---	---	---
R54620	4919, 4952, 4975, 4976	T-9046, T-9047, T-81915, F-83142	---	---	---
R54621	---	---	---	---	---
R54624	---	---	B 265, B 337, B 338	---	---
R54790	4974	T-9047, F-83142	---	---	---
R54810	4915, 4916, 4933, 4955	T-9046, T-9047, T-81556, F-83142	---	---	---
R56080	4908	T-9046	---	---	---
R56210	---	T-9046, T-9047	---	SFA-5.16 (ERTi-15)	A5.16 (ERTi-15)
R56260	4981	T-9047	---	---	---
R56320	4943, 4944	T-9046, T-9047	B 265 grade 9, B 337 grade 9, B 338 grade 9, B 348 grade 9	SFA-5.16 (ERTi-9)	A5.16 (ERTi-9)
R56320	---	---	B 381 grade F-9	---	--
R56321	---	---	---	SFA-5.16 (ERTi-9ELI),	A5.16 (ERTi-9ELI),
R56400	4905, 4906, 4911, 4931, 4934, 4935, 4954, 4965, 4967, 4920, 4928, 4930, 4993	T-9046, T-9047, T-81556, T-81915, F-83142	B 265 grade 5, B 348 grade 5, B 367 grade C-5, B 381 grade F-5	SFA-5.16 (ERTi-5)	A5.16 (ERTi-5)
R56401[a]	4907, 4985, 4991, 4996, 4998	T-9046, T-9047, T-81556, F-83142	---	---	---
R56402	4956	---	---	SFA-5.16 (ERTi-5ELI)	A5.16 (ERTi-5ELI)
R56410	4984, 4986, 4987	---	---	---	---

AMERICAN CROSS REFERENCED SPECIFICATIONS - TITANIUM & TITANIUM ALLOYS (Continued)

UNS	SAE/AMS	MILITARY	ASTM	ASME	AWS
R56430	4912, 4913	T-9046	---	---	---
R56620	4918	T-9046	---	---	---
R56620	4936, 4971, 4978, 4979	T-9047, T-81556, T-81556, F-83142	---	---	---
R56740	4970, 4980	T-9047, T-9047, F-83142	B 337 grade 10, B 338 grade 10, B 348 grade 10, B 381 grade F-10	---	---
R58010	4917, 4959	T-9046, T-9047, F-83142	---	---	---
R58030	4977	T-9046	B 265 grade 10	---	---
R58153	4914	---	---	---	---
R58450	4982	---	---	---	---
R58640	4957, 4958	T-9046, T-9047, F-83142	---	---	---
R58650	4995, 4997	---	---	---	---
R58820	---	T-9046, T-9047, F-83142	---	---	---

a. Also includes former SAE J775 grade XEV-J.

This cross-reference table lists the basic specification or standard number, and since these standards are constantly being revised, it should be kept in mind that they are presented herein as a guide and may not reflect the latest revision.

COMMON NAMES OF TITANIUM & TITANIUM ALLOYS WITH AMS No., UNS No. & FORM

Common Name	AMS - UNS	Form
Ti-Unalloyed/0.26-0.35 O	4900 - R50550	Sheet, Strip, Plate
Ti-Unalloyed/0.36-0.40 O	4901 - R50700	Sheet, Strip, Plate
Ti-Unalloyed/0.19-0.25 O	4902 - R50400	Sheet, Strip, Plate
Ti-6Al-4V	4905 - R56400	Plate
Ti-6Al-4V	4906 - R56400	Sheet, Strip
Ti-6Al-4V ELI	4907 - R56401	Sheet, Strip, Plate
Ti-8Mn	4908 - R56080	Sheet, Strip
Ti-5Al-2.5Sn ELI	4909 - R54521	Sheet, Strip, Plate
Ti-5Al-2.5Sn	4910 - R54520	Sheet, Strip, Plate
Ti-6Al-4V	4911 - R56400	Sheet, Strip, Plate
Ti-15V-3Cr-3Sn-3Al	4914 - R58153	Sheet, Strip
Ti-8Al-1Mo-1V	4915 - R54810	Sheet, Strip, Plate
Ti-8Al-1Mo-1V	4916 - R54810	Sheet, Strip, Plate
Ti-13V-11Cr-3Al	4917 - R58010	Sheet, Strip, Plate
Ti-6Al-6V-2Sn	4918 - R56620	Sheet, Strip, Plate
Ti-6Al-2Sn-4Zr-2Mo	4919 - R54620	Sheet, Strip, Plate
Ti-6Al-4V	4920 - R56400	Forging,
Ti-Unalloyed/0.36-0.40 O	4921 - R50700	Bar, Forging, Ring
Ti-5Al-2.5Sn ELI	4924 - R54521	Bar, Forging, Ring
Ti-5Al-2.5Sn	4926 - R54520	Bar, Forging
Ti-6Al-4V	4928 - R56400	Bar, Forging, Ring
Ti-6Al-4V ELI	4930 - R56401	Bar, Forging, Ring
Ti-6Al-4V ELI	4931 - R56401	Bar, Forging, Ring
Ti-8Al-1Mo-1V	4933 - R54810	Extrusion, Ring
Ti-6Al-4V	4934 - R56400	Extrusion, Ring
Ti-6Al-4V	4935 - R56400	Extrusion, Ring
Ti-6Al-6V-2Sn	4936 - R56620	Extrusion, Ring

COMMON NAMES OF TITANIUM & TITANIUM ALLOYS WITH AMS No., UNS No. & FORM (Continued)		
Common Name	**AMS - UNS**	**Form**
Ti-Unalloyed/0.19-0.25 O	4941 - R50400	Welded Tubing
Ti-Unalloyed/0.19-0.25 O	4942 - R50400	Seamless Tubing
Ti-3Al-2.5V	4943 - R56320	Seamless Tubing
Ti-3Al-2.5V	4944 - R56320	Seamless Tubing
Ti-3Al-2.5V	4945 - R56320	Seamless Tubing
Ti-Unalloyed/0.10-0.18 O	4951 - R50550	Wire
Ti-6Al-2Sn-4Zr-2Mo	4952 - R54620	Wire
Ti-5Al-2.5Sn	4953 - R54520	Wire
Ti-6Al-4V	4954 - R56400	Wire
Ti-8Al-1Mo-1V	4955 - R54810	Wire
Ti-6Al-4V ELI	4956 - R56402	Wire
Ti-3Al-8V-6Cr-4Mo-4Zr	4957 - R58640	Bar, Wire
Ti-3Al-8V-6Cr-4Mo-4Zr	4958 - R58640	Bar, Rod
Ti-13V-11Cr-3Al	4959 - R58010	Wire
Ti-6Al-4V	4965 - R56400	Bar, Forging, Ring
Ti-5Al-2.5Sn	4966 - R54520	Forging,
Ti-6Al-4V	4967 - R56400	Bar, Forging, Ring
Ti-7Al-4Mo	4970 - R56740	Bar, Forging,
Ti-6Al-6V-2Sn	4971 - R56620	Bar, Forging, Ring
Ti-8Al-1Mo-1V	4972 - R54810	Bar, Ring
Ti-8Al-1Mo-1V	4973 - R54810	Forging
Ti-11Sn-5Zr-2Al-1Mo	4974 - R54790	Bar, Forging
Ti-6Al-2Sn-4Zr-2Mo-0.10Si	4975 - R54620	Bar, Ring
Ti-6Al-2Sn-4Zr-2Mo	4976 - R54620	Forging
Ti-6Al-6V-2Sn	4978 - R56620	Bar, Forging, Ring
Ti-6Al-6V-2Sn	4979 - R56620	Bar, Forging, Ring
Ti-6Al-2Sn-4Zr-6Mo	4981 - R56260	Bar, Forging

COMMON NAMES OF TITANIUM & TITANIUM ALLOYS WITH AMS No., UNS No. & FORM (Continued)		
Common Name	**AMS - UNS**	**Form**
Ti-44.5Cb	4982 - R58450	Wire
Ti-10V-2Fe-3Al	4983 - R56410	Forging
Ti-10V-2Fe-3Al	4984 - R56410	Forging
Ti-6Al-4V	4985 - R56401	Casting
Ti-10V-2Fe-3Al	4986 - R56410	Forging
Ti-10V-2Fe-3Al	4987 - R56410	Forging
Ti-6Al-4V	4991 - R56401	Casting
Ti-6Al-4V	4993 - R56400	Powder Compact
Ti-6Al-4V	4994 - R56400	Powder product
Ti-5Al-2Sn-2Zr-4Cr-4Mo	4995 - R58650	Powder Product, Billet
Ti-6Al-4V	4996 - R56401	Powder Product, Billet
Ti-5Al-2Sn-2Zr-4Cr-4Mo	4997 - R58650	Powder
Ti-6Al-4V	4998 - R56401	Powder

BRITISH BSI AEROSPACE SERIES SPECIFICATIONS - TITANIUM & TITANIUM ALLOYS (TA)	
BS	**Title**
2TA 1	Sheet and strip of commercially pure titanium (tensile strength 290-420 MPa)
2TA 2	Sheet and strip of commercially pure titanium (tensile strength 390-540 N/mm^2)
2TA 3	Bar and section for machining of commercially pure titanium (tensile stength 390-540 N/mm^2) (obsolescent)
2TA 4	Forging stock of commercially pure titanium (tensile strength 390-540 N/mm^2) (obsolescent)
2TA 5	Forgings of commercially pure titanium (tensile strength 390-540 N/mm^2) (obsolescent)
2TA 6	Sheet and strip of commercially pure titanium (tensile strength 570-730 N/mm^2)
2TA 7	Bar and section for machining of commercially pure titanium (tensile strength 540-740 N/mm^2)
2TA 8	Forging stock of commercially pure titanium (tensile strength 540-740 N/mm^2)

BRITISH BSI AEROSPACE SERIES SPECIFICATIONS - TITANIUM & TITANIUM ALLOYS (TA) (Continued)	
BS	**Title**
2TA 9	Forgings of commercially pure titanium (tensile strength 540-740 N/mm^2)
2TA 10	Sheet of titanium-aluminium-vanadium alloy (tensile strength 960-1270 MPa)
2TA 11	Bar and section for machining of titanium-aluminium-vanadium alloy (tensile strength 900-1160 MPa) (limiting ruling section 150 mm)
2TA 12	Forging stock of titanium-aluminium-vanadium alloy (tensile strength 900-1160 MPa) (limiting ruling section 150 mm)
2TA 13	Forgings of titanium-aluminum-vanadium alloy (tensile strength 900-1160 MPa) (limiting ruling section 150mm)
2TA 21	Sheet and strip of titanium-copper alloy (tensile strength 540-700 N/mm^2)
2TA 22	Bar and section for machining of titanium-copper alloy (tensile strength 540-770 N/mm^2)
2TA 23	Forging stock of titanium-copper alloy (tensile strength 540-770 N/mm^2)
2TA 24	Forgings of titanium-copper alloy (tensile strength 540-770 N/mm^2)
2TA 28	Forging stock and wire of titanium-aluminium-vanadium alloy (tensile strength 1100-1300 MPa) (limiting ruling section 20 mm) (primarily intended for the manufacture of fasteners complying with the requirements of the 'A' series of BS
TA 38	Bar for machining of titanium-aluminium-molybdenum-tin-silicon-carbon alloy (tensile strength 1250-1420 N/mm^2) (limiting ruling section 25 mm)
TA 39	Forging stock of titanium-aluminium-molybdenum-tin-silicon-carbon alloy (tensile strength 1250-1420 N/mm^2) (limiting ruling section 25 mm)
TA 40	Bar for machining of titanium-aluminium-molybdenum-tin-silicon-carbon alloy (tensile strength 1205-1375 N/mm^2) (limiting ruling section over 25 mm up to and including 75 mm)
TA 41	Forging stock of titanium-aluminium-molybdenum-tin-silicon-carbon alloy (tensile strength 1205-1375 N/mm^2) (limiting ruling section over 25 mm up to and including 75 mm)
TA 42	Forgings of titanium-aluminium-molybdenum-tin-silicon-carbon alloy (tensile strength 1205-1375 N/mm^2) (limiting ruling section over 25 mm up to and including 75 mm)
TA 43	Forging stock of titanium-aluminium-zirconium-molybdenum-silicon alloy (tensile strength 990-1140 N/mm^2) (limiting ruling section 65 mm)
TA 44	Forgings of titanium-aluminium-zirconium-molybdenum-silicon alloy (tensile strength 990-1140 N/mm^2) (limiting ruling section 65 mm)
TA 45	Bar and section for machining of titanium-aluminium-molybdenum-tin-silicon alloy (tensile strength 1100-1280 N/mm^2) (limiting ruling section 25 mm)
TA 46	Bar and section for machining of titanium-aluminium-molybdenum-tin-silicon alloy (tensile strength 1050-1220 N/mm^2) (limiting ruling section over 25 mm up to and including 100 mm)

BRITISH BSI AEROSPACE SERIES SPECIFICATIONS - TITANIUM & TITANIUM ALLOYS (TA) (Continued)	
BS	**Title**
TA 47	Forging stock of titanium-aluminium-molybdenum-tin-silicon alloy (tensile strength 1050-1220 N/mm^2) (limiting ruling section 100 mm)
TA 48	Forgings of titanium-aluminium-molybdenum-tin-silicon alloy (tensile strength 1050-1200 N/mm^2) (limiting ruling section 100 mm)
TA 49	Bar and section for machining of titanium-aluminium-molybdenum-tin-silicon alloy (tensile strength 1000-1200 N/mm^2) (limiting ruling section over 100 mm up to and including 150 mm)
TA 50	Forging stock of titanium-aluminium-molybdenum-tin-silicon alloy (tensile strength 1000-1200 N/mm^2 (limiting ruling section over 100 mm up to and including 150 mm)
TA 51	Forgings of titanium-aluminium-molybdenum-tin-silicon alloy (tensile strength 1000-1200 N/mm^2) (limiting ruling section over 100 mm Up to and including 150 mm)
TA 52	Sheet and strip of titanium-copper alloy (tensile strength 690-920 N/mm^2)
TA 53	Bar and section for machining of titanium-copper alloy (tensile strength 650-880 N/mm^2) (limiting ruling section 75 mm)
TA 54	Forging stock of titanium-copper alloy (tensile strength 650-880 N/mm^2) (limiting ruling section 75 mm)
TA 55	Forgings of titanium-copper alloy (tensile strength 650-880 N/mm^2) (limiting ruling section 75 mm)
TA 56	Plate of titanium-aluminium-vanadium alloy (tensile strength 895-1150 MPa) (maximum thickness 100 mm)
TA 57	Plate of titanium-aluminium-molybdenum-tin-silicon (tensile strength 1030-1220 MPa) (maximum thickness 65 mm)
TA 58	Plate of titanium-copper alloy (tensile strength 520-640 MPa) (maximum thickness 10 mm)
TA 59	Sheet and strip of titanium-alumimum-vanadium alloy (tensile strength 920 MPa to 1180 MPa)
2TA 100	Procedure for inspection and testing of wrought titanium and titanium alloys
A 101	General requirements for titanium bolts
A254-A 259	Hexagonal head titanium alloy bolts 160 000 Lbf/in^2 (1100 MPa)
A266-A 271	Specification for 100° countersunk head titanium alloy bolts 160 000 lbf/in^2 (1100 MPa) with Hi-torque speed drive recesses
M 58	Anodic coating of titanium and titanium alloys by the sulphuric acid process
BSEN 2497	Dry abrasive blasting of titanium and titanium alloys

GERMAN DIN SPECIFICATIONS- TITANIUM & TITANIUM ALLOYS	
DIN	**Title**
1737 Part 1	Filler metals for welding titanium and titanium-palladium alloys; chemical composition, technical delivery conditions
9430	Aerospace; sampling of semi-finished products in light metals; titanium and titanium alloys
17850	Titanium; chemical composition
17851	Titanium alloys; chemical composition
17860	Titanium and titanium alloy plate, sheet and strip; technical delivery conditions
17861	Seamless circular titanium and titanium alloy tubes; technical delivery conditions
17862	Rods of titanium and wrought titanium alloys
17862[a]	Titanium and titanium alloy bars; technical conditions of delivery
17863	Wire of titanium
17864[a]	Titanium and titanium alloys, die forgings and handforgings; technical conditions of delivery
17864	Forging of titanium and wrought titanium alloys (open die and drop forgings)
17865	Titanium and titanium alloy investment castings and rammed graphite castings; technical delivery conditions
17866	Welded circular titanium and titanium alloy tubes; technical deliveryconditions
17869	Material properties of titanium and titanium alloys; additional data
29783	Aerospace; precision castings of titanium and titanium alloys; technical specification
65039	Aerospace; sheets, plates and strips of titanium and titanium alloys; technical specification
65083	Aerospace; heat treatment of castings titanium and titanium alloys
65084	Aerospace; heat treatment of wrought titanium and titanium alloys
65174	Aerospace; round bars of titanium and titanium alloys; dimensions, masses
65179	Aerospace; countersunk head screws, close tolerances, with ribbed Torq-Set (ACR) recess and MJ thread, short thread length, titanium alloy, nominal tensile strength 1100 MPa, for temperatures up to 315°C
65179[b]	Aerospace; countersunk head bolts, close tolerance, with internal offset cruciform ribbed drive and MJ thread, short thread length, titanium alloy, nominal tensile strength 1100 MPa, for temperatures up to 315°C
65197	Aerospace; titanium alloy round bars and wire for screw stock; dimensions, masses
65250	Aerospace; seamless internal pressure tubes in titanium and titanium alloys; dimensions, masses
65265[b]	Aerospace; screws, hexagon, with MJ thread, threaded approximately to head, titanium alloy, nominal tensile strength 1100 MPa, for temperatures up to 315°C

GERMAN DIN SPECIFICATIONS- TITANIUM & TITANIUM ALLOYS (Continued)	
DIN	**Title**
65268[b]	Aerospace; bolts hexagon with MJ thread, short thread, titanium alloy, nominal tensile strength 1100 MPa, for temperatures up to 315°C
65289	Aerospace; countersunk head screws, with ribbed Torq-Set (ACR) recess and MJ thread, short thread length, titanium alloy, nominal strength 1100 MPa, for temperatures up to 315°C
65324	Aerospace; pan head bolts, close tolerance, with ribbed Torq-Set (ACR) recess and short length MJ thread, titanium alloy, nominal tensile strength 1100 MPa, for temperatures up to 315°C
65324[b]	Aerospace; panhead bolts, close tolerance, with internal offset cruciform ribbed drive recess and short length MJ thread, titanium alloy, nominal tensile strength 1100 MPa, for temperatures up to 315°C
65338[b]	Aerospace; hexagon bolts, close tolerance, with short length MJ thread, shoulder, titanium alloy, nominal tensile strength 1100 MPa, for temperatures up to 315°C
65348	Aerospace; seamless internal pressure tubes in titanium and titanium alloys; technical specification
65436	Aerospace; standard quality (STQ) and disc quality (DQ) of titanium and titanium alloys; requirements and tests
65437	Aerospace; processed titanium scrap for reuse in titanium and titanium alloys; requirements and tests
65438	Aerospace; bihexagonal head bolts, close tolerance, with MJ thread, medium thread length, in titanium alloy, nominal tensile strength 1100 MPa, for temperatures up to 315°C
65438[b]	Aerospace; bihexagonal head bolts, close tolerance, with MJ thread, short thread length, in titanium alloy, nominal tensile strength 1100 MPa, for temperatures up to 315°C
65455	Aerospace; seamless tubes in steel, nickel and titanium alloys; ultrasonic inspection
65463[a]	Aerospace; studs with MJ thread, captive ring lock, titanium alloy, nominal tensile strength 1100 MPa, for temperatures up to 315°C
65464	Aerospace; studs with MJ thread, ring locked, titanium alloy, nominal tensile strength 1100 MPa, for temperatures up to 315°C
65464[a]	Aerospace; studs with MJ thread, ring locked, titanium alloy, nominal tensile strength 1100 MPa, for temperatures up to 315°C
65515	Aerospace; cheese head bolts, with internal serration and MJ thread, short thread, titanium alloy, nominal tensile strength 1100 MPa, for temperatures up to 315°C
65517	Aerospace; cheese haed bolts, close tolerance, with internal serration and MJ thread, short thread, titanium alloy, nominal tensile strength 1100 MPa, for temperatures up to 315°C
65518	Aerospace; pan head bolts with ribbed Torq-Set (ACR) recess and MJ thread, short thread, titanium alloy, nominal tensile strength 1100 MPa, for temperatures up to 315°C

GERMAN DIN SPECIFICATIONS- TITANIUM & TITANIUM ALLOYS (Continued)	
DIN	**Title**
65526[b]	Aerospace; hexagon bolts, close tolerance, with short length MJ thread, titanium alloy, nominal tensile strength 1100 MPa, for temperatures up to 315°C
65557	Aerospace; pan head screws with ribbed Torq-Set (ACR) recess and MJ thread, fully threaded, titanium alloy, nominal tensile strength 1100 MPa, for temperatures up to 315°C
65558	Aerospace; countersunk head screws, with ribbed Torq-Set (ACR) recess and MJ thread, fully threaded, titanium alloy, nominal tensile strength 1100 MPa, for temperatures up to 315°C
65596[b]	Aerospace; chemical check analysis limits of titanium and titanium alloys
LN 9297	Aerospace; sheet and plate in titanium and titanium alloys, rolled; dimensions, masses
LN 9425 Part 3	Filler metals for welding magnesium and titanium; welding rods; dimensions, weights
LN 29792	Aerospace; bolts, hexagon, close tolerance, of titanium alloy
LN 29796	Aerospace; pin protruding head, of titanium alloy for shear loads
LN 29798	Aerospace; collars, threaded self-locking for head pins of titanium alloy
LN 65027	Aerospace; hexagon screws of titanium alloy, threaded approximately to head
LN 65047 Part 1	Aerospace; bolts, close tolerance, in titanium alloy, with self-locking collars in aluminum alloy; procurement specification, bolts
LN 65047 Part 2	Aerospace; bolts, close tolerance, in titanium alloy, with self-locking collars in aluminum alloy; procurement specification, collars
LN 65047 Part 3	Aerospace; bolts, close tolerance, in titanium alloy, with self-locking collars in aluminum alloy; technical specification; performance test
LN 65072 Part 1	Joining elements made from titanium alloy; technical specification
WL 3.7000 Part 10	Aerospace; titanium and titanium alloys; index
WL 3.7000 Part 20	Aerospace; titanium and titanium alloys; materials and semi-finished products; survey
WL 3.7024 Part 2	Aerosapce; titanium with approx. 0.1 O; filler metal for welding
WL 3.7024 Part 100	Aerosapce; titanium with approx. 0.1 O; sheet and filler metal for welding; design and production data, conversion
WL3.7034 Part 1	Titanium, unalloyed, with about 0.2 O_2; sheets
WL 3.7034 Part 2	Aerosapce; titanium with approx. 0.2 O; bars and forgingsl
WL 3.7034 Part 3	Titanium, unalloyed, with about 0.2 O_2; filler metal
WL 3.7034 Part 100	Aerosapce; titanium with approx. 0.2 O; sheet, bars, forgings and filler metal for welding; design and production data, conversion
WL 3.7064 Part 1	Titanium, unalloyed, with about 0.3 O_2; sheets
WL 3.7064 Part 2	Titanium, unalloyed, with about 0.3 O_2; rods and forgings

GERMAN DIN SPECIFICATIONS- TITANIUM & TITANIUM ALLOYS (Continued)

DIN	Title
WL 3.7064 Part 100	Aerospace; titanium with approx. 0.3 O; sheet, bars and forgings; design and production data
WL 3.7114 Part 1	Titanium alloy;TiAl5Sn2; sheets, strip and plates
WL 3.7114 Part 2	Titanium alloy; TiAl5Sn2; rods and forgings
WL 3.7114 Part 3	Aerospace; titanium alloy with approx. 5Al-2Sn; filler metal for welding
WL 3.7124 Part 1	Aerospace; titanium alloy with approx. 2Cu; sheet and strips
WL 3.7124 Part 2	Aerospace; titanium alloy with approx. 2Cu; bars and forgings
WL 3.7124 Part 3	Aerospace; titanium alloy with approx. 2Cu; filler metal for welding
WL 3.7124 Part 100	Aerospace; titanium alloy with approx. 2Cu; sheet, strip, bars, forgings and filler metal for welding; design and production data, conversion
WL 3.7144 Part 1	Aerospace; titanium alloy with approx. 6Al-2Sn-4Zr-2Mo; bars and forgings; inactive for new design
WL 3.7144 Part 100	Aerospace; titanium alloy with approx. 6Al-2Sn-4Zr-2Mo; bars and forgings; design and production data, conversion; inactive for new design
WL 3.7154 Part 1	Aerospace; titanium alloy with approx. 6Al-5Zr; bars and forgings
WL 3.7154 Part 100	Aerospace; titanium alloy with approx. 6Al-5Zr; bars and forgings; design and production data, conversion
WL 3.7164 Part 1	Aerospace; titanium alloy with approx. 6Al-4V; sheet and plate
WL 3.7164 Part 2	Aerospace; titanium alloy with approx. 6Al-4V; bars and forgings
WL 3.7164 Part 3	Aerospace; titanium alloy with approx. 6Al-4V; filler metal for welding
WL 3.7164 Part 100	Aerospace; titanium alloy with approx. 6Al-4V; sheet, plate, bars, forgings and filler metal for welding; design and production data, conversion
WL 3.7174 Sheet 1	Titanium alloy with about 6Al-6V-2Sn; sheets, strips, plates, rods and forgings
WL 3.7174 Part 1	Titanium alloy; TiAl6V6Sn2; sheets, strops and plates
WL 3.7174 Part 2	Titanium alloy; 6Al-6V-Sn2; rods and forgings
WL 3.7184 Part 1	Aerospace; titanium alloy with approx. 4Al-4Mo-2Sn; plate
WL 3.7184 Part 2	Aerospace; titanium alloy with approx. 4Al-4Mo-2Sn; bars and forgings
WL 3.7184 Part 100	Aerospace; titanium alloy with approx. 4Al-4Mo-2Sn; plate, bars and forgings; design and production data, conversion
WL 3.7194 Part 1	Aerospace; titanium alloy with approx. 3Al-2.5V, seamless internal pressure tubes
WL 3.7194 Part 2	Aerospace; titanium alloy with approx. 3Al-2.5V; filler for welding

GERMAN DIN SPECIFICATIONS- TITANIUM & TITANIUM ALLOYS (Continued)	
DIN	**Title**
WL 3.7264 Part 1	Aerospace; cast titanium alloy with approx. 6Al-4V; investment casting
WL 3.7264 Part 100	Aerospace; cast titanium alloy with approx. 6Al-4V; investment casting, conversion

a. Draft of a standard or technical rule.

b. Published or withdrawn during 1992; draft of a standard or technical rule.

GERMAN DIN - EUROPEAN HARMONIZED STANDARDS & EC LEGAL PROVISIONS - TITANIUM & TITANIUM ALLOYS	
DIN	**Title**
EN 2497[b]	Aerospace series; dry abrasive blasting of titanium and titanium alloys; german version EN 2497:1989
EN 2500[a] Part 4	Aerospace series; instructions for the drafting and use of metallic material standards; part 4: specific requirements for titanium and titanium alloys
EN 2545[a] Part 1	Aerospace series; titanium and titanium alloy remelting stock and castings; technical specification; part 1: general requirements
EN 2545[a] Part 2	Aerospace series; titanium and titanium alloy remelting stock and castings; technical specification; part 2: remelting stock
EN 2545[a] Part 3	Aerospace series; titanium and titanium alloy remelting stock and castings; technical specification; part 3: pre-productions and production castings
EN 2617[a]	Aerospace; plate in titanium alloys 5 £ a £ 100 mm; dimensions
EN 2858[a] Part 1	Aerospace series; titanium and titanium alloy forging stock and forgings; technical specification; part 1: general requirements
EN 2858[a] Part 2	Aerospace series; titanium and titanium alloy forging stock and forgings; technical specification; part 2: forging stock
EN 2858[a] Part 3	Aerospace series; titanium and titanium alloy forging stock and forgings; technical specification; part 3: pre-production and production forgings
EN 2870[a]	Aerospace series; bolts, bihexagonal normal head, close tolerance normal shank, short thread, in titanium alloy, anodized, MoS_2 lubricated; classification: 1100 MPa (at ambient temperature)/315°C
EN 2884[a]	Aerospace series; bolts, pan head, Torq-Set recess, coarse tolerance shank, short thread, in titanium alloy, anodized; classification: 1100 MPa/315°C
EN 2955[a]	Aerospace series; recycling of titanium and titanium alloy scrap
EN 3242[a]	Aerospace series; pipe coupling 8° 30' in titanium alloy union, welded, threaded

GERMAN DIN - EUROPEAN HARMONIZED STANDARDS & EC LEGAL PROVISIONS - TITANIUM & TITANIUM ALLOYS (Continued)	
DIN	**Title**
EN 3243[a]	Aerospace series; pipe coupling 8° 30' in titanium alloy; ferrule, welded, with dynamic beam seal end
EN 3244[a]	Aerospace series; pipe coupling 8° 30' in titanium alloy; union, double ended
EN 3245[a]	Aerospace series; pipe coupling 8° 30' in titanium alloy; union reducer
EN 3246[a]	Aerospace series; pipe coupling 8° 30' in titanium alloy; union, bulkhead
EN 3247[a]	Aerospace series; pipe coupling 8° 30' in titanium alloy; union, bulkhead, welded
EN 3249[a]	Aerospace series; pipe coupling 8° 30' in titanium alloy; elbow 90°
EN 3250[a]	Aerospace series; pipe coupling 8° 30' in titanium alloy; elbow 90° swivel nut
EN 3251[a]	Aerospace series; pipe coupling 8° 30' in titanium alloy; elbow 90° welded
EN 3252[a]	Aerospace series; pipe coupling 8° 30' in titanium alloy; elbow 90°, swivel nut, welded
EN 3253[a]	Aerospace series; pipe coupling 8° 30' in titanium alloy; elbow 90°, bulkhead
EN 3254[a]	Aerospace series; pipe coupling 8° 30' in titanium alloy; elbow 90°, bulkhead, welded
EN 3255[a]	Aerospace series; pipe coupling 8° 30' in titanium alloy; elbow 45°
EN 3256[a]	Aerospace series; pipe coupling 8° 30' in titanium alloy; elbow 45°, swivel nut, welded
EN 3257[a]	Aerospace series; pipe coupling 8° 30' in titanium alloy; elbow 45°, bulkhead
EN 3258[a]	Aerospace series; pipe coupling 8° 30' in titanium alloy; tee
EN 3260[a]	Aerospace series; pipe coupling 8° 30' in titanium alloy; tee, branch with swivel nut
EN 3261[a]	Aerospace series; pipe coupling 8° 30' in titanium alloy; tee with swivel nut
EN 3262[a]	Aerospace series; pipe coupling 8° 30' in titanium alloy; tee, bulkhead branch
EN 3263[a]	Aerospace series; pipe coupling 8° 30' in titanium alloy; tee, bulkhead end
EN 3264[a]	Aerospace series; pipe coupling 8° 30' in titanium alloy; nut, swivel
EN 3265[a]	Aerospace series; pipe coupling 8° 30' in titanium alloy; nut, union
EN 3266[a]	Aerospace series; nut, bulkhead, in titanium alloy
EN 3267[a]	Aerospace series; washer, bulkhead, in titanium alloy

GERMAN DIN - EUROPEAN HARMONIZED STANDARDS & EC LEGAL PROVISIONS - TITANIUM & TITANIUM ALLOYS (Continued)	
DIN	**Title**
EN 3268[a]	Aerospace series; pipe coupling 8° 30' in titanium alloy; plug, pressure
EN 3269[a]	Aerospace series; pipe coupling 8° 30' in titanium alloy; blind ferrule with dynamic beam end seal
EN 3306[b]	Aerospace series; 100° countersunk normal head, offset cruciform-ribbed recess, threaded to head, in titanium alloy, MoS_2 lubricated; classification: 1100 MPa (at ambient temperature)/315°C
EN 3307[b]	Aerospace series; screws, pan head, offset cruciform-ribbed recess, threaded to head, in titanium alloy, MoS_2 lubricated; classification: 1100 MPa (at ambient temperature)/315°C
EN 3308[b]	Aerospace series; screws, normal hexagonal head, threaded to head, in titanium alloy, MoS_2 lubricated; classification: 1100 MPa (at ambient temperature)/315°C
EN 3381[a]	Aerospace series; bolts, 100° countersunk normal head, offset cruciform-ribbed recess, close tolerance shank, short thread, in titanium, anodized; classification: 1100 MPa/315°C
EN 3424[b]	Aerospace series; rivets, solid, universal head, in titanium alloy Ti45,5Cb; metric series
EN 3441[b]	Aerospace series; titanium TI-P99001; annealed, sheet and strip, hot rolled, a £ 6 mm, 290 MPa £ R_m £ 420 MPa
EN 3442[b]	Aerospace series; titanium TI-P99002; annealed, sheet and strip, hot rolled, a £ 6 mm, 390 MPa £ R_m £ 540 MPa
EN 3443[b]	Aerospace series; titanium TI-P99003; annealed, sheet and strip, hot rolled, a £ 6 mm, 570 MPa £ R_m £ 730 MPa
EN 3456[b]	Aerospace series; titanium alloy TI-P64001; annealed, sheet and strip, hot rolled, a £ 6 mm
EN 3497[b]	Aerospace series; titanium TI-P99001; annealed, sheet and strip, cold rolled, a £ 3 mm, 290 MPa £ R_m £ 420 MPa
EN 3498[b]	Aerospace series; titanium TI-P99002; annealed, sheet and strip, cold rolled, a £ 3 mm, 390 MPa £ R_m £ 540 MPa
EN 3499[b]	Aerospace series; titanium TI-P99003; annealed, sheet and strip, cold rolled, a £ 3 mm, 570 MPa £ R_m £ 730 MPa
EN 3642[a]	Aerospace series; rivets, solid, 100° normal countersunk head with dome, in titanium TI-PO2, anodized
EN 3643[a]	Aerospace series; rivets, solid 100° normal countersunk head, in titanium TI-PO2, anodized
EN 3644[a]	Aerospace series; rivets, solid, universal head, in titanium TI-PO2, anodized
EN 3738[b]	Aerospace series; rivets, solid, 100° normal countersunk head with dome, in titanium alloy Ti 45,5 Cb; metric series
EN 3739[b]	Aerospace series; rivets, solid, 100° normal countersunk head, in titanium alloy Ti 45,5 Cb; metric series
EN 3859[b]	Aerospace series; titanium alloy TI-P19001; annealed, sheet and strip, hot rolled, a £ 6 mm

GERMAN DIN - EUROPEAN HARMONIZED STANDARDS & EC LEGAL PROVISIONS - TITANIUM & TITANIUM ALLOYS (Continued)	
DIN	**Title**
EN 3860[b]	Aerospace series; titanium alloy TI-P19001; annealed, sheet and strip, cold rolled, a £ 3 mm
EN 3870[b]	Aerospace series; titanium alloy TI-P19001; solution treated and aged, sheet and strip, hot rolled, a £ 6 mm
EN 3871[b]	Aerospace series; titanium alloy TI-P19001; solution treated and aged, sheet and strip, cold rolled, a £ 3 mm
EN 6024[b]	Aerospace series; screws 100° countersunk reduced head, offset cruciform recess, close tolerance shank, short thread, in titanium alloy, anodized, MoS_2 lubricated; classification: 1100 MPa (at ambient temperature)/315°C
V EN 2098 Part 2	Aerospace series; inspection and testing requirements for titanium and heat resisting alloy wrought products; part 2: inspection and testing requirements for sheets, strips and plates
V EN 2098 Part 5	Aerospace series; inspection and testing requirements for titanium and heat resisting alloy wrought products; part 5: inspection and testing requirements for wires
V EN 2098 Part 6	Aerospace series; inspection and testing requirements for titanium and heat resisting alloy wrought products; part 6: inspection and testing requirements for bars and wires for fasteners

a. Draft of a standard or technical rule.

b. Published or withdrawn during 1992; draft of a standard or technical rule.

EUROPEAN AECMA[a] SPECIFICATIONS - TITANIUM & TITANIUM ALLOYS	
AECMA	**Title**
prEN 2500 Part 4	Instructions for the drafting and use of metallic material standards - part 4 - specific requirements for titanium and titanium alloys
prEN 2545 Part 1	Titanium and titanium alloy remelting stock and castings - technical specification - part 1 - general requirements
prEN 2545 Part 2	Titanium and titanium alloy remelting stock and castings - technical specification - part 2 - remelting stock
prEN 2545 Part 3	Titanium and titanium alloy remelting stock and castings - technical specification - part 3 - pre-production and production castings
prEN 2808	Anodizing of titanium and titanium alloys
prEN 2870	Bolts, bihexagonal normal head, close tolerance normal shank short thread in titanium alloy, anodized, MoS_2 lubricated classification:1100 MPa (at ambient temperature)/315°C
prEN 2955	Recycling of titanium and titanium alloy scrap
prEN 3248	Pipe coupling 8° 30' in titanium alloy adaptor-reduced pipe end with locking ring
prEN 3270	Pipe coupling 8° 30' in titanium alloy blanking plug with locking ring
prEN 3275	Pipe coupling 8° 30' dynamic beam seal up to 28000 kPa metric series technical specification
prEN 3304	Bolts, 100° countersunk reduced head, offset cruciform recess, close tolerance shank, short thread, in titanium alloy, MoS_2 lubricated classification: 1100 MPa (at ambient temperature)/315°C
prEN 3306	Screws, 100° countersunk normal head, offset cruciform-ribbed recess, threaded to head, in titanium alloy, MoS_2 lubricated classification: 1100 MPa (at ambient temperature)/315°C
prEN 3307	Screws, pan head, offset cruciform recess, threaded to head, in titanium alloy, MoS_2 lubricated classification: 1100 MPa (at ambient temperature)/315°C
prEN 3308	Screws, normal hexagonal head, threaded to head, in titanium alloy, MoS_2 lubricated classification: 1100 MPa (at ambient temperature)/315°C
prEN 3381	Bolts, 100° countersunk normal head, offset cruciform - ribbed recess, close tolerance shank short thread in titanium anodized, MoS_2 lubricated classification: 100 MPa (at ambient temperature)/315°C
prEN 3424	Rivets, solid, universal head, in titanium alloy Ti 45,5 Cb, metric series
prEN 3504	Tube for hydraulic systems titanium and titanium alloys
prEN 3544 Part 1	Titanium and titanium alloy wrought products - technical specification - part 1 - general requirements
prEN 3544 Part 2	Titanium and titanium alloy wrought products - technical specification - part 2 - plate, sheet and strip
prEN 3544 Part 3	Titanium and titanium alloy wrought products - technical specification - part 3 - bar and section
prEN 3544 Part 4	Titanium and titanium alloy wrought products - technical specification - part 4 - seamless tube

EUROPEAN AECMA[a] SPECIFICATIONS - TITANIUM & TITANIUM ALLOYS (Continued)

AECMA	Title
prEN 3544 Part 5	Titanium and titanium alloy wrought products - technical specification - part 5 - wire
prEN 3561	Pipe coupling 8° 30' in titanium alloy ferrule with dynamic beam seal end, welded and reduced at pipe end
prEN 3562	Pipe coupling in titanium alloy elbow 90° double welded
prEN 3563	Pipe coupling in titanium alloy elbow 45° double welded
prEN 3564	Pipe coupling 8° 30' in titanium alloy tee, bulkhead fitting on limb
prEN 3565	Pipe coupling 8° 30' in titanium alloy cross
prEN 3566	Pipe coupling 8° 30' in titanium alloy adaptor with locking ring
prEN 3572	PTFE flexible hose assembly with convoluted inner tube of a nominal pressure up to 6800 kPa and 8° 30' fitting in titanium product standard
prEN 3574	PTFE flexible hose assembly with convoluted inner tube of a nominal pressure up to 6800 kPa and 24° fitting in titanium product standard
prEN3576	PTFE flexible hose assembly of a nominal pressure equal to 10 500 kPa with 8° 30' fitting in titanium product standard
prEN 3578	PTFE flexible hose assembly of a nominal pressure equal to 10 500 kPa with 24° fitting in titanium product standard
prEN 3580	Aerospace series; PTFE flexible hose assembly of a nominal pressure equal to 21 000 kPa with 8° 30' fitting in titanium product standard
prEN 3582	Standard weight PTFE flexible hose assembly of a nominal pressure equal to 21 000 kPa with 24° fitting in titanium product standard
prEN 3584	Lightweight PTFE flexible hose assembly of a nominal pressure equal to 21 000 kPa with 8° 30' fitting in titanium product standard
prEN 3586	Lightweight PTFE flexible hose assembly of a nominal pressure equal to 21 000 kPa with 24° fitting in titanium product standard
prEN 3588	Lightweight PTFE flexible hose assembly of a nominal pressure equal to 28 000 kPa with 8° 30' fitting in titanium product standard
prEN 3590	PTFE flexible hose assembly of a nominal pressure equal to 28 000 kPa with 24° fitting in titanium product standard
prEN 3658	Tube bend radii, for engine application design standard
prEN 3688	T-Ring filler in titanium alloy for welding pipes 14 000 kPa nominal pressure
prEN 3689	T-Ring filler in titanium alloy for welding pipes 28 000 kPa nominal pressure
prEN 3690	Pipe coupling 8° 30' in titanium alloy union bulkhead long
prEN 3691	Pipe coupling 8° 30' in titanium alloy union - bulkhead long welded
prEN 3692	Pipe coupling 8° 30' in titanium alloy elbow 90° - bulkhead long
prEN 3693	Pipe coupling 8° 30' in titanium alloy elbow 90° - bulkhead long welded
prEN 3694	Pipe coupling 8° 30' in titanium alloy tee - bulkhead long branch
prEN 3695	Pipe coupling 8° 30' in titanium alloy tee - reduced one T-End with swivel nut end

EUROPEAN AECMA[a] SPECIFICATIONS - TITANIUM & TITANIUM ALLOYS (Continued)

AECMA	Title
prEN 3724	Bolts, double hexagon head relieved shank, long thread, in titanium alloy TI-P63, MoS_2 coated classification: 1100 MPa (at ambient temperature)/350°C
prEN 3725	Bolts, pan head, 6 lobe recess, long thread, in titanium alloy TI-P63, MoS_2 coated classification: 1100 MPa (at ambient temperature)/350°C
prEN 3738	Rivets, solid, 100° normal countersunk head with dome, in titanium alloy Ti 45,5 Cb, metric series
prEN 3739	Rivets, solid, 100° normal countersunk head, in titanium alloy Ti 45,5 Cb, metric series
prEN 3740	Bolts, shouldered, thin hexagonal head, close tolerance shank, short thread, in titanium alloy, MoS_2 lubricated classification: 1100 MPa (at ambient temperature)/315°C
prEN 3784	Pipe coupling 8° 30' in titanium alloy tee reduced bulkhead branch long
prEN 3785	Pipe coupling 8° 30' in titanium alloy tee reduced bulkhead branch
prEN 3786	Pipe coupling 8° 30' in titanium alloy tee reduced bulkhead long
prEN 3787	Pipe coupling 8° 30' in titanium alloy tee reduced bulkhead
prEN 3818	Bolts in titanium alloy TI-P63 classification: 1100 MPa (at ambient temperature) technical specification
prEN 3851	Nuts, for spherical tube coupling, in titanium alloy TI-P63 use temperature 350°C
prEN 3852	Welding nipple, spherical tube, in titanium alloy TI-P63 use temperature 300°C
prEN 3853	Nipple, spherical tube coupling, in titanium alloy TI-P63 use temperature 300°C
prEN 3854	Welding fende, spherical tube coupling, in titanium alloy TI-P63 use temperature 300°C
prEN 3907	Bolts, double hexagon head, normal shank, long thread, in titanium alloy TI-P63, MoS_2 coated classification: 1100 MPa (at ambient temperature)/350°C
prEN 6024	Bolts, 100° countersunk reduced head, offset cruciform recess, close tolerance shank, short thread, titanium alloy, lubricated classification: 1100 MPa (at ambient temperature)/315°C, inch series
prEN 6024	Screws, 100° countersunk reduced head, offset cruciform recess, close tolerance shank, short thread, in titanium alloy, anodized, MoS_2 lubricated, classification: 1100 MPa (at ambient temperature)/315°C, inch series

a. AECMA - Association Européene des Constructeurs de Matériel Aérospatial.

JAPANESE JIS SPECIFICATIONS - TITANIUM & TITANIUM ALLOYS

JIS	Title
G 3603	Titanium clad steels
H 0511	Testing methods for brinell hardness of titanium sponge
H 4605	Titanium-Palladium alloy sheet, plates and strips
H 4607	Sheets and plates of titanium alloys
H 4630	Titanium pipes and tubes for ordinary piping
H 4631	Titanium tubes for heat exchangers
H 4635	Titanium-Palladium alloy pipes and tubes for ordinary piping
H 4636	Titanium-Palladium alloy pipes and tubes for heat exchanger
H 4650	Titanium rods and bars
H 4655	Titanium-Palladium alloy rods and bars
H 4657	Rods, bars and forgings of titanium alloys
H 4670	Titanium wires
H 4675	Titanium-Palladium alloy wires
H 7001	Glossary of terms used in shape memory alloys
H 7101	Method for Determining the Transformation Temperatures of Shape Memory Alloys
H 7103	Method of fixed temperature tensile test for wires of Ti - Ni shape memory alloys
Z 3107	Methods of radiographic test and classification of radiographs for titanium welds
Z 3331	Titanium and titanium alloy rods and wires for inert gas shielded arc welding

CHEMICAL COMPOSITION OF UNALLOYED (PURE) TITANIUM

UNS	ASTM Spec.	AMS Spec.	MIL Spec.	Grade	C	Fe	H	N	O	Other
R50250	B265, B337, B338, B348, B381[a], F467, F468, F67	---	T-81556,T-81915	1	0.10	0.20	0.015	0.03	0.18	c
R50400	B265, B337, B338, B348, B367[b], B381[a], F467, F468, F67	4902, 4941, 4942	T-9046,T-81556	2	0.10	0.30	0.015	0.03	0.25	c
R50550	B265, B337 B338, B348, B367[b], B381, F67	4900	T-9046,T-81556	3	0.10	0.30	0.015	0.05	0.35	c
R50700	B265, B348, B381, F67, F467, F468	4901, 4921	T-9046, T-9047, T-81556, F-83142	4	0.10	0.50	0.015	0.05	0.40	c

a. ASTM B381 have grade designations as follows: F-1 to F-4; rather than grades 1 to 4, respectively. b. ASTM B367 have grade designations are as follows: C-2 and C-3; rather than grades 2 and 3, respectively. c. Maximum allowable concentration for residual elements shall be 0.1% each and 0.4% maximum total. Single values are maximum. Titanium is the remainder.

CHEMICAL COMPOSITION OF ALLOYED TITANIUM

ASTM Spec.	Gr	UNS	AMS Spec.	MIL Spec.	C	Fe	H	N	O	Other[c]
B265, B348, B367[a],B381[b],F467	5	R56400	4905, 4906, 4911, 4920, 4928, 4930, 4931, 4934, 4935, 4954, 4965, 4967, 4993	T-9046,T-9047, T-81556, T-81915	0.10	0.40	0.015	0.05	0.20	5.5-6.75 Al, 3.5-4.5 V
B265, B348, B367[a], B381[b]	6	R54520	4910, 4926, 4966	T-9046,T-9047, T-81556,T-81915, F-83142	0.10	0.50	0.020	0.05	0.20	4.0-6.0 Al, 2.0-3.0 Sn

CHEMICAL COMPOSITION OF ALLOYED TITANIUM (Continued)

ASTM Spec.	Gr	UNS	AMS Spec.	MIL Spec.	C	Fe	H	N	O	Other[c]
B265, B337 B338, B348, B381[b], F467, F468	7	R52400	---	---	0.10	0.30	0.015	0.03	0.25	0.12-0.25 Pd,
B265, B337, B338, B348, B381[b]	9	R56320	4943, 4944	T-9046, T-9047	0.10	0.25	0.015	0.02	0.15	2.5-3.5Al, 2.0-3.0 V
B265, B337, B338, B348, B381[b]	11	R52250	---	---	0.10	0.20	0.015	0.03	0.18	0.12-0.25 Pd
B265, B337, B338, B348, B381[b]	12	R53400	---	---	0.08	0.30	0.015	0.03	0.25	0.2-0.4 Mo, 0.6-0.9 Ni
B265, B348, B381[b]	13	---	---	---	0.10	0.20	0.015[d]	0.03	0.10	0.04-0.06 Ru, 0.40-0.60 Ni
B265, B348, B381[b]	14	---	---	---	0.010[e]	0.30	0.015[d]	0.03	0.15	0.04-0.06 Ru, 0.40-0.60 Ni
B265, B348, B381[b]	15	---	---	---	0.10	0.30	0.015[d]	0.05	0.25	0.04-0.06 Ru, 0.40-0.60 Ni
B265, B348, B381[b]	16	---	---	---	0.10	0.30	0.015[d]	0.03	0.25	0.04-0.08 Pd
B265, B348, B381[b]	17	---	---	---	0.10	0.20	0.015[d]	0.03	0.18	0.04-0.08 Pd
B265, B348, B381[b]	18	---	---	---	0.10	0.25	0.015[d]	0.05	0.15	0.04-0.08 Pd, 2.5-3.5 Al, 2.0-3.0 V

a. ASTM B367 have grade designations are as follows: C-5, C-6, Ti-Pd 7B, Ti-Pd 8A; rather than grades 5, 6, 7, 8, respectively. b. ASTM B381 have grade designations as follows: F-5 to F-18; rather than grades 5 to 18, respectively. c. Maximum allowable concentration for residual elements shall be 0.1% each and 0.4% maximum total. d. See B348 and B381 for specific hydrogen requirements; hydrogen content listed is for B265. e. Carbon content listed is for B265 and B348; carbon content for B381 is 0.10% maximum. Single value are maximum. Titanium is the remainder.

CHEMICAL COMPOSITION OF TITANIUM BARE ELECTRODE & ROD FILLER METAL[a, b, c]

AWS A5.16 Classification	UNS No.	C	O	H	N	Al	V	Fe	Other
ERTi-1	R50100	0.03	0.10	0.005	0.015		---	0.10	---
ERTi-2	R50120	0.03	0.10	0.008	0.020		---	0.20	---
ERTi-3	R50125	0.03	0.10-0.15	0.008	0.020		---	0.20	---
ERTi-4	R50130	0.03	0.15-0.25	0.008	0.020		---	0.30	---
ERTi-5	R56400	0.05	0.18	0.015	0.030	5.5-6.7	3.5-4.5	0.30	0.005 Y
ERTi-5ELI	R56402	0.03	0.10	0.005	0.012	5.5-6.5	3.5-4.5	0.15	0.005 Y
ERTi-6	R54522	0.08	0.18	0.015	0.050	4.5-5.8	---	0.50	0.005 Y; 2.0-3.0 Sn
ERTi-6ELI	R54523	0.03	0.10	0.005	0.012	4.5-5.8	---	0.20	0.005 Y; 2.0-3.0 Sn
ERTi-7	R52401	0.03	0.10	0.008	0.020	---	---	0.20	0.12-0.25 Pd
ERTi-9	R56320	0.03	0.12	0.008	0.020	2.5-3.5	2.0-3.0	0.25	0.005 Y
ERTi-9ELI	R56321	0.03	0.10	0.005	0.012	2.5-3.5	2.0-3.0	0.20	0.005 Y
ERTi-12	R53400	0.03	0.25	0.008	0.020	---	---	0.30	0.2-0.4 Mo; 0.6-0.9 Ni
ERTi-15	R56210	0.03	0.10	0.005	0.015	5.5-6.5	---	0.15	0.5-1.5 Mo; 1.5-2.5 Cb0.5-1.5 Ta

a. Titanium constitutes the remainder of the composition.

b. Single values are maximum.

c. Residual elements, total, shall not exceed 0.20%, with no single element exceeding 0.05%. See AWS A5.16 specification for more details.

MECHANICAL PROPERTIES OF WROUGHT TITANIUM & TITANIUM ALLOYS

ASTM Standard			Tensile Strength[a]		Yield Strength[a]			Bend Test for ASTM B 265[b]	
B 265	B 337	B 338						Under 0.070 in. in thickness	0.070 to 0.187 in. in thickness
Gr	Gr	Gr	ksi	MPa	ksi	MPa	% El[a]		
1	1	1	35	240	25-45	170-310	24	3T	4T
2	2	2	50	345	40-65	275-450	20	4T	5T
3	3	3	65	450	55-80	380-550	18	4T	5T

MECHANICAL PROPERTIES OF WROUGHT TITANIUM & TITANIUM ALLOYS (Continued)

ASTM Standard			Tensile Strength[a]		Yield Strength[a]		% El[a]	Bend Test for ASTM B 265[b]	
B 265	B 337	B 338						Under 0.070 in. in thickness	0.070 to 0.187 in. in thickness
Gr	Gr	Gr	ksi	MPa	ksi	MPa			
4	---	---	80	550	70-95	485-655	15	5T	6T
5	---	---	130	895	120	830	10[c,d]	9T	10T
6	---	---	120	830	115	795	10[c]	8T	9T
7	7	7	50	345	40-65	275-450	20	4T	5T
9	9[f]	9[f]	90	620	70	485	15[e]	5T	6T
---	9[g]	9[g]	125	860	105	725	10	---	---
---	10[h]	---	100	690	90	620	10	---	---
11	11	11	35	240	25-45	170-310	24	3T	4T
12	12	12	70	483	50	345	18	4T	5T
13	---	---	40	275	25	170	24	3T	4T
14	---	---	60	410	40	275	20	4T	5T
15	---	---	70	483	55	380	18	4T	5T
16	---	---	50	345	40-65	275-450	20	4T	5T
17	---	---	35	240	25-45	170-310	24	3T	4T
18	---	---	90	620	70	483	15[e]	5T	6T

a. Single values are minimum unless otherwise noted.

b. Bend tests applicable to ASTM B 265 only. T equals the thickness of the bend test specimen. Bend tests are not applicable to material over 0.187 in. (4.75 mm) in thickness.

c. For grades 5 and 6 the elongation on materials under 0.025 in. (0.635 mm) in thickness may be obtained only by negotiation.

d. For grade 5, the elongation will be 8% minimum for thicknesses between 0.025 in. and 0.063 in.

e. Elongation for continuous rolled and annealed (strip product from coil) for grade 9 and grade 18 shall be 12% minimum in the longitudinal direction and 8% minimum in the transverse direction. See ASTM B 265 for more details.

f. Properties for material in the annealed condition.

g. Properties for cold-worked and stress-relieved material. h. Properties for material in the solution treated condition.

MECHANICAL PROPERTIES OF WROUGHT TITANIUM & TITANIUM ALLOYS							
ASTM Standard		Tensile Strength[a]		Yield Strength[a]			
B 348 Grade	B 381 Grade	ksi	MPa	ksi	MPa	% El[a]	% RA[a]
1	F-1	35	240	25	170	24	30
2	F-2	50	345	40	275	20	30
3	F-3	65	450	55	380	18	30
4	F-4	80	550	70	483	15	25
5	F-5	130	895	120	825	10	25
6	F-6	120	825	115	795	10	25
7	F-7	50	345	40	275	20	30
9	F-9	90	620	70	483	15	25
11	F-11	35	240	25	170	24	30
12	F-12	70	483	50	345	18	25
13	F-13	40	275	25	170	24	30
14	F-14	60	410	40	275	20	30
15	F-15	70	483	55	380	18	25
16	F-16	50	345	40	275	20	30
17	F-17	35	240	25	170	24	30
18	F-18	90	620	70	483	15	25

a. Single values are minimum.

MECHANICAL PROPERTIES OF CAST TITANIUM & TITANIUM ALLOYS[a]

Casting Standards			Tensile Strength		Yield Strength			Hardness	
ASTM B 367	AMS 4985	AMS 4991	ksi	MPa	Ksi	MPa	% El	Rockwell max	HB max
C-2	---	---	50	345	40	275	15	96 HRB[b]	210[b]
C-3	---	---	65	450	55	380	12	24 HRC[b]	235[b]
C-5	---	---	130	895	120	825	6	39 HRC[b]	365[b]
C-6	---	---	115	795	105	725	8	36 HRC[b]	335[b]
Ti-Pd 7B	---	---	50	345	40	275	15	96 HRB[b]	210[b]
Ti-Pd 8A	---	---	65	450	55	380	12	24 HRC[b]	235[b]
---	6Al - 4V[c]	---	130	895	120	825	6	39 HRC	---
---	6Al - 4V[d]	---	130	895	120	825	6	39 HRC	---
---	6Al - 4V[e]	---	125	860	108	745	4.5	39 HRC	---
---	---	6Al - 4V[d]	130	895	120	825	6	39 HRC[f]	---
---	---	6Al - 4V[e]	127	875	110	760	4.5	39 HRC[f]	---
---	---	6Al - 4V[c]	130	895	120	825	6	39 HRC[f]	---

a. Single values are minimum unless otherwise noted.
b. Supplementary requirement applied only when specified by the purchaser. Values are averages of three tests. See ASTM B 367 for more details.
c. Separately-cast specimens or specimens cut from attached coupons.
d. Specimens cut from castings - designated areas.
e. Specimens cut from castings - non-designated areas.
f. Castings shall not be rejected on the basis of hardness if tensile property requirements of AMS 4991 are met.

MECHANICAL PROPERTIES OF SURGICAL IMPLANT TITANIUM & TITANIUM ALLOYS[a]

ASTM Standards			Tensile Strength		Yield Strength				Bend Test[f]	
F 67	F 136	F 620[e]	ksi	MPa	Ksi	MPa	% El	% RA	Under 0.070 in. in thickness	0.070 to 0.187 in. in thickness
gr 1	---	---	35	240	25-45	170-310	24	---	3T	4T
gr 2	---	---	50	345	40-65	275-450	20	---	4T	5T
gr 3	---	---	65	450	55-80	380-550	18	---	4T	5T
gr 4	---	---	80	550	70-95	483-655	15	---	5T	6T
---	6Al-4V	---	130	896	120[c]	827[c]	10	---	9T	10T
---	6Al-4V	---	125	860	115[d]	795[d]	10	25	---	---
---	---	6Al-4V	---	---	---	---	---	---	---	---

a. Single values are minimum unless otherwise noted. c. Applies to under 0.187 in. (4.75 mm) thickness or diameter.

d. Applies to 0.187 in. (4.75 mm) to 1.75 in. (44.45 mm) inclusive.

e. When specified by the implant manufacturer, the mechanical properties shall be tested and confirmed by the forger, upon completion of forgings, to comply with the minimum mechanical properties as specified by the implant manufacturer. See ASTM F 620 for more details.

f. T equals the thickness of the bend test specimen.

ASME P-No. - BASE METAL TITANIUM & TITANIUM ALLOYS

ASME Specification	Grade	Specified Tensile min, ksi	Nominal Composition	Product Form
P No. 51				
SB-265	1	35	Unalloyed	Plate, Sheet, Strip
	2	50	Unalloyed	Plate, Sheet, Strip
	7	50	Alloyed (0.18Pd)	Plate, Sheet, Strip
	11	35	0.18Pd-Low Fe-Low O_2	Plate, Sheet, Strip
SB-337	1	35	Unalloyed	Seamless & Welded Pipe

ASME P-No. - BASE METAL TITANIUM & TITANIUM ALLOYS (Continued)

ASME Specification	Grade	Specified Tensile min, ksi	Nominal Composition	Product Form
P No. 51 (Continued)				
SB-337	2	50	Unalloyed	Seamless & Welded Pipe
(continued)	7	50	Alloyed (0.18Pd)	Seamless & Welded Pipe
SB-338	1	35	Unalloyed	Seamless & Welded Tube
	2	50	Unalloyed	Seamless & Welded Tube
	7	50	Alloyed (0.18Pd)	Seamless & Welded Tube
SB-348	1	35	Unalloyed	Bar & Billet
	2	50	Unalloyed	Bar & Billet
	7	50	Alloyed (0.18Pd)	Bar & Billet
SB-363	WPT 1	35	Unalloyed	Seamless & Welded Fitting
	WPT 2	50	Unalloyed	Seamless & Welded Fitting
SB-381	F-1	35	Unalloyed	Forging
	F-2	50	Unalloyed	Forging
	F-7	50	Alloyed (0.18Pd)	Forging
P No. 52				
SB-265	3	65	Unalloyed	Plate, Sheet, Strip
	12	70	Alloyed (03.Mo-0.8Ni)	Plate, Sheet, Strip
SB-337	3	65	Unalloyed	Seamless & Welded Pipe
	12	70	Alloyed (03.Mo-0.8Ni)	Seamless & Welded Pipe
SB-338	3	65	Unalloyed	Seamless & Welded Tube
	12	70	Alloyed (03.Mo-0.8Ni)	Seamless & Welded Tube
SB-348	3	65	Unalloyed	Bar & Billet
	12	70	Alloyed (03.Mo-0.8Ni)	Bar & Billet
SB-363	WPT 3	65	Unalloyed	Seamless & Welded Fitting
SB-381	F-3	65	Unalloyed	Forging
	F-12	70	Alloyed (03.Mo-0.8Ni)	Forging

ASME F-No. - WELDING FILLER METAL TITANIUM & TITANIUM ALLOYS		
F No.	ASME Specification No.	AWS Classification No.
51	SFA-5.16	ERTi-1, ERTi-2, ERTi-3, ERTi-4
52	SFA-5.16	ERTi-7, ERTi-11
53	SFA-5.16	ERTi-9, ERTi-9 ELI
54	SFA-5.16	ERTi-12

INTERNATIONAL CROSS REFERENCED SPECIFICATIONS - TITANIUM & TITANIUM ALLOYS						
UNS	SAE/AMS	MILITARY	ASTM	ASME	BS1	DIN
R50250	---	T-81556, T-81915	B 265 grade 1, B 337 grade 1, B 338 grade 1, B 348 grade 1, B 381 grade F-1, F 67 grade 1, F 467 alloy Ti 1, F 468 alloy Ti 1	SB-265 grade 1, SB-337 grade 1, SB-338 grade 1, SB-348 grade 1, SB-381 grade F-1	TA 1	3.7025
R50400	4902, 4941, 4942	T-9046 CP-3, T-81556	B 265 grade 2, B 337 grade 2, B 338 grade 2, B 348 grade 2, B 367 grade C-2, B 381 grade F-2, F 67 grade 2, F 467 alloy Ti 2, F 468 alloy Ti 2	SB-265 grade 2, SB-337 grade 2, SB-338 grade 2, SB-348 grade 2, SB-381 grade F-2	TA 2, 3, 4, 5	3.7035
R50550	4900	T-9046 CP-2, T-81556	B 265 grade 3, B 337 grade 3, B 338 grade 3, B 348 grade 3, B 367 grade C-3, B 381 grade F-3, F 67 grade 3	SB-265 grade 3, SB-337 grade 3, SB-338 grade 3, SB-348 grade 3, SB-381 grade F-3	---	3.7055
R50700	4901	T-9046 CP-1, T-9047 CP-70, F-83142	B 265 grade 4, B 348 grade 4, B 381 grade F-4, F 67 grade 4, F 467 alloy Ti 4, F 468 alloy Ti 4	---	TA 6, 7, 8, 9	3.7065

INTERNATIONAL CROSS REFERENCED SPECIFICATIONS - TITANIUM & TITANIUM ALLOYS

UNS	SAE/AMS	MILITARY	ASTM	ASME	BS1	DIN
R56400	4905, 4906, 4911, 4920, 4928, 4930, 4931, 4934, 4935, 4954, 4965, 4967, 4993	T-9046 AB1/2, T-9047, T-81556, T-81915, F-83142,	B 265 grade 5, B 348 grade 5, B 367 grade C-5, B 381 grade F-5	SFA-5.16 (ERTi-5)	TA 10, 11, 12, TA 28, 56, 59	3.7165

a. It is not practical to directly correlate the various metal designations from country to country, let alone comparing several countries and their metal designations; from the view that chemical composition may be similar, but not identical, and that manufacturing technologies may differ greatly. Consequently, the cross references made in this table are, at best, only listed as a guide to assist in finding comparable metal designations, rather than equivalent metal designations.

UNS - Unified Numbering System; SAE/AMS - Society of Automotive Engineers/Aerospace Material Specifications; MILITARY - U.S. Military Specifications; ASTM - American Society for Testing Materials; ASME - American Society of Mechanical Engineers; BS - British Standards; DIN - Deutsches Institut für Normung.

Chapter 4

REACTIVE & REFRACTORY METALS

The refractory metals group includes niobium (also known as columbium), tantalum, molybdenum, tungsten, and rhenium. The name of this group of metals arises from their very high melting temperatures, which range from 2468 to 3410°C (4474 to 6170°F). The metals and their alloys find specialized applications in the electronics, aerospace, nuclear, and chemical process industries.

Except for niobium, the refractory metals are produced as metal powders which are consolidated by sintering or melting into ingots. These are then processed by combinations of forging, extrusion, and rolling into bar, plate, sheet, foil, and tubing. They can be machined, generally using carbide tools, and a range of techniques is available for their fabrication. Alloys of niobium and tantalum are the most easily fabricated, while more specialized techniques are required for molybdenum and tungsten. Joining is accomplished using electron beam, gas tungsten arc, and resistance welding processes.

Despite their high melting temperatures, the refractory metals react with oxidizing atmospheres at moderate temperatures and this has restricted their use as high temperature materials. In attempts to compensate for this, coatings of the silicide or aluminide type have been developed which enable niobium to be used at temperatures up to about 1650°C (3002°F). For the other refractory metals, coatings developments have been less successful to date, although research continues. Refractory metals and alloys in general have good resistance to corrosion by liquid metals and by acid solutions, and this is responsible for some of their applications.

Niobium (Columbium)

Although most of the niobium produced is used as alloy additions in the steel industry, the combination of relatively light weight and high temperature strength has led to the use of niobium alloys in the aerospace industry. In many cases these applications are only possible when the materials are protected by coatings, for example, a coating of Si-20%Cr-20%Fe applied as a slurry which is subsequently baked to stimulate reaction bonding and diffusion. Niobium alloys such as C-103 (UNS R04295) with 10%Hf and 1%Ti, C-129Y with 10%W, 10%Hf and 0.1%Y, Cb-752 (UNS R04271) with 10%W and 2.5%Zr, and FS-85 (28%Ta, 10%W, 1%Zr) find aerospace applications such as thrust chambers and radiation skirts for rocket and aircraft engines, rocket nozzles, thermal shields, leading edges and nose caps for hypersonic flight vehicles, and guidance structure for glide re-entry vehicles. A Nb-1%Zr alloy (commercial grade UNS R04261, reactor grade UNS R04251) finds applications in the nuclear industry. Several commercial and nuclear grades of unalloyed niobium (UNS R04200, UNS R04210, and UNS R04211) are also available.

Tantalum

The major application of tantalum is in electronics where it is used in the form of porous sintered powder metallurgy electrodes in electrolytic capacitors and as precision foil in foil capacitors, as well as lead wires, seals, and containment cans. Tantalum is also produced in the form of mill products including sheet, plate, rod, bar, and tubing. Tubing can be either seamless or welded and drawn. Tantalum and tantalum-clad steel are used in chemical process applications under many severe conditions including the condensing, reboiling, preheating, and cooling of nitric, hydrochloric, and sulphuric acids. It is used as heating elements and heat shields for furnaces and in specialized aerospace, nuclear, and biomedical applications. Unalloyed tantalum is designated by UNS R05200 (cast), UNS R05210, and UNS R05400 (sintered). Tantalum alloys include Ta-2.5%W (63 metal) which is used in chemical processing for heat exchangers, tower linings, valves, and tubing, and Ta-7.5%W, a powder product which is cold drawn to springs and other components for severe acid service. The alloy Ta-10%W (UNS R05255) finds high temperature applications up to 2480°C (4496°F) in the aerospace industry, for example as hot gas valves and rocket engine components, and also in corrosive and abrasive environments in the chemical process and nuclear industries. Other commercial tantalum alloys include T-111 (8%W, 2%Hf), T-222 (10%W, 10%Hf) and Ta-40%Nb (UNS R05240).

Molybdenum

Like niobium, most molybdenum production goes for use as an alloying element in steels and superalloys; however, some sheet, wire, bar, and tube are produced by forging, rolling and extrusion. Aerospace applications include high temperature structural parts such as nozzles, supports, shields, and pumps. Molybdenum is used for tools in the metalworking industry, for acid service in the chemical process industry, for electrical resistance furnace heating elements, and for numerous components in the electronics, nuclear, metallizing, and glass industries. Unprotected molybdenum oxidizes rapidly above 500°C (932°F) so it cannot be used in oxidizing atmospheres. Commercial alloys include alloy 362 (UNS R03620) which contains 0.5%Ti, the alloy TZC which contains 1%Ti and 0.3%Zr, and the alloy TZM (UNS R03630, which is also known as alloy 363 when arc cast and UNS R03640, or alloy 364 when in powder metallurgical form) which contains 0.5%Ti and 0.1%Zr. Other alloys include HCM (1.1%Hf, 0.07%C), HWM-25 (25%W, 1.0%Hf), and HWM-45 (45%W, 0.9%Hf). Several grades of unalloyed molybdenum are also available.

Tungsten

The main use of tungsten is as tungsten carbide which is discussed in Chapter 5; however, tungsten is also used extensively in metal or alloy form where advantage is taken of its high melting point, density, and elastic modulus. For example, it is used in wire form in lighting, electronic devices, and thermocouples (lamp filaments, electron tube grids, and heating elements were among its original applications). It is also used in counterweights, flywheels, radiation shields, furnace parts, crucibles, and in the metalworking industry. Tungsten mill products can be classified into three groups, based on recrystallization temperature. The first group, including high purity tungsten and its alloys with rhenium or molybdenum, recrystallizes at or below 900°C (1652°F), while the second group, commercial purity (undoped) tungsten, recrystallizes above 1200°C (2192°F). The third group consists of AKS doped tungsten (doped with small amounts of aluminum, potassium, and silicon), doped tungsten alloyed with rhenium, and tungsten containing particles of thoria. These materials recrystallize above 1800°C (3272°F), forming a characteristic microstructure of elongated interlocking grains.

The high strength and creep resistance of tungsten is offset by strong reactivity in air and poor low temperature ductility. Tungsten exhibits a ductile-brittle transition which occurs above 200°C (392°F) in recrystallized material but at lower temperatures after cold work. The retardation of recrystallization which is caused by AKS doping or by additions of a dispersion of thoria particles improves the ductility and also allows the material to retain tensile strength and creep resistance at high temperatures; this is particularly important for applications in filaments for incandescent lamps where sagging is undesirable. In addition to AKS doped tungsten and a range of unalloyed tungsten materials (e.g., UNS R07005 and R07006), commercially available alloys include tungsten containing 1% or 2%ThO_2, tungsten alloyed with 2% or 15%Mo, or with 1.5%, 3% (UNS R07031) or 25%Re. Also available are arc welding electrode materials consisting of tungsten containing either thoria (UNS R07911 with 1%ThO_2 and UNS R07912 with 2%ThO_2), zirconia (UNS R07920 with 0.3%ZrO), ceria (UNS R07932 with 2%CeO_2) or lanthana (UNS R07941 with 1%La_2O_3).

The machinability of tungsten can be a problem, but machinable tungsten heavy-metal alloys bonded with copper, nickel, and iron have been developed. These are produced by liquid phase sintering to create a soft ductile matrix phase in which particles of the tungsten alloy are embedded. Included in this category are the Class 1 alloys containing 6-7%Ni and 3-4%Cu, and a group of alloys containing nickel and iron in the ratio of 4Ni:1Fe (Class 2), 7Ni:3Fe (Class 3), and 1Ni:1Fe (Class 4).

Zirconium

Zirconium, along with metals such as titanium and hafnium, falls into the category of reactive metals. It is similar in many respects chemically and mechanically to titanium, but has both a higher density and a higher melting temperature (6.51 $g.cm^{-3}$ and 1852°C [3366°F] as compared to titanium at 4.51 g cm^{-3} and 1668°C [3034°F]). Zirconium is chemically very reactive, such that in finely divided form it is pyrophoric. This property gives it applications in flashbulbs and as getters for absorbing gases in sealed evacuated devices. However, the oxide formed on bulk zirconium is dense, stable, and protective (self-healing and adherent) so that zirconium and its alloys have excellent corrosion resistance in many media, including superheated water. This, combined with its low capture cross section for thermal neutrons, is responsible for its most important application, namely as a structural (e.g., pressure tubes) and fuel cladding material in water cooled nuclear reactors. It is also used in pressure vessels, pumps, valves, and heat exchangers in chemical processing equipment for severe corrosion environments. Zirconium alloys are processed by most available techniques including forging, rolling, and extrusion as well as by casting. They are generally machinable and are also weldable, most frequently by GTAW (gas tungsten arc welding) and GMAW (gas metal arc welding) techniques using appropriate atmosphere control.

Like titanium, zirconium is allotropic with the hexagonal close-packed (*alpha* phase) crystal structure being stable at low temperatures, and the body centered cubic (*beta*) phase stable above 870°C (1598°F) in the pure metal. As in the case of titanium, particular alloying elements stabilize either the *alpha* or the *beta* phase, and a wide range of microstructures are possible depending on composition and heat treatment. However, the most common commercial zirconium alloys are dilute *alpha* phase alloys, not unlike unalloyed zirconium in metallurgical behaviour. The four commercial alloys used in nuclear applications are reactor grade zirconium (UNS R60001), Zr-2.5%Nb (UNS R60901), Zircaloy-2 (UNS R60802), which contains 1.5%Sn as well as a total of about 0.3% Fe+Cr+Ni, and Zircaloy-4 (UNS R60804), which also contains 1.5%Sn with a total of about 0.35% Fe+Cr. The non-nuclear industrial zirconium alloys are quite similar and include Grade 702, a commercially pure zirconium (UNS R60702), the alloy Grade 704 (UNS R60704), which is similar to the Zircaloys, and two Zr-2.5%Nb alloys known as Grade 705 (UNS R60705) and Grade 706 (UNS R60706).

Cobalt

Cobalt is a metal which is comparable in many of its properties with nickel and iron. For example its melting point, density, and thermal expansion are all close to those of nickel and iron and it is ferromagnetic. Like iron, it is allotropic, its crystal structure below 417°C (783°F) being hexagonal close-packed (hcp) while above this temperature it is face centered cubic (fcc). Cobalt and its alloys find commercial applications as heat-resistant, corrosion-resistant, and wear-resistant alloys and as the matrix component of cemented carbide cutting tools which are discussed in Chapter 5.

Heat-resistant cobalt alloys are used for components in gas turbines and in power generating and chemical processing equipment where their high strength at high temperatures and their resistance to sulphidation are often superior to those of the nickel and nickel-iron based high temperature alloys. Applicable cobalt alloys include the wrought Haynes alloys 25 (also called L-605, UNS R30605) and 188 (UNS R30188) and the casting alloy MAR-M 509. These contain 20-23% chromium and 7-15% tungsten for solution strengthening along with 10-22% nickel to stabilize the fcc phase, and are heat treated to control the distribution of strengthening carbide particles.

The corrosion-resistant cobalt alloys are designed for use where corrosion resistance is required in combination with resistance to wear and high strength over a wide range of temperature. Included in this category are the wrought alloys MP35N (UNS R30035, which contains 20% Cr, 10%Mo, 35%Ni) and Haynes alloy 1233 (UNS R31233, with 26%Cr, 5%Mo, 2%W, 9%Ni). Typical applications are as pump and valve components. The corrosion-resistant cobalt-based alloy Vitallium (30%Cr, 6%Mo in cast form, 20%Cr, 15%W in wrought form), and alloy MP35N are used for biomedical applications such as prosthetic devices and implants as well as dental implants where advantage can be taken of their compatibility with body fluids and tissues.

Cobalt-based wear-resistant alloys are widely used, often as weld overlays (hardfacing alloys), although some are also available as castings or wrought products and some in powder metallurgy versions. They find applications where resistance to wear is required under moderately corrosive and/or elevated temperature environments in the automotive, chemical processing, marine, power generation, pulp and paper, oil and gas, steel, and textile industries. A major application is as weld overlay on valve seating surfaces and they are also used as overlays on bearing surfaces and on knife and shear blades. Wrought applications include erosion shields and guide bars.

The microstructures of these alloys consist basically of particles of a hard carbide phase dispersed throughout a softer matrix, with the amount of carbide exceeding 30% by volume in some cases. The matrix controls resistance to sliding wear (e.g., fretting and galling) and to liquid droplet and cavitation erosion, while resistance to abrasive wear as well as to solid particle and slurry erosion relies on both constituents in a complex manner. Typical compositions include 20-30% chromium with up to 15% tungsten and small amounts of molybdenum, iron, nickel, silicon, and manganese (a few alloys contain substantial additions of nickel, molybdenum, or niobium). These provide solution strengthening and wear resistance along with oxidation and corrosion resistance to the matrix as well as additional strength and wear resistance through carbide precipitation. Carbon contents lie in the range 0.5-3%; the carbide particles are generally of the chromium-rich M_7C_3 type, although in high tungsten alloys the tungsten-rich M_6C type also form. Distribution of carbide particles is not readily controlled in the weld overlay and cast structures, and grain boundary carbides and segregation are common.

A wide range of alloys is available including the Stellite alloys 1, 6, 12, and 21 (UNS R30001, R30006, R30012, and R30021). Of these, Stellite 1 is the hardest, most abrasion resistant, and least ductile, while Stellite 21 is the most corrosion resistant. Haynes alloy 6B is a wrought version of Stellite 6 which has improved properties by virtue of the improved microstructural control which is possible during wrought processing. Other commercial alloys include Havar (UNS R30004), Elgiloy (UNS R30003), and Stellite F (UNS R30002).

The Tribaloy alloys, such as Tribaloy T-400 (UNS R30400) and T-800 have reduced chromium (8% and 17% respectively) and carbon contents and much increased amounts of molybdenum (28%) and silicon (3%), to suppress carbides and stimulate the formation of hard, corrosion-resistant intermetallic "Laves phase" particles which impart excellent abrasion resistance. Some other alloys are designed to be precipitation hardening, including the cobalt-nickel-chromium-molybdenum-tungsten alloys Duratherm 2602 (UNS R30280), Duratherm 477 (UNS R30477), Duratherm 700 (UNS R30700), and wrought Duratherm 600 (UNS R30600), as well as the cobalt-chromium-iron-nickel alloy ARNAVAR (UNS R30007).

SAE/AMS SPECIFICATIONS - REACTIVE & REFRACTORY ALLOYS	
AMS	**Title**
7817	Molybdenum Alloy Sheet, Strip, and Plate, 0.48Ti - 0.09Zr - 0.02C, Arc Cast, Stress Relieved
7819	Molybdenum Alloy Bars, 0.48Ti - 0.09Zr - 0.02C, Arc Cast, Stress Relieved, UNS R03640
7846	Tantalum Alloy Bars and Rods, 90Ta - 10W, Annealed, UNS R05255
7847	Tantalum Alloy Sheet, Strip, and Plate, 90Ta - 10W, UNS R05255
7848	Tantalum Alloy Bars and Rods, 90Ta - 10W, UNS R05255
7849	Tantalum Sheet, Strip, and Plate, Annealed, UNS R05210
7850	Columbium Sheet, Strip, Plate, and Foil, UNS R04211
7851	Foil, sheet, Strip, and Plate, Columbium Alloy, 10W - 2.5Zr, Recrystallized, UNS R04271
7852	Sheet, Strip, and Plate, Columbium Alloy, 10Hf - 1.0Ti, UNS R04295
7855	Columbium Alloy Bars, Rods, and Wire, 10W - 2.5Zr, Recrystallized, UNS R04271
7857	Columbium Alloy Bars, Rods, and Extrusions, 10Hf - 1.0Ti, Recrystallization Annealed, UNS R04295
7875	Powder, Chromium Carbide Plus Nickel-Chromium Alloy, 75Cr2C3 + 25 (80Ni - 20Cr Alloy)
7897	Tungsten Forgings, Pressed, Sintered, and Forged, UNS R07005
7898	Tungsten Sheet, Strip, Plate, and Foil, Pressed, Sintered, and Wrought, UNS R07006

ASTM SPECIFICATIONS - REACTIVE & REFRACTORY ALLOYS	
Molybdenum and Molybdenum Alloys	
B 387	Molybdenum and Molybdenum Alloy Bar, Rod, and Wire
B 386	Molybdenum and Molybdenum Alloy Plate, Sheet, Strip, and Foil
F 289	Molybdenum Wire and Rod for Electronic Applications
F 290	Round Wire for Winding Electron Tube Grid Laterals
Tantalum and Tantalum Alloys	
B 364	Tantalum and Tantalum Alloy Ingots
B 365	Tantalum and Tantalum Alloy Rod and Wire
B 521	Tantalum and Tantalum Alloy Seamless and Welded Tubes
B 708	Tantalum and Tantalum Alloy Plate, Sheet, and Strip

ASTM SPECIFICATIONS - REACTIVE & REFRACTORY ALLOYS (Continued)	
Tungsten	
B 390	Evaluating Apparent Grain Size and Distribution of Cemented Tungsten Carbides
B 482	Preparation of Tungsten and Tungsten Alloys for Electroplating
B 760	Tungsten Plate, Sheet, and Foil
B 777	Tungsten Base, High-Density Metal
F 288	Tungsten Wire for Electron Devices and Lamps
F 290	Round Wire for Winding Electron Tube Grid Laterals
Niobium and Niobium Alloys	
B 391	Niobium and Niobium Alloy Ingots
B 392	Niobium and Niobium Alloy Bar, Rod, and Wire
B 393	Niobium and Niobium Alloy Strip, Sheet, and Plate
B 394	Niobium and Niobium Alloy Seamless and Welded Tubes
B 652	Niobium-Hafnium Alloy Ingots
B 654	Niobium-Hafnium Alloy Foil, Sheet, Strip, and Plate
B 655	Niobium-Hafnium Alloy Bar, Rod, and Wire
Zirconium and Zirconium Alloys	
B 349	Zirconium Sponge and Other Forms of Virgin Metal for Nuclear Application
B 350	Zirconium and Zirconium Alloy Ingots for Nuclear Application
B 351	Hot-Rolled and Cold-Finished Zirconium and Zirconium Alloy Bars, Rod, and Wire for Nuclear Application
B 352	Zirconium and Zirconium Alloy Sheet, Strip, and Plate for Nuclear Application
B 353	Wrought Zirconium and Zirconium Alloy Seamless and Welded Tubes for Nuclear Service
B 493	Zirconium and Zirconium Alloy Forgings
B 494	Primary Zirconium
B 495	Zirconium and Zirconium Alloy Ingots
B 523	Seamless and Welded Zirconium and Zirconium Alloy Tubes
B 550	Zirconium and Zirconium Alloy Bar and Wire
B 551	Zirconium and Zirconium Alloy Strip, Sheet, and Plate
B 653	Seamless and Welded Zirconium and Zirconium Alloy Welding Fittings

ASTM SPECIFICATIONS - REACTIVE & REFRACTORY ALLOYS (Continued)	
Zirconium and Zirconium Alloys (Continued)	
B 658	Seamless and Welded Zirconium and Zirconium Alloy Pipe
B 752	Castings, Zirconium-Base, Corrosion Resistant, for General Application
B 811	Wrought Zirconium Alloy Seamless Tubes for Nuclear Reactor Fuel Cladding

AMERICAN CROSS REFERENCED SPECIFICATIONS - REACTIVE & REFRACTORY ALLOYS

UNS	AMS	ASTM	ASME	AWS
R03600	---	B 386 (360); B 387 (360)	---	---
R03603	---	F 364 (II)	---	---
R03604	---	F 364 (1)	---	---
R03605	7801	---	---	---
R03606	7805	---	---	---
R03610	---	B 386 (361); B 387 (361)	---	---
R03630	7817	B 386 (363); B 387 (363)	---	---
R03640	---	B 386 (364); B 387 (364)	---	---
R03650	---	B 386 (365); B 387 (365)	---	---
R04200	---	B 391; B 392, B 393; B 394	---	---
R04210	---	B 391; B 392; B 393; B 394	---	---
R04211	7850	---	---	---
R04251	---	B 391; B 392; B393; B394	---	---
R04261	---	B 391; B 392 (4); B 393; B 394	---	---
R04271	7851; 7855	---	---	---
R04295	7852; 7857	B 652; B 654; B 655	---	---
R05200	---	B 364; B 365; B 521	---	---
R05210	7849	---	---	---
R05240	---	B 364; B 365; B 521; B 708	---	---
R05255	7846; 7847; 7848	B 364; B 365; B 521; B 708	---	---

AMERICAN CROSS REFERENCED SPECIFICATIONS - REACTIVE & REFRACTORY ALLOYS (Continued)				
UNS	**AMS**	**ASTM**	**ASME**	**AWS**
R05400	---	B 364; B 365; B 521; B 708	---	---
R07005	7897	F 288 (1A, 1B); F 290	---	---
R07006	7898	---	---	---
R07030	---	F 73 (electronic grade)	---	---
R07900	---	---	SFA-5.12 (EWP)	A5.12 (EWP)
R07911	---	---	SFA-5.12 (EWTh-1)	A5.12 (EWTh-1)
R07912	---	---	SFA-5.12 (EWTh-2)	A5.12 (EWTh-2)
R07920	---	---	SFA-5.12 (EWZr-1)	A5.12 (EWZr-1)
R07932	---	---	SFA-5.12 (EWCe-2)	A5.12 (EWCe-2)
R07941	---	---	SFA-5.12 (EWLa-1)	A5.12 (EWLa-1)
R19800	7901; 7902	---	---	---
R19920	7900	---	---	---
R60001	---	B 349; B3 50; B 351; B 352; B 353	---	---
R60701	---	B 494; B 550	---	---
R60702	---	B 493; B 495; B 523; B 550; B 551; B 653; B 658; B 752 (702C)	SB493; SB494; SB495; SB523; SB550; SB551; SFA-5.24 (ERZr2)	A5.24 (ERZ r2)
R60703	---	B 494; B 495	---	---
R60704	---	B 493; B 495; B 523; B 550; B 551; B 653; B 658; B 752 (704C)	SFA-5.24 (ERZr3)	A5.24 (ERZr3)
R60705	---	B 493; B 495; B 523; B 550; B 551; B 653; B 658; B 752 (705C)	---	---
R60706	---	B 495; B 551	---	---
R60707	---	---	SFA-5.24 (ERZr4)	A5.24 (ERZr4)
R60802	---	B 350; B 351; B 352; B 353	---	---
R60804	---	B 350; B 351; B 352; B 353	---	---
R60901	---	B 350; B 351; B 352; B 353	---	---
R60904	---	B 353	---	---

AMERICAN CROSS REFERENCED SPECIFICATIONS - REACTIVE & REFRACTORY ALLOYS (Continued)

a. This cross-reference table lists the basic specification or standard number, and since these standards are constantly being revised, it should be kept in mind that they are presented herein as a guide and may not reflect the latest revision.

COMMON NAMES OF REACTIVE & REFRACTORY ALLOYS WITH UNS No.
Molybdenum, Unalloyed, R03600
Molybdenum Sealing Alloy, R03601
Molybdenum Sealing Alloy, R03602
Molybdenum, R03603
Molybdenum, R03604
Molybdenum Metal, R30605
Molybdenum Metal, R03606
Molybdenum Unalloyed, R03610
Molybdenum Alloy, R03620
Molybdenum Alloy, R03630
Molybdenum Alloy, R03640
Molybdenum Unalloyed, Low Carbon, R03650
Niobium (Columbium), Unalloyed, Reactor Grade, R04200
Niobium (Columbium), Unalloyed, Commercial Grade, R04210
Niobium (Columbium), Unalloyed, Commercial Grade, R04211
Niobium (Columbium) - 1% Zirconium, Reactor Grade, R04251
Niobium (Columbium) - 1% Zirconium, Commercial Grade, R04261
Niobium (Columbium) Alloy, 10W-2.5Zr, R04271
Niobium (Columbium) Alloy, R04295
Tantalum, Unalloyed, Cast, R05200
Tantalum Metal, R05210
Tantalum - 40 Nb (Cb) Alloy, R05240
90 Tantalum - 10 Tungsten, R05255

COMMON NAMES OF REACTIVE & REFRACTORY ALLOYS WITH UNS No. (Continued)
Tantalum, Unalloyed, Sintered, R05400
Tungsten, R07005
Tungsten Metal, R07006
Tungsten, R07030
Tungsten-Rhenium, R07031
Tungsten, R07080
Tungsten, R07100
Tungsten Arc Welding Electrode, R07900
Tungsten-Thorium Alloy Arc Welding Electrode, R07911
Tungsten-Thorium Alloy Arc Welding Electrode, R07912
Tungsten-Zirconium Alloy Arc Welding Electrode, R07920
Tungsten-Cerium Alloy Arc Welding Electrode, R07932
Tungsten-Lanthanum Alloy Arc Welding Electrode, R07941
Zirconium Unalloyed Reactor Grade, R60001
Zirconium, Unalloyed, R60701
Zirconium, Unalloyed, R60702
Zirconium Alloy, R60703
Zirconium Alloy, R60704
Zirconium Alloy, R60705
Zirconium Alloy, R60706
Zirconium Alloy (2.5 Cb) Welding Filler Metal, R60707
Zirconium Alloy Reactor Grade, R60802
Zirconium Alloy Reactor Grade, R60804
Zirconium Alloy Reactor Grade, R60901
Zirconium Alloy Reactor Grade, R60902
Zirconium - 0.65 Nb, R60904

CHEMICAL COMPOSITION OF REACTIVE & REFRACTORY ALLOYS

UNS	CHEMICAL COMPOSITION
R03600	Mo rem C 0.010-0.040 Fe 0.010 max N 0.0010 max Ni 0.005 max O 0.0030 max Si 0.10 max
R03601	Mo 99.90 min C 0.04 max Fe 0.01 max N 0.001 max Ni 0.01 max O 0.005 max Si 0.01 max W 0.02 max Other each 0.005 max
R03602	Mo 99.90 min C 0.005 max Fe 0.01 max N 0.002 max Ni 0.01 max O 0.008 max Si 0.01 max W 0.02 max Other each 0.005 max
R03603	Mo 99.90 min Al 0.015 max C 0.0015 max Fe 0.001 max H 0.001 max N 0.001 max O 0.175 max Si 0.035 max Sn 0.0025 max W 0.02 max Other each 0.005 max, Ca 0.005 max, K 0.015 max
R03604	Mo 99.90 min Al 0.015 max C 0.005 max Fe 0.01 max H 0.001 max N 0.002 max O 0.008 max Si 0.01 max Sn 0.0025 max W 0.02 max Other each 0.005 max, Ca 0.005 max, K 0.015 max
R03605	Mo 99.90 min C 0.030 max Fe 0.020 max N 0.0010 max Ni 0.010 max O 0.0030 max Si 0.010 max
R03606	Mo 99.95 min C 0.030 max Fe 0.008 max H 0.0005 max N 0.002 max Ni 0.002 max O 0.0015 max Si 0.008 max
R03610	Mo rem C 0.010 max Fe 0.010 max N 0.0020 max Ni 0.005 max O 0.0070 max Si 0.010 max
R03620	Mo rem C 0.010-0.040 Fe 0.010 max N 0.0010 max Ni 0.005 max O 0.0030 max Si 0.010 max Ti 0.40-0.55
R03630	Mo rem C 0.010-0.040 Fe 0.010 max N 0.0010 max Ni 0.005 max O 0.0030 max Si 0.010 max Ti 0.40-0.55 Zr 0.06-0.12
R03640	Mo rem C 0.010-0.040 Fe 0.010 max N 0.0020 max Ni 0.005 max O 0.030 max Si 0.005 max Ti 0.40-0.55 Zr 0.06-0.12
R03650	Mo rem C 0.010 max Fe 0.010 max N 0.0010 max Ni 0.005 max O 0.0030 max Si 0.010 max
R04200	Mo 0.005 max C 0.01 max Fe 0.005 max H 0.001 max Hf 0.01 max N 0.01 max Nb rem Ni 0.005 max O 0.015 max Si 0.005 max Ta 0.1 max W 0.03 max Zr 0.01 max
R04210	Mo 0.005 max C 0.01 max Fe 0.01 max H 0.001 max Hf 0.01 max N 0.01 max Nb rem Ni 0.005 max O 0.025 max Si 0.005 max Ta 0.2 max W 0.05 max Zr 0.01 max
R04211	Nb rem C 0.005 max Fe 0.010 max H 0.002 max N 0.010 max O 0.030 max Si 0.005 max Ta 0.10 max Ti 0.005 max Zr 0.010 max Other each 0.010 max, total 0.15 max
R04251	Nb rem C 0.01 max Fe 0.005 max H 0.001 max Hf 0.01 max Mo 0.005 max N 0.01 max Ni 0.005 max O 0.015 max Si 0.005 max Ta 0.1 max W 0.03 max Zr 0.8-1.2
R04261	Nb rem C 0.01 max Fe 0.01 max H 0.001 max Hf 0.01 max Mo 0.005 max N 0.01 max Ni 0.005 max O 0.025 max Si 0.005 max Ta 0.2 max W 0.05 max Zr 0.8-1.2
R04271	Nb rem C 0.030 max Fe 0.02 max H 0.001 max N 0.010 max O 0.020 max Si 0.02 max Ta 0.15 max Ti 0.01 max W 9.00-11.00 Zr 2.00-3.00
R04295	Nb rem C 0.015 max H 0.0015 max Hf 9-11 N 0.010 max O 0.020 max Ta 0.500 max Ti 0.7-1.3 W 0.500 max Zr 0.700 max

CHEMICAL COMPOSITION OF REACTIVE & REFRACTORY ALLOYS (Continued)	
UNS	**CHEMICAL COMPOSITION**
R05200	Ta rem C 0.01 max Cb 0.05 max Fe 0.01 max H 0.001 max Mo 0.01 max N 0.01 max Ni 0.01 max O 0.015 max Si 0.005 max Ti 0.01 max W 0.03 max
R05210	Ta 99.85 min C 0.0075 max Cb 0.100 max Co 0.002 max Fe 0.01 max H 0.0010 max Mo 0.01 max N 0.0075 max Ni 0.005 max O 0.015 max Si 0.005 max Ti 0.005 max W 0.03 max Zr 0.01 max Other total 0.15 max
R05240	Ta rem C 0.010 max Cb 35.0-42.0 H 0.0015 max Fe 0.010 max Mo 0.020 max Ni 0.010 max Ni 0.010 max O 0.020 max Si 0.005 max Ti 0.010 max W 0.050 max Trace impurities 0.00 max
R05255	Ta rem C 0.01 max Cb 0.10 max Fe 0.01 max H 0.001 max Mo 0.01 max N 0.01 max Ni 0.01 max O 0.015 max Si 0.005 max Ti 0.01 max W 9.0-11.0
R05400	Ta rem C 0.01 max Cb 0.05 max Fe 0.01 max H 0.001 max Mo 0.01 max N 0.015 max Ni 0.01 max O 0.03 max Si 0.005 max Ti 0.01 max W 0.03 max
R60001	Zr rem Al 0.0075 max B 0.00005 max C 0.027 max Cd 0.00005 max Co 0.0020 max Cr 0.020 max Cu 0.0050 max Fe 0.150 max H 0.0025 max Hf 0.010 max Mn 0.0050 max N 0.0065 max Ni 0.0070 max Si 0.0120 max Ti 0.0050 max U 0.00055 max W 0.010 max
R60701	Zr+Hf 99.5 min C 0.05 max H 0.005 max Hf 4.5 max N 0.025 max Other Fe+Cr 0.05 max
R60702	Zr+Hf 99.2 min C 0.05 max H 0.005 max Hf 4.5 max N 0.025 max Other Fe+Cr 0.2 max
R60703	Zr+Hf 98.0 min Hf 4.5 max
R60704	Zr+Hf 97.5 min C 0.05 max H 0.005 max Hf 4.5 max N 0.025 max Sn 1.00-2.00 Other Fe+Cr 0.20-0.40
R60705	Zr+Hf 95.5 min C 0.05 max Cb 2.0-3.0 H 0.005 max Hf 4.5 max N 0.025 max O 0.18 max
R60706	Zr+Hf 95.5 min C 0.050 max Cb 2.0-3.0 H 0.005 max Hf 4.5 max N 0.025 max O 0.16 max Other Fe+Cr 0.2 max
R60707	Zr+Hf 95.5 min C 0.05 max Cb 2.0-3.0 Fe 0.20 max includes Cr H 0.005 max Hf 4.5 max N 0.025 max
R60802	Zr rem Al 0.0075 max B 0.00005 max C 0.027 max Cd 0.00005 max Co 0.0020 max Cr 0.05-0.15 Cu 0.0050 max Fe 0.07-0.20 H 0.0025 max Hf 0.010 max Mn 0.0050 max N 0.0065 max Ni 0.05-0.08 Si 0.020 max Sn 1.20-1.70 Ti 0.0050 max U 0.00035 max W 0.010 max Other Fe+Cr+Ni 0.18-0.38
R60804	Zr rem Al 0.0075 max B 0.00005 max C 0.027 max Cd 0.00005 max Co 0.0020 max Cr 0.07-0.13 Cu 0.0050 max Fe 0.18-0.24 H 0.0025 max Hf 0.010 max Mn 0.0050 max N 0.0065 max Ni 0.0070 max Si 0.0120 max Sn 1.20-1.70 Ti 0.0050 max U 0.00035 max W 0.010 max Other Fe+Cr 0.28-0.37
R60901	Zr rem Al 0.0075 max B 0.00005 max C 0.027 max Cb 2.40-2.80 Cd 0.00005 max Co 0.0020 max Cr 0.020 max Cu 0.0050 max Fe 0.150 max H 0.0025 max Hf 0.010 max Mn 0.0050 max N 0.0065 max Ni 0.0070 max O 0.09-0.13 Si 0.0120 max Ti 0.0050 max U 0.00035 max W 0.010 max

CHEMICAL COMPOSITION OF REACTIVE & REFRACTORY ALLOYS (Continued)	
UNS	**CHEMICAL COMPOSITION**
R60902	Zr rem Al 0.0075 max B 0.00005 max C 0.030 max Cb 2.4-2.8 Cd 0.00005 max Co 0.0020 max Cr 0.020 max Cu 0.3-0.7 Fe 0.15 max H 0.0025 max Hf 0.010 max Mg 0.0020 max Mn 0.0050 max N 0.0065 max Ni 0.0070 max O 0.08-0.12 Pb 0.013 max Si 0.012 max Sn 0.0050 max Ta 0.020 max Ti 0.0050 max U 0.00035 max V 0.0050 max W 0.010 max
R60904	Zr remAl 0.0075 max B 0.00005 max Cb 2.50-2.80 Co 0.0020 max Cu 0.0050 max Fe 0.0650 max H 0.0020 max Hf 0.0050 max Mg 0.0020 max Mn 0.0050 max Mo 0.0050 max N 0.0065 max Ni 0.0035 max Pb 0.0050 max Si 0.0120 max Sn 0.0100 max Ta 0.0100 max Ti 0.0050 max U 0.00035 max V 0.0050 max W 0.0050 max

Chapter 5

CEMENTED CARBIDES

These materials are composites consisting of particles of a hard, brittle, nonmetallic refractory material, typically carbides but sometimes carbonitrides or nitrides, in a metallic binder (matrix) phase, typically cobalt but sometimes nickel, iron, or their alloys. The matrix phase, which imparts ductility, toughness, and thermal conductivity to the composite, typically occupies 10-15% of the volume. The outer surface may or may not be coated for improved surface properties. The original material in this category was WC-Co; however, there has been a long history of development which has led to many other formulations. Applications include metal cutting, mining, construction, rock drilling, metal forming, structural components, and wear parts, but of these metal cutting is the most important, accounting for approximately half of production. The steps involved in manufacture of these materials include powder preparation, then compaction or consolidation by pill or hydrostatic pressing or extrusion, followed by sintering, and finally forming. During sintering, the metallic binder melts and draws the carbide particles together. Pressure can be applied during the sintering, for example this can be carried out by hot isostatic pressing in a vacuum. The properties of sintered hardmetals are profoundly influenced by microstructure, including the compositions of the phases, and the sizes, shapes, and distributions of carbide particles. One of the problems which can arise is that when there is a deficiency of carbon, some of the original WC particles can dissolve, and particles of a mixed cobalt-tungsten carbide ("*eta* phase") can form, giving rise to serious embrittlement.

Carbides produced for use in hardmetals include WC, TiC, TaC, NbC, and solid solution mixed carbides such as WC-TiC, WC-TiC-TaC, and WC-TiC-(Ta,Nb)C.

The basic WC-Co alloys, referred to as straight grades, have excellent resistance to abrasive wear and find many applications in metal cutting. Although alloys ranging between 3% and 30%Co are produced, those most widely used in machining typically contain 3-12% Co and WC particles with grain sizes between 0.5 and 5 micrometers (0.00002 in.).

When more toughness or edge strength is required, finer (submicron) grain size carbide particles are used, for example in indexable inserts and solid carbide drilling and milling tools. The fine grain size is achieved by additions of other carbides, such as vanadium, tantalum, niobium, or chromium carbide. Higher Co contents, up to 30%, and larger carbide grain sizes, up to 15 micrometers (0.00006 in.) are used in tools, dies, and wear parts subjected to shock loading. Mixed WC-TiC-Co hardmetals were developed for machining steel, where WC-Co materials are subject to chemical attack. Here the TiC is in solid solution with the WC, and is used in amounts up to 18% of the total WC-TiC-Co alloy; higher proportions of TiC cause the carbides to be excessively brittle. Further developments have led to these WC-TiC-Co alloys being replaced by more advanced alloys containing in addition to WC and Co, both TiC and solid solution (Ta,Nb)C. These are called complex grades, multigrades, or steel-cutting grades. Their microstructures contain angular WC particles in a cobalt matrix, as do the straight grades, but in addition they contain rounded particles of solid solution WC-TiC or WC-TiC-(Ta,Nb)C grains. The need for even more advanced alloys has been partially met by the development of surface coatings for these alloys.

Widespread use is now being made of hardmetals to which a surface coating has been applied in order to combine a tougher substrate with a more wear resistant surface layer. Initially surface layers of TiC were applied, first by sintering and then later by chemical vapour deposition (CVD). These have now been superseded by multilayer coatings consisting of combinations of TiC, TiCN, TiN, or more recently Al_2O_3, Hf, Ta, and Zr carbides and nitrides, and many other complex compounds in as many as 13 different layers. The total thickness of surface coatings is typically about 10 micrometers (0.00004 in.), and even the thickest coatings rarely exceed 15 micrometers (0.00006 in.). Production with a cobalt enriched zone beneath the surface coating improves performance in some cases and permits use of higher cutting speeds and feed rates in machining. Coating by physical vapour deposition (PVD), at lower temperatures than is possible with CVD, has also become common; here the formation of brittle phases is avoided and fine grain size can be obtained in the coating. Furthermore, it is possible to obtain sharper edges using the PVD technique, and coatings as thin as 3 micrometers (0.00001 in) can be produced.

There is no universally recognized classification scheme for hardmetals, but two application-oriented systems are in use. The C-Grade (U.S.) system, based on the General Motors Buick system, classifies according to application category (Table 1) with no reference to material or properties; hence, materials in the same category produced by different manufacturers can exhibit different properties and performance. Both machining and non-machining applications are included. The same limitations apply to the more recent ISO codes. ISO Recommendation R513 divides machining grades into three colour-coded groups: straight WC grades (letter K, colour red); alloyed WC grades (letter M, colour yellow); and highly alloyed grades (letter P, colour blue). Within each grade, a number is assigned giving the approximate position in the range between maximum hardness and maximum toughness (shock resistance). These are shown with their applications in Table 2. It must be noted that with this system the coding is assigned to a given material by its manufacturer and that quality control cannot be applied. The ISO R513 has been revised to take into account the increasing use of coated hardmetals, as well as the superhard ceramics, polycrystalline diamond, and cubic boron nitride.

Table 1 American Cemented Carbide Classifications

Buick Code	**IndustryCode**	**Application**
Machining - cast iron, nonferrous & nonmetallic materials		
TC-1	C-1	Roughing cuts
TC-2	C-2	General purpose
TC-3	C-3	Light finishing
TC-4	C-4	Precision machining
Machining - steel		
TC-5	C-5	Roughing cuts
TC-6	C-6	General purpose
TC-7	C-7	Light finishing
TC-8	C-8	Precision machining
Wear Surface		
TC-9	C-9	No shock
TC-10	C-10	Light shock
TC-11	C-11	Heavy shock
Impact		
TC-12	C-12	Light
TC-13	C-13	Medium
TC-14	C-14	Heavy
Miscellaneous		
---	C-15	Light cut hot flash weld removal
---	C-15A	Heavy cut hot flash weld removal
---	C-16	Rock bits
---	C-17	Cold Header dies
---	C-18	Wear at elevated temperatures and/or resistance to chemical reactions
---	C-19	Radioactive shielding, counterbalances and kinetics applications

Table 2 ISO Std. 513 Classification of Cutting Tool Materials

Material Categories	Color Code	ISO	Material Machined
Ferrous metals with long chips	Blue	P01	Steel, steel castings
		P10	Steel, steel castings
		P20	Steel, steel castings, malleable cast iron with long chips
		P30	Steel, steel castings, malleable cast iron with long chips
		P40	Steel, steel castings with sand inclusion and cavities
		P50	Steel, steel castings of medium or low tensile strength, with sand inclusion and cavities
Ferrous metals with long or short chips and nonferrous metals	Yellow	M10	Steel, steel castings, manganese steel, grey cast iron, alloy cast iron
		M20	Steel, steel castings, austenitic or manganese steel, grey cast iron
		M30	Steel, steel castings, austenitic steel, grey cast iron, high temperature resistant alloys
		M40	Mild free-cutting steel, low tensile steel, nonferrous metals, light alloys
Ferrous metals with short chips, nonferrous metals and nonmetallic materials	Red	K01	Very hard grey cast iron, chilled castings over 85 Shore, high silicon-aluminum alloys, hardened steel, highly abrasive plastics, hard cardboard, ceramics
		K10	Grey cast iron over 220 Brinell, malleable cast iron with short chips, hardened steel, silicon-aluminum alloys, copper alloys, plastic, glass, hard rubber, hard cardboard, porcelain, stone
		K20	Grey cast iron over 220 Brinell, nonferrous metals: copper, brass, aluminum
		K30	Low hardness grey cast iron, low tensile steel, compressed wood
		K40	Soft wood or hard wood, nonferrous metals

COATED CEMENTED CARBIDES CROSS REFERENCE

Trade Name	Company Name	Grade	Layer Composition	Coating Depth, μm	ISO Code(s)
Allenite Gold	Edgar Allen Tools Ltd	EM100	TiC+Ti(C,N)+TiN	12	M05-M20/K05-K30
		EM100	TiC+TiN	5-10	M05-M20/K05-K30
		ES200	TiC+Ti(C,N)+TiN	12	P10-P25
		ES201	TiC+TiN	5-10	P10-P25
		ES350	TiC+Ti(C,N)+TiN	12	P30-P40
		ES351	TiC+TiN	5-10	P30-P40
ANC	American National Carbide Co	AOX	$TiC+Al_2O_3$	8	P15-P35/K15-K30
		CAD	$TiC+Al_2O_3+TiN$	10	P15-P35/K15-K30
		CAM	$TiC+Al_2O_3+TiN$	10	P25-P35/K25-K35
		TEX	Ti(C,N)+TiC+TiN	8	P30-P40/K30-K40
		TEXO	TiN	4	P30-P40/K30-K40
		TNT	Ti(C,N)+TiC+TiN	8	P20-P35/K20-K30
		TNTO	TiN	4	P20-P35/K20-K30
		TOX	$TiC+Al_20_3$	8	P25-P35/K25-K35
Arsenal	Arsenal Company	BGC015	$TiC+Al_2O_3$	7+1.5	P05-P30
		BGC135	TiC	5	P25-P45/M15-M30
		BGC315	TiC	5	M15-M20/K05-K25
		BGC1025	TiC	7	P10-P35
Atrax	Atrax	A6G	TiN+Ti(C,N)+TiC+TiN	8	M05-M20/K05-K20
		A7G	TiN+Ti(C,N)+TiC+TiN	8	M05-M15/K05-K15
		A62G	TiN+Ti(C,N)+TiC+TiN	8	M05-M25/K05-K25
		T50G	TiN+Ti(C,N)+TiC+TiN	8	P10-P30
		T55B	$TiN+Ti(C,N)+Al_2O_3$	8	P20-P45/M30-M40/K30-K40
		T55G	TiN+Ti(C,N)+TiC+TiN	8	P20-P45
		T70B	$TiN+Ti(C,N)+Al_2O_3$	8	P05-P25/M05-M20/K05-K20
		T70G	TiN+Ti(C,N)+TiC+TiN	8	P05-P25

COATED CEMENTED CARBIDES CROSS REFERENCE (Continued)

Trade Name	Company Name	Grade	Layer Composition	Coating Depth, μm	ISO Code(s)
Atrax (con't)	Atrax	T76G	TiN+Ti(C,N)+TiC+TiN	8	P10-P25
Bohlerit Royal	Bohlerit GmH & Co KG	111	TiC	4	K05-K20
		121	TiC	4	P10-P35/K05-K20
		131	TiC	4.5	P25-P45
		321	TiC+TiN+Ti(C,N)	8.5	P10-P35/M10-M20
		331	TiC+TiN+TiC+TiN	8.5	M15-M20
		341	TiC+TiN+TiC+TiN	11.5	P20-P50/M20-M40
		421	TiC+Al_2O_3+TiN	8	P05-P30/K05-K20
		432	TiC+Al_2O_3+TiN	7	P25-P45/M15-M25
		610	TiC+Al_2O_3+TiN	6	K05-K15
		615	TiC+Al_2O_3+TiN	6	K05-K20
		625	TiN+TiC+TiN	4.5	P10-P30
		630	TiN+TiC+TiN	4.5	P15-P35
		635	TiN+TiC+TiN	4.5	P15-P35
		640	TiN+TiC+TiN	3	P20-P35
Bonastre	Bonastre SA	HK10	TiC+Al_2O_3+TiN	8	M10-M20/K01-K20
		HK20	TiC	5	M15-M25/K10-K35
		HK25	TiN	3	M10-M25/K10-K30
		SK10	TiC+Al_2O_3+TiN	8	P05-P35/M10-M30/K05-K30
		SK15	TiC+TiN	8	P10-P30/M10-M25/K10-K25
		SK25	TiN	3	P10-P35/M15-M35
		SK30	TiC+TiN	8	P20-P35/M15-M35
		SK35	TiN	3	P25-P40
		SK40	TiC	5	P25-P40/M25-M40
Carbex	Carbex SA	CNX25	composite TiC	8-10	P10-P30/M10-M30
		CNX35	composite TiC	3-4	P20-P40/K10-K25

COATED CEMENTED CARBIDES CROSS REFERENCE (Continued)

Trade Name	Company Name	Grade	Layer Composition	Coating Depth, µm	ISO Code(s)
Carbex (continued)	Carbex SA	CNX45	composite TiC	3-4	P30-P50/M20-M40
		CNX725	TiN	2-3	P10-P35/M10-M30
		CNX745	TiN	2-3	P30-P50/M20-M40
		CS130	TiC	5-6	P20-P40
		CS310	TiC	5-6	K05-K20
		RW2110	TiC+TiN	8-10	P10-P35/K10-K20
		RW2115	TiC+Al_2O_3+TiN	8-10	P05-P30/M10-M30
		RW2310	TiC+TiN	8-10	P30-P50/M20-M40
Carboloy	Carboloy Inc	550	TiN+TiC+TiN	9	P20-P45
		560	TiN+TiC+Al_2O_3+TiN	10	P10-P35/K10-K20
		570	Al_2O_3	6	P05-P25/M10-M20/K05-K20
		T25M	Ti(C,N)+TiC+TiN	5	P05-P40/M30-M40/K30-K40
		TP05	TiC+Al_2O_3+TiN	9	P05/K10
		TP10	TiC+Al_2O_3+TiN	11	P10/K15
		TP15	TiC+Ti(C,N)+Al_2O_3+TiN	7	P05-P25/M05-M20/K05-K20
		TP20	TiC+Al_2O_3+TiN	7	P10-P35/M20
		TP30	TiC+Ti(C,N)+TiN	12	P30/M30
		TP40	Ti(C,N)+TiC+TiN	5	P40/M40
		TP301	Ti(C,N)+TiC+TiN	4	P30/M40
		TP401	TiC+Al_2O_3+TiN	7	P40/M40
Carmet	Carmet Company	7000	TiN+TiC+Al_2O_3+TiN	12	P05-P25/M10-M20/K05-K20
		9443	TiN+TiC+TiN	7	M10-M20/K10-K20
		9720	TiN+TiC+TiN	7	P10-P20
		9721	TiN+TiC+TiN	7	P20-P40
		9740	TiN+TiC+TiN	7	P25-P50

COATED CEMENTED CARBIDES CROSS REFERENCE (Continued)

Trade Name	Company Name	Grade	Layer Composition	Coating Depth, μm	ISO Code(s)
CRH	Pulvi Metalurgia Rossi	K155	TiC+TiN	5	K10-K20
		KTIN	TiC+TiN	4	K10+K20
		P156	TiC-TiN	5	P10-P30
		PTIC	TiC	4	P20-P30
		PTIN	TiC+TiN	4	P10-P30
		R15	TiC+TiN+Al_2O_3	5	P10-P20/K10-K20
		R157	TiC+TiN+Al_2O_3	7	P10-P30/K10-K20
Denitool	H Deni AG	DC15	TiC+TiN	7-8	P05-P30/M10-M20/K05-K20
		DC25	TiC+Ti(C,N)+TiN	9-11	P10-P30/M15-M25/K10-K25
		DP25	TiN	2-3	P10-P20/M10-M25
		DP35	Ti(C,N)	2-3	P10-P20/M10-M20
		DX10	Ti(C,N)+TiN	7-8	P05-P15/M10-M15/K05-K15
		DX20	TiN	2-3	P05-P20/M10-M20/K05-K15
		DX30	Ti(C,N)	2-3	P05-P10/M10-M15/K05-K10
Dijet	Dijet Industrial Co Ltd	JC220	TiC+Al_2O_3	5-10	P10-P30/M10-M30/K20-K40
		JC330	TiC+Al_2O_3+TiN	5-10	P20-P40/M20-M40
		JC1211	TiC+Al_2O_3	5-10	P01-P15/M05-M15/K01-K15
		JC1231	TiC+Al_2O_3	5-10	P05-P20/M05-M20/K05-K20
		JC1341	TiC+Ti(C,N)+TiN	5-10	P15-P35/M10-M30
		JC1361	TiC+Ti(C,N)+TiN	5-10	P20-P40/M20-M40
		JC3512	TiN	2-4	P01-P20/M05-M20/K01-K20
		JC3562	TiN	2-4	P10-P30/M10-M30/K10-K30
DoALL	DoAll Company	DO40	TiC	5	P20-P30
		DO42	TiC	5	P10-P20
		DO44	TiC	5	P25-P30
		DO46	TiC	5	K10-K20

COATED CEMENTED CARBIDES CROSS REFERENCE (Continued)

Trade Name	Company Name	Grade	Layer Composition	Coating Depth, μm	ISO Code(s)
DoALL (continued)	DoAll Company	DO50	$TiC+Al_2O_3+TiN$	10	P05-P30/M10-M25/K05-K20
		DO51	TiC+Ti(C,N)+TiN	7	P25-P45/M30-M40
		DO52	TiC+Ti(C,N)+TiN	7	P10-P35/M20-M30
Duracarb	Duracarb	DC120	TiN	---	P15-P30
		DC150	TiN	---	P20-P30
		DC710	TiC+Ti(C,N)+TiN	8	P05-P25/K05-K25
		DC720	TiC+Ti(C,N)+TiN	8	K05-K20
		DC725	Ti(C,N)	6	P15-P35/K10-K30
		DC735	TiN+TiC+TiN	6	P25-P45
Duramet	Duramet Corporation	DU227	TiN+TiC+TiN	4-8	K20
		DU577	TiN+TiC+TiN	4-8	P40
		DU607	TiN+TiC+TiN	4-8	P30
		DU811P	TiN	2-3	P30-P50
		DU817	TiN+TiC+TiN	4-8	P30-P50
		DU857	TiN+TiC+TiN	4-8	P30
		DU859	$TiC+Al_2O_3$	3-5	P20-P30
Fansteel VR/Wesson	Fansteel VR/Wesson	623	TiC+Ti(C,N)+TiN	5	P40-P50
		630	TiN+TiC	5	M10-M30/K10-K25
		633	TiC+Ti(C,N)+TiN	7	M10-M30/K10-K25
		643	TiC+Ti(C,N)+TiN	7	P30-P45
		650	TiN+TiC	5	P30-P45/M30-M40
		653	TiN+TiC+TiN	7	P25-P45/M30-M40
		655	$TiN+TiC+Al_2O_3+TiN$	10	P20-P40/M25-M35
		660	TiN+TiC	5	P10-P35/K15-K25
		663	TiC+Ti(C,N)+TiN	7	P10-P35/M20-M30

COATED CEMENTED CARBIDES CROSS REFERENCE (Continued)

Trade Name	Company Name	Grade	Layer Composition	Coating Depth, μm	ISO Code(s)
Fansteel VR/Wesson (continued)	Fansteel VR/Wesson	680	$TiN+TiC+Al_2O_3+TiN$	10	P05-P30/K05-K20/M10-M25
		689	$TiN+TiC+Al_2O_3+TiN+TiC$	---	P01-P10/K01-K15
		690	$TiN+TiC+Al_2O_3+TiN$	7	P01-P10/K01-K15
Greenleaf	Greenleaf Corporation	G1	$TiC+Al_2O_3$	---	P05-P35/M10-M20/K15
		G2	$TiC+Al_2O_3$	---	M05-M15/K01-K20
		GA6	TiC-Ti(C,N)+TiN	---	P10-P35/M10-M20
		GA56	TiC+Ti(C,N)+TiN	---	P35-P50/M10-M40/K10-K35
		GA60	TiC+Ti(C,N)+TiN	---	P15-P35/M10-M40
		Ti5	TiC	---	P20-P45/M15-M30
		Ti6	TiC	---	P10-P45/M10-M20/K10
Groovall	US Tool & Cutter Co	C2/Al2O3	Al_2O_3	---	K01-K20
		C2/TiC	TiC	---	K05/K20
		C2/TiN	TiN	---	K05-K20
		C5/Al2O3	Al_2O_3	---	P05-P20
		C5/TiC	TiC	---	P10-P20
		C5/TiN	TiN	---	P10-P20
		C7/Al2O3	Al_2O_3	---	P01-P10
		C7/TiC	TiC	---	P01-P10
		C7/TiN	TiN	---	P01-P10
Walmet	Walmet Inc	A60	$TiC+Al_2O_3$	---	P01-P20/K10-K20
		P2	TiC+TiN	10	K01-K20
		P5	TiC+TiN	10	P05-P30
		P47	TiC+TiN	10	P01-P20
		P52	TiC+TiN	10	P01-P20/K10-K20

COATED CEMENTED CARBIDES CROSS REFERENCE (Continued)

Trade Name	Company Name	Grade	Layer Composition	Coating Depth, µm	ISO Code(s)
Walmet (continued)	Walmet Inc	P54	TiC+TiN	10	P20-P40
		P57	TiC+TiN	10	P10-P35
Hardmet	Nippon Hardmetal Co Ltd	NC115	---	8	P10-P30/M10-M20/K10-K30
		NC22H	---	8	K01-K20
		NC110	---	8	P01-P20/K10-K20
		NC120	---	8	P10-P40/M10-M30
		NC210	---	8	P01-P20/K10-K20
		NCA2H	---	8	K01-K20
		NCA20	---	8	P10-P40/M10-M30
		NPN10	---	2	P01-P20/K10-K20
		NPN20	---	2	P10-P40/M10-M30
Hertel	Hertel AG Werkzeuge & Hartstoffe	13E	TiN+Ti(C,N)+Al_2O_3+TiN	8	P05-P25/M05-M15/K05-K20
		CD2	TiN+Ti(C,N)+TiN	1.5	P30-P45/K15-K25
		CF2	TiN+Ti(C,N)+TiN	5	P10-P35/M15-M30
		CF3	TiN+Ti(C,N)+TiN	5	K05-K20
		CM2	TiN+Ti(C,N)+TiN	8	P10-P30/M10-M20/K10-K20
		CM3	TiN+Ti(C,N)+TiN	8	P15-P40/M15-M25/K15-K25
		CM4	TiN+Ti(C,N)+TiN	8	P20-P45/M20-M30/K20-K25
Hertel	Hertel Cutting Techologies	13E	TiN+Ti(C,N)+Al_2O_3+TiN	10	P05-P25/M05-M15/K05-K20
		CD2	TiN+Ti(C,N)+TiN	1.5	P30-P45/K15-K25
		CF2	TiN+Ti(C,N)+TiN	5	P10-P35/M15-M30
		CF3	TiN+Ti(C,N)+TiN	5	K10-K30
		CM2	TiN+Ti(C,N)+TiN	8	P10-P30/M10-M20/K05-K20
		CM3	TiN+Ti(C,N)+TiN	8	P15-P40/M15-M25/K15-K25
		CM4	TiN+Ti(C,N)+TiN	8	P20-P45/M20-M30/K20-K25
		CNC	Ti(C,N)+TiC+TiN	6	P15-P40

COATED CEMENTED CARBIDES CROSS REFERENCE (Continued)

Trade Name	Company Name	Grade	Layer Composition	Coating Depth, µm	ISO Code(s)
Hertel (continued)	Hertel Cutting Techologies	CNCM	Ti(C,N)+TiC+TiN	6	P20-P30
		CNCT	Ti(C,N)+TiC+TiN	6	P40-P50/K30-K40
		CNCW	Ti(C,N)+TiC+TiN	6	K05-K15
		CP1	TiN+Ti(C,N)+Al_2O_3	9	P05-P20/K05-K10
		Roxide	Ti(C,N)+Al_2O_3	10	P10-P40/K10-K30
		Roxide-M	Ti(C,N)+Al_2O_3	10	P30-P40/K20-K30
		Roxide-W	Ti(C,N)+Al_2O_3	10	K01-K20
Hitachi	Hitachi Tool Engineering	HC500	Al_2O_3+TiC	7	P01/K01-K10
		HC512	Al_2O_3+TiC+Ti(C,N)	7	P10-P20/M20/K20
		HC514	Al_2O_3+TiC	7	P30/M30/K30
		HC516	Al_2O_3+TiC	7	P40/M40
		HC730	Al_2O_3+TiC	7	K01
		HC830	TiN	3	K01
		HC831	TiN	3	K10-K20
		HC842	TiN	3	P20/M20
		HC843	TiN	3	P30/M30
		HC844	TiN	3	P40/M40
		HC5000	Al_2O_3+TiC+Ti(C,N)	12	K01-K20
Impero	Impero SpA	SK05	TiC+Al_2O_3+TiN	9	K05-K20
		SK3	TiN	3	P01-P15
		SK8	TiN	3	P20-P40
		SK10	TiC+Al_2O_3	9	P05-P30/M10-M20/K05-K20
		SK15	TiC+TiN	6	P10-P30/M10-M30
		SK20	TiC+Al_2O_3	6	P20-P30

COATED CEMENTED CARBIDES CROSS REFERENCE (Continued)

Trade Name	Company Name	Grade	Layer Composition	Coating Depth, µm	ISO Code(s)
Impero (continued)	Impero SpA	SK30	TiC+TiN	5	P25-P40/M10-M40
		SK45	TiC+Al_2O_3+TiN	9	P25-P45/M15-M25
		SK50	TiN+TiC+TiN	3	P30-P50/M20-M40
		SK90	TiC+Al_2O_3+TiN	9	K05-K30/M10-M30
		SKM	TiN	3	P10-P40/M20-M30
Iscar	Iscar Ltd	IC220	TiN	3	K10-K20
		IC250	TiN	3	P20-P30
		IC428	Composite + Al_2O_3	7	K05-K20
		IC520M	Ti(C,N)	5	P15-P30/K10-K25
		IC540	Ti(C,N)	6	P20-P40
		IC635	TiN+TiC+TiN	3	P30-P50
		IC656	TiC+Ti(C,N)+TiN	8	P20-P40
		IC805	TiC+Al_2O_3+TiN	11	P05-P35/K05-K25
ISG	ISG Szerszam es Porkohaszati Gyar	EK1	TiC	5-8	K10-K20
		EN1	TiC	5-8	P20-P40/M20-M30
		ENN1	TiC	5-8	P30-P30
		ES1	TiC	5-8	P10-P30/M10-M20
		TK14	TiC+Al_2O_3	6-8	K10-K20
		TK123	TiC+Ti(C,N)+TiN	6-8	K10-K30
		TKM	Ti(C,N)+TiC+TiN	4-6	K10-K20
		TN123	TiC+Ti(C,N)+TiN	6-8	P20-P40/M20-M30
		TNM23	Ti(C,N)+TiC+TiN	4-6	P25-P40
		TS14	TiC+Al_2O_3	6-8	P10-P20/K10-K20
		TS123	TiC+Ti(C,N)+TiN	6-8	P10-P30/M10-M20
		TS143	TiC+Al_2O_3+TiN	6-8	P10-P30/K10-K20

COATED CEMENTED CARBIDES CROSS REFERENCE (Continued)

Trade Name	Company Name	Grade	Layer Composition	Coating Depth, μm	ISO Code(s)
ISG (continued)	ISG Szerszam es Porkohaszati Gyar	TSM23	Ti(C,N)+TiC+TiN	4-6	P15-P25
Kennametal	Kennametal Inc	KC210	TiC+Ti(C,N)+TiN	10	K05-K30
		KC250	TiC+Ti(C,N)+TiN	10	M30-M45/K25-K35
		KC710	TiN	3	P15-P25/M15-M25
		KC720	TiN	3	P25-P45/M30-M40/K25-K35
		KC729M	TiN+Ti(C,N)(CVD)+TiN(PVD)	8	P10-P30
		KC730	TiN	3	M05-M15/K05-K15
		KC740	TiN	3	P05-P15
		KC810	TiC+Ti(C,N)+TiN	10	P10-P30/M15-M35
		KC820	TiC+Ti(C,N)+TiN	10	P25-P35/M10-M25
		KC840	TiC+Ti(C,N)+TiN	10	P35-P45
		KC850	TiC+Ti(C,N)+TiN	10	P25-P45/M25-M45
		KC910	TiC+Al_2O_3	9	P01-P20/M05-M20/K01-K15
		KC950	TiC+Al_2O_3+TiN	10	P05-P25/M10-M25/K10-K20
		KC990	Ti(C,N)+Al_2O_3+TiN+Al_2O_3+ TiN+Al_2O_3+Ti(C,N)+TiN	11	P05-P25/M10-M25/K05-K25
		KC992M	Ti(C,N)+Al_2O_3	4	K10-K25
Komet	Komet Stahlhalter-und Werkzeugfabrik Robert Breuning GmbH	BK/BK60	TiC+Ti(C,N)+TiN	---	P10-P30
		BK1/BK61	TiC+Al_2O_3	---	K05-K15
		BK3/BK65	TiC+Ti(C,N)+TiN	---	P15-P35
		BK4/BK64	TiC+Ti(C,N)+TiN	---	P20-P40
		BK65	TiC+Ti(C,N)+TiN	---	---
		BK66	TiC+Ti(C,N)+TiN	---	P25-P40

COATED CEMENTED CARBIDES CROSS REFERENCE (Continued)

Trade Name	Company Name	Grade	Layer Composition	Coating Depth, µm	ISO Code(s)
Komet (continued)	Komet Stahlhalter-und Werkzeugfabrik Robert Breuning GmbH	BK85	TiN	---	P25
		BK86	TiN	---	P35
KT/Korea Tungsten	Korea Tungsten Mining Co Ltd	KT150	$TiC+Al_2O_3+TiN$	5+3+1	P05-P20/M10/K05-K20
		KT200	$TiC+Al_2O_3$	6+1	P10-P30/M10-M20/K10-K25
		KT250	$TiC+Al_2O_3+TiN$	4+3+1	P15-P35/M15-M25
		KT300	TiC+Ti(C,N)+TiN	4+2+2	P10-P35/M10-M20/K15-K35
		KT350	TiC+Ti(C,N)+TiN	3+2+2	P25-P40/M20-M30
		KT650	Ti(C,N)+TiC+TiN	4+1+1	K10-K30
Manchester	Manchester Tool Company	GC	TiN+TiC+TiN	---	---
		M40	TiN	---	---
		M50	TiN	---	---
		PTN	TiN+TiC+TiN	---	---
Mircona	Mircona AB	ALC150	$TiC+Al_2O_3+TiN$	10	P25-P45/M15-M25
		ALC250	$TiC+Al_2O_3$	12	K05-K25
		ALC300	$TiC+Al_2O_3$	4	P05-P25/M10-M20/K05-K15
		ALC350	$TiC+Al_2O_3+TiN$	10	P05-P25/K05-K20
		TNC100	TiC+Ti(C,N)+TiN	5	P45/M20-M40
		TNC150	TiC+Ti(C,N)+TiN	10	P25+P40
		TNC250	TiC+Ti(C,N)+TiN	10	P10-P25/M10-M20/K20-K30
Mitsubishi	Mitsubishi Materials Corporation	F515	$TiC+Al_2O_3$	4	K10-K20
		U66	$TiC+Al_2O_3$	8	P01-P10/K01-K10
		U77	$TiC+Al_2O_3$	8	P10-P30/K10-K20

COATED CEMENTED CARBIDES CROSS REFERENCE (Continued)

Trade Name	Company Name	Grade	Layer Composition	Coating Depth, μm	ISO Code(s)
Mitsubishi (continued)	Mitsubishi Materials Corporation	U88	$TiC+Al_2O_3$	8	P20-P40
		U505	Composite + $TiC+Al_2O_3$	8	P01-P10/K01-K10
		U510	Composite + $TiC+Al_2O_3$	8	K01-K20
		U610	Composite + $TiC+Al_2O_3$	8	P01-P20/K01-K20
		U625	Composite + $TiC+Al_2O_3$	8	P10-P30
Newpro	Newcomer Products Inc	NA02	$TiC+Al_2O_3$	10	P10-P25/M10-M25/K01-K20
		NN55	TiC+TiN	10	P10-P30/M10-M30
		NN60	TiC+TiN	10	P10-P25/M10-M30
		NP92	TiC+TiN	10	P15-P45/M20-M40/K15-K40
		NP94	$TiC+TiN+TiC+Al_2O_3$	20	P10-P45/M20-M40/K10-K40
		NP1000	TiC+TiN	10	P10-P35/M10-M30/K05-K20
		NP2000	$TiC+Al_2O_3$	10	P05-P20/M05-M20/K01-K20
		NP4000	$TiC+TiN+Al_2O_3$	20	P01-P25/M05-M25/K01-K25
		NT56	TiC	7	P05-P15/M10-M20
		PV52	TiN	3	P20-P40/M10-M35
North American	North American Carbide	111-5	TiC+TiN	6-8	P10-P30/K10-K30
		111-54	TiC+TiN	6-8	P30-P40/K30-K40
		111-N	TiC+TiN	6-8	P20-P40/K20-K40
		112	TiC	6-8	---
		113-5	$TiC+Al_2O_3$	6-8	P10-P30/K20-K40
		113-N	$TiC+Al_2O_3$	6-8	P20-P40/K20-K40
		114	TiN	2-4	P25/K25
		115-N	$TiC+TiN+Al_2O_3$	6-8	P20-P40/K20-K40

COATED CEMENTED CARBIDES CROSS REFERENCE (Continued)

Trade Name	Company Name	Grade	Layer Composition	Coating Depth, µm	ISO Code(s)
Poldi Diadur	Pramet Sumperk	015P	Ti(C,N)	4-8	P10-P20
		020P	Ti(C,N)	4-8	P20-P30
		210K	Ti(C,N)+TiC+Al_2O_3	4-8	K10-K20
		320P	Ti(C,N)+TiC+Al_2O_3+TiN	4-8	P10-P30/M10-M20
		515P	Ti(C,N)+TiC+Ti(C,N)+TiN	4-8	P15-P20/M10-M15
		520P	Ti(C,N)+TiC+Ti(C,N)+TiN	4-8	P15-P30
		525P	Ti(C,N)+TiC+Ti(C,N)+TiN	4-8	P25-P35/M15-M20/K10-K20
		530P	Ti(C,N)+TiC+Ti(C,N)+TiN	4-8	P15-P35/M10-M30/K15-K30
		535P	Ti(C,N)+TiC+Ti(C,N)+TiN	4-8	P30-P40
RTW	Rogers Tool Works	027	TiN+Ti(C,N)+TiC+TiN	8	M05-M20/K05-K20
		714	TiN+Ti(C,N)+TiC+TiN	8	P05-P25
		716	TiN+Ti(C,N)+TiC+TiN	8	P10-P30
		718	TiN+Ti(C,N)+TiC+TiN	8	P10-P25
		725	TiN+Ti(C,N)+TiC+TiN	8	P10-P30
		731	TiN+Ti(C,N)+TiC+TiN	8	P01-P10
		755	TiN+Ti(C,N)+TiC+TiN	8	P20-P45
		914	TiN+Ti(C,N)+Al_2O_3	8	P05-P25/M05-M20/K05-K20
		916	TiN+Ti(C,N)+Al_2O_3	8	P10-P30/M10-M30/K10-K30
		918	TiN+Ti(C,N)+Al_2O_3	8	P05-P25/M10-M30/K10-K30
		925	TiN+Ti(C,N)+Al_2O_3	8	P10-P30/M15-M35/K15-K35
		931	TiN+Ti(C,N)+Al_2O_3	8	P01-P10/M05-M15/K05-K20
		955	TiN+Ti(C,N)+Al_2O_3	8	P20-P45/M30-M40/K30-K40
		CX37	TiN+Ti(C,N)+TiC+TiN	8	P40-P50
		CX47	TiN+Ti(C,N)+TiC+TiN	8	P40-P50

COATED CEMENTED CARBIDES CROSS REFERENCE (Continued)

Trade Name	Company Name	Grade	Layer Composition	Coating Depth, µm	ISO Code(s)
Sandvik Coromant	AB Sandvik Coromant	Coronite	TiN	3	---
		GC015	TiC+Al_2O_3	7	P05-P35/K05-K20
		GC135	TiC	5	P25-P45/M15-M20
		GC215	Ti(C,N)	8	P10-P40/M10-M25
		GC225	TiN+TiC+TiN	3	P10-P30
		GC235	TiN+TiC+TiN	3	P30-P50/M20-M40
		GC315	TiC	5	M15-M20/K05-K25
		GC320	TiC+Al_2O_3	4	K10-K25
		GC415	TiC+Al_2O_3+TiN	8	P05-P30/K05-K20
		GC425	TiC+TiN	8	P10-P35/M10-M20
		GC435	TiC+Al_2O_3+TiN	8	P25-P45/M15-M25
		GC1020	TiN	3	M10-M30
		GC1025	TiC	6	P10-P35/K05-K20
		GC3015	TiC+Al_2O_3	11	K05-K20
		GCA	Ti(C,N)+TiN	5	P10-P35
Seco	Seco Tools AB	550	TiN+TiC+TiN	9	P20-P45
		560	TiN+TiC+Al_2O_3+TiN	10	P10-P35/K10-K20
		570	Al_2O_3	6	P05-P25/M10-M20/K05-K20
		CP30	Ti(C,N)+TiC+TiN	4	P10-P20
		T15M	TiC+Ti(C,N)+Al_2O_3+TiN	7	K05-K15
		T25M	Ti(C,N)+TiC+TiN	5	P05-P40/M30-M40/K30-K40
		TP05	TiC+Al_2O_3+TiN	9	P05/K10
		TP10	TiC+Al_2O_3+TiN	11	P10/K15
		TP15	TiC+Ti(C,N)+Al_2O_3+TiN	7	P05-P25/M05-M20/K05-K20

COATED CEMENTED CARBIDES CROSS REFERENCE (Continued)

Trade Name	Company Name	Grade	Layer Composition	Coating Depth, μm	ISO Code(s)
Seco (continued)	Seco Tools AB	TP20	$TiC=Al_2O_3+TiN$	7	P10-P35/M20
		TP30	TiC+Ti(C,N)+TiN	12	P30/M30
		TP35	TiC	5	P20-P35/M15-M30
		TP40	Ti(C,N)+TiC+TiN	5	P40/M40
		TP45	TiC	4	P35-P50/M30-M40/K40
		TP301	Ti(C,N)+TiC+TiN	4	P30/M40
		TP401	$TiC+Al_2O_3+TiN$	7	P40/M40
		TX10	$TiC+Al_2O_3+TiN$	7	P05-P15/M05-M15/K05-K15
SKF	SKF Tools AB	QM15	TiN	---	M05-M20/K01-K25
		QM25	TiN	---	P10-P35
		QT09	TiC	5	M10-M15/K05-K20
		QT11	Al_2O_3	---	K05-K20
		QT15	TiN	---	P05-P30/M10-M25/K05-K20
		QT25	TiC+TiN	7-8	P05-P35/M10-M40/K05-K20
		QT27	TiC-TiN	5-7	P25-P40/M10-M40
		QT35	TiC	5	P25-P40/M15-M30
Sumitomo	Sumitomo Electric Industries Ltd	AC05	$TiC+Al_2O_3$	8	P01-P10/K01-K15
		AC10	$TiC+Al_2O_3$	8	P05-P25
		AC10G	$TiC+Al_2O_3$	8	K05-K20
		AC15	$Special+Al_2O_3+TiN$	8	P10-P30
		AC25	$Special+Al_2O_3+TiN$	8	P15-P45
		AC105	$TiC+Al_2O_3+TiN$	10	P01-P05/K01-K10
		AC211	$Special+Al_2O_3+TiN$	4	K10-K20
		AC225	$Special+Al_2O_3+TiN$	4	P15-P45

COATED CEMENTED CARBIDES CROSS REFERENCE (Continued)

Trade Name	Company Name	Grade	Layer Composition	Coating Depth, μm	ISO Code(s)
Sumitomo (continued)	Sumitomo Electric Industries Ltd	AC305	TiN+TiC+TiN	4	K05-K15
		AC325	TiN+TiC+TiN	4	P10-P30
		AC330	TiN+TiC+TiN	4	P20-P30
		AC720	TiC	8	P20-P30
		AC815	TiN+TiC+TiN	8	P15-P25
		T12Z	TiN+TiC+TiN	4	P05-P15
		T110Z	TiN+TiC+TiN	4	P01-P15/K01-K10
		T130Z	TiN+TiC+TiN	4	P05-P25
Teledyne	Teledyne Firth Sterling	CC44	Al_2O_3	8	P15-P35
		CC46	Al_2O_3	8	K05-K30
		HN+	HfN	8	P05-P20
		MP21	TiN+Al_2O_3+TiN+Ti(C,N)	10	K05-K20
		MP26	---	10	P10-P25/K20-K30
		MP51	TiN+Al_2O_3+TiN+Ti(C,N)	10	P20-P40/K30-K40
		MP62	TiN+Al_2O_3+TiN+Ti(C,N)	---	P10-P35/M15-M30
		TP21	TiN	3	P10-P20/K05-K20
		TC41M	---	5	P30-P40/K40
Tizit Goldmaster	Plansee Tizit GmbH	Gm15	TiC+Ti(C,N)+TiN	10-12	P05-P20/K05-K20
		Gm25	TiC+Ti(C,N)+TiN	10-12	P10-P30/K10-K30
		Gm26	TiC+Ti(C,N)+TiN	5	P10-P30/K10-K30
		Gm35	TiC+Ti(C,N)+TiN	10-12	P20-P40
		Gm36	TiC+Ti(C,N)+TiN	5	P20-P40
		Gm40	TiC+Ti(C,N)+TiN	5	P25-P40
		Gm43	TiC+Ti(C,N)+TiN	3	P25-P40

COATED CEMENTED CARBIDES CROSS REFERENCE (Continued)

Trade Name	Company Name	Grade	Layer Composition	Coating Depth, μm	ISO Code(s)
Tizit Starmaster (continued)	Plansee Tizit GmbH	Gm176	TiC+Ti(C,N)+TiN	5	P15
		Gm306	TiC+Ti(C,N)+TiN	5-6	P40-P50/M40
		Sr16	TiC+Ti(C,N)+KS+ZS	7	K10
		Sr17	TiC+Ti(C,N)+KS+ZS	8	P05-P25/K05-K20
		Sr117	TiC+Ti(C,N)+TiN+KS+ZS	10-12	P10-P20
		Sr127	TiC+Ti(C,N)+TiN+KS+ZS	10-12	P15-P35/K10-K30
		Sr137	TiC+Ti(C,N)+TiN+KS+ZS	10-12	P30-P40
Tool-Flo	Tool-Flo Manufacturing	A53	$TiC+Al_2O_3$	---	P10-P20
		AG53	$TiC+Al_2O_3+TiN$	---	P10-P20
		G2	TiN+TiC+TiN	---	K20
		G4	TiN+TiC+TiN	---	K30-K40
		G5	TiN+TiC+TiN	---	P25-P30
		G6	TiN+TiC+TiN	---	P15-P20
		G7	TiN+TiC+TiN	---	P10
		G25	TiN+TiC+TiN	---	K05-K15
		G44	TiN+TiC+TiN	---	P35-P45
		G50	TiN+TiC+TiN	---	P25-P30
		G53	TiN+TiC+TiN	---	P25-P30
		G54	TiN+TiC+TiN	---	P35-P40
		GP2	TiN	---	K20
		GP4	TiN	---	P35-P45
		GP6	TiN	---	P15-P20
		GP7	TiN	---	P10
		GP25	TiN	---	K05-K15
		N5	TiN	---	P25-P30
Tungaloy	Toshiba Tungaloy Co Ltd	PEM	TiN	3	(drills)

COATED CEMENTED CARBIDES CROSS REFERENCE (Continued)

Trade Name	Company Name	Grade	Layer Composition	Coating Depth, µm	ISO Code(s)
Tungaloy (continued)	Toshiba Tungaloy Co Ltd	PG2	TiN	3	K10-K20
		PK56	TiN	3	P30-P40
		T221	(Ti,Ta)(N,O)	3	K01-K20
		T260	(Ti,Ta)(N,O)	3	P20-P30/M20-M40
		T313V	---	2	P20-P30
		T370	TiC	2	P20-P30/M20-M30/K01-K30
		T530	TiC+Ti(C,N)+TiN	10	K10-K20
		T553	TiC	7	P10-P30
		T801	TiC+Al(O,N)	8	K01-K10
		T802	TiC+Al(O,N)	8	P10-P20/M10-M20/K01-K20
		T803	TiC+Al(O,N)	8	P10-P30/M10-M30/K10-K20
		T813	Ti(C,N,O)+Al(O,N)	8	P10-P40/M10-M30/K10-K30
		T821	Ti(C,N,O)+Al(O,N)	8	K01-K10
		T822	Ti(C,N,O)+Al(O,N)	8	P01-P20/M10
		T823	Ti(C,N,O)+Al(O,N)	8	P10-P30/M10-M30/K10-K20
		T841	TiC+Al(O,N)	12	M10-M20/K01-K10
		T842	TiC+Al(O,N)	12	M10-M30/K01-K20
Ultra-Met	Ultra-Met Manufacturing	ZCB	TiN	8	P25-P40/M20-M40
		ZCC	TiN	8	P10-P15
		ZCK	TiN	8	K30
		ZCJ	TiN	8	P10-P30/M15-M30/K15-K30
		ZCS	TiN	8	K15-K30
		ZCU	TiN	8	P25-P40/M20-M40
		ZDC	---	8	P05-P15/K05-K10
		ZDF	---	8	K20-K30
		ZDI	---	8	P05-P20/M10-M20/K10-K20
		ZDS	---	8	K05-K20

COATED CEMENTED CARBIDES CROSS REFERENCE (Continued)

Trade Name	Company Name	Grade	Layer Composition	Coating Depth, µm	ISO Code(s)
Ultra-Met (continued)	Ultra-Met Manufacturing	ZFY	TiN	3	P20-P30/K20-K30
		ZSF	TiN	8	K20-K30
Valenite	Valenite Inc	902	TiN	---	K05-K20
		905	TiN	---	P20-P30
		907	TiN	---	P05-P15
		SV3	---	---	P01-P30/M10-M30/K05-K20
		SV4	TiN	---	P10-P40/M20-M40/K10-K40
		SV200	TiC+TiN	---	P20-P50/K20-K40
		V01	---	---	P05-P20/M10-M20/K05-K15
		V1N	TiN	---	P30-P40/M30-M40/K20-K40
		VN8	TiN	---	P10-P30/M15-M20/K10-K20
		VX8	TiC+TiN	---	P15-P30/M15-M30/K10-K20
		V88	TiC	---	P01-P30/M10-M30/K01-K30
		V05	---	---	P01-P30/M05-M20/K01-K30
Vandurit	Metalloceramica Vanxetti	V111	---	4+4	P05-P25/M05-M25/K05-K25
		V134	TiC+Ti(C,N)+TiN	8-10	P25-P40/M25-M40/K25-K40
		V635F	Ti(C,N)+TiN	3+1	P10-P30
		VMT	TiC+Ti(C,N)+TiN	8-10	P20-P35/M20-M35
		VOR	TiC+Ti(C,N)+TiN	8-10	P15-P30/M15-M30
		VRC	TiC	6	K15-K30
Walter	Walter Hartmetall GmbH	WTA21	---	5	P05-P20/M10-M25/K05-K20
		WTA31	---	5	P05-P20/M10-M25/K05-K20
		WTA33	---	12	P05-P20/M10-M25/K05-K20
		WTA41	---	5	P15-P30/M20-M35/K15-K30
		WTA43	---	12	P15-P30/M20-M35/K15-K30
		WTL14	Ti(C,N)+TiN	5	P15-P35/K15-K35
		WTL41	Ti(C,N)+TiN	5	P15-P35/M15-M20

COATED CEMENTED CARBIDES CROSS REFERENCE (Continued)

Trade Name	Company Name	Grade	Layer Composition	Coating Depth, μm	ISO Code(s)
Walter (continued)	Walter Hartmetall GmbH	WTL71	Ti(C,N)+TiN	5	P35-P40/K30-K40
		WTL74	Ti(C,N)+TiN	5	P35-P40/K30-K40
		WTL82	Ti(C,N)+TiN	5	M10-M15/K10-K30
		WTN33	Ti(C,N)+TiN	12	P05-P20
		WTN41	Ti(C,N)+TiN	5	P15-P30
		WTN43	Ti(C,N)+TiN	12	P15-P30
		WTN53	Ti(C,N)+TiN	12	P25-P40
Widia	Krupp Widia GmbH	Widadur TN25	TiC+Ti(C,N)+TiN	10	P25
		Widadur TN25M	TiC+Ti(C,N)+TiN	4	P25
		Widadur TN35	TiC+Ti(C,N)+TiN	10	P30
		Widadur TN35M	TiC+Ti(C,N)+TiN	4	P30
		Widadur TN250	TiC+Ti(C,N)+TiN	10	P25
		Widadur TN450	TiC+Ti(C,N)+TiN	8	P30
		Widalon HK15	multilayer Al(O,N)	8	K15
		Widalon HK15M	multilayer Al(O,N)	5	K15
		Widalon HK35	multilayer Al(O,N)	8	M25
		Widalon HK150	Ti(C,N)+Al_2O_3	8	K15
		Widalon TK15	multilayer Al(O,N)	8	P15
		Widaplas TPC15	TiN	3	P15
		Widaplas TPC25	TiN	3	P25
		Widaplas TPC35	TiN	3	P30
		Widianit CN1000	Al_2O_3	1-2	K20
		Widianit CN2000	Al_2O_3	1-2	K20
Wolframcarb	Wolframcarb SpA	333X	TiC+Ti(C,N)+TiN	5	P10-P30
		666/S	TiC+TiN	2	P30-P45
		666X	TiC+Ti(C,N)+TiN	5	P30-P45

COATED CEMENTED CARBIDES CROSS REFERENCE (Continued)

Trade Name	Company Name	Grade	Layer Composition	Coating Depth, μm	ISO Code(s)
Wolframcarb (continued)	Wolframcarb SpA	A55	$TiC+TiN+Al_2O_3+TiN$	5-6	P01-P20/K01-K10
		A66	$TiC+Al_2O_3$	5	P30-P45/K30-K35
		A99	$TiC+Al_2O_3$	5	P20-P30/K20-K30
		A355	$TiC+TiN+Al_2O_3+TiN$	5-6	P10-P30/K20-K30
		AK33	$TiC+Al_2O_3$	5	K01-K10
		AK44	$TiC+Al_2O_3$	5	K10-K20
		G166X	TiC+Ti(C,N)+TiN	5	P30-P45
		T125	TiC+TiN	2	P20-P25
		T140	TiC+TiN	2	P30-P40
		W30X	TiC+TiN	2	P10-P30

a. This cross-reference table lists the ISO Code and company-related information; since this information is constantly being revised, it should be kept in mind that it is presented herein as a guide and may not reflect the latest revisions.

Chapter 6

LEAD, TIN & ZINC ALLOYS

LEAD

The origins of metallic lead lie in the far distant past. Lead ores are relatively easy to reduce, and hence the metal has been known and used for both ornamental and structural purposes probably since before 5000 B.C. Today the uses of lead and its alloys are based on its unique combination of properties, among which are high density, castability, formability, conductivity, thermal expansion, and corrosion resistance combined with low tensile and creep strength, hardness, elastic stiffness, and melting point. The low creep strength is a serious limitation to the structural applications of lead, but this can be overcome by design or by the use of inserts or supports. Battery applications represent the most important uses of lead and its alloys, but they are also used as shielding, bearing materials, type metal, cable sheathing, and solders. These applications are discussed in more detail below.

Primary lead production comes from the smelting of the mineral galena (lead sulphide, PbS). The ore is concentrated, using flotation to separate the galena from associated minerals. The concentrate is sintered and roasted, then smelted in a blast furnace to produce an impure lead bullion which is purified in a sequence of processes to remove most of the residual copper, antimony, tin, arsenic, precious metals, and zinc. The final step is often an electrolytic refining. Purities as high as 99.99% are readily obtainable.

Lead is also extensively recycled. Scrap processing accounts for more than half of lead production, giving it the highest recycling rate of all the metals.

Pure lead has a face-centered cubic crystal structure, and this combined with its low melting temperature (327°C, 621°F) account for its high ductility. Compared to most of the common metals, it has a high density, its specific gravity being 11.3.

Categories

The commercial grades of lead and its alloys are shown in Table 1 along with the corresponding UNS designations. The initial letter in the UNS designations, "L", refers to low melting metals and alloys (as such, the L category also includes cadmium, lithium, tin, and their alloys). Cast and wrought lead alloys are available but the UNS categories do not distinguish between these.

Table 1 Commmercial Grades of Lead With UNS No.

Commercial purity lead	UNS L50000-L50099
Lead-silver alloys	UNS L50100-L50199
Lead-arsenic alloys	UNS L50300-L50399
Lead-barium alloys	UNS L50500-L50599
Lead-calcium alloys	UNS L50700-L50899
Lead-cadmium alloys	UNS L50900-L50999
Lead-copper alloys	UNS L51100-L51199
Lead-indium alloys	UNS L51500-L51599
Lead-lithium alloys	UNS L51700-L51799
Lead-antimony alloys	UNS L52500-L53799
Lead-tin alloys	UNS L54000-L55099
Lead-strontium alloys	UNS L55200-L55299

Commercial Purity Lead

The four grades of commercial purity lead are "chemical lead" (UNS L51120) and "acid-copper" lead (UNS L51121), both of which contain a minimum of 99.9% lead, along with "pure lead" also called "corroding lead" (UNS L50042) and "common lead" (UNS L50045) which contain a minimum of 99.94% lead. Corroding lead is named not for its corrosion resistance but rather for a process to which it was subjected in order to produce products such as oxides, pigments, and other chemicals. Chemical lead contains copper and silver impurities which confer on it an improved corrosion resistance and strength, so that it is particularly appropriate for use in the chemical industries.

Lead Alloys

The most common alloying elements in lead are antimony, tin, arsenic, and calcium. Antimony is added in amounts ranging up to 25%, but most commonly 2-5%, in order to improve the strength for applications such as storage battery grids, pipe, and sheet. Calcium in amounts below 0.2% has a similar hardening effect on lead, and is sometimes added along with small amounts of tin and/or aluminum. Tin not only strengthens lead but also improves its casting and wetting properties, for uses as solders and as type metal for printing. Arsenic is also a strengthening addition to lead, both alone and in combination with antimony. Some lead alloys, including antimonial lead and lead-calcium alloys, exhibit pronounced precipitation hardening and can be considered heat treatable.

Battery Applications

The primary applications of lead and its alloys are as grid plates, posts, and connector straps in lead-acid storage batteries. Grid plates function in a sulphuric acid environment under repeated cycles of charging and discharging, hence corrosion resistance is important. Both cast and wrought plates are used, typically lead-calcium, lead-calcium-tin, or antimonial lead alloys. For example, positive plates for automotive batteries may be made from antimonial lead (1.5-4%Sb plus minor amounts of tin, arsenic, etc.) including alloys UNS L52760, L52765, L52770, and L52840) while negative plates may be cast calcium-lead alloys (up to 0.1%Ca) with minor amounts of tin and/or aluminum (UNS L50775, L50780, and L50790) or without the tin/aluminum addition (UNS L50760 and L50770). Grids for industrial batteries usually are lead containing 5-8%Sb (e.g., UNS L53135). Other alloys for battery applications include the lead-strontium alloys (e.g., UNS L55260).

Type Metal

Lead-based materials are also used for type in the printing industry. These are generally lead-antimony alloys, often with tin and copper as well. Here the good castability and low melting temperature of lead are further improved by the antimony and tin. Furthermore, the antimony and copper increase the hardness, while the tin reduces brittleness and improves the ability to reproduce fine detail. The different printing processes (electrotype, stereotype, linotype, monotype, foundry type) require different type characteristics, and hence different alloys are used. For example electrotype has relatively low alloy content (2-4%Sn, 2-3%Sb, e.g., UNS L52730 and L52830), since it is not required to resist wear, while linotype (3-5%Sn, 11-12%Sb, e.g., UNS L53425 and L53455), stereotype (6-8%Sn, 13-15%Sb; e.g., UNS L53530 and L53575), and monotype (7-12%Sn, 15-24%Sb; e.g., UNS L53685 and L53750) are increasingly highly alloyed for increased hardness and wear resistance. Foundry type, which is repeatedly reused without remelting and can be subjected to high pressures in forming molds, has up to 2% copper added for even greater hardness (e.g., UNS L53710 and L53780).

Cable Sheathing

Lead sheathing is often extruded around electrical power and communication cables for protection against moisture, corrosion, and mechanical damage. Alloys used include chemical lead, antimonial lead, arsenical lead, and lead-calcium alloys, some with additions of copper or tellurium. In all cases, the total amount of alloying element remains small (less than 1%) so that the sheathing can be readily extruded around the cable. Examples include the arsenical lead alloy UNS L50310, the antimonial leads UNS L52520 and L52535), and the calcium leads UNS L50710, L50712, and L50725.

Shielding and Damping

Lead and its alloys find many applications as shielding materials. Its high density provides it with good radiation shielding properties against x-rays and gamma rays, and it is frequently used for this purpose as a lining in concrete structures. The combination of high density, low stiffness, and high damping capacity gives lead excellent vibration and sound absorbing properties. It is widely used to isolate equipment and structures from vibration, for example as lead-steel composite pads under column footings in buildings. For sound absorption it is used either alone or in composites with polymer foams or sheets. Lead foils, either alone or as a sandwich rolled between thin layers of tin, are used as moisture barriers in the construction industry and as oxygen barriers on wine bottles.

Corrosion Resistant Sheet and Pipe

In addition, lead and its alloys are used in sheet form in corrosion-resistant applications, for example in the construction industry for roofing and flashing, in pans below shower and bath stalls, and in flooring. Sheet and piping made from chemical lead (UNS L51120) and alloys such as antimonial lead (e.g., the 6%Sb alloy UNS L53125) are also used in the chemical industry as well as in plumbing and water distribution and waste systems, with the more highly alloyed material used where resistance to creep and erosion is important.

Corrosion resistance can be imparted to mild steel sheet or plate by coating it with a lead-tin (3-15%Sn) alloy. This is known as a terne coating and the product is called terne sheet or terne plate.

Solders

Lead-tin alloys are very widely used as solders, since their low melting temperatures allow joining without damage to heat-sensitive materials. Depending on the composition, melting temperatures between 182° and 315°C (360 and 599°F) are possible, and commercial solders are available across the full range from pure tin to pure lead. The very low-tin solders (e.g., 5%Sn or 5/95 solder, UNS L54320) are used for coating, sealing and joining and at service temperatures as high as 120°C (248°F). Solders with higher amounts of tin (e.g., 10/90, UNS L54520; 15/85, L54560; 20/80, L54710) are used in the automotive industry, both for radiator joining applications and for body damage repairs. Solders 40/60 (UNS L54915) and 50/50 (UNS L55030) are general purpose solders used in the automotive, electrical and electronic, and construction industries, in the latter case for roofing seams. Some lead-tin based solders also contain additional alloying elements, notably antimony and silver for improved fatigue and creep resistance and improved performance at high and low temperatures. Examples include the lead-tin-silver alloys UNS L54525, L54750, and L54855, and the lead-tin-antimony alloys UNS L54211, L54321, and L54905.

Bearing Materials

Many bearing materials are lead-based alloys; these are often referred to as lead-base babbitt metals. There are two basic groups of these alloys, one being lead-tin-antimony often with arsenic as well, the other being lead-calcium-tin often with other alkaline earth metals. The first group is the most common and includes four arsenical alloys covered by ASTM B23 (Alloy 7, UNS L53585; Alloy 8, UNS L53565; Alloy 13, UNS L53346, and the most common, Alloy 15, UNS L53620). Alloy 15 is used in automotive engines, often as a continuously cast bimetallic strip (steel and babbitt metal). For service in applications such as rolling mill bearings, a higher (3%) arsenic alloy is used. Another lead-tin-antimony alloy, known as SAE 16 (UNS L52860), is used for automotive applications as an overlay, cast onto a porous sintered non-ferrous matrix which is bonded to steel. Alloys in the second group of babbitt metals, the lead-calcium-tin alloys, are used for railway applications and some diesel engine bearings.

Miscellaneous Applications

In addition to the applications discussed above, there are a number of miscellaneous uses for which the properties of lead alloys make them suitable materials. For example, the high density and excellent castability of lead make it a useful material for counterweights. Lead-silver and lead-calcium-tin alloys in the form of rolled sheet find uses as anodes for electroplating and electrowinning (e.g., UNS L50110, L50120, L50730, and L53120). Unalloyed lead and alloys with tin, silver (e.g., UNS L50140), or antimony are used as anodes for cathodic protection, for example on ships and offshore rigs; these are used in cast or extruded bar form or as supported sheet. Fusible alloys, i.e., alloys with especially low melting points, in some cases below 100°C (212°F), are used in automatic sprinkler systems, as electrical fuses, and as boiler plugs. These are often lead-based alloys containing tin and bismuth, including UNS L54755, L54830, and L54930.

TIN

Tin has been used as an alloying element in bronze for at least 6000 years and has been known as a metal in its own right for at least 3500 years. It is characterized by a low melting point (232°C, 450°F), an attractive appearance, softness, and corrosion resistance. The major applications of tin are in coatings, mainly tinplated steel sheet, and as a component of solders, but other uses include bearing materials and pewter. Tin and its alloys are used in cast and wrought forms and as a powder.

Tin is obtained primarily from the smelting of the mineral cassiterite (SnO_2) mined from low-grade placer deposits or veins in Brazil, Southeast Asia, Bolivia, and Australia. The high density of cassiterite permits it to be readily beneficiated, and the concentrate is then smelted in a reverberatory furnace, refined by a remelting step, and further refined if necessary by liquating or electrolytic refining.

Tin exists normally as *beta* tin or "white tin" with a body-centered tetragonal crystal structure. Below 13°C (56°F) the equilibrium structure is *alpha* tin or "grey tin" which is a nonmetallic cubic phase. The transformation from white tin to grey tin is usually associated with disintegration or blistering problems which result from the volume increase; however, grey tin does not normally form. The transformation is difficult to initiate and impurities inhibit it further, so this factor does not usually have to be taken into consideration.

Tin is a reasonably dense metal with a specific gravity of 7.28, very slightly greater than zinc, but much less than lead. As is the case for lead, the tensile strength of tin is not as important a design criterion as its creep strength, since it can readily undergo creep at room temperature.

Unalloyed Tin

Commercially pure tin is available in several grades, but the most important is Grade A tin (UNS L13008) which contains a minimum of 99.8%Sn. Uses include pewter and tinplate foil. It is also used as a lining material in contact with high purity water, for example in distillation plants. The corrosion resistance of tin is excellent in near-neutral solutions, for example dilute solutions of weak alkalis, but tin does corrode in strong acids and bases and in aerated aqueous solutions.

Tinplate

Tinplate was originally produced by dipping iron or steel sheet into a bath of molten tin, but this process has been largely replaced by electroplating techniques. Coatings on steel sheet include pure tin, as well as alloys including tin-zinc (brass), tin-cadmium, and tin-copper (bronze). Coating thicknesses are almost always less than 2 micrometers (80 microinches) with overall tinplate thickness of 0.15 to 0.60 mm (0.006 to 0.024 in.). Electrolytic tinplate is used mainly for containers for food products, motor oil, and many other materials. Hot dip tin coatings are used for component leads, bonding layers in bearings, and in the food handling and processing industries.

Solders

Solder materials cover the entire composition range of the lead-tin system from pure tin to pure lead. Alloys containing up to 50%Sn have been discussed above in the Lead section. These include the widely used general purpose solders 40/60 and 50/50 which contain 40% and 50%Sn respectively. The roles of tin in solder include lowering the melting temperature range as well as increasing the wetting and adhesion of the solder to a wide range of base materials. Relatively high tin solders are used in the electrical and electronics industries because of their higher strength and electrical conductivity. These include the 60%Sn solders UNS L13600 and L13601 and the eutectic (63%Sn) solders UNS L13630 and L13631. Tin alloy solders often contain zinc, antimony, or silver for higher strength or higher temperature applications. These typically contain more than 95%Sn, for example the tin-antimony solders UNS L13940 and L13950 and the tin-silver solders UNS L13960 and L13961. Unalloyed tin and very high tin alloys with a maximum of 0.2%Pb (UNS L13940, L13950, and L13960) are used for soldering side seams of some food product containers and in numerous other applications where the presence of lead could be a hazard.

Pewter

The material known as pewter is traditionally a lead-tin based alloy; however, modern pewters are tin alloys containing up to 8%Sb and up to 3%Cu in order to avoid the toxicity and surface brightness problems of the older lead-bearing pewters. Pewter is used in both cast (e.g., UNS L13911) and wrought (e.g., UNS L13912) forms. Its high formability permits the use of a wide range of shaping processes including spinning, rolling, hammering, and drawing, often without the necessity for annealing steps during fabrication. Pewter is used for such products as trays, dishes, vases, and candlesticks.

Bearing Alloys

Bearing alloys which are rich in lead (lead-base babbitt metals) have been discussed above in the Lead section. However, many bearing alloys are tin-rich; these are usually alloyed with antimony and copper for increased hardness, tensile strength, and fatigue resistance. Examples include UNS L13840, L13870, and L13910. Other tin-base bearing alloys include intermediate lead-tin babbitt alloys and aluminum-tin bearing alloys. The latter find applications in contact with hardened steel or ductile iron crankshafts, where they are capable of withstanding higher loading and provide better fatigue resistance than tin-base or lead-base bearing alloys. These are used in high-duty engine applications including connecting rod bearings, crosshead bearings, and thrust bearings.

Miscellaneous Applications

"Spotted metals," tin-lead alloys in the form of cast strip, are used for organ pipes; these contain 20% to 90% tin depending on the tone desired. Tin-based die casting alloys, some of which are also used for sleeve bearings, contain antimony and copper (UNS L13913 and L13820) and sometimes lead (UNS L13650). A tin-antimony alloy (8%Sb), known as white metal, is used for jewelry applications. Collapsible tubes made of tin containing 0.4%Cu are used for packaging (e.g., artist paint tubes), and tin foil containing 8%Zn is used for food packaging.

Tin in powder form is used in sintered components whose primary constituent is bronze or iron powder. The role of the tin is to create a liquid which can function as a carrier phase for liquid-phase sintering. Tin-lead solder powders are also used in powder compacts which are to be bonded by warm compression. Tin powder also finds applications as spray-deposited coatings for corrosion resistance, bearing repairs, and metallizing of non-conducting components.

ZINC

Zinc has been used since before Roman times as a component of brass, but this early brass was produced not from metallic zinc but from zinc oxide, which in turn was obtained from calamine (zinc carbonate) or roasted sphalerite (zinc sulphide). Metallic zinc was known in Europe and in China by about 1600 A.D., but was clearly known in India at least a few hundred years earlier.

The production of metallic zinc is complicated by the fact that the temperatures necessary to reduce zinc from its ores are higher than the boiling point of zinc, so that zinc is produced as a vapour which readily oxidizes during cooling unless a reducing environment is maintained during cooling. Hence, zinc smelting traditionally involved distillation processes, but now is carried out by hydrometallurgy, with leaching of zinc concentrate being followed by electrowinning. The cathode zinc is then remelted and cast into ingots or slabs. There are also pyrometallurgical processes for zinc extraction, notably the Imperial Smelting process which involves a blast furnace and a lead splash condenser for recovering zinc from the blast furnace gas stream.

Pure zinc has a hexagonal close-packed (hcp) crystal structure, a melting temperature of 420°C (788°F), and a specific gravity of 7.14 at room temperature.

The major use of zinc is as a coating on steel, the process being known as galvanizing and the product as galvanized steel. Zinc and zinc alloys are also widely used as casting alloys, especially for die casting, and to a lesser extent in wrought form as sheet, extrusions, drawn wire, and forgings.

Nomenclature

Zinc and its alloys are categorized under the UNS system using the prefix letter Z. Other designations follow various ASTM standards in utilizing simpler nomenclatures (e.g., die casting alloy no. 2). These are described in more detail below in connection with various applications of zinc and its alloys.

Zinc is available in several grades in slab form, intended for remelting especially for galvanizing. These grades include, in order of decreasing purity, Special High Grade zinc (min. 99.99%Zn, UNS Z13001), High Grade zinc (min. 99.90%Zn, UNS Z15001), and Prime Western zinc (min. 98.0%Zn, UNS Z19001). The higher purity grades are used for continuous and electrolytic galvanizing and for alloy production, both in brassmaking and in zinc die casting alloys, where control of impurity levels is critical. For after-fabrication hot dip galvanizing, zinc purity is not as important and the Prime Western grade is satisfactory or even advantageous as its lead content (1%Pb) has benefits in the galvanizing process.

Galvanizing

Zinc has the ability to cathodically protect steel against corrosion by acting as a sacrificial anode, so that the zinc corrodes itself in preference to the steel and thereby protects it. Galvanized steel has good resistance to corrosion in many environments, and galvanized products cover a wide range, from small fasteners to large structures. Zinc is readily applied to steel as a surface coating, the most common techniques being hot dip galvanizing (dipping into a molten zinc bath), electroplating, metallizing (spraying molten zinc onto the surface), and mechanical galvanizing (forcing zinc powder onto the surface under pressure). Of these, hot dip galvanizing remains the most common. Typically, continuously hot dip galvanized steel strip with a 25 micron (0.001 in.) thick layer of Zn containing 0.2%Al finds applications in the automotive, appliance, and building industries. Variations on this process include the galvannealing process where a post-galvanizing heat treatment converts the zinc layer to a layer of iron-zinc alloy, and the galvalume process which creates a zinc-aluminum alloy coating for improved corrosion resistance.

Fabricated components or structures are hot dip galvanized in a bath of Prime Western or other lead-bearing zinc, the resultant coating having a thickness typically of 85-100 microns (0.0034-0.004 in.). Galvanized structural components include electricity and microwave transmission towers, highway signs and guard rails, pipe, and reinforcing bar.

Electrogalvanizing is carried out by a continuous high speed process, creating a smooth thin coating. This is finding applications in the automotive industry.

Application of a zinc coating by thermal spraying (metallizing) is utilized when a heavy coating of zinc is desired, typically in field application and for repair of existing coatings. Pure zinc (UNS Z13001) or zinc-15%Al alloy are typical sprayed coatings. Mechanical galvanizing, by tumbling the components in a drum with zinc dust, chemicals, and glass beads, finds applications where high temperatures, such as those encountered in hot dipping cannot be tolerated, for example fasteners manufactured from high strength steels.

Sacrificial Anodes

Zinc plates are used as sacrificial anodes, especially for ships hulls, ballast tanks, and underwater structures where the surface area is too large for galvanizing to be appropriate. Alloys used include a relatively high purity zinc (Type II Anodes, UNS Z13000) and alloys containing small amounts of aluminum (Type I, UNS Z32120 and Type III, UNS Z32121).

Zinc Alloy Castings

Zinc and its alloys have low melting points, high fluidity in the liquid state, and do not require fluxing or protective atmospheres, thus making them excellent casting alloys. They are utilized for all forms of castings including pressure die castings, permanent mold castings, sand castings, investment castings, and centrifugal castings. Of these, pressure die castings are the most important. Most zinc casting alloys are based on the zinc-aluminum system, which exhibits a eutectic point at about 5%Al and 382°C (720°F). Many of the die casting alloys are hypoeutectic, solidifying with a microstructure consisting of proeutectic dendrites of zinc-rich *eta* phase in a eutectic matrix. The hypereutectic alloys have aluminum-rich *beta* or *alpha* dendrites in the eutectic matrix. Because of the high cooling rates associated with die casting, stabilization heat treatments, e.g., 70°-100°C (158°-212°F) for 3-20 hours, are sometimes necessary to prevent small changes in dimensions and properties which result from long-term aging.

Zinc alloy castings are readily machinable, can be joined by soldering or brazing, have good sound and vibration damping properties, especially at high temperatures, and are used also as bearings and bushings under low speed high load conditions. They have good corrosion resistance, which can be improved by surface treatments such as chromating and anodizing; they can also be painted or plated for improved appearance. Zinc alloy castings are extensively used in the automotive industry, for example for carburetor and fuel pump bodies and various frames, levers, brackets, and moldings. They are also used for hardware components in computers, business machines, vending machines, and hand tools, as well as for components of domestic appliances, motor housings, locks, and clocks.

Each of the common die casting alloys is available with two similar specifications, differing slightly in composition; one is a specification for castings, the other with slightly lower impurity limits is a specification for ingots to be remelted for casting. The four common hypoeutectic casting alloys, all of which contain about 4%Al, are referred to as Alloys

2, 3, 5, and 7, or ASTM designations AC43A, AG40A, AC41A, and AG40B respectively. Alloy No. 3 (UNS Z33520 castings, UNS Z33521 ingots) is a very widely used alloy, with an excellent combination of strength and castability at a low cost. Alloy No. 7 (UNS Z33523 castings, UNS Z33522 ingots) is a high purity version of No. 3, with improved castability and ductility. Alloy No. 5 (UNS Z33531 castings, UNS Z33530 ingots) contains about 1%Cu for improved strength and creep resistance but decreased ductility, so that more care is necessary in any post-casting operations. Alloy No. 2 (UNS Z33541 castings, UNS Z33540 ingots) contains close to 3%Cu, giving it the highest strength and creep resistance of the hypoeutectoid alloys. It also has good bearing properties, but suffers from reduced ductility and some dimensional instability.

The hypereutectoid alloys include alloys ZA-8, ZA-12, and ZA-27 which contain approximately 8, 12, and 27%Al respectively. ZA-8 (UNS Z35636 castings, UNS Z35635 ingots) has properties similar to those of alloy No. 2, but with improved dimensional stability and higher strength. It is the only hypereutectoid alloy which can be cast using hot chamber die casting machines, as the hypoeutectoid alloys can. ZA-12 (UNS Z35631 castings, UNS Z35630 ingots) has good castability and a low density, along with excellent bearing and wear properties. ZA-27 (UNS Z35841 castings, UNS Z35840 ingots) has the highest strength, lowest density, and best sound, and vibration damping properties; however, it is not as easily castable. The ZA alloys can also be gravity cast, using sand or permanent molds. The coarser microstructures of the gravity cast alloys give them properties which are slightly different from those of the equivalent pressure die cast alloys. For example, strength is lower and creep resistance higher. Of the three ZA alloys, ZA-8 is most used with ferrous permanent molds, ZA-12 with graphite molds, and ZA-27 with sand molds.

Also included among the gravity cast alloys are the Kirksite alloys which are used for cast two-piece dies for sheet metal forming and polymer injection molding. These alloys, which are similar in composition to die casting alloy No. 2, are referred to as Kirksite I or forming die alloy A (UNS Z35543) and Kirksite II or forming die alloy B (UNS Z35542).

Slush casting is used for the production of hollow castings where interior surface quality is unimportant (e.g., hollow lamp bases). Zinc alloys used for this purpose include the slush casting alloys A (UNS Z34510) and B (UNS Z30500).

Wrought Zinc Alloys

Zinc alloys are rolled to strip, sheet, and foil which can be subsequently formed by deep drawing, spinning, roll forming, coining, and impact extrusion. A number of zinc alloys are applied in wrought form, including alloys containing small amounts of copper (UNS Z40330 and Z44330), copper plus titanium (UNS Z41320), or lead and cadmium (UNS Z21210, Z21220, and Z21540). Wrought zinc is used in roofing and flashing, and after forming as dry cell battery cans, cups, cases, gaskets, and other electrical and hardware components. An alloy of zinc containing approximately 22%Al has superplastic forming properties, which can be removed by a subsequent heat treatment, permitting complex forming operations followed by strengthening.

Zinc can be formed into rod and then drawn into wire, for example for use in thermal spray applications (metallizing). Forged zinc alloys include Korloy 2573, which contains 15%Al, and a zinc-copper-titanium alloy (Korloy 3130).

SAE/AMS SPECIFICATIONS - SOLDER, BABBITT & LEAD ALLOYS	
AMS	**Title**
4750	Solder, Tin-Lead, UNS L54950
4751	Solder, Tin-Lead, Eutectic, 63Sn - 37Pb, UNS L13630
4755	Solder, Lead-Silver, 94Pb - 5.5Ag, UNS L50180
4756	Solder, 97.5Pb - 1.5Ag - 1Sn, UNS L50131
4800	Bearings, Babbitt, 91Sn - 4.5Sb - 4.5Cu, UNS L13910
7721	Lead Alloy, Sheet and Extrusions, Special Property Material, 93Pb - 6.5Sb - 0.5Sn, As Fabricated, UNS L53131

ASTM SPECIFICATIONS - LEAD, TIN & ZINC ALLOYS	
ASTM	**Title**
Lead & Tin Alloys	
B 23	White Metal Bearing Alloys (Known Commercially as Babbitt Metal)
B 29	Pig Lead
B 32	Solder Metal
B 102	Lead- and Tin-Alloy Die Castings
B 339	Pig Tin
B 560	Modern Pewter Alloys
B 749	Lead and Lead Alloy Strip, Sheet, and Plate Products
B 774	Low Melting Point Alloys
Zinc & Zinc Alloys	
B 6	Zinc
B 69	Rolled Zinc
B 86	Zinc-Alloy Die Castings
B 240	Zinc Alloys in Ingot Form for Die Castings
B 327	Aluminum-Alloy Hardeners Used in Making Zinc Die-Casting Alloys
B 418	Cast and Wrought Galvanic Zinc Anodes
B 669	Zinc-Aluminum Alloys in Ingot Form for Foundry and Die Castings

ASTM SPECIFICATIONS - LEAD, TIN & ZINC ALLOYS (Continued)

ASTM	Title
Zinc & Zinc Alloys (Continued)	
B 750	Zinc-5% Aluminum-Mischmetal Alloy (UNS Z38510) in Ingot Form for Hot-Dip Coatings
B 791	Zinc-Aluminum Alloy Foundry and Die Castings
B 792	Zinc Alloys in Ingot Form for Slush Casting
B 793	Zinc Casting Alloy Ingot for Sheet Metal Forming Dies

AMERICAN CROSS REFERENCED SPECIFICATIONS - LEAD, TIN & ZINC ALLOYS

UNS	AMS	ASTM	FED	SAE	MIL SPEC
Lead & Tin Alloys					
L01900	---	B 440	---	---	---
L05120	---	---	QQ-T-390	---	---
L13008	---	B 339 (A)	---	---	---
L13600	---	---	QQ-S-571 (Sn60)	---	---
L13630	4751	B 32 (Sn63)	QQ-S-571 (Sn63)	---	---
L13650	---	B 102 (PY1815A)	---	---	---
L13700	---	B 32 (Sn70)	QQ-S-571 (Sn70)	---	---
L13820	---	B 102 (YC135A)	---	---	---
L13840	---	B 23 (3)	QQ-T-390	---	---
L13870	---	B 23 (11)	---	---	---
L13890	---	B 23 (2)	QQ-T-390	---	---
L13910	4800	B 23 (1)	QQ-T-390	---	---
L13911	---	B 560 (1)	---	---	---
L13912	---	B 560 (2)	---	---	---
L13913	---	B 102 (CY44A)	---	---	---
L13940	---	---	QQ-S-571 (Sb5)	---	---
L13950	---	B 32 (Sb5)	---	---	---

AMERICAN CROSS REFERENCED SPECIFICATIONS - LEAD, TIN & ZINC ALLOYS (Continued)					
UNS	AMS	ASTM	FED	SAE	MIL SPEC
Lead & Tin Alloys (Continued)					
L13961	---	---	QQ-S-571 (Sn96)	---	---
L13963	---	B 560 (3)	---	---	---
L50042	---	B 29 (Corroding Lead); B 749	---	---	---
L50045	---	B 29 (Common Lead)	---	---	---
L50050	---	---	QQ-L-171	---	---
L50065	---	---	QQ-C-40 (AA, C)	---	---
L50070	---	---	QQ-L-201	---	---
L50080	---	---	QQ-L-171	---	---
L50132	---	B 32 (Ag1.5)	QQ-S-571 (Ag 1.5)	---	---
L50151	---	B 32 (Ag2.5)	QQ-S-571 (Ag2.5)	---	---
L51120	---	B 29 (Chemical lead); B749	---	---	---
L51121	---	B 29 (Copper Bearing Lead); B 749	QQ-L-171 (C); QQ-L-201 (C)	---	---
L51123	---	B 749	QQ-L-201 (D)	---	---
L51124	---	---	QQ-C-40 (D)	---	---
L51180	---	---	---	J460, No. 485	---
L53131	7721	---	---	---	---
L53340	---	B 102 (Y10A)	---	---	---
L53345	---	---	---	J460 (13)	---
L53346	---	B 23 (13)	---	---	---
L53560	---	B 102 (YT155A)	---	---	---
L53565	---	B 23 (8)	---	---	---
L53585	---	B 23 (7)	---	---	---
L53620	---	B 23 (15)	---	---	---
L54210	---	B 32 (Sn2)	---	---	---
L54250	---	---	---	J473 (No. 9B)	---
L54322	---	B 32 (Sn5)	QQ-S-571 (Sn5)	---	---

AMERICAN CROSS REFERENCED SPECIFICATIONS - LEAD, TIN & ZINC ALLOYS (Continued)

UNS	AMS	ASTM	FED	SAE	MIL SPEC
Lead & Tin Alloys (Continued)					
L54370	---	---	---	J460, No. 190	---
L54510	---	---	---	J460, No. 19	---
L54520	---	B 32 (Sn10A)	---	---	---
L54525	---	B 32 (Sn10B)	QQ-S-571 (Sn10)	---	---
L54555	---	---	---	J473 (No. 6B)	---
L54560	---	B 32 (Sn15)	---	---	---
L54610	---	---	---	J473 (No. 5B)	---
L54711	---	B 32 (Sn20A)	QQ-S-571 (Pb80)	---	---
L54712	---	B 32 (Sn20B)	QQ-S-571 (Sn20)	---	---
L54721	---	B 32 (Sn25A)	---	---	---
L54722	---	B 32 (Sn 25B)	---	---	---
L54815	---	---	---	J473 (No. 3B)	---
L54821	---	B 32 (Sn30A)	QQ-S-571 (Pb70)	---	---
L54822	---	B 32 (Sn30B)	QQ-S-571 (Sn30)	---	---
L54851	---	B 32 (Sn35A)	QQ-S-571 (Pb65)	---	---
L54852	---	B 32 (Sn35B)	QQ-S-571 (Sn35)	---	---
L54905	---	---	---	J473 (No. 2B)	---
L54915	---	B 32 (Sn40A)	---	---	---
L54916	---	---	QQ-S-571 (Sn40)	---	---
L54918	---	B 32 (Sn40B)	---	---	---
L54935	---	B 774 (291-325)	---	---	---
L54940	---	---	---	J473 (No. 1B)	---
L54950	4750	---	---	---	---
L54951	---	B 32 (Sn45)	---	---	---
L55031	---	B 32 (Sn50)	QQ-S-571 (Sn50)	---	---

AMERICAN CROSS REFERENCED SPECIFICATIONS - LEAD, TIN & ZINC ALLOYS (Continued)					
UNS	**AMS**	**ASTM**	**FED**	**SAE**	**MIL SPEC**
Zinc & Zinc Alloys					
Z13000	---	B 418 (Zinc Anodes Type II)	---	---	---
Z13001	---	B 6 (Special High Grade)	---	---	---
Z15001	---	B 6 (High Grade)	---	---	---
Z19001	---	B 6 (Prime Western)	---	---	---
Z21210	---	B 69 (Zn-0.08Pb)	---	---	---
Z21220	---	B 69 (Zn-0.06Pb-0.06Cd)	---	---	---
Z21540	---	B 69 (Zn-0.3Pb-0.3Cd)	---	---	---
Z30500	---	B 792	---	---	---
Z32120	---	B 418 (Zinc Anodes - Type I)	---	---	---
Z32121	---	---	---	---	MIL-A-18001
Z33520	4803	B 86 (AG40A)	QQ-Z-363	J468 (903) (	---
Z33521	---	B 240 (AG40A)	---	J468 (903)	---
Z33522	---	B 240	---	---	---
Z33523	---	B 86	---	---	---
Z34510	---	B 792	---	---	---
Z35530	---	B 240	---	J468 (925)	---
Z35531	---	B 86	QQ-Z-363	J468 (925)	---
Z35540	---	B 240	---		---
Z35541	---	B 86 (AC43A)	---		---
Z35542	---	B 793	---	J469 (921)	MIL-I-7068 (II)
Z35543	---	B 793	---	---	MIL-I-7068 (I)
Z35630	---	B 669 (ZA-12)	---	---	---
Z35631	---	B 791	---	---	---
Z35635	---	B 669 (ZA-8)	---	---	---
Z35636	---	B 791	---	---	---
Z35840	---	B 669 (ZA-27)	---	---	---

AMERICAN CROSS REFERENCED SPECIFICATIONS - LEAD, TIN & ZINC ALLOYS (Continued)

UNS	AMS	ASTM	FED	SAE	MIL SPEC
Zinc & Zinc Alloys (Continued)					
Z35841	---	B 791	---	---	---
Z38510	---	B 750	---	---	---
Z40330	---	B 69 (Zn-0.8Cu)	---	---	---
Z41320	---	B 69 (Zn-0.8Cu-0.15Ti)	---	---	---
Z44330	---	B 69 (Zn-1Cu)	---	---	---
Z45330	---	B 69 (Zn-1Cu-0.01Mg)	---	---	---

a. This cross-reference table lists the basic specification or standard number, and since these standards are constantly being revised, it should be kept in mind that they are presented herein as a guide and may not reflect the latest revision.

COMMON NAMES OF LEAD & TIN ALLOYS WITH UNS No.

UNS	Common Name	UNS	Common Name
L05120	Bearing Alloy White Metal (Babbitt Metal)	L53131	Lead Alloy
L13840	Bearing Alloy White Metal (Babbitt Metal)	L51123	Lead Alloy Grade D
L13870	Bearing Alloy White Metal (Babbitt Metal)	L51124	Lead Alloy Grade D
L13910	Bearing Alloy, White Metal (Babbitt Metal)	L13963	Pewter, Modern
L13890	Bearing Alloy, White Metal (Babbitt Metal)	L13911	Pewter Casting Alloys, Modern
L53345	Bearing Alloy, Lead-Base	L13912	Pewter Sheet Alloy, Modern
L53346	Bearing Alloy, Lead-Base	L13008	Pig Tin
L53565	Bearing Alloy, White Metal , Lead-Base	L54905	Solder Alloy, Antimonial
L53585	Bearing Alloy, White Metal , Lead-Base	L54250	Solder Alloy
L53620	Bearing Alloy, Lead-Base	L54555	Solder Alloy
L51180	Bearing Alloy, Copper-Lead	L54610	Solder Alloy
L54370	Bearings, Plated Overlay for	L54815	Solder Alloy
L54510	Bearings, Plated Overlay for	L54821	Solder Alloy
L53340	Die Casting Alloy, Lead-Base	L54822	Solder Alloy

COMMON NAMES OF LEAD & TIN ALLOYS WITH UNS No. (Continued)

UNS	Common Name	UNS	Common Name
L53560	Die Casting Alloy, Lead-Base	L54851	Solder Alloy
L54935	Fusible Alloy	L54852	Solder Alloy
L50042	Lead, Corroding	L54916	Solder Alloy
L50045	Lead, Common	L54918	Solder Alloy
L50050	Lead, Grade A	L54940	Solder Alloy
L50065	Lead, Grade AA, or Grade C	L54951	Solder Alloy
L50070	Lead, Remelted, Grade B	L55031	Solder Alloy
L50080	Lead, Grade B	L54711	Solder Alloy 20B
L51120	Lead, Chemical	L54712	Solder Alloy 20C
L51121	Lead, Copper Bearing	L54721	Solder Alloy 25B
L54722	Solder Alloy 25C	L54915	Solder, 40/60
L50131	Solder Alloy-Grade 1.5S	L54950	Solder, 45/55
L50150	Solder Alloy-Grade 2.5S	L55030	Solder, 50/50
L50132	Solder Alloy-Grade Ag1.5	L13600	Solder, Tin-Lead
L50151	Solder Alloy-Grade Ag2.5	L13601	Solder, Tin-Lead
L50180	Solder Alloy-Grade Ag5.5	L13630	Solder, Tin-Lead
L54210	Solder, 2% Tin	L13631	Solder, Tin-Lead
L54211	Solder, 2% Tin Antimonial	L13700	Solder, Tin-Lead
L54320	Solder, 5/95	L13701	Solder, Tin-Lead
L54321	Solder, 5% Tin Antimonial	L13940	Solder, Tin-Antimony
L54322	Solder, SN5	L13950	Solder, Tin-Antimony
L54520	Solder, 10/90	L13960	Solder, Tin-Silver
L54525	Solder, 88-10-2	L13961	Solder, Tin-Silver
L54560	Solder, 15/85	L13965	Solder, Tin-Silver
L54720	Solder, 25/75	L13650	Tin Die Casting Alloy
L54820	Solder, 30/70	L13820	Tin Die Casting Alloy
L54850	Solder, 35/65	L13913	Tin Die-Casting Alloy

COMMON NAMES OF ZINC & ZINC ALLOYS WITH UNS No.

UNS	Common Name	UNS	Common Name
Z13000	Zinc Anodes (Type II)	Z35531	Zinc Alloy (AC41A) Die Casting
Z13001	Zinc Metal	Z35540	Zinc Alloy (AC43A) Ingot Form
Z15001	Zinc Metal	Z35541	Zinc Alloy (AC43A) Die Casting
Z19001	Zinc Metal	Z35542	Zinc Alloy (Kirksite II or B) Forming Die
Z21210	Zinc Rolled	Z35543	Zinc Alloy (Kirksite I or A) Forming Die
Z21220	Zinc Alloy Rolled	Z35630	Zinc-Aluminum Alloy (ZA-12) Ingot Form
Z21540	Zinc Alloy Rolled	Z35631	Zinc-Aluminum Alloy (ZA-12) Casting
Z30500	Zinc Alloy (B) Slush Casting	Z35635	Zinc-Aluminum Alloy (ZA-8) Ingot Form
Z32120	Zinc Anodes (Type I)	Z35636	Zinc-Aluminum Alloy (ZA-8) Casting
Z32121	Zinc Anodes (Type III)	Z35840	Zinc-Aluminum Alloy (ZA-27) Ingot Form
Z33520	Zinc Alloy (AG40A) Die Casting	Z35841	Zinc-Aluminum Alloy (ZA-27) Casting
Z33521	Zinc Alloy (AG40A) Ingot Form	Z38510	Zinc Alloy Ingot Form
Z33522	Zinc Alloy (AG40B) Ingot Form	Z40330	Zinc Alloy, Rolled
Z33523	Zinc Alloy (AG40B) Die Casting	Z41320	Zinc Alloy, Rolled
Z34510	Zinc Alloy (A) Slush Casting	Z44330	Zinc Alloy, Rolled
Z35530	Zinc Alloy (AC41A) Ingot Form	Z45330	Zinc Alloy, Rolled

CHEMICAL COMPOSITION OF LEAD & TIN ALLOYS

UNS	Chemical Composition
L01900	Cd 99.90 min Ag 0.01 max As 0.003 max Cu 0.015 max Pb 0.025 max Sb 0.001 max Sn 0.01 max Zn 0.035 max Other Ti 0.003 max
L05120	Pb 83.0-88.0 Al 0.005 max As 0.20 max Cu 0.50 max Fe 0.10 max Sb 8.0-10.0 Sn 4.0-6.0 Zn 0.005 max Other total 0.75 max
L13008	Sn 99.80 min As 0.05 max Bi 0.015 max Cd 0.001 max Cu 0.04 max Fe 0.015 max Fe 0.015 max Pb 0.05 max S 0.01 max Sb 0.04 max Zn 0.005 max Other Ni+Co 0.01 max
L13600	Sn 60 nom Pb 40 nom Al 0.005 max As 0.03 max Bi 0.25 max Co 0.08 max Fe 0.02 max Sb 0.12 max Zn 0.005 max
L13601	Sn 60 nom Pb 40 nom Al 0.005 max As 0.03 max Bi 0.25 max Cu 0.08 max Fe 0.02 max Sb 0.20-0.50 Zn 0.005 max
L13630	Sn 63 nom Pb 37 nom Al 0.005 max As 0.03 max Bi 0.25 max Cu 0.08 max Fe 0.02 max Sb 0.12 max Zn 0.005 max

CHEMICAL COMPOSITION OF LEAD & TIN ALLOYS (Continued)	
UNS	**Chemical Composition**
L13631	Sn 63 nom Pb 37 nom Al 0.005 max As 0.03 max Bi 0.25 max Cu 0.08 max Fe 0.02 max Sb 0.20-0.50 Zn 0.005 max
L13650	Sn 64-66 Pb 17-19 Al 0.01 max As 0.15 max Cu 1.5-2.5 Fe 0.08 max Sb 14-16 Zn 0.01 max
L13700	Sn 70 nom Pb 30 nom Al 0.005 max As 0.003 max Bi 0.25 max Cu 0.08 max Fe 0.02 max Sb 0.20 nom Zn 0.005 max
L13701	Sn 70 nom Pb 30 nom Al 0.005 max As 0.03 max Bi 0.25 max Cu 0.08 max Fe 0.02 max Sb 0.20-0.50 Zn 0.005 max
L13820	Sn 80-84 Sb 12-14 Al 0.01 max As 0.08 max Cu 4-6 Fe 0.08 max Pb 0.35 max Zn 0.01 max
L13840	Sn 83.0-85.0 Sb 7.5-8.5 Al 0.005 max As 0.10 max Bi 0.08 max Cd 0.05 max Cu 7.5-8.5 Fe 0.08 max Pb 0.35 max Zn 0.005 max
L13870	Sn 86.0-89.0 Sb 6.0-7.5 Al 0.005 max As 0.10 max Bi 0.08 max Cd 0.05 max Cu 5.0-6.5 Fe 0.08 max Pb 0.50 max Zn 0.005 max
L13890	Sn 88.0-90.0 Sb 7.0-8.0 Al 0.005 max As 0.10 max Bi 0.08 max Cd 0.05 max Cu 3.0-4.0 Fe 0.08 max Pb 0.35 max Zn 0.005 max
L13910	Sn 90.0-92.0 Sb 4.0-5.0 Al 0.005 max As 0.10 max Bi 0.08 max Cd 0.05 max Cu 4.0-5.0 Fe 0.08 max Pb 0.35 max Zn 0.005 max
L13911	Sn 90-93 Sb 6-8 As 0.05 max Cu 0.25-2.0 Fe 0.015 max Pb 0.05 max Zn 0.005 max
L13912	Sn 90-93 Sb 5-7.5 As 0.05 max Cu 1.5-3.0 Fe 0.015 max Pb 0.05 max Zn 0.005 max
L13913	Sn 90-92 Sb 4-5 Al 0.01 max As 0.08 max Cu 4-5 Fe 0.08 max Pb 0.35 max Zn 0.01 max
L13940	Sn 94 min Sb 4.0-6.0 Al 0.03 max As 0.06 max Cd 0.03 max Cu 0.08 max Fe 0.08 max Pb 0.20 max Zn 0.03 max Other total 0.03 max
L13950	Sn 95 nom Sb 4.5-5.5 Al 0.005 max As 0.05 max Bi 0.15 max Cu 0.08 max Fe 0.04 max Pb 0.20 max Zn 0.005 max
L13960	Sn 96 nom Ag 3.6-4.4 Al 0.005 max As 0.05 max Bi 0.15 max Cu 0.08 max Fe 0.02 max Pb 0.20 max Sb 0.20-0.50 Zn 0.005 max
L13961	Sn rem Ag 3.6-4.4 As 0.05 max Cd 0.005 max Cu 0.20 max Pb 0.10 max Zn 0.005 max
L13963	Sn 95-98 As 0.05 max Cu 1.0-2.0 Fe 0.015 max Pb 0.05 max Sb 1.0-3.0 Zn 0.005 max
L13965	Sn 96.5 nom Ag 3.3-3.7 Al 0.005 max As 0.05 max Bi 0.15 max Cu 0.08 max Fe 0.02 max Pb 0.20 max Sb 0.20-0.50 Zn 0.005 max
L50042	Pb 99.94 min Ag 0.0015 max Bi 0.050 max Cu 0.0015 max Fe 0.002 max Zn 0.001 max Other Cu+Ag 0.0025 max, As+Sb+Sn 0.002 max
L50045	Pb 99.94 min Ag 0.005 max Bi 0.050 max Cu 0.0015 max Fe 0.002 max Zn 0.001 max Other As+Sb+Sn 0.002 max
L50050	Pb 99.90 min Other total other elements 0.10 max
L50065	Pb 99.7 min Sb 0.02 max
L50070	Pb 39.5 min Bi 0.025 max (Optional)
L50080	Pb 95.0 min Other total other elements 5.0 max
L50131	Pb 97.5 nom As 1.3-1.7 Al 0.005 max As 0.02 max Bi 0.25 max Cu 0.08 max Fe 0.02 max Sn 0.75-1.25 Zn 0.005 max
L50132	Pb rem Ag 1.3-1.7 Al 0.005 max As 0.02 max Bi 0.25 max Cd 0.001 max Cu 0.30 max Fe 0.02 max Sb 0.40 max Sn 0.75-1.25 Zn 0.005 max Other total of all others 0.08 max

CHEMICAL COMPOSITION OF LEAD & TIN ALLOYS (Continued)	
UNS	**Chemical Composition**
L50150	Pb 97.5 nom Ag 2.3-2.7 Al 0.005 max As 0.02 max Bi 0.25 max Cu 0.08 max Fe 0.02 max Sn 0.25 max Zn 0.005 max
L50151	Pb rem Ag 2.3-2.7 Al 0.005 max As 0.02 max Bi 0.25 max Cd 0.001 max Cu 0.30 max Fe 0.02 max Sb 0.40 max Sn 0.25 max Zn 0.005 max Other total of all others 0.03 max
L50180	Pb rem Ag 5.0-6.0 Al 0.005 max As 0.02 max Bi 0.25 max Cd 0.001 max Cu 0.30 max Fe 0.002 max Cd 0.001 max Cu 0.30 max Fe 0.002 max Sb 0.40 max Sn 0.25 max Zn 0.005 max Other total of all others 0.03 max
L51120	Pb 99.90 min Ag 0.002-0.02 Bi 0.005 max Cu 0.04-0.08 Fe 0.002 max Zn 0.001 max Other As+Sb+Sn 0.002 max
L51121	Pb 99.90 min Ag 0.020 max Bi 0.025 max Cu 0.04 -0.08 Fe 0.002 max Zn 0.001 max Other As+Sb+Sn 0.002 max
L51123	Pb 99.85 min Ag 0.020 max Bi 0.025 max Cu 0.04-0.08 Fe 0.002 max Te 0.035-0.060 Zn 0.001 max Other As+Sb+Sn 0.002 max
L51124	Pb 99.82 min Ag 0.020 max Bi 0.025 max Cu 0.04-0.08 Fe 0.002 max Sn 0.016 max Te 0.035-0.055 Zn 0.001 max Other Sb+As 0.002 max
L51180	Cu rem Fe 0.35 max Pb 44.0-58.0 Sn 1.0-5.0 Other total of others 0.45 max, others, each 0.15 max
L53131	Pb rem Sb 6.0-7.0 Sn 0.25-0.75
L53340	Pb 89-91 Sb 9.25-10.75 As 0.15 max Cu 0.50 max Zn 0.01 max
L53345	Pb rem Sb 9.0-11.0 Sn 5.0-7.0 Al 0.005 max As 0.25 max Bi 0.10 max Cd 0.05 max Cu 0.50 max Zn 0.005 max Other total of others 0.20 max
L53346	Pb rem Sb 9.5-10.5 Sn 5.5-6.5 Al 0.005 max As 0.25 max Bi 0.10 max Cd 0.05 max Cu 0.50 max Fe 0.10 max Zn 0.005 max
L53560	Pb 79-81 Sb 14-16 Sn 4-6 Al 0.01 max As 0.15 max Cu 0.50 max Zn 0.01 max
L53565	Pb rem Sb 14.0-16.0 Sn 4.5-5.5 Al 0.005 max As 0.30-0.60 Bi 0.10 max Cd 0.05 max Cu 0.50 max Fe 0.10 max Zn 0.005 max
L53585	Pb rem Sb 14.0-16.0 Sn 9.3-10.7 Al 0.005 max As 0.30-0.60 Bi 0.10 max Cd 0.05 max Cu 0.50 max Fe 0.10 max Zn 0.005 max
L53620	Pb rem Sb 14.5-17.5 Sn 0.8-1.2 Al 0.005 max As 0.8-1.4 Bi 0.10 max Cd 0.05 max Cu 0.6 max Fe 0.10 max Zn 0.005 max
L54210	Pb 98 nom Sb 0.12 max Sn 1.5-2.5 Al 0.005 max As 0.02 max Bi 0.25 max Cu 0.08 max Fe 0.02 max Zn 0.005 max
L54211	Pb 98 nom Sb 0.20-0.50 Sn 1.5-2.5 Al 0.005 max As 0.02 max Bi 0.25 max Cu 0.08 max Fe 0.02 max Zn 0.005 max
L54250	Pb rem Sb 4.90-5.40 Sn 2.50-2.75 Al 0.005 max As 0.40-0.60 Bi 0.25 max Cu 0.08 max Fe 0.02 max Zn 0.005 max Other other elements, total 0.08 max
L54320	Pb 95 nom Sn 4.5-5.5 Al 0.005 max As 0.02 max Bi 0.25 max Cu 0.08 max Fe 0.02 max Sb 0.12 max Zn 0.005 max
L54321	Pb 95 nom Sn 4.5-5.5 Al 0.005 max As 0.02 max Bi 0.25 max Cu 0.08 max Fe 0.02 max Sb 0.20-0.50 Zn 0.005 max
L54322	Pb rem Sn 4.5-5.5 Ag 0.015 max Al 0.005 max As 0.02 max Bi 0.25 max Cd 0.001 max Cu 0.08 max Fe 0.02 max Sb 0.50 max Zn 0.005 max Other total of all others 0.08 max
L54370	Pb rem Sn 5.0-9.0 Other other elements, total 3.5 max

CHEMICAL COMPOSITION OF LEAD & TIN ALLOYS (Continued)	
UNS	**Chemical Composition**
L54510	Pb rem Sn 8.0-12.0 Other other elements, total 3.5 max
L54520	Pb 90 nom Sn 10 nom Al 0.005 max As 0.02 max Bi 0.25 max Cu 0.08 max Fe 0.02 max Sb 0.20-0.50 Zn 0.005 max
L54525	Pb rem Sn 9.0-11.0 Ag 1.7-2.4 Al 0.005 max As 0.02 max Bi 0.03 max Cd 0.001 max Cu 0.08 max Sb 0.20 max Zn 0.005 max Other total of all others 0.10 max
L54555	Pb rem Sn 14.0-15.0 Sb as specified 2.75 max Al 0.005 max Bi 0.25 max Cu 0.08 max Fe 0.02 max Zn 0.005 max Other other elements, total 0.08 max
L54560	Pb 85 nom Sn 15 desired Al 0.005 max As 0.02 max Bi 0.25 max Cu 0.08 max Fe 0.02 max Sb 0.20-0.50 Zn 0.005 max
L54610	Pb rem Sn 19.0-20.0 Sb 1.25-1.75 Al 0.005 max Bi 0.25 max Cu 0.08 max Fe 0.02 max Zn 0.005 max Other elements, total 0.08 max
L54711	Pb 80 nom Sn 20 desired Al 0.005 max As 0.02 max Bi 0.25 max Cu 0.08 max Fe 0.02 max Sb 0.20-0.50 Zn 0.005 max
L54712	Pb 79 nom Sn 20 desired Al 0.005 max As 0.02 max Bi 0.25 max Cu 0.08 max Fe 0.02 max Sb 0.8-1.2 Zn 0.005 max
L54720	Pb 75 nom Sn 25 nom Al 0.005 max As 0.02 max Bi 0.25 max Cu 0.08 max Fe 0.02 max Sb 0.25 max Zn 0.005 max
L54721	Pb 75 nom Sn 25 desired Al 0.005 max As 0.02 max Bi 0.25 max Cu 0.08 max Fe 0.02 max Sb 0.20-0.50 Zn 0.005 max
L54722	Pb 73.7 nom Sn 25 desired Al 0.005 max As 0.02 max Bi 0.25 max Cu 0.08 max Fe 0.02 max Sb 1.1-1.5 Zn 0.005 max
L54815	Pb rem Sn 29.0-30.0 Al 0.005 max Bi 0.25 max Cu 0.08 max Fe 0.02 max Sb 0.75-1.25 Zn 0.005 max Other other elements, total 0.08 max
L54820	Pb 70 nom Sn 30 desired Al 0.005 max As 0.02 max Bi 0.25 max Cu 0.08 max Fe 0.02 max Sb 0.25 max Zn 0.005 max
L54821	Pb 70 nom Sn 30 desired Al 0.005 max As 0.02 max Bi 0.25 max Cu 0.08 max Fe 0.02 max Sb 0.20-0.50 Zn 0.005 max
L54822	Pb 68.4 nom Sn 30 desired Al 0.005 max As 0.02 max Bi 0.25 max Cu 0.08 max Fe 0.02 max Sb 1.4-1.8 Zn 0.005 max
L54850	Pb 65 nom Sn 35 nom Al 0.005 max As 0.02 max Bi 0.25 max Cu 0.08 max Fe 0.02 max Sb 0.25 max Zn 0.005 max
L54851	Pb 65 nom Sn 35 nom Al 0.005 max As 0.02 max Bi 0.25 max Cu 0.08 max Fe 0.02 max Sb 0.20-0.50 Zn 0.005 max
L54852	Pb 63.2 nom Sn 35 nom Al 0.005 max As 0.02 max Bi 0.25 max Cu 0.08 max Fe 0.02 max Sb 1.6-2.0 Zn 0.005 max
L54905	Pb rem Sn 38.0-38.5 Al 0.005 max Bi 0.25 max Cu 0.08 max Fe 0.02 max Sb 1.5-2.00 Zn 0.005 max Other other elements, total 0.08 max
L54915	Pb 60 nom Sn 40 nom Al 0.005 max As 0.02 max Bi 0.25 max Cu 0.08 max Fe 0.02 max Sb 0.12 max Zn 0.005 max
L54916	Pb 60 nom Sn 40 nom Al 0.005 max As 0.02 max Bi 0.25 max Cu 0.08 max Fe 0.02 max Sb 0.20-0.50 Zn 0.005 max
L54918	Pb 58 nom Sn 40 nom Al 0.005 max As 0.02 max Bi 0.25 max Cu 0.08 max Fe 0.02 max Sb 1.8-2.4 Zn 0.005 max
L54935	Pb 43 nom Sn 43 nomBi 14 nom
L54940	Pb rem Sn 43.0-43.5 Sb 1.5-2.00 Al 0.005 max Bi 0.25 max Cu 0.08 max Fe 0.02 max Zn 0.005 max Other other elements, total 0.08 max
L54950	Pb 55 nom Sn 45 nom Sb 0.12 max Al 0.005 max As 0.03 max Bi 0.25 max Cu 0.08 max Fe 0.02 max Zn 0.005 max

CHEMICAL COMPOSITION OF LEAD & TIN ALLOYS (Continued)

UNS	Chemical Composition
L54951	Pb 55 nom Sn 45 nom Sb 0.20-0.50 Al 0.005 max As 0.03 max Bi 0.25 max Cu 0.08 max Fe 0.02 max Zn 0.005 max
L55030	Pb 50 nom Sn 50 nom Sb 0.12 max Al 0.005 max As 0.03 max Bi 0.25 max Cu 0.08 max Fe 0.02 max Zn 0.005 max
L55031	Pb 50 nom Sn 50 nom Sb 0.20-0.50 Al 0.005 max As 0.03 max Bi 0.25 max Cu 0.08 max Fe 0.02 max Zn 0.005 max

CHEMICAL COMPOSITION OF ZINC ALLOYS

UNS	Chemical Composition
Z13000	Zn rem Al 0.005 max Cd 0.003 max Cu 0.002 max Fe 0.0014 max Pb 0.003 max
Z13001	Zn 99.99 minAl 0.002 max Cd 0.003 max Cu 0.002 max Fe 0.003 max Pb 0.003 max Sn 0.001 max
Z15001	Zn 99.90 min Al 0.01 max Cd 0.02 max Fe 0.02 max Pb 0.03 max
Z19001	Zn 98.0 min Al 0.01 max Cd 0.20 max Cu 0.20 max Fe 0.05 max Pb 0.5-1.4
Z21210	Zn rem Al 0.001 max Cd 0.005 max Cu 0.001 max Fe 0.012 max Pb 0.10 max Sn 0.001 max
Z21220	Zn rem Al 0.001 max Cd 0.05-0.08 Cu 0.005 max Fe 0.012 max Pb 0.05-0.10 Sn 0.001 max
Z21540	Zn rem Al 0.001 max Cd 0.25-0.45 Cu 0.005 max Fe 0.002 max Pb 0.25-0.50
Z30500	Zn rem Al 5.25-5.75 Cd 0.005 max Cu 0.1 max Fe 0.10 max Pb 0.007 max Sn 0.005 max
Z32120	Zn rem Al 0.10-0.5 Cd 0.025-0.07 Cu 0.005 max Fe 0.005 max Pb 0.006 max Other 0.1 max
Z32121	Zn rem Al 0.1-0.5 Cd 0.025-0.15 Cu 0.005 max Fe 0.005 max Pb 0.006 max Si 0.125 max
Z33520	Zn rem Al 3.5-4.3 Cd 0.004 max Cu 0.25 max Fe 0.100 max Mg 0.02-0.05 Pb 0.005 max Sn 0.003 max
Z33521	Zn rem Al 3.9-4.3 Cd 0.003 max Cu 0.10 max Fe 0.075 max Pb 0.004 max Mg 0.025-0.05 Sn 0.002 max
Z33522	Zn rem Al 3.9-4.3 Cd 0.0020 max Cr 0.02 max Cu 0.10 max Fe 0.075 max Mg 0.010-0.020 Mn 0.05 max Ni 0.005-0.020 Pb 0.0020 max Si 0.035 max Sn 0.0010 max
Z33523	Zn rem Al 3.5-4.3 Cd 0.0020 max Cr 0.02 max Cu 0.25 max Fe 0.075 max Mg 0.005-0.020 Mn 0.5 max Ni 0.005-0.020 Pb 0.0030 max Si 0.035 max Sn 0.0010 max
Z34510	Zn rem Al 4.5-5.0 Cd 0.005 max Cu 0.2-0.3 Fe 0.100 max Pb 0.007 max Sn 0.005 max
Z35530	Zn rem Al 3.9-4.3 Cd 0.003 max Cu 0.75-1.25 Fe 0.075 max Pb 0.004 max Mg 0.03-0.06 Sn 0.002 max
Z35531	Zn rem Al 3.5-4.3 Cd 0.004 max Cu 0.75-1.25 Fe 0.100 max Mg 0.03-0.08 Pb 0.005 max Sn 0.003 max
Z35540	Zn rem Al 3.9-4.3 Cd 0.003 max Cu 2.6-2.9 Fe 0.075 max Pb 0.004 max Mg 0.025-0.050 Sn 0.002 max

CHEMICAL COMPOSITION OF ZINC ALLOYS (Continued)	
UNS	**Chemical Composition**
Z35541	Zn rem Al 3.5-4.3 Cd 0.004 max Cu 2.5-3.0 Fe 0.100 max Mg 0.020-0.050 Pb 0.005 max Sn 0.003 max
Z35542	Zn rem Al 3.9-4.3 Cd 0.003 max Cu 2.5-2.9 Fe 0.075 max Mg 0.02-0.05 Pb 0.003 max Sn 0.001 max
Z35543	Zn rem Al 3.5-4.5 Cd 0.005 max Cu 2.5-3.5 Fe 0.100 max Mg 0.02-0.10 Pb 0.007 max Sn 0.005 max
Z35630	Zn rem Al 10.8-11.5 Cd 0.005 max Cu 0.50-1.2 Fe 0.065 max Mg 0.020-0.030 Pb 0.005 max Sn 0.002 max
Z35631	Zn rem Al 10.5-11.5 Cd 0.004 max Cu 0.5-1.2 Fe 0.075 max Pb 0.006 max Mg 0.015-0.030 Sn 0.003 max
Z35635	Zn rem Al 8.2-8.8 Cd 0.005 max Cu 0.8-1.3 Fe 0.10 max Mg 0.020-0.030 Pb 0.005 max Sn 0.002 max
Z35636	Zn rem Al 8.0-8.8 Cd 0.006 max Cu 0.8-1.3 Fe 0.075 max Pb 0.006 max Mg 0.015-0.030 Sn 0.003 max
Z35840	Zn rem Al 25.5-28.0 Cd 0.005 max Cu 2.0-2.5 Fe 0.072 max Mg 0.012-0.020 Pb 0.005 max Sn 0.002 max
Z35841	Zn rem Al 25.0-28.0 Cd 0.006 max Cu 2.0-2.5 Fe 0.075 max Pb 0.006 max Mg 0.010-0.020 Sn 0.003 max
Z38510	Zn rem Al 4.2-6.2 Cd 0.005 max Cu 0.1 max Fe 0.075 max Mg 0.05 max Pb 0.005 max Sb 0.002 max Si 0.015 max Sn 0.002 max Ti 0.02 max Zr 0.02 max Other Ce+La 0.03-0.10 Other reach 0.02 max, total 0.04 max
Z40330	Zn rem Al 0.005 max Cd 0.02 max Cu 0.70-0.90 Fe 0.01 max Pb 0.02 max Ti 0.02 max
Z41320	Zn rem Al 0.001 max Cd 0.05 max Cu 0.50-1.50 Fe 0.012 max Pb 0.10 max Sn 0.001 max Ti 0.12-0.50
Z44330	Zn rem Al 0.001 max Cd 0.005 max Cu 0.85-1.25 Fe 0.012 max Pb 0.10 max Sn 0.001 max
Z45330	Zn rem Al 0.001 max Cd 0.04 max Cu 0.85-1.25 Fe 0.015 max Mg 0.006-0.016 Pb 0.15 max Sn 0.001 max

PHYSICAL PROPERTIES OF WHITE METAL BEARING ALLOYS - ASTM B 23

Alloy No.	Johnson's Apparent Elastic Limit[a]				Ultimate Tensile Strength in Compression[b]				Brinell Hardness[c]	
	psi 68°F	MPa 20°C	psi 68°F	MPa 20°C	psi 68°F	MPa 20°C	psi 212°F	MPa 100°C	68°F 20°C	212°F 100°C
1	2450	16.9	1050	7.2	12 850	88.6	6950	47.9	17.0	8.0
2	3350	23.1	1100	7.6	14 900	102.7	8700	60.0	24.5	12.0
3	5350	36.9	1300	9.0	17 600	121.3	9900	68.3	27.0	14.5
7	2500	17.2	1350	9.3	15 650	107.9	6150	42.4	22.5	10.5
8	2650	18.3	1200	8.3	15 600	107.6	6150	42.4	20.0	9.5
15	---	---	---	---	---	---	---	---	21.0	13.0

a. Johnson's apparent elastic limit is taken as the unit stress at the point where the slope of the tangent to the curve is 2/3 times its slope at the origin.
b. The ultimate strength values were taken as the unit load necessary to produce a deformation of 25% of the length of the specimen.
c. These values are the average Brinell number of three impressions on each alloy using a 10-mm ball and 500-kg load applied for 30 s.

PHYSICAL PROPERTIES OF WHITE METAL BEARING ALLOYS - ASTM B 23

Alloy No.	Melting Point		Temperature of Complete Liquefaction		Proper Pouring Temperature	
	°F	°C	°F	°C	°F	°C
1	433	223	700	371	825	441
2	466	241	669	354	795	424
3	464	240	792	422	915	491
7	464	240	514	268	640	338
8	459	237	522	272	645	341
15	479	248	538	281	662	350

PHYSICAL PROPERTIES OF BULK SOLDERS

Alloy Composition[a]	Melting Temperature °C	Solidus Temperature °C	Liquidus Temperature °C	Density Mg m^{-3}	Coefficient of Linear Thermal Expansion x 10^{-6} K^{-1}	Hardness
100Sn	231.9	---	---	7.29	23.5	3.9 HB
99Sn-1Cu	---	227	255	7.31	---	9 HB
98Sn-2Ag	---	221[b]	230[b]	---	---	---
97Sn-3Cu	---	227[b]	310	7.34	---	11 HB
96.5Sn-3.5Ag	221	---	---	10.38	---	14.8 HV
95Sn-5Sb	---	234	240	7.250	---	15 HB
60Sn-40Pb	---	183	188	8.520	23.9	16 HV
50Sn-50Pb	---	183	216	8.89	23.4	14 HV
42Sn-58Bi	138	---	---	8.72	13.78	22 HB
40Sn-60Pb	---	183	234	9.280	24.58	12 HV
10Sn-90Pb	---	268	301	10.50	27.98	10 HV
62Sn-36Pb-2Ag	---	177	189	8.50	27.0	---
50Sn-32Pb-18Cd	145	---	---	8.5	13.2	14 HB
40Sn-58Pb-2Sb	---	185	231	9.23	24.3	---
34Sn-42Pb-24Bi	---	99.5	146	---	---	---
5Sn-93.5Pb-1.5Ag	---	296	301	11.10	---	---
1Sn-97.5Pb-1.5Ag	309	---	---	11.28	---	13 HB
95Pb-5Ag	---	304	365	---	---	---

a. By weight. b. Estimated.

Chapter 7

PRECIOUS METALS

The term precious metals is taken to include silver, gold, and the six platinum group metals - platinum, palladium, ruthenium, rhodium, osmium, and iridium. Gold and silver have been known and used for many millennia, as has platinum to a lesser extent. The others are recent discoveries. These metals and their alloys are used in the electronics, jewelry, dentistry, coinage, textile, automotive, aerospace, ceramic, and chemical industries. They all have high densities (specific gravities ranging from silver at 10.5 to iridium at 22.6), relatively good corrosion resistance, good electrical conductivity and light reflectivity, and relatively high cost. Despite the latter factor, they are often economically acceptable because of their properties and performance in service. Scrap values are high and recycling is common.

Products available include rod and wire drawn as fine as 7.5 microns (0.0003 in) diameter for some alloys, and rolled products such as sheet, strip, ribbon, and foil (in some cases as thin as 2.5 microns (0.0001 in). Tubing of various cross-sections is also available for various precious metal alloys in the form of seamless tube for small sizes down to 0.4 mm (0.16 in) outside diameter, and as seamed tubing for larger sizes up to 75 mm (3 in) inside diameter and in less ductile alloys. Powder products are produced for applications in the electronic industry, including inks and films, and for applications as protective coatings.

Base metal alloys clad with precious metals (via mechanical and thermal bonding) are available in a range of product forms including sheet, wire, and tubing. Precious metal coatings can be applied using such techniques as vacuum metallizing and sputtering, electroplating, and thermal decomposition (firing).

The purities of silver and gold are designated in terms of several systems which do not apply to other metals. Fineness is used to designate purity in parts per thousand by weight; hence, 995 fine gold is 99.5% gold, and 925 fine silver (sterling silver) is 92.5% silver. Gold bullion, as traded, is at least 995 fine and silver bullion at least 999 fine. Gold purity for jewelry applications is specified in karats, where 1 karat is 1/24th part of gold; hence, for example, 18 karat gold is 75% gold.

SILVER

Silver is characterized by its bright appearance, high thermal and electrical conductivities, reflectivity, and very high malleability. It is resistant to corrosion in many organic acids and in sodium and potassium hydroxide, but it is susceptible to corrosion in mineral acids. It resists oxidation at room temperature but is susceptible to attack by sulphur. Although its major use is in photographic emulsions, it finds wide applications as brazing filler metal alloys ("silver solder"), as well as in electrical contacts, bearings, jewelry, and tableware, batteries, mirror backings, dental alloys, catalysts, and coinage. Silver-clad base metals (copper, brass, nickel, iron) are also used, in particular in the electrical and chemical industries.

Silver and the other precious metals are described by the UNS system in a single category with the prefix letter P. Other than for unalloyed silver, the designation given is P07xxy, where xx gives the approximate silver content of the alloy in percent; hence, P07650 contains about 65%Ag. Refined silver is available at several levels of purity from 99.99% (UNS P07010) to 99.90% (UNS P07020).

Brazing Alloys

A very wide range of silver based brazing filler metal alloys is described by various standards, notably ASME SFA5.8 and AWS A5.8, which give alloy designations in the form BAg-x. These alloys are characterized by low melting temperatures and the ability to wet the solid base metals; hence, they are suitable filler metals for brazing steel, cast iron, stainless steel, and copper alloys to themselves and each other. They are also used in the brazing of some reactive and refractory metals, although the latter are restricted to low temperature use.

These brazing alloys are generally silver-copper-zinc alloys, some with additions of cadmium, manganese, nickel, or tin. The most widely used are the alloys BAg-1 (UNS P07450 which contains 45%Ag, 15%Cu, 16%Zn, 24%Cd) and BAg-1a (UNS P07500 containing 50%Ag, 15%Cu, 16%Zn, 18%Cd). The cadmium additions give these alloys particularly low melting temperatures, narrow melting temperature ranges, and high fluidities. Other brazing alloys include BAg-2 (UNS P07350 with 35%Ag, 26%Cu, 21%Zn, 18%Cd) and BAg-2a (UNS P07300 with 30%Ag, 27%Cu, 23%Zn, 20%Cd) which contain less silver and are therefore cheaper but have higher melting temperatures and are less fluid. For food processing applications, where toxicity must be considered, it is necessary to utilize brazing alloys which are free of cadmium; these alloys include BAg-4 (UNS P07400 with 40%Ag, 30%Cu, 28%Zn, 2%Ni), BAg-5 (UNS P07453 with 45%Ag, 30%Cu, 25%Zn), BAg-20 (UNS P07301 with 30%Ag, 38%Cu, 32%Zn), and BAg-28 (UNS P07401 with 40%Ag, 30%Cu, 28%Zn, 2%Sn). For the brazing of stainless steel, the nickel-bearing alloy BAg-3 (UNS P07501 with 50%Ag, 15%Cu, 16%Zn, 16%Cd, 3%Ni) is most commonly used, but many other silver-based brazing alloys are applicable as well.

The vacuum brazing grades have maximum contents specified for a number of impurities, notably cadmium, phosphorus, and zinc which have high vapour pressures. Typical examples include BVAg-0 (UNS P07017 with 99.95% Ag) and BVAg-8 (UNS P07727 with 72%Ag, 28%Cu). A few brazing alloys contain lithium in amounts less than 0.5%; brazing with these alloys, such as BAg-8a (UNS P07723 with 72%Ag, 27%Cu, 0.4%Li) and BAg-19 (UNS P07925 with 92%Ag, 8%Cu, 0.2%Li), can be performed without the use of a flux, since the alloys are self-fluxing in dry, non-oxidizing protective atmospheres.

Electrical Contacts

Silver is a useful material for medium- and heavy-duty electrical contacts, particularly when alloyed with a dispersion of cadmium oxide. Here, advantage is taken of the high thermal and electrical conductivity and low surface contact resistance of silver, and the ability of the particles of cadmium oxide to prevent sticking and welding and to minimize arc erosion. Under light-duty conditions (low voltage, low current) these materials are not suitable because of the tendency of silver to react with sulphur in the atmosphere to form a sulphide surface layer. The cadmium oxide dispersion can be formed either by powder metallurgical processing or by internal oxidation (e.g., the Ag-15%Cd alloy). Silver alloys containing approximately 0.25% each of magnesium, nickel, and in some cases copper, can also be dispersion strengthened by internal oxidation for use as electrical contact materials. Other silver-based alloys used for electrical contact purposes include fine silver, for low current applications, and alloys of silver with copper, palladium, platinum, or gold. These alloying elements increase the hardness, but decrease the electrical conductivity. Silver is also used in the form of an electroplated coating on electrical connection plugs and sockets.

Jewelry and Tableware Alloys, Coinage

The high reflectivity of silver makes it particularly attractive for applications in jewelry and tableware. Strength, hardness, and wear resistance are obtained through the use of alloys, in particular sterling silver, which contains 92.5%Ag and 7.5%Cu. There are two grades of sterling silver - the standard grade (UNS P07931) and silversmiths grade (P07932), which has closer limits on the allowable silver and copper contents. Sterling silver is heat treatable by precipitation hardening, for example by solution treatment at 720°C (1328°F) followed by aging for 1 hr at 300°C (572°F) but this is not a normal operation.

Silver alloys containing copper (e.g., Ag-10%Cu) are traditional coinage alloys.

Dental Alloys

Silver-tin-mercury and silver-tin-copper-mercury alloys (amalgams) find wide use in dentistry because of their attractive mechanical properties and the small dimensional changes which occur during their "setting." Typically, particles containing 65%Ag min, 29%Sn max, and 6%Cu max are mixed with a similar weight of mercury to form the amalgam. The setting process occurs as mercury diffuses in the alloy and reacts with the constituents so as to transform from the liquid state into solid intermetallic compounds such as Ag_2Hg_3 or Sn_7Hg. These, along with compounds such as Ag_3Sn and Cu_6Sn_5, make up the final filling material.

Miscellaneous Applications

Silver also finds use as a catalyst in chemical reactions such as the oxidation of ethylene and the production of formaldehyde from methanol, and as a crucible material for molten NaOH. It is used in engine bearings where advantage is taken of its lubricating properties, its high thermal conductivity, and its low solubility in iron. It makes an excellent reflector for visible and infrared light, and can be applied to glass or ceramics by a vacuum sublimation metallizing process. In the form of a silver-silver oxide composite, it finds specialized application as a battery plate material.

GOLD

Gold is a metal with a bright, pleasing yellow colour which can be substantially modified by alloy addition. It has very high malleability, thermal and electrical conductivity, and excellent corrosion resistance, including resistance to oxidation and sulphidation (i.e., to tarnish). This combination of properties has given gold and its alloys traditional applications in the jewelry, dental, electrical, and electronics industries and as brazing filler metals. Gold and gold alloys are used in the form of bulk materials, and also as surface coatings applied by a wide range of techniques.

Gold and its alloys are given UNS designations in the precious metals and alloys group, specifically designations of the form P00xxx where xxx refers to specific alloys. Thus gold with a minimum purity of 99.99% is categorized as UNS P00010, P00015, or P00016 depending on impurities and their limits, while UNS P00020 and P00025 are refined golds with minimum gold contents of 99.95% and 99.5% respectively.

Jewelry

Gold and its alloys are the most popular of the jewelry metals, as they have been for thousands of years. Jewelry alloys are classified by their karat ratings and described by their colours. The most widely used alloys are 14 k gold (58%Au), 18 k gold (75%Au) and 10 k gold (42%Au). The elements which make up the balance of these alloys are silver and the base metals copper, zinc, and nickel, with the colour of the alloy being determined by which elements are present and their proportions. Thus, increasing silver content changes the colour of gold from yellow to greenish yellow to white, copper increases the redness, nickel increases whiteness, and zinc acts as a decolourizer. The jewelry alloys which result from these additions include the green, yellow, red, and white golds. At the18 k level, the green golds are gold-silver alloys (e.g., UNS P00280) and red golds are gold-copper alloys (UNS P00285), while the gold-silver-copper alloys are yellow (UNS P00250, P00255, P00260). The red 18 k gold alloy is difficult to work because it undergoes an ordering transformation in the solid state which increases its strength and lowers its ductility. The 10 k and 14 k gold alloys are based on the gold-silver-copper system (e.g., 14 k yellow golds UNS P00180, P00190, P00200, P00220); the properties of these alloys are dependent mainly on the ratios of silver to copper, and not on the gold contents. The white golds are either gold-silver-copper alloys to which 5-12% zinc is added (e.g., 10 k alloy UNS P00125) or else alloys in the gold-nickel-copper system, also containing zinc (e.g., 14 k alloys UNS P00150, P00160). The gold-nickel-copper white golds are harder but are susceptible to cracking when annealed after light working ("firecracking").

In addition, there are several ways of producing jewelry with a gold surface and a base metal interior. The expression "gold filled" applies to situations in which the weight of the coating (minimum 10 k gold) is at least 5% of the total weight. If the coating weight is less than this, the expression "rolled gold plate" is used. These coatings are applied by soldering, brazing, welding, or are applied mechanically. Gold and gold alloy electrodeposits are also used; in some cases alloying is accomplished by electrodeposition of successive layers of different materials followed by heating to permit alloying by diffusion.

Electronic Applications

Gold finds applications in the electronic industry in printed circuit boards, connectors, keyboard contacts, and miniaturized circuitry, applications where high conductivity and chemical and metallurgical stability are required. Gold and gold alloys are used in sliding electrical contacts for low current service; typical alloys include those containing about 25%Ag and 6-8%Pt (UNS P00691 and P00692). Other electrical contact gold alloys include UNS P00710 (15%Cu, 8%Pt, 5%Ag), UNS P00750 (22%Ag, 3%Ni), and UNS P00901 (Au-10%Cu). Gold in the form of a surface electrodeposit finds electronic uses include surfacing of plug-type electrical connectors and of high frequency conductors for use in environments where silver corrodes.

Brazing Alloys

Gold brazing alloys are used for joining jet and rocket engine components made from nickel- or cobalt-based superalloys or stainless steel. Here, joint ductility is high since alloying with the base metal is minimized. They are also used for joining hermetically sealed vacuum tube components where the low vapour pressure is advantageous. The common brazing alloys include gold-copper alloys such as BAu-1 (UNS P00375 with 38%Au, 62%Cu), BAu-2 (UNS P00800 with 80%Au and 20%Cu), and BAu-3 (UNS P00350 with 35%Au, 62%Cu and 3%Ni), gold-nickel alloys like BAu-4 (UNS P00820 with 82%Au, 18%Ni), and gold-nickel-palladium alloys like BAu-5 (UNS P00300 with 30%Au, 36%Ni and 34%Pd) and BAu-6 (UNS P00700 with 70%Au, 22%Ni and 8%Pd). As in the case of silver-base brazing alloys, some of these have BVAu equivalents designed specifically for vacuum brazing applications with restricted contents of volatile impurities. Examples include BVAu-2 (UNS P00807), BVAu-4 (UNS P00827), the gold-nickel-palladium alloy BVAu-7 (UNS P00507), and the gold-palladium alloy BVAu-8 (UNS P00927) with 92%Au and 8%Pd).

Miscellaneous Applications

Gold is a good reflector of radiation in the red and infrared part of the spectrum, and thus finds applications as coatings for infrared reflectors and in heating and drying devices including thermal barrier windows, space vehicle components, and space suits. Gold alloys are also used extensively in dentistry for crown and bridge alloys, inlays, and porcelain fused to metal alloys. Many compositions are used, typically gold-silver alloys with additions of copper, zinc, and palladium. The gold alloy 70%Au-30%Pt is used for spinnerettes in rayon production in severe corrosion service. This is also a high melting point solder for platinum.

PLATINUM

Platinum is the most abundant of the platinum group metals, and it is also the most widely used. Applications are based on the corrosion resistance, high melting point, ductility, electronic properties, and overall appearance. Platinum remains bright when heated in air at all temperatures up to the melting point and is soluble only in acids which generate free chlorine, such as aqua regia. Unalloyed platinum is available at two purity levels, UNS P04980 (min. 99.80%Pt) and P04995 (min. 99.95%Pt). Purity better than 99.9%Pt is required for laboratory ware and for electrical contacts. Even higher purity platinum is used for thermocouples and resistance thermometers. Platinum and its alloys are used for jewelry, as a brazing alloy for tungsten, as sensing elements for gas analyzers, as containers for various chemicals at high temperatures, and as various types of electrodes. Platinum alloys are excellent catalysts for oxidation and hydrogenation reactions and in petroleum reforming. Catalysts, in the form of platinum-palladium alloys for the automotive industry, account for the largest use of platinum.

Platinum-palladium alloys are also used in jewelry and for electrodes and electrical contacts. Platinum-iridium alloys up to 30%Ir are used for laboratory ware, jewelry, electrical contacts and electrodes, and as fine tubing. Platinum-rhodium alloys (up to 40%Rh, with Pt-10%Rh being the most popular) are used for applications at high temperatures in oxidizing conditions, in addition to being used as a catalyst, as a furnace element material (at 10%, 20% or 40%Rh), and as standard thermocouple alloys (at 10%, 13%, and 30%Rh). Platinum-rhodium alloys are also used for crucibles and dies in the glass industry, and as spinnerettes for synthetic fibers and bushings in the extrusion of fiberglass. Platinum-ruthenium alloys (5-11%Ru) are used for electrical contacts, for jewelry, and for hypodermic needles. The very strong alloy Pt-15%Rh-6%Ru is used in wire form in many applications including wire-wound electrical instruments. Pt-8%W wire is used widely in electrical, electronic, and biochemical applications. Pt-10%Ni has outstanding high temperature strength. Pt-23%Co (i.e., 50 atomic %Co) is a permanent magnet material with an unusually high coercive force. The creep resistance of platinum and its alloys at high temperatures can be much improved by using alloys containing a dispersion of zirconia; hence, the term ZGS (Zirconia-Grain-Stabilized) platinum and platinum alloys. These find applications in such areas as the glass industry where it is used for bushings, stirrers, and carrying apparatus.

PALLADIUM

Palladium is similar to platinum in its appearance, ductility, and strength but it is lower in melting point and in corrosion/tarnish resistance and much lower in density. Overall, it is the second most widely used of the platinum group elements with wide applications as a material for electrical contacts. In the automotive industry palladium is employed as a component of catalyst alloys, and palladium alloy catalysts also have applications in the chemical and pharmaceutical industries. Other uses of palladium alloys are found in medical and dental materials, in jewelry, and in brazing alloys. It can be electroplated, electroformed, and deposited by electroless methods.

Commercial purity palladium with a minimum 99.8%Pd (UNS P03980) is used in electrical contacts for such light duty purposes as relays in telephone systems. Palladium-silver alloys also find applications as electrical contact materials, and the Pd-40%Ag alloy is used as well for precision resistance wires, and as a diffusion membrane for separation of hydrogen from other gases. Electrical contacts are also made from palladium-copper (40%Cu) alloy and from palladium-silver-copper alloys such as UNS P03440 (38%Ag, 16%Cu, 1%Pt) and UNS P03350 (30%Ag, 14%Cu, 10%Pt, 10%Au). The latter alloys can be precipitation hardened, making them useful in contact applications where sliding wear resistance or spring properties are required. Some palladium-silver-gold alloys are solution strengthened but there are others which are precipitation hardened. These are used in dental applications and in the form of cladding for severe corrosion service, particularly in halogen acids. Alloys used for brazing include palladium-cobalt alloys, such as the vacuum brazing alloy BVPd-1 (UNS P03657 with 65%Pd-35%Co). Pd-4.5%Ru is a standard jewelry alloy, in both cast and wrought forms, and palladium-ruthenium alloys up to 12%Ru are also used for electrical contacts.

RHODIUM, IRIDIUM, RUTHENIUM, OSMIUM

These platinum group metals are all hard white metals which have little ductility at room temperature, but all except osmium can be worked at elevated temperatures. They are all used primarily as alloying elements in other precious metals, usually platinum or palladium, but some have applications in their own right. In general, they have excellent corrosion resistance, accompanied by high melting temperature and high density.

Rhodium is available in unalloyed form (e.g., UNS P05980, P05981 and P05982) and finds commercial applications, many of which are based on its very bright appearance (high specular reflectivity). For example, it is used as a plating on other white jewelry metals for whiteness and wear resistance. It remains bright under all atmospheric conditions at ordinary temperatures and is resistant to high temperature oxidation and to corrosion by aqua regia.

Iridium has a combination of excellent corrosion resistance and high temperature strength, giving it applications as a crucible material for melting nonmetallic materials up to 2100°C (3812°F). It is also used as a neutron absorber and as a gamma ray source. Iridium has the highest density of any metal (22.65 g/cm^3) and is the only metal which can be used for short periods at 2000°C (3632°F) in air.

Ruthenium and osmium are mainly used as alloying elements in other precious metals, where they provide hardness and corrosion resistance. Osmium-ruthenium alloys are used in specialized instrument applications such as galvanometer pivots.

COMMON NAMES OF PRECIOUS METALS WITH UNS No.

UNS	Common Name	UNS	Common Name
P00140	10-Karat Green Gold Jewelry Alloy	P00225	14-Karat Yellow Gold Jewelry Alloy
P00145	10-Karat Red Gold Jewelry Alloy	P00280	18-Karat Green Gold Jewelry Alloy
P00125	10-Karat White Gold Jewelry Alloy	P00285	18-Karat Red Gold Jewelry Alloy
P00130	10-Karat White Gold Jewelry Alloy	P00270	18-Karat White Gold Jewelry Alloy
P00135	10-Karat White Gold Jewelry Alloy	P00275	18-Karat White Gold Jewelry Alloy
P00105	10-Karat Yellow Gold Jewelry Alloy	P00250	18-Karat Yellow Gold Jewelry Alloy
P00110	10-Karat Yellow Gold Jewelry Alloy	P00255	18-Karat Yellow Gold Jewelry Alloy
P00115	10-Karat Yellow Gold Jewelry Alloy	P00260	18-Karat Yellow Gold Jewelry Alloy
P00120	10-Karat Yellow Gold Jewelry Alloy	P00016	Gold
P00230	14-Karat Green Gold Jewelry Alloy	P00100	Gold Alloy (990) Gold
P00235	14-Karat Green Gold Jewelry Alloy	P00350	Gold Brazing Alloy
P00170	14-Karat Red Gold Jewelry Alloy	P00375	Gold Brazing Alloy
P00150	14-Karat White Gold Jewelry Alloy	P00500	Gold Brazing Alloy
P00155	14-Karat White Gold Jewelry Alloy	P00700	Gold Brazing Alloy
P00160	14-Karat White Gold Jewelry Alloy	P00800	Gold Brazing Alloy
P00165	14-Karat White Gold Jewelry Alloy	P00820	Gold Brazing Alloy
P00032	14-Karat Yellow Gold Jewelry Alloy	P00807	Gold Brazing Alloy (Vacuum Grade)
P00175	14-Karat Yellow Gold Jewelry Alloy	P00827	Gold Brazing Alloy (Vacuum Grade)
P00180	14-Karat Yellow Gold Jewelry Alloy	P00691	Gold Electrical Contact Alloy
P00185	14-Karat Yellow Gold Jewelry Alloy	P00692	Gold Electrical Contact Alloy
P00190	14-Karat Yellow Gold Jewelry Alloy	P00710	Gold Electrical Contact Alloy
P00195	14-Karat Yellow Gold Jewelry Alloy	P00750	Gold Electrical Contact Alloy
P00200	14-Karat Yellow Gold Jewelry Alloy	P00901	Gold Electrical Contact Alloy
P00205	14-Karat Yellow Gold Jewelry Alloy	P00927	Gold-Palladium Brazing Alloy (Vacuum Grade)
P00210	14-Karat Yellow Gold Jewelry Alloy	P00300	Gold-Palladium-Nickel Brazing Alloy
P00215	14-Karat Yellow Gold Jewelry Alloy	P00507	Gold-Palladium-Nickel Brazing Alloy (Vacuum Grade)
P00220	14-Karat Yellow Gold Jewelry Alloy	P00900	Gold-Silver Alloy

COMMON NAMES OF PRECIOUS METALS WITH UNS No. (Continued)

UNS	Common Name	UNS	Common Name
P00580	Gold: 14 Karat Yellow Gold - New U.S. (After September 1981	P07560	Silver Brazing Alloy
P00560	Gold: 14 Karat Yellow Gold - Old U.S. (Prior to October 1981)	P07563	Silver Brazing Alloy
P03300	Palladium Alloy (Paliney No. 7, "W" Alloy 3350, No. 226 Alloy)	P07600	Silver Brazing Alloy
P03350	Palladium Electrical Contact Alloy	P07650	Silver Brazing Alloy
P03440	Palladium Electrical Contact Alloy	P07700	Silver Brazing Alloy
P03657	Palladium-Cobalt Brazing Alloy (Vacuum Grade)	P07720	Silver Brazing Alloy
P00010	Refined Gold	P07723	Silver Brazing Alloy
P00015	Refined Gold	P07925	Silver Brazing Alloy
P00020	Refined Gold	P07450	Silver Brazing Alloy (BAg-1)
P00025	Refined Gold	P07540	Silver Brazing Alloy (BAg-13)
P06100	Refined Iridium	P07500	Silver Brazing Alloy (BAg-1a)
P03980	Refined Palladium	P07350	Silver Brazing Alloy (BAg-2)
P03995	Refined Palladium	P07301	Silver Brazing Alloy (BAg-20)
P04980	Refined Platinum	P07630	Silver Brazing Alloy (BAg-21)
P04995	Refined Platinum	P07490	Silver Brazing Alloy (BAg-22)
P05981	Refined Rhodium	P07850	Silver Brazing Alloy (BAg-23)
P05982	Refined Rhodium	P07505	Silver Brazing Alloy (BAg-24)
P07015	Refined Silver	P07200	Silver Brazing Alloy (BAg-25)
P07020	Refined Silver	P07250	Silver Brazing Alloy (BAg-26)
P07010	Refined Silver	P07251	Silver Brazing Alloy (BAg-27)
P05980	Rhodium	P07401	Silver Brazing Alloy (BAg-28)
P07016	Silver	P07300	Silver Brazing Alloy (BAg-2a)
P07253	Silver Brazing Alloy	P07501	Silver Brazing Alloy (BAg-3)
P07351	Silver Brazing Alloy	P07252	Silver Brazing Alloy (BAg-33)
P07402	Silver Brazing Alloy	P07380	Silver Brazing Alloy (BAg-34)
P07454	Silver Brazing Alloy	P07400	Silver Brazing Alloy (BAg-4)
P07502	Silver Brazing Alloy	P07453	Silver Brazing Alloy (BAg-5)

COMMON NAMES OF PRECIOUS METALS WITH UNS No. (Continued)

UNS	Common Name	UNS	Common Name
P07503	Silver Brazing Alloy (BAg-6)	P07728	Silver Brazing Alloy (Vacuum Grade)
P07507	Silver Brazing Alloy (Vacuum Grade)	P07017	Silver Brazing Alloy (Vacuum Grade) (BVAg-0)
P07587	Silver Brazing Alloy (Vacuum Grade)	P07900	Silver Electrical Contact Alloy
P07607	Silver Brazing Alloy (Vacuum Grade)	P07932	Silver, Sterling Silver (Silversmiths Grade)
P07627	Silver Brazing Alloy (Vacuum Grade)	P07931	Silver, Sterling Silver (Standard Grade)
P07687	Silver Brazing Alloy (Vacuum Grade)	P07547	Silver-Palladium Brazing Alloy (Vacuum Grade)
P07727	Silver Brazing Alloy (Vacuum Grade)		

AMERICAN CROSS REFERENCED SPECIFICATIONS - PRECIOUS METALS

UNS	AMS	ASME	AWS	ASTM	MIL SPEC
P00010	---	---	---	B562 (99.995)	---
P00015	---	---	---	B562 (99.99)	---
P00016	---	---	---	F72	---
P00020	7731	---	---	B562 (99.95)	---
P00025	---	---	---	B562 (99.5)	---
P00300	4785	SFA5.8 (BAu-5)	A5.8 (BAu-5)	---	---
P00350	---	SFA5.8 (BAu-3)	A5.8 (BAu-3)	---	---
P00375	---	SFA5.8 (BAu-1)	A5.8 (BAu-1)	---	---
P00500	4784	---	---	---	---
P00507	---	SFA5.8 (BVAu-7)	A5.8 (BVAu-7)	---	---
P00691	---	---	---	B522 (I)	---
P00692	---	---	---	B522 (II)	---
P00700	4786	SFA5.8 (BAu-6)	A5.8 (BAu-6)	---	---
P00710	---	---	---	B541	---
P00750	---	---	---	B477	---
P00800	---	SFA5.8 (BAu-2)	A5.8 (BAu-2)	---	---

AMERICAN CROSS REFERENCED SPECIFICATIONS - PRECIOUS METALS (Continued)

UNS	AMS	ASME	AWS	ASTM	MIL SPEC
P00807	---	SFA5.8 (BVAu-2)	A5.8 (BVAu-2)	---	---
P00820	4787	SFA5.8 (BAu-4)	A5.8 (BAu-4)	---	---
P00827	---	SFA5.8 (BVAu-4)	A5.8 (BVAu-4)	---	---
P00901	---	---	---	B596	---
P00927	---	SFA5.8 (BVAu-8)	A5.8 (BVAu-8)	---	---
P03300	7735	---	---	---	---
P03350	---	---	---	B540	---
P03440	---	---	---	B563	---
P03657	---	SFA5.8 (BVPd-1)	A5.8 (BVPd-1)	---	---
P03980	---	---	---	B589 (99.80)	---
P03995	---	---	---	B589 (99.95)	---
P04980	---	---	---	B561	---
P04995	---	---	---	B561 (99.95)	---
P05980	---	---	---	B616 (99.80)	---
P05981	---	---	---	B616 (99.90)	---
P05982	---	---	---	B616 (99.95)	---
P06100	---	---	---	B671	---
P07010	---	---	---	B413 (99.99)	---
P07015	---	---	---	B413 (99.95)	MIL-S-13282
P07016	---	---	---	F106 (TB Ag)	---
P07017	---	SFA5.8 (BVAg-0)	A5.8 (BVAg-0)	---	---
P07020	---	---	---	B413 (99.90)	---
P07200	---	SFA5.8 (BAg-25)	A5.8 (BAg-25)	---	---
P07250	---	SFA5.8 (BAg-26)	A5.8 (BAg-26)	---	---
P07251	---	SFA5.8 (BAg-27)	A5.8 (BAg-27)	---	---
P07252	---	SFA5.8 (BAg-33)	A5.8 (BAg-33)	---	---
P07253	---	SFA5.8 (BAg-37)	A5.8 (BAg-37)	---	---

AMERICAN CROSS REFERENCED SPECIFICATIONS - PRECIOUS METALS (Continued)

UNS	AMS	ASME	AWS	ASTM	MIL SPEC
P07300	---	SFA5.8 (BAg-2a)	A5.8 (BAg-2a)	---	---
P07301	---	SFA5.8 (BAg-20)	A5.8 (BAg-20)	---	---
P07350	4768	SFA5.8 (BAg-2)	A5.8 (BAg-2)	---	MIL-B-7883 (BAg-2)
P07351	---	SFA5.8 (BAg-35)	A5.8 (BAg-35)	---	---
P07380	---	SFA5.8 (BAg-34)	A5.8 (BAg-34)	---	---
P07400	---	SFA5.8 (BAg-4)	A5.8 (BAg-4)	---	---
P07401	---	SFA5.8 (BAg-28)	A5.8 (BAg-28)	---	---
P07402	4762	---	---	---	---
P07450	4769	SFA5.8 (BAg-1)	A5.8 (BAg-1)	---	MIL-B-7883 (BAg-1)
P07453	---	SFA5.8 (BAg-5)	A5.8 (BAg-5)	---	---
P07454	---	SFA5.8 (BAg-36)	A5.8 (BAg-36)	---	---
P07490	---	SFA5.8 (BAg-22)	A5.8 (BAg-22)	---	---
P07500	4770	SFA5.8 (BAg-1a)	A5.8 (BAg-1a)	---	MIL-B-7883 (BAg-1a)
P07501	4771	SFA5.8 (BAg-3)	A5.8 (BAg-3)	---	MIL-B-7883 (BAg-3)
P07502	---	---	---	---	---
P07503	---	SFA5.8 (BAg-6)	A5.8 (BAg-6)	---	---
P07505	4788	SFA5.8 (BAg-24)	A5.8 (BAg-24)	---	---
P07507	---	SFA5.8 (BVAg-6c)	A5.8 (BVAg-6c)	---	---
P07540	4772	SFA5.8 (BAg-13)	A5.8 (BAg-13)	---	MIL-B-7883 (BAg-13)
P07547	---	SFA5.8 (BVAg-32)	A5.8 (BVAg-32)	---	---
P07560	4765	SFA5.8 (BAg-13a)	A5.8 (BAg-13a)	---	---
P07563	---	SFA5.8 (BAg-7)	A5.8 (BAg-7)	---	---
P07587	---	SFA5.8 (BVAg-31)	A5.8 (BVAg-31)	---	---
P07600	4773	SFA5.8 (BAg-18)	A5.8 (BAg-18)	---	---
P07607	---	SFA5.8 (BVAg-18)	A5.8 (BVAg-18)	---	---
P07627	---	SFA5.8 (BVAg-29)	A5.8 (BVAg-29)	---	---
P07630	4774	SFA5.8 (BAg-21)	A5.8 (BAg-21)	---	---

AMERICAN CROSS REFERENCED SPECIFICATIONS - PRECIOUS METALS (Continued)

UNS	AMS	ASME	AWS	ASTM	MIL SPEC
P07650	---	SFA5.8 (BAg-9)	A5.8 (BAg-9)	---	---
P07687	---	SFA5.8 (BVAg-30)	A5.8 (BVAg-30)	---	---
P07700	---	SFA5.8 (BAg-10)	A5.8 (BAg-10)	---	---
P07720	---	SFA5.8 (BAg-8)	A5.8 (BAg-8)	---	MIL-B-7883 (BAg-8)
P07723	---	SFA5.8 (BAg-8a)	A5.8 (BAg-8a)	---	MIL-B-7883 (BAg-8a)
P07727	---	SFA5.8 (BVAg-8)	A5.8 (BVAg-8)	---	---
P07728	---	SFA5.8 (BVAg-8b)	A5.8 (BVAg-8b)	---	---
P07850	4766	SFA5.8 (BAg-23)	A5.8 (BAg-23)	---	MIL-B-7883 (BAg-23)
P07900	---	---	---	B617	---
P07925	4767	SFA5.8 (BAg-19)	A5.8 (BAg-19)	---	MIL-B-7883 (BAg-19)

a. This cross-reference table lists the basic specification or standard number, and since these standards are constantly being revised, it should be kept in mind that they are presented herein as a guide and may not reflect the latest revision.

CHEMICAL COMPOSITIONS OF PRECIOUS METALS

UNS	Chemical Composition
P00010	Au 99.995 min Ag 0.001 max Bi 0.001 max Cr 0.0003 max Cu 0.001 max Fe 0.001 max Mg 0.001 max Mn 0.0003 max Pb 0.001 max Pd 0.001 max Si 0.001 max Sn 0.001 max
P00015	Au 99.99 min As 0.003 max Bi 0.002 max Cr 0.0003 max Cu 0.005 max Fe 0.002 max Mg 0.003 max Mn 0.0003 max Ni 0.0003 max Pb 0.002 max Pd 0.005 max Si 0.005 max Sn 0.001 max
P00016	Au 99.99 min Other Ag+Cu 0.009 max, each, other impurity 0.003 max
P00020	Au 99.95 min Ag 0.035 max Cu 0.02 max Fe 0.005 max Pb 0.005 max Pd 0.02 max Other AG+Cu 0.04 max
P00025	Au 99.5 min
P00032	Au 58.33 (-0.3) Cu 31.08 (+/- 1.0) rem Ag 4.00 (+/- 0.5) Fe 0.05 (+/- 0.5) Ni 0.10 (+/- 0.5) Si 0.01 (+/- 0.5) Zn 6.43 (+/- 0.5)
P00100	Au 99.0 min Ti 1.0 (0.8 min)
P00105	Au 41.817 (-0.3) Cu 38.5 (rem) Zn 12.68 (+/- 1.0)Ag 5.85 (+/- 0.5) Ni 1.15 (+/- 0.5)
P00110	Au 41.70 (-0.3) Ag 11.66 (+/- 1.0) Zn 5.83 (+/- 0.5) B 0.02 Cu 40.81 (rem) Si 0.03

CHEMICAL COMPOSITIONS OF PRECIOUS METALS (Continued)	
UNS	**Chemical Composition**
P00115	Au 41.70 (-0.3) Cu 43.8 (rem) Zn 9.00 (+/- 0.5) Ag 5.50 (+/- 0.5)
P00120	Au 41.70 (-0.3) Cu 48.00 (rem) Ag 6.60 (+/- 0.5) Zn 3.70 (+/- 0.5)
P00125	Au 41.817 (-0.3) Cu 46.683 (rem) Zn 10.35 (+/- 1.0) Ni 1.15 (+/- 0.5)
P00130	Au 41.70 (-0.3) Cu 32.82 (rem) Ni 17.08 (+/- 1.0) Zn 8.40 (+/- 0.5)
P00135	Au 41.66 (-0.3) Cu 29.20 (rem) Ni 15.05 (+/- 1.0) Zn 12.12 (+/- 0.5) Co 1.97 (+/- 0.5)
P00140	Ag 48.90 (+/- 1.0) Au 41.70 (-0.3) Cu 9.05 (rem) Zn 0.35
P00145	Au 41.70 (-0.3) Cu 55.48 (rem) Ag 2.82 (+/- 0.5)
P00150	Au 58.484 (-0.3) Cu 24.00 (rem) Ni 8.00 (+/- 0.5) Zn 7.52 (+/- 0.5) Ag 2.00 (+/- 0.5)
P00155	Au 58.33 (-0.3) Cu 22.10 (rem) Ni 10.80 (+/- 1.0) Zn 8.77 (+/- 0.5)
P00160	Au 58.33 (-0.3) Cu 23.47 (rem) Ni 12.21 (+/- 1.0) Zn 5.99 (+/- 0.5)
P00165	Au 58.33 (-0.3) Cu 28.32 (rem) Ni 8.55 (+/- 0.5) Zn 4.80 (+/- 0.5)
P00170	Au 58.33 (-0.3) Cu 39.59 (rem) Ag 2.08 (+/- 0.5)
P00175	Au 58.33 (-0.3) Zn 6.63 (+/- 0.5) Ag 5.00 (+/-0.5) B 0.02 Cu 30.00 (rem) Si 0.03
P00180	Au 58.33 (-0.3) Cu 24.97 (rem) Ag 16.50 (+/- 1.0) Zn 0.20
P00185	Au 58.33 (-0.3) Cu 26.58 (rem) Ag 7.46 (+/- 0.5) Zn 6.63 (+/- 0.5) Ni 1.00
P00195	Au 58.33 (-0.3) Cu 29.67 (rem) Ag 10.00 (+/- 1.0) Zn 2.00 (+/- 0.5)
P00200	Au 58.33 (-0.3) Cu 20.17 (rem) Ag 21.20 (+/- 1.0 Zn 0.30
P00205	Au 58.33 (-0.3) Cu 29.19 (rem) Ag 8.31 (+/- 0.5) Zn 4.17 (+/- 0.5)
P00210	Au 58.33 (-0.3) Cu 31.24 (rem) Ag 4.00 (+/- 0.5) Zn 6.43 (+/- 0.5)
P00215	Au 58.33 (-0.3) Ag 24.78 (+/- 1.0) Cu 16.75 (rem) Zn 0.14
P00220	Au 58.33 (-0.3) Ag 21.20 (+/- 1.0) Cu 20.17 (rem) Zn 0.30
P00225	Au 58.33 (-0.3) Cu 31.24 (rem) Zn 6.43 (+/- 0.5) Ag 4.00 (+/- 0.5)
P00230	Au 58.33 (-0.3) Ag 32.50 (+/- 1.0) Cu 8.97 (rem) Zn 0.20 (+/- 0.5)
P00235	Au 58.33 (-0.3) Ag 35.00 (+/- 1.0) Cu 6.47 (rem) Zn 0.20
P00250	Au 75.00 (-0.3) Ag 13.00 (+/- 1.0) Cu 12.00 (rem)
P00255	Au 75.00 (-0.3) Ag 15.00 (+/- 1.0) Cu 10.00 (rem)
P00260	Au 75.15 Ag 19.00 (+/- 1.0) (-0.3) Cu 2.85 (rem) Zn 3.00 (+/- 0.5)

CHEMICAL COMPOSITIONS OF PRECIOUS METALS (Continued)

UNS	Chemical Composition
P00270	Au 75.15 (-0.3) Ni 12.00 (+/- 1.0) Cu 9.85 (rem) Zn 3.00 (+/- 0.5)
P00275	Au 75.00 (-0.3) Ni 17.80 (+/- 1.0) Cu 1.73 (rem) Zn 5.47 (+/- 0.5)
P00280	Au 75.00 (-0.3) Ag 22.50 (+/- 1.0) Cu 2.50 (rem)
P00285	Au 75.00 (-0.3) Cu 20.00 (rem) Ag 5.00 (+/- 0.5)
P00300	Au 29.5-30.5 Ni 35.5-36.5 Pd 33.5-34.5 Other 0.15 max
P00350	Au 34.5-35.5 Cu rem Ni 2.5-3.5 Other 0.15 max
P00375	Au 37.0-38.0 Cu rem Other 0.15 max
P00500	Au 49.5-50.5 Ni 24.5-25.5 Pd 24.5-25.5
P00507	Au 49.5-50.5 C 0.005 max Cd 0.001 max Co 0.06 max Ni 24.5-25.5 P 0.002 max Pb 0.002 max Pd rem Zn 0.001 max
P00560	Au 56.25-56.55 Cu 32.30-33.30 Ag 3.70-4.70 Fe 0.05 max Ni 0.10 max Si 0.01 max Zn 6.00-7.50 Other all elements except Au, Ag, Cu, Zn: max 0.15
P00580	Ag 3.0-4.0 Au 58.03-58.63 Cu 30.74-31.74 Fe 0.05 max Ni 0.10 max Si 0.01 max Zn 5.68-7.18 Other all elements except Au, Ag, Cu, Zn: max 0.15
P00691	Ag 23.5-26.5 Au 68.0-70.0 Pt 5.0-7.0
P00692	Ag 24.5-25.5 Au 68.5-69.5 Pt 5.5-6.5 S 0.01 max
P00700	Au 69.5-70.5 Ni 21.5-22.5 Pd 7.5-8.5 Other 0.15 max
P00710	Ag 4.0-5.0 Au 70.5-72.5 Cu 13.5-15.5 Pt 8.0-9.0 Zn 0.7-1.3
P00750	Ag 21.4-22.6 Au 74.2-75.8 Ni 2.6-3.4
P00800	Au 79.5-80.5 Cu rem
P00807	Au 79.5-80.5 C 0.005 max Cd 0.001 max Cu rem P 0.002 max Pb 0.002 max Zn 0.001 max Other 0.15 max
P00820	Au 81.5-82.5 Ni rem Other 0.15 max
P00827	Au 81.5-82.5 C 0.005 max Cd 0.001 max Ni rem P 0.002 max Pb 0.002 max Zn 0.001 max
P00900	Ag 3.0-10.0 Au 90.0 min
P00901	Au 89.0-91.0 Cu 9.0-11.0
P00927	Au 91.0-93.0 C 0.005 max Cd 0.001 max P 0.002 max Pb 0.002 max Pd rem Zn 0.001 max
P03300	Pd 34.0-36.0 Ag 29.0-31.0 Au 9.5-10.5 Cu 13.5-14.5 Pt 9.5-10.5 Other elements 0.01 max total
P03350	Pd 34.0-36.0 Ag 29-31 Au 9.5-10.5 Cu 13.5-14.5 Pt 9.5-10.5 Zn 0.6-1.2 Other total 0.1 max
P03440	Pd 43.0-45.0 Ag 37.0-39.0 Cu 15.5-16.5 Ni 0.8-1.2 Pt 0.8-1.2

CHEMICAL COMPOSITIONS OF PRECIOUS METALS (Continued)

UNS	Chemical Composition
P03657	Pd 64.0-66.0 Co rem C 0.005 max Cd 0.001 max Ni 0.06 max P 0.002 max Pb 0.002 max Zn 0.001 max
P03980	Pd 99.80 min Ir 0.05 max Pt 0.15 max Rh 0.10 max Other Ru 0.05 max
P03995	Pd 99.95 min Ag 0.005 max Al 0.005 max Au 0.01 max Ca 0.005 max Co 0.001 max Cr 0.001 max Cu 0.005 max Fe 0.005 max Mg 0.005 max Mn 0.001 max Ni 0.005 max Pb 0.005 max Sb 0.002 max Si 0.005 max Sn 0.005 max Zn 0.0025 max Other Pt+Rh+Ru+Ir 0.03 max
P04980	Pt 99.80 min Ir 0.05 max Rh 0.10 max Other Pd 0.15 max Ru 0.05 max
P04995	Pt 99.5 min Ag 0.005 max Al 0.005 max As 0.005 max Au 0.01 max Bi 0.005 max Cd 0.005 max Cr 0.005 max Cu 0.01 max Ir 0.015 max Fe 0.01 max Mg 0.005 max Mn 0.005 max Mo 0.01 max Ni 0.005 max Pb 0.005 max Rh 0.03 max Sb 0.005 max Si 0.01 max Sn 0.005 max Te 0.005 max Zn 0.005 max Other Ca 0.005 max, Ru 0.01 max, Pd 0.02 max
P05980	Rh 99.8 min As 0.01 max Bi 0.01 max Cd 0.01 max Ir 0.01 max Fe 0.01 max Pb 0.01 max Pd 0.05 max Pt 0.01 max Ru 0.05 max Sn 0.01 max Zn 0.01 max
P05981	Rh 99.90 min Ag 0.02 max Al 0.01 max As 0.005 max Au 0.01 max B 0.005 max Bi 0.005 max Cd 0.005 max Co 0.005 max Cr 0.01 max Cu 0.01 max Fe 0.01 max Ir 0.05 max Mg 0.01 max Mn 0.005 max Ni 0.01 max Pb 0.01 max Pt 0.05 max Sb 0.005 max Si 0.01 max Sn 0.01 max Te 0.01 max Zn 0.01 max Other Pd 0.05 max, Ru 0.05 max, Ca 0.01 max
P05982	Rh 99.95 min Ag 0.005 max Al 0.005 max As 0.003 max Au 0.003 max B 0.001 max Bi 0.005 max Cd 0.005 max Co 0.001 max Cr 0.005 max Cu 0.005 max Fe 0.003 max Ir 0.02 max Mg 0.005 max Mn 0.005 max Ni 0.003 max Pb 0.005 max Pt 0.02 max Sb 0.003 max Si 0.005 max Sn 0.003 max Te 0.005 max Zn 0.003 max Other Pd 0.005 max, Ru 0.01 max, Ca 0.005 max
P06100	Ir 99.80 min As 0.01 max Bi 0.01 max Cd 0.01 max Fe 0.01 max Pb 0.02 max Pt 0.10 max Rh 0.15 max Si 0.02 max Sn 0.01 max Other Pd 0.05 max, Ru 0.05 max
P07010	Ag 99.99 min Bi 0.0005 max Cu 0.010 max Fe 0.001 max Pb 0.001 max Se 0.0005 max Te 0.0005 max Other Pd 0.001 max
P07015	Ag 99.95 min Bi 0.001 max Cu 0.04 max Fe 0.002 max Pb 0.015 max
P07016	Ag 99.95 min Cu 0.05 max P 0.002 max Other total volatile impurities 0.010 max
P07017	Ag 99.95 min C 0.005 max Cd 0.001 max Cu 0.05 max P 0.002 max Pb 0.002 max Zn 0.001 max
P07020	Ag 99.90 min Bi 0.001 max Cu 0.08 max Fe 0.002 max Pb 0.025 max Other Ag+Cu 99.95 min
P07200	Ag 19.0-21.0 Cu 39.0-41.0 Mn 4.5-5.5 Zn 33.0-37.0
P07250	Ag 24.0-26.0 Cu 37.0-39.0 Mn 1.5-2.5 Ni 1.5-2.5 Zn 31.0-35.0
P07251	Ag 24.0-26.0 Cd 12.5-14.5 Cu 34.0-36.0 Zn 24.5-28.5
P07252	Ag 24.0-26.0 Cd 16.5-18.5 Cu 29.0-31.0 Zn 26.5-28.5
P07253	Ag 24.0-26.0 Cu 39.0-41.0 Sn 1.5-2.5 Zn 31.0-35.0

CHEMICAL COMPOSITIONS OF PRECIOUS METALS (Continued)

UNS	Chemical Composition
P07300	Ag 29.0-31.0 Cd 19.0-21.0 Cu 26.0-28.0 Zn 21.0-25.0
P07301	Ag 29.0-31.0 Cu 37.0-39.0 Zn 30.0-34.0
P07350	Ag 34.0-36.0 Cd 17.0-19.0 Cu 25.0-27.0 Zn 19.0-23.0
P07351	Ag 34.0-36.0 Cu 31.0-33.0 Zn 31.0-35.0
P07380	Ag 37.0-39.0 Cu 31.0-33.0 Sn 1.5-2.5 Zn 26.0-30.0
P07400	Ag 39.0-41.0 Cu 29.0-31.0 Ni 1.5-2.5 Zn 26.0-30.0
P07401	Ag 39.0-41.0 Cu 29.0-31.0 Sn 1.5-2.5 Zn 26.0-30.0
P07402	Ag 39.0-41.0 Cu 29.0-32.0 Zn 28.0-32.0 Other 0.15 max
P07450	Ag 44.0-46.0 Cd 23.0-25.0 Cu 14.0-16.0 Zn 14.0-18.0
P07453	Ag 44.0-46.0 Cu 29.0-31.0 Zn 23.0-27.0
P07454	Ag 44.0-46.0 Cu 26.0-28.0 Sn 2.5-3.5 Zn 23.0-27.0
P07490	Ag 48.0-50.0 Cu 15.0-17.0 Mn 7.0-8.0 Ni 4.0-5.0 Zn 21.0-25.0
P07500	Ag 49.0-51.0 Cd 17.0-19.0 Cu 14.5-16.5 Zn 14.5-18.5
P07501	Ag 49.0-51.0 Cd 15.0-17.0 Cu 14.5-16.5 Ni 2.5-3.5 Zn 13.5-17.5
P07502	Ag 49.0-51.0 Cd 9.0-11.0 Cu 17.0-19.0 Zn 20.0-24.0
P07503	Ag 49.0-51.0 Cu 33.0-35.0 Zn 14.0-18.0
P07505	Ag 49.0-51.0 Cu 19.0-21.0 Ni 1.5-2.5 Zn 26.0-30.0
P07507	Ag 49.0-51.0 Cu rem C 0.005 max Cd 0.001 max P 0.002 max Pb 0.002 max Zn 0.001 max
P07540	Ag 53.0-55.0 Cu rem Ni 0.5-1.5 Zn 4.0-6.0
P07547	Ag 53.0-55.0 Pd rem C 0.005 max Cd 0.001 max Cu 20.0-22.0 P 0.002 max Pb 0.002 max Zn 0.001 max
P07560	Ag 55.0-57.0 Cu rem Ni 1.5-2.5
P07563	Ag 55.0-57.0 Cu 21.0-23.0 Sn 4.5-5.5 Zn 15.0-19.0
P07587	Ag 57.0-59.0 Pd rem C 0.005 max Cd 0.001 max Cu 31.0-33.0 P 0.002 max Pb 0.002 max Zn 0.001 max
P07600	Ag 59.0-61.0 Cr rem Sn 9.5-10.5
P07607	Ag 59.0-61.0 Cu rem C 0.005 max Cd 0.001 max P 0.002 max Pb 0.002 max Sn 9.5-10.5 Zn 0.001 max
P07627	Ag 60.5-62.5 Cu rem C 0.005 max Cd 0.001 max P 0.002 max Pb 0.002 max In 14.0-15.0 Zn 0.001 max
P07630	Ag 62.0-64.0 Cu 27.5-29.5 Ni 2.0-3.0 Sn 5.0-7.0

CHEMICAL COMPOSITIONS OF PRECIOUS METALS (Continued)

UNS	Chemical Composition
P07650	Ag 64.0-66.0 Cu 19.0-21.0 Zn 13.0-17.0
P07687	Ag 67.0-69.0 Cu rem C 0.005 max Cd 0.001 max P 0.002 max Pb 0.002 max Pd 4.5-5.5 Zn 0.001 max
P07700	Ag 69.0-71.0 Cu 19.0-21.0 Zn 8.0-12.0
P07720	Ag 71.0-73.0 Cu rem
P07723	Ag 71.0-73.0 Cu rem Li 0.25-0.50
P07727	Ag 71.0-73.0 Cu rem C 0.005 max Cd 0.001 max P 0.002 max Pb 0.002 max Zn 0.001 max
P07728	Ag 70.5-72.5 Cu rem C 0.005 max Cd 0.001 max P 0.002 max Pb 0.002 max Ni 0.3-0.7 Zn 0.001 max
P07850	Ag 84.0-86.0 Mn rem
P07900	Ag 89.6-91.0 Cu 9.0-10.4 Al 0.005 max Cd 0.05 max Fe 0.05 max Ni 0.01 max P 0.02 max Pb 0.03 max Zn 0.06 max
P07925	Ag 92.0-93.0 Cu rem Li 0.15-0.30
P07931	Ag 92.10-93.50 Cu 6.50-7.90 Cd 0.05 max Fe 0.05 max Pb 0.03 max Zn 0.06 max Other 0.06 max
P07932	Ag 92.50-93.50 Cu 6.50-7.50 Cd 0.05 max Fe 0.05 max Pb 0.03 max Zn 0.06 max Other 0.06 max

Chapter 8

COPPER & COPPER ALLOYS

Copper is one of the first metals used by humanity. Initially, the naturally occurring native copper was exploited but since at least 5000 BC copper has been produced as a smelted product. It was likely the first metal to be smelted from its ore as well as the first metal used in alloyed form. The copper-tin alloy bronze has been used since at least 3500 BC and the copper-zinc alloy brass since before Roman times. Copper and its alloys remain to this day among the most important engineering materials as a result of a combination of properties which include resistance to corrosion, outstanding electrical and thermal conductivity, attractive appearance, strength, ductility, and ease of fabrication and joining.

Copper is produced by the smelting of sulphide minerals. Since copper ores are generally low-grade, with copper contents well below 1%, they are first subjected to beneficiation to produce a concentrate. This concentrate is smelted, usually via pyrometallurgical processes, to impure "blister copper." Blister copper is fire refined to tough pitch copper, which contains at least 99.5%Cu with a significant content of copper oxide. Tough pitch copper is normally cast into anodes and refined electrolytically to produce electrolytic copper (≥99.95%Cu), also known as cathode copper, which is remelted and cast into shapes for further fabrication. Alloying can be carried out as desired during this final remelting. Some copper is extracted by hydrometallurgical methods, producing a copper-bearing aqueous solution from which refined copper is extracted by electrowinning. Processing of copper-bearing scrap is very common, with more than half of refined copper produced coming from recycled scrap.

About half of the copper consumption is processed in wire mills, for production of electrical conductors (wire and cable). Another 40% goes to brass mills where both unalloyed and alloyed copper products are produced. The unalloyed copper and high-copper alloys are typically produced as tube for plumbing and air conditioning, as well as bus bar and roofing sheet. The main copper alloy products are free-cutting brass rod and brass strip, with other products including plate, sheet, tube, bar, forgings, extrusions and mechanical wire. Minor amounts of copper are consumed in foundries, as powder metallurgy products and in other metallurgical and chemical industries.

The major markets for copper and copper alloy products are, in decreasing order of importance, the building industry, electrical and electronic products, industrial machinery and equipment, transportation, and consumer and general products. Both cast and wrought products are used.

Pure copper has a face-centered cubic (fcc) crystal structure, and is a very soft material, with a ductility so high that it can be subjected to very large changes in shape in a single reduction pass. It can be strengthened by solid solution strengthening and by cold working, with reasonably high strength levels being achieved in some alloys. Formability decreases with increasing cold work. Many copper-based materials are solid solutions, with single phase (face-centered cubic) microstructures. In a few alloys, a dispersion of particles of an intermetallic phase can form, providing a grain refining and strengthening effect which results in improved formability for a given strength level. In UNS C63800 and C68800, for example, these intermetallic particles result from the presence of cobalt.

Most copper-based alloys are not strengthened by heat treatment but there are some important exceptions. Precipitation hardening can give very high strengths to the beryllium-coppers (e.g., UNS C17000, C17200, C17300, C17500, and C17510), the zirconium-coppers (e.g. UNS C15000 and C15100), the chromium-coppers (e.g., UNS C18200, 18400, and 18500), some copper-nickel-phosphorus alloys (e.g., UNS C19000 and C19100) and the copper-nickel-silicon alloys UNS C64700 and C70250. A few other alloys, including UNS C71900 (copper-nickel-chromium) can be hardened by spinodal decomposition. Certain aluminum bronzes, notably those containing more than 9%Al, can be hardened by quenching from above a critical temperature to form a martensitic microstructure. Control of the quenching rate or tempering after the quench gives more control over mechanical properties, as is the case in ferrous martensites. Some alloys have additions of lead, sulphur, or tellurium to improve machinability, but these generally have detrimental effects on hot workability (lead in high-zinc brass is an exception as discussed below). Other alloying additions are made to deoxidize the copper; phosphorus in tin bronzes and silicon in chromium-coppers are examples.

Corrosion Resistance

Copper and its alloys are relatively resistant to general corrosion in most environments, but can be susceptible to stress corrosion cracking (e.g., brass in ammonia, amines, mercury compounds or cyanide). Some copper-based alloys in some environments suffer from hydrogen embrittlement by the steam reaction (e.g., tough pitch copper in reducing atmospheres). Brasses containing more than 15%Zn are also susceptible to dezincification, in which zinc is selectively removed by corrosion from the surface of the material, leaving a porous layer of copper and corrosion products.

Weldability

Copper and its alloys are most frequently welded using GTAW (gas tungsten arc welding), or its pulsed variant, especially for thin sections, since high localized heat input is important in materials with high thermal conductivity. In thicker sections, GMAW (gas metal arc welding) is preferred. The weldability varies among the different alloys for a variety of reasons, including the occurrence of hot cracking in the leaded (free-machining) alloys and unsound welds in alloys containing copper oxide. Tin and zinc both reduce the weldability of copper alloys. The presence in the alloy of residual phosphorus is beneficial to weldability, since it combines with absorbed oxygen, thereby preventing the formation of copper oxide in the weld. Resistance welding is also widely used, particularly in alloys with low thermal conductivity. Most copper alloys can be brazed satisfactorily. Oxygen-bearing coppers can be subject to gassing and embrittlement, particularly in oxyacetylene welding.

Nomenclature

Copper and its alloys were formerly given designations according to the Copper Development Association (CDA) system of three digits. This CDA system was used as the basis for the Unified Numbering System (UNS) designations, shown in Table 1. For copper and its alloys, the UNS system has come into common usage by all specification writing organizations.

The traditional system of temper designations referred to the amount of reduction in cold work, stated in Brown & Sharpe (B&S) gage numbers for rolled sheet and drawn wire. The current system defined in ASTM B 601 is more comprehensive; this assigns an alphanumeric code to each of the standard descriptive temper designations. Here cold worked tempers are defined by the letter H and 2 digits where the digits refer either to the amount of cold work or to the specific product. Other letter designations, which are followed by digits used for many purposes, include: HR (cold worked and stress relieved), M (as manufactured), O and OS (annealed), TB (solution treated), TD (solution treated and cold worked), TF (solution treated and precipitation hardened), TM (mill hardened), TQ (quench hardened), and WH (welded and drawn).

Table 1 UNS Designations For Copper & Copper Alloys	
UNS Designations - Wrought Alloys	
C10100-C15760	Coppers (>99%Cu)
C16200-C19600	High-copper alloys (>96%Cu)
C21000-C28000	Brasses (Cu-Zn)
C31200-C38500	Leaded brasses (Cu-Zn-Pb)
C40400-C48600	Tin brasses (Cu-Zn-Sn-Pb)
C50100-C52400	Phosphor bronzes (Cu-Sn-P)
C53400-C54400	Leaded phosphor bronzes (Cu-Sn-Pb-P)
C55180-C55284	Copper-phosphorus and copper-silver-phosphorus alloys (Cu-P-Ag)
C60800-C64210	Aluminum bronzes and aluminum-silicon bronzes (Cu-Al-Ni-Fe-Si-Sn)
C64700-C66100	Silicon bronzes (Cu-Si-Sn)
C66400-C69900	Other copper-zinc alloys
C70100-C72950	Copper-nickels (Cu-Ni-Fe)
C73500-C79800	Nickel silvers (Cu-Ni-Zn)
UNS Designations - Cast Alloys	
C80100-C81200	Coppers (>99%Cu)
C81400-C82800	High-copper alloys (>94%Cu)
C83300-C84800	Red and leaded red brasses (Cu-Zn-Sn-Pb, 75-89%Cu)
C85200-C85800	Yellow and leaded yellow brasses (Cu-Zn-Sn-Pb, 57-74%Cu)
C86100-C86800	Manganese bronzes and leaded manganese bronzes (Cu-Zn-Mn-Fe-Pb)
C87300-C87800	Silicon bronzes, silicon brasses (Cu-Zn-Si)
C90200-C94500	Tin bronzes and leaded tin bronzes (Cu-Sn-Zn-Pb)
C94700-C94900	Nickel-tin bronzes (Cu-Ni-Sn-Zn-Pb)
C95200-C95900	Aluminum bronzes (Cu-Al-Fe-Ni)
C96200-C96800	Copper-nickels (Cu-Ni-Fe)
C97300-C97800	Nickel silvers (Cu-Ni-Zn-Pb-Sn)
C98200-C98800	Leaded coppers (Cu-Pb)
C99300-C99750	Miscellaneous alloys

Copper and its alloys are discussed in this chapter in order of the UNS system general categories, i.e., wrought alloys first, beginning with coppers and high-copper alloys and moving through brasses, bronzes, copper-nickels and nickel silvers, followed by cast alloys in a similar order. It is important to note that although the terms brass and bronze can be clearly defined as copper-zinc and copper-tin alloys respectively, the common names are not always consistent with these definitions. In particular, the name "bronze" is applied to many different copper-based alloys. Hence, for example, Cu-10%Zn is called "commercial bronze" despite the fact that it is clearly a brass and is so categorized in the CDA and UNS designation systems.

Wrought Alloys - Coppers (UNS C10100-C15760)

Unalloyed copper is widely used, mainly because of its high electrical conductivity, but also to take advantage of its attractive appearance, its corrosion resistance, ease of fabrication, and reasonable strength. These "coppers" are classified according to their oxygen and impurity contents since these have strong effects on electrical conductivity. Oxygen is normally present as copper oxide particles, often in the copper-copper oxide eutectic which forms interdendritically during ingot casting. Oxygen also forms a fine dispersion of blowholes which prevent pipe cavities from forming during ingot solidification. These are welded together during hot working.

The electrical conductivity, which is reported as %IACS (International Annealed Copper Standard), is affected by both oxygen and impurity atoms. Impurity atoms in solid solution degrade the conductivity, but if they can react with oxygen to form oxides they have a less detrimental effect. On the other hand, oxygen in the form of copper oxide can create problems when the material is heated above 400°C (752°F) in atmospheres containing hydrogen. The hydrogen atoms diffuse into the copper and reduce copper oxide particles to create steam under high pressure which can cause rupture along grain boundaries. This "steam embrittlement" creates problems during annealing, brazing, or welding in a reducing atmosphere.

In general, coppers have good resistance to corrosion by non-oxidizing acids unless the acids are aerated. Coppers also resist caustic solutions, saline solutions, and natural and process waters, including ground water. They are not used in oxidizing environments such as nitric acid, ammonia, ferric chloride, acid chromate solutions, mercury salts, perchlorates, or persulphates.

UNS C11000, the most common of the electrical coppers, is called electrolytic tough pitch (ETP) copper. It contains 0.02-0.04% oxygen, and less than 50 ppm total metallic impurities including sulphur. This gives it a conductivity of 101%IACS in annealed form. It is used for building sheet products, gaskets, radiators, fasteners, and chemical process equipment in addition to numerous electrical applications (wire, busbar, stranded conductors, switches, etc.).

There are many variations on ETP copper, including the oxygen-free coppers produced by remelting electrolytic copper under carefully controlled reducing atmospheres. Oxygen-free electronic copper (UNS C10100) and oxygen-free copper (UNS C10200) are used for electrical and electronic applications where their high conductivity (101%IACS) and high ductility are required.

UNS C12500 is fire refined tough pitch (FRTP) copper, i.e., deoxidized copper which has not been electrolytically refined. It contains 500 to 3000 ppm of cuprous oxide, and is used for architectural, automotive, and electrical applications which are not demanding enough to require ETP copper.

Phosphorus is added to some alloys to tie up the oxygen as phosphorus pentoxide. If the amount of phosphorus is just sufficient to do this without leaving excess phosphorus in solid solution, these alloys retain high electrical conductivity (e.g., UNS C10800 and C12000 at greater than 90%IACS). Alternatively, excess phosphorus may be added (e.g., deoxidized high-phosphorus (DHP) copper UNS C12200) so that there is phosphorus available to provide deoxidation during hot working or annealing; this also increases the weldability. However, the electrical conductivity is lowered by the excess phosphorus, so that this alloy finds its major use as piping and tubing.

Small additions (a few tenths of a percent at most) of elements such as cadmium (UNS C11100, C14300) or silver (UNS C10400, C10500, C10700, C11300, C11400, C11500, C11600, C12900) are made to oxygen-free, ETP and FRTP coppers in order to impart resistance to softening at slightly elevated temperatures, for example so that they can withstand the conditions encountered during soldering of automotive radiators or semiconductor packaging. Silver is especially useful in this respect as small amounts raise the recrystallization temperature markedly, while having little effect on electrical conductivity. The zirconium-coppers, such as UNS C15000 with 0.15%Zr, are heat-treatable by precipitation hardening and retain strength at temperatures as high as 450°C (842°F).

The machinability of these coppers is poor because of their high ductility, but some compositions have been developed specifically to counteract this deficiency. UNS C14500 and C14700 contain small additions of tellurium and sulphur, respectively, to make them free-machining for use in screw machine products requiring some combination of conductivity, corrosion resistance, colour, etc., for example electrical and plumbing components and fittings.

High-Copper Alloys (UNS C16200-C19600)

These materials, sometimes referred to as dilute copper alloys, contain more than 96%Cu generally with more than 0.7% impurities and additions, including beryllium, cadmium, chromium, and iron. Many of the alloys in this classification are heat-treatable by precipitation hardening. The first stage of this process involves solution treating at temperatures above 750°C (1382°F) to take the alloying elements into solid solution, then quenching to create a supersaturated solid solution. This is then given an aging treatment at 260-500°C (500-932°F) to allow a fine dispersion of strengthening particles to precipitate from solution. The coarseness of the dispersion of particles determines the strength and ductility. If aging is carried out too long or at too high a temperature, the particles become too coarse, the strength falls below the maximum possible (the "peak aged" condition), and the alloy is said to be "overaged." The particles which precipitate from solution are generally intermetallic phases or non-equilibrium transition phases which are their precursors. For example, in the beryllium-copper system the precipitating intermetallic phase is (Cu,Co)Be, a mixed copper-cobalt beryllide.

The most important of the high-copper alloys are beryllium-copper alloys, which can be precipitation hardened to very high strength levels. There are two commercially important groups of beryllium-copper alloys, with different beryllium contents as well as additions of cobalt or nickel. One group, the "red" alloys, which includes alloys UNS C17500 and C17510, contains approximately 0.2-0.7%Be with additions of nickel or cobalt totaling 1.4-2.7%. These alloys maintain a moderate electrical conductivity (50%IACS) and yield strengths as high as 895 MPa (130 ksi) and find applications as fasteners and connectors, springs, switch parts, and spot welding tips. The higher alloyed "gold" alloys, such as UNS C17000 and C17200, contain 1.4-2.7%Be, which gives them their shiny lustrous colour, plus about 0.25%Co. These are even stronger than the "red" alloys, with yield strength as high as 1380 MPa (200 ksi). For enhanced machinability, lead additions can be made (UNS C17300) but electrical conductivity is as low as 20%IACS. These alloys are used for applications requiring high strength combined with good formability, for example in bellows, valves, pumps, rolling mill parts, and nonsparking safety tools in addition to electrical and electronic components. The beryllium-copper alloys are available as rod, bar, flat products, tubing, billets, and forgings and extrusions.

Among other precipitation-hardening alloys in this series are UNS C18200 and C18400 which are chromium-copper alloys with excellent cold workability and good hot workability. These find applications in welding equipment. The copper-nickel-phosphorus alloys UNS C19000 and C19100 are also precipitation hardenable.

High-copper alloys which are not heat-treatable include the leaded free-machining copper UNS C18700 (Cu-1%Pb) and the copper-iron-phosphorus alloys UNS C19200, C19210, C19400, C19500, and C19700. Many of these latter alloys are selected for use where a combination of high strength and conductivity is required. Other attractive properties of individual alloys in this group include formability and resistance to softening and stress corrosion. They are used for electrical and electronic components; UNS C19200 and C19210 are also used for tubing in automotive, heat exchanger, and air conditioning applications.

Brasses (UNS C21000-C28000)

The brasses are a very large family of alloys based on the copper-zinc system. Commercial brasses contain up to about 40% zinc, and some also contain small amounts (up to about 5%) of alloying elements such as tin, aluminum, silicon, manganese, and lead. Alloys containing up to above 30%Zn are solid solutions; these are referred to as α-solid solutions, hence the brasses are known as α-brasses. These solid solution alloys are not heat-treatable but are solid solution strengthened and can be strengthened further by cold work. Among them are the very widely used Cu-30%Zn alloy known as cartridge brass, which has the optimum combination of strength and ductility, maintaining the high formability of copper.

In these alloys increasing zinc content increases the strength and ductility, while the colour changes from red through gold to green-yellow. The ductility reaches a maximum at about 30%Zn. All of these alloys have excellent cold workability and alloys up to about 20%Zn have good hot workability as well, provided that the lead content is very low. They are used in some cases after plating with chromium or nickel

If the zinc content is higher than the solubility limit, e.g. 40%Zn, the microstructures consist of a mixture of the α-solid solution with a body-centered cubic solid solution known as the β phase; hence, alloys such as this are known as α-β brasses. The most common of these alloys is Cu-40%Zn, known as Muntz metal (UNS C28000). In fact, during cooling of these alloys, the β phase transforms at about 465°C (869°F) to an ordered β' phase, which has higher strength and much lower ductility than the β phase. The strengths of these alloys increase with increasing zinc content up to about 45%Zn. However, the presence of the β phase reduces the ductility of these brasses and makes them difficult to cold work, although it does make them more machinable. These alloys can be readily hot worked since at hot working temperatures they consist entirely of β phase. For example, a 40%Zn alloy is single phase β above 760°C (1400°F).

Another consequence of the phase behaviour of these α-β alloys is that they are heat-treatable since at high temperature the β phase is stable while at lower temperatures the stable state is a two-phase α-β mixture. Thus if the alloy is quenched from above 760°C (1400°F) to α temperature in the range 500-700°C (932-1292°F), particles of the α phase can be made to precipitate in the β matrix, which becomes β' on further cooling. Although the β' phase is too brittle to be useful, alloys in which it co-exists with precipitated α phase have useful properties, depending on the precise heat treatment and resultant microstructure.

This series of plain copper-zinc brasses includes alloys with zinc contents ranging upward from 5%. The 5%Zn alloy UNS C21000 is known as gilding metal. UNS C22000 and C22600, containing 10% and 12.5%Zn, are called commercial bronze and jewelry bronze respectively, despite being brasses. UNS C23000, which contains 15%Zn, is known as red brass; UNS C24000, containing 20%Zn, is called low brass; while cartridge brass, UNS C26000, contains 30%Zn. Yellow brass, UNS C26800 and C27000, contains approximately 35%Zn while C28000, Muntz metal, contains 40%Zn.

The lower zinc alloys are used for jewelry and coinage applications and as a base for gold plate and enamel. The mid-range zinc alloys are also suitable for these applications but in addition find usage in architectural, hardware, fasteners, and munitions applications. The higher zinc material is used in applications where greater strength is important. For example cartridge brass is very widely used in such applications as architectural grillwork, automotive radiator cores and tanks, lamp fixtures, locks, plumbing fittings, and pump cylinders. Muntz metal with its attractive hot workability and machinability is used in architectural panels, in heavy plate for structural applications, and as bolting, valve stems, and forgings.

Leaded Brasses (UNS C31200-C38500)

This is a series similar to the brasses described above but with small lead contents in order to make them machinable. Lead is insoluble in the solid solution matrix of these alloys. It exists in the microstructure as small inclusions of metallic lead, which act as chip breakers during machining, thus making them free-machining. This has little effect on strength; however, since these lead particles melt at low temperatures, their presence dramatically reduces the hot workability of the alloys.

These alloys are referred to as medium leaded brasses when their lead content is about 1%, high leaded at 2%, and extra high leaded at 2.5%Pb. UNS C31400 is commercial bronze (9%Zn) with 2%Pb; a variant UNS C31600 also contains 1%Ni. These are used for hardware and electrical applications. UNS C33000, C33200, C33500, and C34000 contain 32-35%Zn with 0.5 -2%Pb and are used when both machinability and cold working properties are required, as in plumbing, hardware, and munitions applications. Leaded alloys containing slightly higher zinc content (up to about 37% but still solid solution alloys), such as UNS C34200, C35000, C35600, and C36000 are used for products such as gears, pinions, wheels, key and lock components, valve stems, and parts for clocks and meters.

Higher zinc contents cause the structure to consist of α and β phases as in the unleaded brasses discussed above. Copper alloys UNS C36500 (uninhibited leaded Muntz metal) contains about 40%Zn with 0.6%Pb. The addition of about 0.1% of arsenic, antimony, or phosphorus serves to provide high resistance to dezincification, giving applications as tube sheets in condensers and heat exchangers. UNS C37000 (free-cutting Muntz metal) has 39%Zn and 1%Pb for use in automatic screw machine parts. Architectural bronze, UNS C38500 (40%Zn, 3%Pb), is used for such architectural components as trim, as well as hardware.

Although the presence of lead is normally seriously detrimental to hot workability, as discussed above, the leaded high zinc brasses can be hot worked because the lead goes into solution in the β phase which is predominant at high working temperatures. Thus UNS C37700, forging brass (38%Zn, 2%Pb), is the standard against which the forging qualities of all other copper alloys are judged.

Tin Brasses (UNS C40400-C48600)

This series of alloys includes material with zinc ranging from 3% to 40% and tin ranging from 0.5% to about 2% (a few as high as 5%). The presence of small amounts of tin in these alloys improves their strengths, as well as their corrosion resistance in sea water. For this reason cartridge brass (30%Zn) containing 1% tin is called "admiralty metal," while Muntz metal (40%Zn) containing 1% tin is called "naval brass." The leaded alloys in this series, as in the UNS C3xxxx series, contain lead for improved machinability, but at the expense of hot workability.

Alloys in this series containing <15%Zn find applications as rolled strip, bar, and sheet for such purposes as electrical switch springs, terminals and connectors. Examples are UNS C41500 (7%Zn, 2%Sn), C42200 (11%Zn, 1%Sn), and C43400 (14%Zn, 1%Sn). UNS C43500 (18%Zn, 1%Sn), with its higher strength is used as strip and tubing, for example in musical instruments. Small amounts of arsenic, antimony and phosphorus can be added to the higher zinc alloys in this series, as in the UNS C3xxxx series, in order to inhibit dezincification. This creates inhibited admiralty metal, which contains 28%Zn, 1%Sn, and 0.06% of either arsenic (UNS C44300, arsenical admiralty metal), antimony (UNS C44400, antimonial admiralty metal), or phosphorus (UNS C44500, phosphorized admiralty metal). Applications include tubes and other components of heat exchangers, condensers, and distillation equipment. Naval brass (39%Zn, 1%Sn) is available in uninhibited form (UNS C46400) or inhibited by the same three elements (UNS C46500, C46600, 46700, respectively), for use in marine hardware, shafts, fittings, and fasteners. Uninhibited naval brass is also available as a medium- or high-leaded alloy (UNS C48200 and C48500, respectively) with improved machinability for hardware and screw machine products.

Phosphor Bronzes (UNS C50100-C52400) and Leaded Phosphor Bronzes (UNS C53400-54400)

Phosphor bronzes are copper-tin alloys with small amounts of phosphorus added to improve ingot castability and to act as a deoxidizer. These alloys are stronger than the brasses and can be cold worked to even higher strength. They also have good resistance to wear and to sea water corrosion as well as to stress corrosion. Their cold formabilities are good, but not as good as brass, and they have a poor capacity for hot forming.

In the copper-tin system, the maximum solubility of tin in copper is about 16% as compared to almost 40%Zn in the copper-zinc system. However, non-equilibrium microstructures are common in copper-tin alloys, since many of the phase transformations are very sluggish. For example, the phase diagram predicts that the ε phase should be present at room temperature; however, this is not observed in alloys containing less than 5%Sn, and commonly annealed alloys of 8%Sn are single phase solid solutions. When the solubility limit is exceeded, it is often the δ phase which is observed at room temperature, even though the phase diagram predicts that this phase is unstable below 350°C (662°F). The hardness of the δ phase contributes to wear resistance of the alloys and improves their suitability for use as bearing materials.

If phosphorus is present in excess of that required for deoxidation, it occurs as the hard intermetallic compound Cu3P, which can have a strengthening effect. Most of the phosphor bronzes contain about 0.2%P, which is sufficient for purposes of deoxidation, while tin contents range up to above 10%. Most alloys are produced as flat products, bar, rod, wire, and tubing. Common alloys include UNS C50500 (1.3%Sn), which is used in such applications as electrical hardware, and UNS C51000 (5%Sn) and C51100 (4%Sn) which are used for architectural, electrical, fasteners, and industrial components. For more severe service conditions, higher tin alloys are employed, including UNS C52100 (8%Sn) and C52400 (10%Sn). Examples of specific applications include bridge bearing and expansion plates and fittings, Bourdon tubes, fasteners, chemical hardware, and textile machinery components.

In addition to these lead-free alloys, there is a free-machining phosphor bronze, UNS C54400, which contains 4% of each of tin, lead, and zinc. This is used for bearings, bushings, gears, and screw machine products.

Aluminum Bronzes (UNS C60800-C64210)

These materials have good strength properties, with good resistance to wear and corrosion; the latter property is due to the formation of a protective surface film of aluminum oxide. This gives them resistance to non-oxidizing mineral acids as long as dissolved oxygen or oxidizing agents are not present in quantity. As well, they are suitable for service in alkalis, neutral salts, non-oxidizing acid salts, and many organic acids. They have good resistance to corrosion by water, including sea water, although resistance is less in softened water. They are susceptible to stress corrosion cracking in moist ammonia and mercury compounds.

The commercial alloys in this system contain 5-13%Al with minor amounts of nickel, iron, silicon, and cobalt. In these alloys, a range of microstructures can be obtained depending on composition and heat treatment. Alloys containing 5-8%Al are solid solution alloys similar to the α-brasses. Aluminum is a highly effective solid solution strengthening agent, comparable to tin and much more effective than zinc; hence, these alloys have good strength (increasing with aluminum content) as well as good toughness, cold workability, and corrosion resistance. Examples include UNS C60800 (5%Al) which is used in the form of seamless tubing for heat exchangers and condensers, and UNS

C61000 (8%Al) which is produced as rod and wire for bolts, shafts and pump parts. It is also used as a weld overlay on steel for improved wear resistance. Alloys UNS C61300 (7%Al with 3%Fe and 0.3%Sn) and C61400 (7%Al with 2%Fe) are both used as seamless tubing and pipe as well as rod, bar, plate, and sheet for various condenser and heat exchanger applications. The microstructures of the latter two alloys exhibit precipitates of an iron-rich phase in the solid solution.

At higher aluminum contents the alloys become heat-treatable, with duplex microstructures consisting of mixtures of the α-solid solution with metastable β phase particles, for example UNS C62300 (10%Al, 3%Fe) and C62400 (11%Al, 3%Fe). In addition, precipitates of an iron-rich phase may also be present. The precise microstructural features depend on the heat treatment, including the cooling rate. Above about 9.5%Al, a eutectoid decomposition can occur during slow cooling, as the high temperature β phase decomposes to a lamellar mixture of α-solid solution and the intermetallic γ_2 phase. This lamellar structure is comparable to that of pearlite in the iron-carbon system and in fact is sometimes referred to by that name. The eutectoid composition in this system is 11.8%Al, so hypereutectoid or hypoeutectoid alloys (containing more or less aluminum than the eutectoid amount) develop slow cooled microstructures consisting of pearlite with proeutectoid γ_2 phase and α phase, respectively. One commercial hypereutectoid alloy is UNS C62500 (13%Al, 4%Fe) which has low ductility and impact resistance because of the presence of the brittle γ_2 phase, but finds applications as wear strips and as metal forming dies and rolls.

The brittleness of the γ_2 phase makes its presence undesirable in these alloys, although it does improve machinability. However, as in the iron-carbon system, quenching from the β phase region causes the microstructure to transform to needles of a β'-martensite phase which is metastable and hence does not appear on the phase diagram. Unlike iron-carbon martensite, this martensite does not have a particularly high strength; however, it can be strengthened appreciably by tempering in the range 400-650°C (752-1202°F). The microstructure after tempering consists of a fine dispersion of the γ_2 phase in an α phase matrix. This provides a good combination of strength and ductility, which can be controlled by controlling the aging time and temperature.

Additions of up to 5%Fe and 5%Ni render these alloys especially strong and tough, with excellent high temperature oxidation and corrosion resistance. This provides applications as high strength fasteners, gear wheels, pump parts, and structural members. Alloys available include UNS C63000 (10%Al, 5%Ni, 3%Fe) and C63200 (9%Al, 5%Ni, 4%Fe).

UNS C63800 (3%Al, 2%Si, 0.4%Co) is a high strength aluminum bronze in which the presence of cobalt is responsible for strengthening by a dispersion of intermetallic particles. This alloy also has excellent resistance to crevice corrosion and to high temperature oxidation and is used for springs, switch parts, contacts, glass sealing, and porcelain enamelling.

Silicon Bronzes (UNS C64700-C66100)

These alloys contain 1-3%Si, with small additions of manganese and iron to improve properties. They are not heat treatable but can be cold worked to high strength levels. Their combination of corrosion resistance with high strength and toughness compares favourably to that of low-carbon steels.

Low-silicon bronze UNS C65100 (1.5%Si), available in bar, rod, wire, and shapes as well as tubular and flat products, is used for aircraft hydraulic pressure lines, heat exchanger tubes, and marine and industrial hardware and fasteners. The high-silicon bronze UNS C65500 (3%Si) is used for similar applications as well as for chemical process equipment and marine propeller shafts.

Copper-Nickels (UNS C70100-C72950)

Nickel has complete solid solubility in copper so that a continuous range of solid solution copper-nickel alloys is possible, all with the fcc crystal structure. The cupronickel alloys containing 10-30%Ni have moderate strength provided by the nickel which also improves the oxidation and corrosion resistance of copper. These alloys have good hot and cold formability and are produced as flat products, pipe, rod, tube, and forgings. The common alloys are UNS C70600 (10%NI), C71000 (20%Ni), and C71500 (30%Ni). These find applications as plates and tubes for evaporators, condensers, and heat exchangers as well as for saltwater piping. The alloy UNS C71900 (30%Ni, 3%Cr) is similar to UNS C71500 but the chromium content permits it to undergo spinodal decomposition during slow cooling. This produces a two-phase microstructure, giving higher strength for a given amount of cold work. The tin-bearing copper-nickel UNS C72500 (10%Ni, 2.5%Sn) has excellent hot and cold formability in a wide range of forming processes, as well as excellent resistance to sea water. It is used for relay and switch springs, connectors, sensing bellows, and as a brazing alloy.

Nickel Silvers (UNS C73500-C79800)

Although these alloys are referred to as German silvers or nickel silvers, they contain no silver; they are copper-nickel-zinc alloys containing zinc in the range 17-27% along with 8-18%Ni. The use of the name silver arises from their colour, which ranges from soft ivory to silvery white as the nickel content is increased. Most of these alloys are single phase solid solutions, and have medium strengths with good cold workability. Corrosion resistance is excellent, but some are subject to dezincification and they can be susceptible to stress corrosion cracking as well.

Commercial alloys include UNS C74500 (25%Zn, 10%Ni), C75200 (17%Zn, 18%Ni), C75400 (20%Zn, 15%Ni), C75700 (23%Zn, 12%Ni), and C77000 (27%Zn, 18%Ni). Specific applications include hardware, fasteners, optical and camera parts, etching stock, and hollowware. They are particularly appropriate for articles which require a smooth surface for plating. The alloy UNS C78200 (25%Zn, 8%Ni) also contains 2%Pb to provide free-machining properties, making it suitable for use as key blanks and watch parts.

Cast Alloys

The cast copper alloys are generally comparable to the wrought alloys, however the ranges of allowable impurity and alloying elements are often larger in castings, as no consideration need be given to their effects on cold or hot workability. On the other hand, consideration must be given to castability, which is related to solidification shrinkage characteristics and the temperature range over which freezing occurs. Foundry alloys are conveniently classified as high-shrinkage or low-shrinkage alloys. Among the common high-shrinkage alloys are the manganese bronzes, aluminum bronzes, and silicon bronzes; the red brasses are low-shrinkage alloys. High-shrinkage alloys can produce high grade castings, but careful design is necessary to avoid internal shrinkage porosity and cracks.

Pure copper is extremely difficult to cast, with the potential for surface cracks and internal cavities. Small additions of alloying elements improve the casting characteristics, but for property improvement, larger amounts of alloying elements are necessary. In particular, for higher strength cast alloys, it is necessary to select aluminum bronzes, manganese bronzes, silicon bronzes, or some nickel silvers.

Sand casting is appropriate for all copper alloys. Permanent mold casting and die casting are applicable to tin bronzes, silicon bronzes, aluminum bronzes, and manganese bronzes, as well as to yellow brasses.

Investment casting is also applicable to many of these alloys, while plaster casting is most suitable for low-lead alloys. Most casting alloys are readily machinable, especially those containing lead particles in a copper-based solid solution. The most difficult alloys to machine are the high strength manganese bronzes and aluminum bronzes with high iron or nickel content.

Coppers (UNS C80100-C81200) and High Copper Alloys (UNS C81400-82800)

The high purity copper casting alloys, such as UNS C81100 (99.7%Cu), are soft and low in strength, but can be used in applications requiring electrical and thermal conductivity along with resistance to corrosion and oxidation. However, many high copper alloys are available which give better strength while retaining good conductivity. These are typically heat-treatable alloys containing more than 96%Cu with small amounts of chromium or beryllium and various combinations of silicon, cobalt, and nickel. Examples include the precipitation-hardenable chromium coppers UNS C81400 (0.8%Cu, 0.06%Be) and C81500 (1%Cr), used as electrical and/or thermal conductors which are also structural members. The heat-treatable low-beryllium casting coppers, containing about 0.5%Be, in which the silver content provides improved surface conductivity, includes UNS C82000 (2.5%Co, 0.5%Be) which is used for circuit breaker and switch gear parts and continuous casting molds, and UNS C82200 (1.5%Ni, 0.5%Be) whose uses include brake drums and clutch rings. In addition, these three low beryllium alloys find applications as components of welding equipment, such as electrodes, holders, tips, etc.

The high-strength precipitation-hardenable high-beryllium alloys often include cobalt for grain size control and silicon for improved castability. These alloys include UNS C82500 (2%Be, 0.5%Co, 0.25%Si) which is especially suited for investment casting and finds applications as safety (non-sparking) tools, machine components, and structural parts. Several modifications to this alloy are available, including UNS C82400 (1.7%Be, 0.3%Co), a lower cost, high toughness alloy which is used in marine applications, UNS C82800 (2%Be, 0.5%Co, 0.25%Si) which is a highly castable alloy used for molds for forming of plastics as well as for precision cast parts for the communications, textile, aerospace, and business machine industries, and beryllium-copper 21C (2%Be, 1%Co) which is desirable for thin section castings because of the grain-refining effect of the cobalt.

Red Brasses/Leaded Red Brasses (UNS C83300-C84800) and Yellow Brasses/Leaded Yellow Brasses (UNS C85200-C85800)

The red brass casting alloys contain copper in the range of 75 - 89%, giving them a more reddish colour than the higher-zinc alloys which are called yellow brasses. The leaded red brasses are readily castable. These include the most common of the copper casting alloys UNS C83600 (5%Sn, 5%Pb, 5%Zn), also known as "ounce metal," and its modification UNS C83800 (4%Sn, 6%Pb, 7%Zn). These alloys have moderate strength, ductility and corrosion resistance with good machinability and castability, and are used for plumbing goods, general household and machinery hardware and fixtures, hydraulic and steam valves, impellers, injectors, and statuary. As the zinc content of these cast alloys increases, the cost and the mechanical properties both decrease. The alloys UNS C84400 (3%Sn, 7%Pb, 9%Zn) and C84800 (2.5%Sn, 6.5%Pb, 15%Zn), with their higher zinc contents, are sometimes referred to as "semi-red" brasses, while the leaded yellow brasses include UNS C85200 (1%Sn, 3%Pb, 24%Zn), C85400 (1%Sn, 3%Pb, 29%Zn), and C85700/C85800 (1%Sn, 1%Pb, 35%Zn). These are used for low pressure valves and fittings, as well as general hardware, ornamental castings, and especially as plumbing fittings and fixtures.

All of these red, semi-red, and yellow brasses are suitable for chromium or nickel plating after appropriate surface treatment.

Manganese Bronzes and Leaded Manganese Bronzes (UNS C86100-C86800)

These alloys, along with the aluminum bronzes discussed below, are higher strength materials than the cast brasses, with some having strengths well above 700 MPa (100 ksi) and with good castability and machinability. In fact, the manganese bronzes are better described as modified high-strength yellow brasses, with zinc contents in the range of 24-40%, along with 0.5-4%Mn, and varying amounts of nickel, tin, aluminum, and iron. The tin additions are made to enhance resistance to dezincification, and some also contain lead additions for improved machinability, but these additions are often detrimental to strength and ductility. The manganese bronzes are not heat treatable, but are available with a range of strengths. They find applications as marine propellers and fittings, ball bearing races, architectural components, and high strength structural components such as bridge trunnions, gears, and bearings. The lower-zinc, higher strength commercial manganese bronzes include the alloys UNS C86100/86200 (24%Zn, 3%Fe, 5%Al, 4%Mn) and C86300 (26%Zn, 3%Fe, 3%Al, 4%Mn). The higher-zinc alloys include UNS C86500 (39%Zn, 1.3%Fe, 1%Al, 0.5%Mn), as well as the leaded manganese bronze UNS C86400 (37%Zn, 1%Sn, 1%Pb,1%Fe, 1%Al, 0.5%Mn), and the nickel-manganese bronze UNS C86800 (35%Zn, 0.5%Sn, 3%Ni, 2%Fe, 1%Al, 3%Mn).

Silicon Bronzes and Silicon Brasses (UNS C87300-C87800)

Although these alloys generally have higher strengths than the red brasses, their applications are often based on superior corrosion resistance under particular service conditions. The silicon bronze alloy known as Everdur (UNS C87300 - 4%Si, 1%Mn) is used as a substitute for tin bronze in bearings, bells, pump and valve components, and in statuary and art castings. The silicon brasses cover a range of zinc contents. Low- and moderate-zinc varieties including UNS C87600 (4.5%Si, 5.5%Zn), C87610 (4%Si, 4%Zn), and C87500/C87800 (4%Si, 14%Zn) find applications similar to those of UNS C87300.

Tin Bronzes and Leaded Tin Bronzes (UNS C90200-C94500)

Cast copper-tin alloys can sustain higher tin contents than the wrought tin bronzes, which are limited to about 10%Sn for reasons of workability. Many cast alloys contain less tin than this, along with additions of zinc, nickel, and lead, but alloys with tin contents up to 20% are available. Tin bronzes are used when higher strength, leak tightness or corrosion resistance are desired or at higher temperatures than the red or semi-red brasses can tolerate. The maximum operating temperature is about 290°C (554°F) for bronzes with tin contents of less than about 8%; above this tin content the precipitation of a brittle tin-rich intermetallic phase limits the operating temperature to below about 260°C (500°F). Nickel additions to tin bronzes increase density and leak tightness, as well as making the alloy heat treatable above 5%Ni. An example is UNS C94700 (5%Sn, 2%Zn, 5%Ni).

One of the major applications for cast tin bronzes, leaded and unleaded, is as bearing materials. Alloys with high tin and phosphorus (up to 1%) contents such as UNS C91100 (16%Sn) and C91300 (19%Sn) are used in bridge turntables, where loads are high and rotational motion is slow. The phosphorus is normally added for deoxidation and strengthening along with, in some cases, nickel for control of the grain size and the distribution of lead particles.

The cast unleaded tin bronzes include UNS C90300 (8%Sn, 4%Zn), C90500 (10%Sn, 2%Zn), sometimes known as gun metal, and the gear bronzes UNS C90700 (11%Sn) and C91700 (12%Sn, 1.5%Ni). These are used as pump and valve components, as well as bearings, bushings, and gears.

A wide variety of leaded tin bronzes is available. Lead at levels up to 3% is added to tin bronzes to increase machinability, but it also decreases ductility so lower tin content is sometimes used in compensation. Common leaded tin bronzes include steam bronze (also known as Navy M bronze) UNS C92200 (6%Sn, 1.5%Pb, 4.5%Zn), Navy G bronze UNS C92300 (8%Sn, 1%Pb, 4%Zn), and UNS C92600 (10%Sn, 1%Pb, 2%Zn). The high lead tin bronzes, such as UNS C93200 (7%Sn, 7%Pb, 7%Zn), C93400 (8%Sn, 8%Pb), and C93700 (10%Sn, 10%Pb) are used where a soft bearing or bushing metal is required, at low loads and speeds, especially where lubrication may be deficient. Where loads are higher and lubrication conditions more problematic, or where corrosion or dusty atmospheres inhibit lubrication, even higher lead alloys are used, including UNS C93800 (7%Sn, 15%Pb) and C94300 (5%Sn, 25%Pb). This situation may apply for example, in bearings for railway and mine locomotives and rail cars or equipment for crushing plants in the stone and cement industries.

Aluminum Bronzes (UNS C95200-C95900)

Like the wrought aluminum bronzes discussed above, these are high strength alloys, containing aluminum along with iron, manganese, and nickel. As for wrought alloys, those with aluminum contents between 9.5% and 11.5%, along with iron, are heat treatable by quenching and tempering to strengths well above 700 MPa (100 ksi). They have high ductility and toughness compared to the tin bronzes, and can be used at higher temperatures than the manganese bronzes.

Like tin and manganese bronzes, aluminum bronzes find applications as bearing and bushing alloys. The lower (8-9%) aluminum alloys are used for light-duty or high-speed machinery, while increasingly higher aluminum contents have higher hardness properties suitable for heavy duty applications such as rolling mill bearings, screwdown nuts, guides and aligning plates, and dies.

The higher strength properties of the aluminum bronzes are accompanied by less desirable behaviour in respect to other properties. The high aluminum alloys are generally less resistant to corrosion because they contain aluminum-rich phases which can be susceptible to preferential attack in some environments. They are unsuitable for use in oxidizing acids. Their machinabilities range from moderate to difficult.

Commercially available non-heat-treatable aluminum bronzes include UNS C95200 (9%Al, 3%Fe) which is used for acid-resisting pumps, gears, and valve seats, and the manganese-aluminum bronze UNS C95700 (8%Al, 3%Fe, 2%Ni, 12%Mn) which is used for propellers, impellers, and marine fittings. The heat-treatable alloys include UNS C95400/C95410 (11%Al, 4%Fe) and C95500 (11%Al, 4%Fe, 4%Ni), which are used for valve guides and seats, bushings, gears, and impellers. Alloy UNS C95800 (9%Al, 4%Fe, 4%Ni, 1%Mn) has higher resistance to cavitation and sea water fouling and so is especially suited for marine applications such as propeller blades and hubs.

Copper-Nickels (UNS C96200-C96800), Nickel Silvers (UNS C97300-97800) and Miscellaneous Alloys (UNS C99300-99750)

The common cast copper nickels are the solid solution alloys UNS C96200 (10%Ni) and C96400 (30%Ni) which are used for their sea water corrosion resistance as centrifugal and sand castings for valves, pump components, etc. A heat-treatable beryllium cupronickel, UNS C96600 (30%Ni, 0.5%Be), has about twice the strength of C96400 while maintaining sea water corrosion resistance for applications as pressure housings, pump and valve bodies, and line fittings for marine service. For improved castability, a higher-beryllium cupronickel 72C (30%Ni, 1.2%Be) is also available with applications in the plastic tooling industry. Leaded nickel silver alloys are used for investment, centrifugal, permanent mold, and sand castings for hardware fittings, valves, statuary, and ornamental castings. Alloys include UNS C97300 (12%Ni, 20%Zn, 10%Pb, 2%Sn), C97600 (20%Ni, 8%Zn, 4%Pb, 4%Sn), and C97800 (25%Ni, 2%Zn, 1.5%Pb, 5%Sn).

Alloys developed for resistance to dezincification and/or dealuminification under corrosion conditions are UNS C99400 (2%Ni, 2%Fe, 1%Al, 1%Si, 3%Zn) and the higher strength equivalent alloy UNS C99500 (4%Ni, 4%Fe, 1%Al, 1%Si, 1%Zn). Uses include valves, wheels, gears, and electrical parts for mining equipment and the outboard marine industry.

SAE/AMS SPECIFICATIONS - COPPER & COPPER ALLOYS

AMS	Title
4500	Copper Sheet, Strip, and Plate, Soft Annealed, UNS C11000
4501	Copper Sheet, Strip, and Plate, Oxygen-Free, Light Cold Rolled, UNS C10200
4505	Brass Sheet, Strip, and Plate, 70Cu - 30Zn, Annealed (O61), UNS C26000
4507	Brass Sheet, Strip, and Plate, 70Cu - 30Zn, Annealed, UNS C26000
4508	Brass Sheet, Laminated, 70Cu - 30Zn, Surface Bonded
4510	Phosphor Bronze Sheet, Strip, and Plate, 94.5Cu - 5.0Sn - 0.19P, Spring Temper (H08), UNS C51000
4511	Copper-Beryllium Alloy Castings, 97Cu - 2.1Be - 0.52(Co + Ni) - 0.28Si, Solution and Precipitation Heat Treated (TF00), UNS 82500
4520	Bronze Strip, 88.5Cu - 4.0Sn - 4.0Pb - 3.0Zn - 0.26P, Cold Rolled, Half Hard (H02), UNS C54400
4533	Copper-Beryllium Alloy Bars and Rods, 98Cu - 1.9Be, Solution and Precipitation Heat Treated (TF00, formerly AT), UNS C17200
4534	Copper-Beryllium Alloy, Bars and Rods, 98Cu - 1.9Be, Solution Heat Treated, Cold Worked, and Precipitation Heat Treated (TH04, formerly HT), UNS C17200
4535	Copper-Beryllium Alloy, Mechanical Tubing, 98Cu - 1.9Be, Solution and Precipitation Heat Treated (TF00, formerly AT), UNS C17200
4553	Brass Tubing, Seamless, 85Cu - 15Zn, UNS C23000
4554	Brass Tubing, Seamless, 66.5Cu - 33Zn - 0.45Pb, Annealed, UNS C33000
4555	Brass Tubing, Seamless, 66.5Cu - 32.5Zn - 0.48Pb, Light Annealed (O50), UNS C33000
4558	Brass Tubing, Seamless, 66.5Cu - 31.5Zn - 1.6Pb, Drawn Temper (H58), UNS C33200
4590	Extrusions, Nickel-Aluminum Bronze, Martensitic, 78.5Cu - 10.5Al - 5.1Ni - 4.8Fe, Solution Heat Treated and Tempered (T050), UNS C63020
4602	Copper Bars, Rods, and Shapes, Oxygen-Free, Hard Temper (H04), UNS C10200
4610	Brass Bars and Rods, Free-Cutting, 61.5Cu - 3.1Pb - 35Zn, Half Hard (H02), UNS C36000
4611	Brass Bars and Rods, Naval, 60.5Cu - 38.7Zn - 0.8Sn, Half Hard (H02), UNS C46400
4612	Brass Rods and Bars, Naval Brass, 60.5Cu - 38.5Zn - 0.75Sn, Hard Temper (H04), UNS C46400
4614	Brass Forgings, Free Cutting, 60Cu - 2.0Pb - 37.5Zn, As Forged (M10), UNS C37700
4616	Silicon Bronze Bars, Rods, Forgings, and Tubing, 92Cu - 3.2Si - 2.8Zn - 1.5Fe, Stress Relieved, UNS C65620
4625	Phosphor Bronze Bars, Rods, and Tubing, 95Cu - 5Sn, Hard Temper, UNS C51000
4633	Bronze, Aluminum Silicon, Rods, Bars, and Forgings, 90Cu - 7.0Al - 1.8Si, Drawn and Stress Relieved (HR50), UNS C64200
4634	Aluminum Bronze Bars, 90.5Cu - 7.5Al - 1.9Si, Stress Relieved, UNS C64200
4635	Aluminum Bronze Bars, Rods, and Forgings, 87Cu - 9Al - 3Fe, Stress Relieved, UNS C62300

SAE/AMS SPECIFICATIONS - COPPER & COPPER ALLOYS (Continued)	
AMS	**Title**
4640	Aluminum Bronze Bars, Rods, Shapes, Tubes, and Forgings, 81.5Cu - 10.0Al - 4.8Ni - 3.0Fe, Annealed, UNS C63000
4650	Copper-Beryllium Alloy Bars, Rods, Shapes, and Forgings, 98Cu - 1.9Be, Solution Heat Treated TB00 (A), UNS C17200
4651	Copper-Beryllium Alloy Bars and Rods, 98Cu - 1.9Be (CDA 172), Hard Temper, UNS C17200
4700	Copper Wire, Bare, High Purity, UNS C10100
4701	Copper Wire, Oxygen-Free, 99.95(Cu+Ag), Annealed, UNS C10200
4710	Brass Wire, Tinned, 65Cu - 35Zn, Annealed (061), UNS C27000
4712	Brass Wire, 65Cu - 35Zn, Annealed (061), UNS C27000
4713	Wire, Brass, 65Cu - 35Zn, Eighth-Hard (H00), UNS C27000
4720	Wire, Phosphor Bronze, 94Cu - 5.0Sn - 0.19P, Cold Drawn, Spring Temper (H08), UNS C51000
4725	Copper-Beryllium Alloy Wire, 98Cu - 1.9Be, Solution Heat Treated, UNS C17200
4732	Copper-Nickel Alloy Wire and Ribbon, 77.5Cu - 22.5Ni, UNS C71110
4740	Copper Powder, 99.0Cu minimum, As Fabricated
4805	Bearings, Sintered Metal Powder, 89Cu - 10Sn, Oil Impregnated
4820	Bearings, Leaded Copper, 70Cu - 28.5Pb, Steel Back, UNS C98400
4824	Bearings, Babbitt-Coated Leaded Bronze, Steel Back, UNS C98200
4827	Bearings, Leaded Bronze, 80Cu - 10Pb - 10Sn, Steel Back
4842	Castings, Leaded Bronze, Sand and Centrifugal, 80Cu - 10Sn - 9.5Pb, As Cast, UNS C93700
4845	Tin Bronze Castings, Sand and Centrifugal, 87.5Cu - 10Sn - 2Zn, As Cast, UNS C90500
4855	Leaded Red Brass Castings, Sand and Centrifugal, 85Cu - 5.0Sn - 5.0Pb - 5.0Zn (CDA 83600), As Cast, UNS C83600
4860	Manganese Bronze Castings, Sand and Centrifugal, 58Cu - 39Zn - 1.2Fe - 1.0Al - 0.80Mn (CDA 86500), As Cast, UNS C86500
4862	Manganese Bronze Castings, Sand and Centrifugal, 63Cu - 24Zn - 6.2Al - 3.8Mn - 3.0Fe (CDA 86300), High Strength, As Cast, UNS C86300
4870	Aluminum Bronze Castings, Centrifugal and Chill, 85Cu - 11Al - 3.6Fe, As Cast, UNS C95420
4871	Aluminum Bronze Castings, Centrifugal and Chill, 85Cu - 11Al - 3.6Fe, Solution Heat Treated and Tempered, UNS C95420
4873	Aluminum Bronze Castings, Sand, 85Cu - 11Al - 3.6Fe, Solution Heat Treated and Tempered, UNS C95420
4880	Aluminum Bronze Castings, Centrifugal and Continuous Cast, 81.5Cu - 10.3Al - 5.0Ni - 2.8Fe, Solution Heat Treated and Tempered, UNS C95510
4881	Nickel-Aluminum-Bronze Castings, Martensitic, Sand and Centrifugal, 78Cu - 11Al - 5.1Ni - 4.8Fe, Solution Heat Treated and Tempered, UNS C95520

SAE/AMS SPECIFICATIONS - COPPER & COPPER ALLOYS (Continued)	
AMS	**Title**
4890	Copper-Beryllium Alloy Castings, Investment, 2Be - 0.4Co - 0.3Si, UNS C82500
7322	Sealing Rings, Cast Tin Bronze, 80Cu - 19Sn, As Cast, Rockwell B85-92, UNS C91300

ASTM SPECIFICATIONS - COPPER & COPPER ALLOYS	
Copper and Copper Alloys for Electron Devices and Electronic Applications and Electrical Conductors	
ASTM	**Title**
B 1	Hard-Drawn Copper Wire
B 2	Medium-Hard-Drawn Copper Wire
B 3	Soft or Annealed Copper Wire
B 8	Concentric-Lay-Stranded Copper Conductors, Hard, Medium-Hard or Soft
B 9	Bronze Trolley Wire
B 33	Tinned Soft or Annealed Copper Wire for Electrical Purposes
B 47	Copper Trolley Wire
B 48	Soft Rectangular and Square Bare Copper Wire for Electrical Conductors
B 49	Hot-Rolled Copper Redraw Bar for Electrical Purposes
B 68	Oxygen-Free Copper in Wrought Forms for Electron Devices
B 96	Electronic Grade Alloys of Copper and Nickel in Wrought Forms
B 105	Hard-Drawn Copper Alloy Wires for Electrical Conductors
B 116	Figure-9 Deep-Grooved and Figure-8 Copper Trolley Wire for Industrial Haulage
B 172	Rope-Lay Stranded Copper Conductors Having Bunch-Stranded Members, for Electrical Conductors
B 173	Rope-Lay Stranded Copper Conductors Having Concentric-Stranded Members, for Electrical Conductors
B 174	Bunch-Stranded Copper Conductors for Electrical Conductors
B 187	Copper Bus Bar, Rod, and Shapes
B 188	Seamless Copper Bus Pipe and Tube
B 193	Resistivity of Electrical Conductor Materials
B 226	Cored, Annular, Concentric-Lay-Stranded Copper Conductors

ASTM SPECIFICATIONS - COPPER & COPPER ALLOYS (Continued)

Copper and Copper Alloys for Electron Devices and Electronic Applications and Electrical Conductors (Continued)

ASTM	Title
B 272	Flat Copper Products with Finished (Rolled or Drawn) Edges (Flat Wire and Strip)
B 286	Copper Conductors for Use in Hookup Wire for Electronic Equipment
B 372	Seamless Copper and Copper-Alloy Rectangular Waveguide Tube
B 451	Copper Foil, Strip, and Sheet for Printed Circuits
B 470	Bonded Copper Conductors for Use in Hookup Wires for Electronic Equipment
B 496	Compact Round Concentric-Lay-Stranded Copper Conductors
B 624	High-strength, High-Conductivity Copper-Alloy Wire for Electronic Application
B 738	Fine-Wire Bunch-Stranded and Rope-Lay Bunch Stranded Copper Conductors for Use as Electrical Conductors
F 68	Oxygen-Free Copper in Wrought Forms for Electron Devices
F 96	Electronic Grade Alloys of Copper and Nickel in Wrought Forms
Copper Refinery Products	
B 5	Electrolytic Tough-Pitch Copper - Refinery Shapes
B 115	Electrolytic Cathode Copper
B 170	Oxygen-Free Electrolytic Copper - Refinery Shapes
B 216	Tough-Pitch Fire-Refined Copper for Wrought Products and Alloys - Refinery Shapes
B 224	Coppers
B 379	Phosphorized Coppers - Refinery Shapes
B 623	Tough-Pitch Fire-Refined High-Conductivity Copper - Refinery Shapes
Copper and Copper Alloy Plate, Sheet, Strip, and Rolled Bar	
B 19	Cartridge Brass Sheet, Strip, Plate, Bar, and Disks
B 36/B 36M	Brass Plate, Sheet, Strip, and Rolled Bar
B 96	Copper-Silicon Alloy Plate, Sheet, Strip, and Rolled Bar for General Purposes and Pressure Vessels
B 96M	Copper-Silicon Alloy Plate, Sheet, Strip, and Rolled Bar for General Purposes (Metric)
B 100	Rolled Copper-Alloy Bearing and Expansion Plates and Sheets for Bridge and Other Structural Uses
B 101	Lead-Coated Copper Sheet and Strip for Building Construction
B 103/B 103M	Phosphor Bronze Plate, Sheet, Strip, and Rolled Bar

ASTM SPECIFICATIONS - COPPER & COPPER ALLOYS (Continued)	
Copper and Copper Alloy Plate, Sheet, Strip, and Rolled Bar (Continued)	
ASTM	**Title**
B 121/B 121M	Leaded Brass Plate, Sheet, Strip, and Rolled Bar
B 122	Copper-Nickel-Tin Alloy, Copper-Nickel-Zinc Alloy (Nickel Silver), and Copper-Nickel Alloy Plate, Sheet, Strip, and Rolled Bar
B 129	Cartridge Brass Cartridge Case Cups
B 130	Commercial Bronze Strip for Bullet Jackets
B 131	Copper Alloy Bullet Jacket Cups
B 152	Copper Sheet, Strip, Plate, and Rolled Bar
B 152M	Copper Sheet, Strip, Plate, and Rolled Bar (Metric)
B 169	Aluminum Bronze Plate, Sheet, Strip, and Rolled Bar
B 169M	Aluminum Bronze Plate, Sheet, Strip, and Rolled Bar (Metric)
B 171	Copper-Alloy Plate and Sheet for Pressure Vessels, Condensers, and Heat Exchangers
B 171M	Copper-Alloy Plate and Sheet for Pressure Vessels, Condensers, and Heat Exchangers (Metric)
B 194	Copper-Beryllium Alloy Plate, Sheet, Strip, and Rolled Bar
B 248	General Requirements for Wrought Copper and Copper-Alloy Plate, Sheet, Strip, and Rolled Bar
B 248M	General Requirements for Wrought Copper and Copper-Alloy Plate, Sheet, Strip, and Rolled Bar (Metric)
B 291	Copper-Zinc Manganese Alloy (Manganese Brass) Sheet and Strip
B 370	Copper Sheet and Strip for Building Construction
B 422	Copper-Aluminum-Silicon-Cobalt Alloy, Copper-Nickel-Aluminum-Silicon Alloy, Copper-Nickel-Silicon-Magnesium Alloys, and Copper-Nickel-Aluminum-Magnesium Alloy Sheet and Strip
B 432	Copper and Copper Alloy Clad Steel Plate
B 451	Copper Foil, Strip, and Sheet for Printed Circuits and Carrier Tapes
B 465	Copper-Iron Alloy Plate, Sheet, Strip, and Rolled Bar
B 506	Copper-Clad Stainless Steel Sheet and Strip for Building Construction
B 508	Copper Alloy Strip for Flexible Metal Hose
B 534	Copper-Cobalt-Beryllium Alloy and Copper-Nickel-Beryllium Alloy Plate, Sheet, Strip, and Rolled Bar
B 569	UNS No. C26000 Brass Strip in Narrow Widths and Light Gage for Heat-Exchanger Tubing
B 591	Copper-Zinc-Tin Alloys Plate, Sheet, Strip, and Rolled Bar

ASTM SPECIFICATIONS - COPPER & COPPER ALLOYS (Continued)	
Copper and Copper Alloy Plate, Sheet, Strip, and Rolled Bar (Continued)	
ASTM	**Title**
B 592	Copper-Zinc-Aluminum-Cobalt Alloy Plate, Sheet, Strip, and Rolled Bar
B 638	Copper and Copper Alloy Solar Heat Absorber Panels
B 694	Copper, Copper Alloy, and Copper-Clay Stainless Steel Sheet and Strip for Electrical Cable Shielding
B 740	Copper-Nickel-Tin Spinodal Alloy Strip
B 747	Copper Zirconium Alloy Sheet and Strip
B 754	Measuring and Recording the Deviations from Flatness in Copper and Copper Alloy Strip
B 768	Copper-Cobalt-Beryllium Alloy Strip and Sheet
Copper and Copper Alloy Rod, Bar, and Shapes, and Die Forgings	
B 16	Free-Cutting Brass Rod, Bar, and Shapes for Use in Screw Machines
B 16M	Free-Cutting Brass Rod, Bar, and Shapes for Use in Screw Machines (Metric)
B 21	Naval Brass Rod, Bar, and Shapes
B 21M	Naval Brass Rod, Bar, and Shapes (Metric)
B 98	Copper-Silicon Alloy Rod, Bar, and Shapes
B 98M	Copper-Silicon Alloy Rod, Bar, and Shapes (Metric)
B 124	Copper and Copper Alloy Forging Rod, Bar, and Shapes
B 124M	Specification for Copper and Copper-Alloy Forging Rod, Bar, and Shapes (Metric)
B 133	Copper Rod, Bar, and Shapes
B 133M	Copper Rod, Bar, and Shapes (Metric)
B 138	Manganese Bronze Rod, Bar, and Shapes
B 138M	Manganese Bronze Rod, Bar, and Shapes (Metric)
B 139	Phosphor Bronze Rod, Bar, and Shapes
B 139M	Phosphor Bronze Rod, Bar, and Shapes (Metric)
B 140	Copper-Zinc-Lead (Leaded Red Brass or Hardware Bronze) Rod, Bar, and Shapes
B 140M	Copper-Zinc-Lead (Leaded Red Brass or Hardware Bronze) Rod, Bar and Shapes (Metric)
B 150	Aluminum Bronze Rod, Bar, and Shapes
B 150M	Aluminum Bronze Rod, Bar, and Shapes (Metric)

ASTM SPECIFICATIONS - COPPER & COPPER ALLOYS (Continued)	
Copper and Copper Alloy Rod, Bar, and Shapes, and Die Forgings (Continued)	
ASTM	**Title**
B 151	Copper-Nickel-Zinc Alloy (Nickel Silver) and Copper-Nickel Rod and Bar
B 151M	Copper-Nickel-Zinc Alloy (Nickel Silver) and Copper-Nickel Rod and Bar (Metric)
B 187	Copper Bus Bar, Rod, and Shapes
B 196	Copper-Beryllium Alloy Rod and Bar
B 196M	Copper-Beryllium Alloy Rod and Bar (Metric)
B 249	General Requirements for Wrought Copper and Copper-Alloy Rod, Bar, and Shapes
B 249M	General Requirements for Wrought Copper and Copper-Alloy Rod, Bar, and Shapes (Metric)
B 283	Copper and Copper-Alloy Die Forgings (Hot-Pressed)
B 301	Free-Cutting Copper Rod and Bar
B 301M	Free-Cutting Copper Rod and Bar (Metric)
B 371	Copper-Zinc-Silicon Alloy Rod
B 411	Copper-Nickel-Silicon Alloy Rod and Bar
B 441	Copper-Cobalt-Beryllium (UNS No. C17500) and Copper-Nickel-Beryllium (UNS No. C17510) Rod and Bar
B 453	Copper-Zinc-Lead Alloy (Leaded Brass) Rod
B 453M	Copper-Zinc-Lead Alloy (Leaded Brass) Rod (Metric)
B 455	Copper-Zinc-Lead Alloy (Leaded Brass) Extruded Shapes
B 570	Copper-Beryllium Alloy Forgings and Extrusions
F 467	Nonferrous Nuts for General Use
F 468	Nonferrous Bolts, Hex Cap Screws, and Studs for General Use
Copper, Copper Alloy, and Copper-Clad Wire	
B 99	Copper-Silicon Alloy Wire for General Purposes
B 99M	Copper-Silicon Alloy Wire for General Purposes (Metric)
B 134	Brass Wire
B 159	Phosphor Bronze Wire
B 159M	Phosphor Bronze Wire (Metric)
B 189	Lead-Coated and Lead-Alloy-Coated Soft Copper Wire for Electrical Purposes

ASTM SPECIFICATIONS - COPPER & COPPER ALLOYS (Continued)	
Copper, Copper Alloy, and Copper-Clad Wire (Continued)	
ASTM	**Title**
B 197	Copper-Beryllium Alloy Wire
B 197M	Copper-Beryllium Alloy Wire (Metric)
B 206	Copper-Nickel-Zinc Alloy (Nickel Silver) Wire and Copper-Nickel Alloy Wire
B 206M	Copper-Nickel-Zinc Alloy (Nickel Silver) Wire and Copper-Nickel Alloy Wire (Metric)
B 229	Concentric-Lay-Stranded Copper and Copper-Clad Steel Composite Conductors
B 246	Tinned Hard-Drawn and Medium-Hard-Drawn Copper Wire for Electrical Purposes
B 250	General Requirements for Wrought Copper-Alloy Wire
B 250M	General Requirements for Wrought Copper-Alloy Wire (Metric)
B 272	Flat Copper Products with Finished (Rolled or Drawn) Edges (Flat Wire and Strip)
B 298	Silver-Coated Soft or Annealed Copper Wire
B 355	Nickel-Coated Soft or Annealed Copper Wire
B 412	Copper-Nickel-Silicon Alloy Wire
B 42	Seamless Copper Pipe, Standard Sizes
Copper and Copper Alloy Pipe And Tube	
B 43	Seamless Red Brass Pipe, Standard Sizes
B 68	Seamless Copper Tube, Bright Annealed
B 68M	Seamless Copper Tube, Bright Annealed (Metric)
B 75	Seamless Copper Tube
B 75M	Seamless Copper Tube (Metric)
B 88	Seamless Copper Water Tube
B 88M	Seamless Copper Water Tube (Metric)
B 111	Copper and Copper-Alloy Seamless Condenser Tubes and Ferrule Stock
B 111M	Copper and Copper-Alloy Seamless Condenser Tubes and Ferrule Stock (Metric)
B 135	Seamless Brass Tube
B 135M	Seamless Brass Tube (Metric)
B 153	Expansion (Pin Test) of Copper and Copper Alloy Pipe and Tubing

ASTM SPECIFICATIONS - COPPER & COPPER ALLOYS (Continued)	
Copper and Copper Alloy Pipe And Tube (Continued)	
ASTM	**Title**
B 188	Seamless Copper Bus Pipe and Tube
B 251	General Requirements for Wrought Seamless Copper and Copper-Alloy Tube
B 251M	General Requirements for Wrought Seamless Copper and Copper-Alloy Tube (Metric)
B 280	Seamless Copper Tube for Air Conditioning and Refrigeration Field Service
B 302	Threadless Copper Pipe
B 306	Copper Drainage Tube (DWV)
B 315	Seamless Copper Alloy Pipe and Tube
B 359	Copper and Copper-Alloy Seamless Condenser and Heat Exchanger Tubes with Integral Fins
B 359M	Copper and Copper-Alloy Seamless Condenser and Heat-Exchanger Tubes with Integral Fins (Metric)
B 360	Hard-Drawn Copper Capillary Tube for Restrictor Applications
B 395	U-Bend Seamless Copper and Copper-Alloy Heat-Exchanger and Condenser Tubes
B 395M	U-Bend Seamless Copper and Copper-Alloy Heat-Exchanger and Condenser Tubes (Metric)
B 428	Angle of Twist in Rectangular and Square Copper and Copper Alloy Tube
B 447	Welded Copper Tube
B 466	Seamless Copper-Nickel Pipe and Tube
B 466M	Seamless Copper-Nickel Pipe and Tube (Metric)
B 467	Welded Copper-Nickel Pipe
B 469	Seamless Copper Alloy Tubes for Pressure Applications
B 543	Welded Copper and Copper-Alloy Heat-Exchanger Tube
B 543M	Welded Copper and Copper-Alloy Heat-Exchanger Tube (Metric)
B 552	Seamless and Welded Copper-Nickel Tubes for Water Desalting Plants
B 587	Weld Brass Tube
B 608	Welded Copper-Alloy Pipe
B 640	Welded Copper and Copper Alloy Tube for Air Conditioning and Refrigeration Service
B 641	Seamless and Welded Copper Distribution Tube (Type D)
B 642	Welded Copper Alloy UNS No. C21000 Water Tube

ASTM SPECIFICATIONS - COPPER & COPPER ALLOYS (Continued)	
Copper and Copper Alloy Pipe And Tube (Continued)	
ASTM	**Title**
B 643	Copper-Beryllium Alloy Seamless Tube
B 687	Brass, Copper, and Chromium-Plated Pipe Nipples
B 698	Seamless and Welded Copper and Copper Alloy Plumbing Pipe and Tube (Classification)
B 706	Seamless Copper Alloy (UNS No. C69100) Pipe and Tube
B 716	Welded Copper Water Tube
B 716M	Welded Copper Water Tube (Metric)
B 732	Evaluating the Corrosivity of Solder Fluxes for Copper Tubing Systems
B 743	Seamless Copper Tube in Coils
B 813	Liquid and Paste Fluxes for Soldering Applications of Copper and Copper Tube
B 819	Seamless Copper Tube for Medical Gas Systems
E 243	Electromagnetic (Eddy-Current) Testing of Seamless Copper and Copper-Alloy Tubes
Copper Alloy Ingot and Castings	
B 22	Bronze Castings for Bridges and Turntables
B 30	Copper-Base Alloys in Ingot Form
B 61	Bronze Castings, Steam or Valve
B 62	Composition Bronze or Ounce Metal Castings
B 66	Bronze Castings for Steam Locomotive Wearing Parts
B 67	Car and Tender Journal Bearings, Lined
B 148	Aluminum-Bronze Sand Castings
B 176	Copper Alloy Die Castings
B 208	Preparing Tension Test Specimens for Copper-Base Alloys for Sand Permanent Mold, Centrifugal, and Continuous Castings
B 271	Copper-Base Alloy Centrifugal Castings
B 369	Copper-Nickel Alloy Castings
B 427	Gear Bronze Alloy Castings
B 505	Copper-Base Alloy Continuous Castings
B 584	Copper Alloy Sand Castings for General Applications

ASTM SPECIFICATIONS - COPPER & COPPER ALLOYS (Continued)	
Copper Alloy Ingot and Castings (Continued)	
ASTM	**Title**
B 644	Copper Alloy Addition Agents
B 763	Copper Alloy Sand Castings for Valve Application
B 770	Copper-Beryllium Alloy Sand Castings for General Applications
B 806	Copper Alloy Permanent Mold Castings for General Applications
B 824	General Requirements for Copper Alloy Castings
Copper Classifications Systems	
B 224	Coppers

AWS WELDING FILLER METAL SPECIFICATIONS - COPPER & COPPER ALLOYS	
AWS	**Title**
A5.6	Copper and Copper Alloy Covered Electrodes
A5.7	Copper and Copper Alloy Bare Welding Rods and Electrodes
A5.27	Copper and Copper Alloy Gas Welding Rods

AMERICAN CROSS REFERENCED SPECIFICATIONS - COPPER & COPPER ALLOYS

UNS	AMS	MIL SPEC	ASTM	ASME	AWS
C10100	4700	MIL-W-85; MIL-W-3318; MIL-B-18907	B1; B3; B8; B33, B48; B49; F68; B75; B111 (OFE); B133; B152; B172; B173; B174; B187; B188; B189; B224 (OFE); B246; B272; B280; B286, B298; B355; B359; B432; B447; B451; B470; B496; B640; B641; B698 (OFE)	---	---
C10200	4501; 4602; 4701	MIL-W-85; MIL-B-2029; MIL-W-6712; MIL-B-18907; MIL-T-24107	B1; B2; B3; B8; F9; B33; B42; B48; B49; B68; B75; B88; B111; B133; B152; B172; B173; B174; B187; B188; B189; B224 (OF); B640; B641; B687; B698 (OF); B716; B738; B743	SB42; SB75, SB111; SB133; SB152; SB359; SB395	---
C10300	---	MIL-T-3235; MIL-B-20292	B42; B68; B75; B88; B111; B133; B152; B187; B188; B224 (OFXLP); B280; B302; B306; B359; B372; B379 (OFXLP); B395; B432; B447; B638 (OFXLP); B640; B641; B687; B698 (OFXLP); B716; B743	---	---
C10400	---	MIL-W-3318; MIL-B20292	B1; B2; B3; B49; B133; B152; B187; B188; B189; B224 (OFS); B246; B272; B298; B355; B506	SB152	---
C10500	---	---	B1; B2; B3; B49; B133; B152; B187; B188; B189; B224 (OFS); B246; B272; B298; B355; B506	SB152	---
C10700	---	MIL-W-3318; MIL-B-19231; MIL-B-20292	B1; B2; B3; B49; B133; B152; B187; B188; B189; B224 (OFS); B246; B272; B298; B355; B506	SB152	---
C10800	---	MIL-B-20292; MIL-T-24107	B42; B68; B75; B88; B111; B133; B152; B224 (OFLP); B280; B302; B306; B359; B360; B379; B432; B447; B543; B638 (OFLP); B640; B641; B687; B698 (OFLP); B716; B743	---	---

AMERICAN CROSS REFERENCED SPECIFICATIONS - COPPER & COPPER ALLOYS (Continued)					
UNS	**AMS**	**MIL SPEC**	**ASTM**	**ASME**	**AWS**
C11000	4500	MIL-W-6712; MIL-C-12168; MIL-B-20292	B1; B2; B3; B8; B33; B47; B48; B49; B101; B116; B124; B133; B152; B172; B173; B174; B187; B188; B189; B224 (ETP); B226; B228; B229; B246; B272; B283; B286; B298; B355; B370; B447; B451; B470; B496; B506; B566; B638; B694; B738; F467 (110); F468 (110)	SB133	---
C11010	---	---	B224 (RHC)	---	---
C11020	---	---	B224 (FRHC)	---	---
C11030	---	---	B224 (CRTP)	---	---
C11040	---	---	B49	---	---
C11100	---	---	B224 (ETP)	---	---
C11300	---	MIL-B-20292	B1; B2; B3; B5 (STP); B8; B33; B47; B48; B49; B116; B33; B48; B152; B172; B173; B174; B187; B188; B189; B224 (ETP); B226; B228; B229; B246; B272; B286; B298; B355; B442 (STP); B470; B496; B506; B623 (STP)	---	---
C11400	---	MIL-B-20292	B1; B2; B3; B4; B5 (STP); B8; B47; B48; B49; B116; B152; B172; B173; B174; B187; B188; B189; B224 (ETP); B226; B229; B246; B272; B286; B298; B355; B442 (STP); B470; B496; B506; B623 (STP)	---	---
C11500	---	---	B1; B2; B3; B5 (STP); B8; B33; B47; B49; B116; B172; B173; B174; B189; B224 (ETP); B226; B228; B229; B246; B272; B286; B298; B355; B442 (STP); B470; B496; B623 (STP)	---	---
C11600	---	MIL-B-19231; MIL-B-20292	B1; B2; B3; B5 (STP); B8; B33; B47; B48; B49; B116; B152; B172; B173; B174; B187; B188; B189; B224 (ETP); B226; B229; B246; B272; B286; B298; B355; B442 (STP); B470; B496; B506; B623 (STP)	---	---

AMERICAN CROSS REFERENCED SPECIFICATIONS - COPPER & COPPER ALLOYS (Continued)

UNS	AMS	MIL SPEC	ASTM	ASME	AWS
C12000	---	MIL-W-85; MIL-T-3235; MIL-B-18907; MIL-B-20292; MIL-T-24107	B5 (DLP); B42; B68; B75; B88; B111; B133; B152; B187; B188; B224 (DLP); B447; B451; B506; B623 (DLP); B638; B640; B641; B687; B698 (DLP); B716; B743	SB42; SB75; SB111; SB133; SB359; SB395	---
C12100	---	MIL-B-20292	B224 (DLPS)	---	---
C12200	---	MIL-T-3235; MIL-B-18907; MIL-B-20292; MIL-T-22214; MIL-T-24107	B5 (DHP); B42; B68; B75; B88; B101; B111; B133; B152; B224 (DHP); B272; B280; B302; B306; B359; B360; B370; B379 (OFLP); B395; B432; B442 (DHP); B447; B506; B543; B623 (DHP); B638; B640; B641; B687; B698 (DHP); B716; B743	B16.22; B16.29; SB42; SB75; SB111; SB133; SB152; SB359; SB395; SB543	---
C12300	---	---	B5 (DHPS); B152; B224 (DHPS); B442 (DHPS); B506; B623 (DHPS)	SB152	---
C12900	---	MIL-B-20292	B224 (FRSTP)	---	---
C14180	---	MIL-B-7883(BCu-1)	---	SFA5.8 (BCu-1)	A5.8 (BCu-1)
C14181	---	---	---	SFA5.8 (BCu-1x)	A5.8 (BCu-1x)
C14200	---	MIL-T-24107	B5 (DPA); B75; B111; B133; B152; B216 (DPA); B224 (DPA); B359; B379 (DPA); B395; B442 (DPA); B447; B623 (DPA)	SB75; SB111; SB133; SB359; SB395	---
C14420	---	---	B152	---	---
C14500	---	---	B5 (DPTE); B124; B216 (DPTE); B224 (DPTE); B283; B301; B442 (DPTE); B623 (SPTE)	B16.22	
C14510	---	---	B301	---	---
C14520	---	---	B301	---	---
C14700	---	---	B124; B224; B283; B301	B16.22	---
C15000	---	---	B224	---	---
C15100	---	---	B747	---	---
C16200	---	---	B105; B624	---	---

AMERICAN CROSS REFERENCED SPECIFICATIONS - COPPER & COPPER ALLOYS (Continued)

UNS	AMS	MIL SPEC	ASTM	ASME	AWS
C16500	---	---	B9; B105	---	---
C17000	---	---	B194; B196; B570	---	---
C17200	4530; 4532; 4533; 4534; 4535; 4650; 4651; 4725	MIL-C-21657	B194; B196; B197; B570; B643	---	---
C17300	---	---	B196; B197	---	---
C17410	---	---	B768	---	---
C17500	---	MIL-C-81020	B441; B534	---	---
C17510	---	MIL-C-81021	B441; B534	---	---
C18200	---	MIL-C-19311	F9	---	---
C18700	---	---	B301	---	---
C18980	---	MIL-C-19654 (MIL-RCu-2)	---	SFA5.7 (ERCu); SFA5.27 (RCu)	A5.7 (ERCu); A5.27 (RCu)
C19200	---	---	B111; B359; B395; B465; B469; B638; B640; B698	SB111; SB359; SB395; B640	---
C19400	---	---	B465; B543; B638; B640; B694; B698	SB543	---
C19500	---	---	B465	---	---
C21000	---	---	B36; B134; B587; B642; B698 (Gilding-95%)	---	---
C22000	---	MIL-W-85; MIL-C-3383; MIL-W-6712; MIL-B-18907; MIL-B-20292	B36; B130; B131; B134; B135; B372; B587; B694	---	---
C22600	---	---	B36	---	---
C23000	---	MIL-T-20168	B36; B43; B111; B134; B135; B359; B395; B543; B587; B687; B698 (Red Brass-85%)	B16.22; B16.29; SB43; SB111; SB359; SB395; SB543	---

AMERICAN CROSS REFERENCED SPECIFICATIONS - COPPER & COPPER ALLOYS (Continued)					
UNS	**AMS**	**MIL SPEC**	**ASTM**	**ASME**	**AWS**
C24000	---	---	B36; B134	---	---
C26000	4505; 4507; 4508; 4555	MIL-C-10375; MIL-T-20219; MIL-S-22499	B19; B36; B129; B134; B135; B569; B587	---	---
C26800	---	MIL-W-6712	B36; B587	---	---
C27000	4710; 4712; 4713	---	B134; B135; B587; F467; F468	---	---
C27200	---	---	B36; B135; B587	---	---
C27400	---	---	B134	---	---
C28000	---	---	B111; B135	SB111	---
C31400	---	MIL-V-18436	B140	---	---
C31600	---	MIL-V-18436	B140	---	---
C32000	---	---	B140	---	---
C33000	4555	MIL-T-46072	B135	---	---
C33200	4558	MIL-T-46072	B135	---	---
C33500	---	---	B121; B453	---	---
C34000	---	---	B121; B453	---	---
C34200	---	---	B121	---	---
C34500	---	---	B453	---	---
C35000	---	---	B121; B453	---	---
C35300	---	---	B121; B453	---	---
C35600	---	---	B121; B453	---	---
C36000	4610	MIL-V-18436	B16	---	---
C36500	---	---	B124; B171; B283; B432	SB171	---
C37000	---	MIL-T-46072	B135	---	---
C37700	4614	---	B124; B283	SB283	---
C38000	---	---	B455	---	---

AMERICAN CROSS REFERENCED SPECIFICATIONS - COPPER & COPPER ALLOYS (Continued)

UNS	AMS	MIL SPEC	ASTM	ASME	AWS
C38500	---	---	B455	---	---
C40500	---	---	B591	---	---
C41100	---	MIL-B-13501	B105; B508; B591	---	---
C41300	---	---	B591	---	---
C41500	---	---	B591	---	---
C42200	---	---	B591	---	---
C42500	---	---	B591	---	---
C43000	---	---	B591	---	---
C43400	---	---	B591	---	---
C44300	---	---	B111; B135; B171; B359; B395; B432; B543	SB111; SB171; SB359; SB395; SB543	---
C44400	---	---	B111; B171; B359; B395; B432; B543	SB111; SB171; SB359; SB395; SB543	---
C44500	---	---	B111; B171; B359; B395; B432; B543	SB111; SB171; SB359; SB395; SB543	---
C46200	---	---	B21; F68; F467; F468	---	---
C46400	4611; 4612	MIL-W-6712	B21; B124; B171; B283; B432; F467; F468	SB171	---
C46500	---	---	B171; B432	SB171	---
C47000	---	MIL-B-7883 (RBCuZn-A)	---	SFA5.8 (RBCuZn-A)	A5.8 (RBCuZn-A); A5.27 (RBCuZn-A)
C47940	---	---	B21	---	---
C48200	---	---	B21; B124; B283	---	---
C48500	---	---	B21; B124; B283	---	---
C50100	---	---	B105	---	---
C50200	---	---	B105	---	---
C50500	---	---	B9; B105; B508	---	---
C50700	---	---	B105	---	---

AMERICAN CROSS REFERENCED SPECIFICATIONS - COPPER & COPPER ALLOYS (Continued)					
UNS	**AMS**	**MIL SPEC**	**ASTM**	**ASME**	**AWS**
C51000	4510; 4625; 4720	MIL-T-3595; MIL-W-6712; MIL-B-13501	B100; B103; B139; B159; F467 (510); F468 (510)	---	---
C51100	---	---	B100; B103	---	---
C51800	---	---	---	SFA5.7 (ERCuSn-A); SFA5.21 (ERCuSn-A)	A5.7 (ERCuSn-A); A5.21 (ERCuSn-A)
C52100	---	---	B103; B139; B159	SFA5.7 (ERCuSn-C)	A5.7 (ERCuSn-C)
C52400	---	---	B103; B139; B159	SFA5.21 (ERCuSn-D)	A5.21 (ERCuSn-D)
C53400	---	---	B103; B139	---	---
C54400	4520	---	B103; B139	---	---
C55180	---	---	---	SFA5.8 (BCuP-1)	A5.8 (BCuP-1)
C55181	---	---	---	SFA5.8 (BCuP-2)	A5.8 (BCuP-2)
C55280	---	---	---	SFA5.8 (BCuP-6)	A5.8 (CVuP-6)
C55281	---	MIL-B-7883 (BCuP-3)	---	SFA5.8 (BCuP-3)	A5.8 (BCuP-3)
C55282	---	---	---	SFA5.8 (BCuP-7)	A5.8 (BCuP-7)
C55283	---	---	---	SFA5.8 (BCuP-4)	A5.8 (BCuP-4)
C55284	---	MIL-B-7883 (BCuP-5)	---	SFA5.8 (BCuP-5)	A5.8 (BCuP-5)
C60800	---	---	B111; B359; B395	SB111; SB359; SB395	---
C61000	---	---	---	SB169; SFA5.7 (ERCuAl-Al)	A5.7 (ERCuAl-Al)
C61300	---	---	B111; B150; B169; B171; B315; B432; B608; F467 (613);F468 (613)	---	---
C61400	---	---	B111; B150; B169; B171; B315; B432; B608; F467; F468	SB150; SB169; SB171	---

AMERICAN CROSS REFERENCED SPECIFICATIONS - COPPER & COPPER ALLOYS (Continued)

UNS	AMS	MIL SPEC	ASTM	ASME	AWS
C61800	---	MIL-W-6712; MIL-E-23765/3 (MIL-CuAl-A2)	---	SFA5.7 (ERCuAl-A2)	A5.7 (ERCuAl-A2)
C61900	---	---	B124; B150; B283	---	---
C62300	4635	MIL-B-16166	B124; B150; B283	SB150	---
C62400	---	---	B150	SFA5.7 (ERCuAl-A3)	A5.7 (ERCuAl-A3)
C63000	4640	MIL-B-16166	B124; B150; B171; B283; F467; F468	SB150; SB171	---
C63200	---	---	B124; B150; B283	---	---
C63280	---	MIL-E-23765/3 (MIL-CuNiAl)	---	SFA5.7 (ERCuNiAl)	A5.7 (ERCuNiAl)
C63380	---	MIL-E-23765/3 (MIL-CuMnNiAl)	---	SFA5.7 (ERCuMnNiAl)	A5.7(ERCuMnNiAl)
C63800	---	---	B422	---	---
C64200	4631	---	B124; B150; B283; F467; F468	SB150; SB283	---
C64210	---	---	B124; B150; B283	---	---
C64700	---	---	B411; B412	---	---
C65100	---	---	B96; B98; B99; B105; B315; B432; F467; F468	SB98; SB315	---
C65400	---	---	B96	---	---
C65500	4615; 4665	MIL-T-8231	B96; B98; B99; B100; B105; B124; B283; B315; B432; F467; F468	SB96; SB98; SB315	---
C65600	---	MIL-E-23765/3 (MIL-CuSi)	---	SFA5.7 (ERCuSi-A)	A5.7 (ERCuSi-A); A5.27 (RCuSi-A)
C66100	---	---	B98; F467; F468	SB98	---
C66400	---	---	B694	---	---
C66410	---	---	B694	---	---
C66700	---	---	B291	---	---
C67000	---	---	B138	---	---

AMERICAN CROSS REFERENCED SPECIFICATIONS - COPPER & COPPER ALLOYS (Continued)					
UNS	**AMS**	**MIL SPEC**	**ASTM**	**ASME**	**AWS**
C67500	---	---	B124; B138; B283; F467; F468	---	---
C68000	---	---	---	---	A5.27 (RCuZn-B)
C68100	---	---	---	SFA5.27 (RBCuZn-C)	A5.8 (RBCuZn-C); A5.27 (RBCuZn-C)
C68700	---	---	B111; B359; B395; B543	SB111, SB359; SB395; SB543	---
C68800	---	---	B592	---	---
C69100	---	---	B706	---	---
C69400	---	---	B371	---	---
C69430	---	---	B371	---	---
C69700	---	---	B371	---	---
C69710	---	---	B371	---	---
C70200	---	MIL-B-18907; MIL-B-20292	---	---	---
C70400	---	---	B111; B359; B395; B466; B543	SB111; SB359; SB466; SB543	---
C70600	---	MIL-T-15005; MIL-C-15726; MIL-T-16420; MIL-T-22214	B111; B122; B151; B171; B359; B395; B432; B466; B467; B469; B543; B552; B608	SB111; SB171; SB359; SB395; SB402; SB466; SB467; SB543	---
C70690	---	---	F96	---	---
C71000	---	---	B111; B122; B206; B359; B395; B466; B543; B694; F467; F468	SB111; SB359; SB395; SB466	---
C71500	---	MIL-T-15005; MIL-C-15726; MIL-T-16420; MIL-T-22214	B111; B122; B171; B359; B395; B432; B466; B467; B543; B552; B608; F467; F468	SB111; SB171; SB359; SB395; SB402; SB466; SB467; SB543	---
C71580	---	---	F96	---	---

AMERICAN CROSS REFERENCED SPECIFICATIONS - COPPER & COPPER ALLOYS (Continued)

UNS	AMS	MIL SPEC	ASTM	ASME	AWS
C71581	---	MIL-E-21562 (RN7,EN67,RN67); MIL-I-23413 (MIL-67)	---	SFA5.7 (ERCuNi); SFA5.27 (RCuNi); SFA5.30 (IN67)	A5.7 (ERCuNi); A5.27 (RCuNi); A5.30 (IN67)
C71590	---	---	F96	---	---
C71640	---	---	B111; B543; B552	---	---
C72200	---	---	B111; B122; B171; B359; B395; B466; B543; B552	---	---
C72500	---	---	B122	---	---
C72700	---	---	B740	---	---
C72900	---	---	B740	---	---
C73500	---	---	B122	---	---
C74000	---	---	B122	---	---
C74500	---	---	B122; B151; B206	---	---
C75200	---	---	B122; B151; B206; B458 (1)	---	---
C75700	---	---	B151; B206	---	---
C76200	---	---	B122	---	---
C76400	---	---	B151; B206	---	---
C77000	---	---	B122; B151; B206	---	---
C77300	---	---	---	SFA5.7 (RBCuZn-D); SFA5.27 (RBCuZn-D)	A5.8 (RBCuZn-D) A5.27 (RBCuZn-D)
C77400	---	---	B124; B283	---	---
C79200	---	---	B151; B206	---	---
C81400	---	---	B770	---	---
C82000	---	---	B30 (10C); B770	---	---
C82200	---	---	B30 (3C, 14C); B770	---	---
C82400	---	---	B30 (165C, 165CT); B770	---	---

AMERICAN CROSS REFERENCED SPECIFICATIONS - COPPER & COPPER ALLOYS (Continued)

UNS	AMS	MIL SPEC	ASTM	ASME	AWS
C82500	4890	MIL-C-11866 (17); MIL-C-22087 (10)	B30 (20C, 20CT); B770	---	---
C82510	---	---	B30 (21C); B770	---	---
C82600	---	---	BS0 (245C, 245CT); B770	---	---
C82800	---	---	B30 (275C, 275CT); B770	---	---
C83400	---	MIL-B-46066	---	---	---
C83450	---	---	B30; B584; B763	---	---
C83600	4855	MIL-C-11866 (25); MIL-V-18436	B30 (4A); B62; B271; B505; B584 (formerly B145)	B16.15; B16.18; B16.23; B16.24; B16.26; B16.32; SB62	---
C83800	---	---	B30; B271; B505; B584 (formerly 145); B763	B16.15; B16.18; B16.23	---
C84200	---	---	B30; B505	---	---
C84400	---	---	B30; B271; B505; B584 (formerly B145); B763	B16.15; B16.18; B16.23; B16.26; B16.32	---
C84800	---	---	B30; B271; B505; B584 (formerly B145); B763	---	---
C85200	---	---	B30; B271; B584 (formerly B146); B763	---	---
C85400	---	---	B30; B271; B584 (formerly B146); B763	---	---
C85700	---	---	B30; B176; B271; B584; B763	---	---
C85800	---	---	B30; B176	---	---
C86200	---	---	B30; B271; B505; B584 (formerly B147); B763	---	---
C86300	4862	MIL-C-11866 (21)	B22; B30; B271; B505; B584 (formerly B147); B763	---	---
C86400	---	---	B30; B271; B505; B584 (formerly B132, B147); B763	---	---
C86500	4860	---	B30; B176; B271; B505; B584 (formerly B147); B763	---	---
C86700	---	---	B30; B271; B584 (formerly B132); B763	---	---
C87300	---	MIL-C-11866 (Comp 19)	B30; B271; B584; B763	---	---
C87400	---	---	B30; B271; B584 (formerly B198); B763	---	---

AMERICAN CROSS REFERENCED SPECIFICATIONS - COPPER & COPPER ALLOYS (Continued)					
UNS	**AMS**	**MIL SPEC**	**ASTM**	**ASME**	**AWS**
C87500	---	---	B30; B271; B584 (formerly B198); B763; B806	---	---
C87600	---	---	B30; B584; B763	---	---
C87610	---	---	B30; B584; B763	---	---
C87800	---	---	B30; B176; B806	---	---
C90300	---	MIL-C-11866 (26)	B30; B271; B505; B584 (formerly B143); B763	---	---
C90500	4845	---	B22; B30; B271; B505; B584 (formerly B143); B763	---	---
C90700	---	---	B30; B427; B505	---	---
C90800	---	---	B30; B427	---	---
C91000	---	---	B30; B505	---	---
C91100	---	---	B22; B30	---	---
C91300	7322	---	B22; B30; B505	---	---
C91600	---	---	B30; B427	---	---
C91700	---	---	B30; B427	---	---
C92200	---	MIL-V-17547	B30; B271; B505; B584 (formerly B143)	SB61; SB584	---
C92300	---	---	B30; B271; B505; B584 (formerly B143); B763	---	---
C92500	---	---	B30; B505	---	---
C92600	---	---	B584	---	---
C92700	---	---	B30; B505	---	---
C92800	7320	---	B30; B505	---	---
C92900	---	---	B30; B427; B505	---	---
C93200	---	---	B30; B271; B505; B584 (formerly B144); B763	---	---
C93400	---	---	B30; B505	---	---
C93500	---	---	B30; B271; B505; B584 (formerly B144); B763	---	---
C93600	---	---	B30; B271; B505	---	---
C93700	4842; 4827	---	B30; B271; B505; B584 (formerly B144); B763	SB584	---
C93800	---	---	B30; B66; B271; B505; B584; B763	---	---
C93900	---	---	B30; B505	---	---

AMERICAN CROSS REFERENCED SPECIFICATIONS - COPPER & COPPER ALLOYS (Continued)

UNS	AMS	MIL SPEC	ASTM	ASME	AWS
C94000	---	---	B30; B505	---	---
C94100	---	---	B30; B67; B505	---	---
C94300	---	---	B30; B66; B271; B505; B584 (formerly B144); B763	---	---
C94400	---	---	B30; B66	---	---
C94500	---	---	B30; B66	---	---
C94700	---	---	B30, B505; B584 (formerly B292); B763	---	---
C94800	---	---	B30; B505; B584 (formerly B292); B763	---	---
C94900	---	---	B30; B584 (formerly B292); B763	---	---
C95200	---	---	B30; B148; B271; B505; B763	B16.24; SB148; SB505; SB271	---
C95300	---	MIL-C-11866 (22)	B30; B148; B271; B505; B763; B806	---	---
C95400	---	MIL-C-11866 (Comp 23)	B30; B148; B271; B505; B763; B806	SB148; SB271	---
C95410	---	---	B30; B148; B271; B505; B763; B806	---	---
C95420	4870; 4871; 4872; 4873	---	---	---	---
C95500	---	---	B30; B148; B271; B505; B763; B806	---	---
C95510	4880	---	---	---	---
C95520	4881	---	---	---	---
C95600	---	---	B30; B148; B763	---	---
C95700	---	---	B30; B148; B505	---	---
C95800	---	MIL-B-24480	B30; B148; B271; B505; B763; B806	---	---
C95900	---	---	B148; B505	---	---
C96200	---	MIL-V-18436	B30; B369	---	---
C96400	---	---	B30; B369; B505	---	---
C96700	---	---	B30 (72C); B770	---	---

AMERICAN CROSS REFERENCED SPECIFICATIONS - COPPER & COPPER ALLOYS (Continued)

UNS	AMS	MIL SPEC	ASTM	ASME	AWS
C96800	---	---	B30; B584	---	---
C97300	---	---	B30; B271; B505; B584 (formerly B149); B763	---	---
C97600	---	MIL-V-18436	B30; B271; B505; B584 (formerly B149); B763	SB584	---
C97800	---	MIL-V-18436	B30; B271; B505; B584 (formerly B149); B763	---	---
C98200	4824	---	---	---	---
C98400	4822	MIL-B-13506	---	---	---
C99400	---	---	B30; B763	---	---
C99500	---	---	B30; B763	---	---
C99700	---	---	B30; B176	---	---
C99750	---	---	B30; B176	---	---

This cross-reference table lists the basic specification or standard number, and since these standards are constantly being revised, it should be kept in mind that they are presented herein as a guide and may not reflect the latest revision.

COMMON NAMES OF COPPER & COPPER ALLOYS WITH UNS No.

Common Name	UNS	Common Name	UNS
Admiralty, Antimonial (Cu-Zn-Sn)	C44400	Aluminum Bronze	C63000
Admiralty, Arsenical (Cu-Zn-Sn)	C44300	Aluminum Bronze	C63200
Admiralty, Phosphorized (Cu-Zn-Sn)	C44500	Aluminum Bronze	C63800
Aluminum Brass, Arsenical	C68700	Aluminum Bronze	C64200
Aluminum Bronze	C60800	Aluminum Bronze D	C61400
Aluminum Bronze	C61000	Aluminum Silicon Bronze	C64210
Aluminum Bronze	C61300	Architectural Bronze	C38500
Aluminum Bronze	C61800	Beryllium Copper	C17000
Aluminum Bronze	C61900	Beryllium Copper	C17200
Aluminum Bronze	C62300	Beryllium Copper	C17500
Aluminum Bronze	C62400	Beryllium Copper with Lead	C17300

COMMON NAMES OF COPPER & COPPER ALLOYS WITH UNS No. (Continued)

Common Name	UNS	Common Name	UNS
Brass	C27200	Cast Copper Alloy	C99500
Brass Alloy	C47940	Cast Copper Alloy	C99700
Bronze, Low Fuming	C68100	Cast Copper Alloy	C99750
Bronze, Low Fuming (Nickel)	C68000	Cast Copper Nickel Beryllium Alloy	C96700
Cadmium Copper	C16200	Cast Copper-Nickel	C96200
Cadmium Copper Alloy	C16500	Cast Copper-Nickel	C96400
Cartridge Brass, 70%	C26000	Cast Copper-Nickel Alloy	C96800
Cast Aluminum Bronze	C95200	Cast High-Leaded Tin Bronze	C94400
Cast Aluminum Bronze	C95300	Cast High-Leaded Tin Bronze	C94500
Cast Aluminum Bronze	C95400	Cast High-Leaded Tin Bronze	C93200
Cast Aluminum Bronze	C95410	Cast High-Leaded Tin Bronze	C93400
Cast Aluminum Bronze	C95420	Cast High-Leaded Tin Bronze	C93500
Cast Aluminum Bronze	C95500	Cast High-Leaded Tin Bronze	C93600
Cast Aluminum Bronze	C95510	Cast High-Leaded Tin Bronze	C93700
Cast Aluminum Bronze	C95520	Cast High-Leaded Tin Bronze	C93720
Cast Aluminum Bronze	C95600	Cast High-Leaded Tin Bronze	C93800
Cast Aluminum Bronze	C95700	Cast High-Leaded Tin Bronze	C93900
Cast Aluminum Bronze	C95800	Cast High-Leaded Tin Bronze	C94000
Cast Aluminum Bronze	C95900	Cast High-Leaded Tin Bronze	C94100
Cast Beryllium Copper	C82000	Cast High-Leaded Tin Bronze	C94300
Cast Beryllium Copper, 165C	C82400	Cast High-Leaded Tin Bronze	C94320
Cast Beryllium Copper, 20C	C82500	Cast High-Leaded Tin Bronze	C94330
Cast Beryllium Copper, 245C	C82600	Cast Leaded Copper	C98200
Cast Beryllium Copper, 275C	C82800	Cast Leaded Copper	C98400
Cast Beryllium Copper, 35B, 35C	C82200	Cast Leaded Copper	C98820
Cast Chromium Copper, 70C	C81400	Cast Leaded Copper	C98840
Cast Copper Alloy	C99400	Cast Leaded Manganese Bronze	C86400

COMMON NAMES OF COPPER & COPPER ALLOYS WITH UNS No. (Continued)

Common Name	UNS	Common Name	UNS
Cast Leaded Manganese Bronze	C86700	Cast Semi-Red Brass	C84800
Cast Leaded Red Brass	C83600	Cast Silicon Brass	C87400
Cast Leaded Red Brass	C83800	Cast Silicon Brass	C87500
Cast Leaded Tin Bronze	C87610	Cast Silicon Brass	C87600
Cast Leaded Tin Bronze	C92200	Cast Silicon Bronze	C87800
Cast Leaded Tin Bronze	C92300	Cast Tin Bronze	C91600
Cast Leaded Tin Bronze	C92500	Cast Tin Bronze	C91700
Cast Leaded Tin Bronze	C92600	Cast Tin Bronze	C90300
Cast Leaded Tin Bronze	C92700	Cast Tin Bronze	C90500
Cast Leaded Tin Bronze	C92800	Cast Tin Bronze	C90700
Cast Leaded Tin Bronze	C92900	Cast Tin Bronze	C90800
Cast Leaded Yellow Brass	C85200	Cast Tin Bronze	C91000
Cast Leaded Yellow Brass	C85400	Cast Tin Bronze	C91100
Cast Leaded Yellow Brass	C85700	Cast Tin Bronze	C91300
Cast Leaded Yellow Brass	C85800	Chemically Refined Tough Pitch Copper	C11030
Cast Manganese Bronze	C86300	Chromium Copper	C18200
Cast Manganese Bronze	C86500	Commercial Bronze, 90%	C22000
Cast Manganese Bronze	C86200	Copper Alloy	C66400
Cast Nickel-Silver	C97300	Copper Alloy	C68800
Cast Nickel-Silver	C97600	Copper Alloy	C69100
Cast Nickel-Silver	C97800	Copper Beryllium Alloy	C17510
Cast Nickel-Tin Bronze	C94700	Copper Brazing Filler Metal	C14180
Cast Nickel-Tin Bronze	C94800	Copper Silicon	C65400
Cast Nickel-Tin Bronze	C94900	Copper Silicon Alloy	C87300
Cast Red Brass	C83400	Copper Vacuum Grade Brazing Filler Metal	C14181
Cast Semi-Red Brass	C84200	Copper Welding Filler Metal	C18980
Cast Semi-Red Brass	C84400	Copper-Beryllium	C82510

COMMON NAMES OF COPPER & COPPER ALLOYS WITH UNS No. (Continued)

Common Name	UNS	Common Name	UNS
Copper-Beryllium Alloy (Beryllium-Copper)	C17410	Electronic/Electrowon Tough Pitch Copper	C11040
Copper-Nickel	C71580	Fine Refined High Conductivity Tough Pitch Copper	C11020
Copper-Nickel	C71640	Fire-Refined Tough Pitch Copper with Silver (FRSTP)	C12900
Copper-Nickel	C72200	Forging Brass	C37700
Copper-Nickel	C72500	Free Cutting Brass	C36000
Copper-Nickel	C70200	Free Cutting Muntz Metal (Cu-Zn-Pb)	C37000
Copper-Nickel Tin Spinodal Alloy	C72700	Gilding, 95% (Copper/Zinc)	C21000
Copper-Nickel Tin Spinodal Alloy	C72900	High Copper Alloy	C19200
Copper-Nickel Welding Filler Metal	C71581	High Copper Alloy	C19400
Copper-Nickel, 10%	C70600	High Copper Alloy	C19500
Copper-Nickel, 20%	C71000	High Leaded Brass (Tube)	C33200
Copper-Nickel, 30%	C71500	High Leaded Brass, 62%	C35300
Copper-Nickel, 5%	C70400	High Leaded Brass, 64-½%	C34200
Copper-Nickel, 70Cu-30Ni	C71590	High Silicon Bronze A	C65500
Copper-Nickel, 90Cu-10Ni	C70690	Jewelry Bronze, 87-½%	C22600
Copper-Phosphorus Brazing Filler Metal	C55180	Leaded Brass	C34500
Copper-Phosphorus Brazing Filler Metal	C55181	Leaded Brass	C38000
Copper-Tellurium	C14520	Leaded Brass Alloy	C35600
Copper-Tellurium Alloy	C14510	Leaded Commercial Bronze	C31400
Copper-Tellurium-Tin Alloy	C14420	Leaded Commercial Bronze (Nickel Bearing)	C31600
Cu-Ag-P Brazing Filler Metal	C55280	Leaded Copper	C18700
Cu-Ag-P Brazing Filler Metal	C55281	Leaded Muntz Metal, Uninhibited (Cu-Zn-Pb)	C36500
Cu-Ag-P Brazing Filler Metal	C55282	Leaded Nickel silver	C79200
Cu-Ag-P Brazing Filler Metal	C55283	Leaded Red Brass	C32000
Cu-Ag-P Brazing Filler Metal	C55284	Leaded Red Brass	C83450
Electrolytic Tough Pitch Anneal Resistant	C11100	Leaded Silicon Brass	C69700
Electrolytic Tough Pitch Copper (ETP)	C11000	Low Brass, 80%	C24000

COMMON NAMES OF COPPER & COPPER ALLOYS WITH UNS No. (Continued)

Common Name	UNS	Common Name	UNS
Low Leaded Brass	C33500	Nickel Silver	C77400
Low Leaded Brass (Tube)	C33000	Nickel Silver, 55-18	C77000
Low Silicon Bronze B	C65100	Nickel Silver, 65-10	C74500
Manganese Aluminum Bronze Welding Filler Metal	C63380	Nickel Silver, 65-12	C75700
Manganese Brass	C66700	Nickel Silver, 65-18	C75200
Manganese Bronze	C67300	Oxygen-Free Copper (OF)	C10200
Manganese Bronze	C67400	Oxygen-Free Copper with Silver (OFS)	C10400
Manganese Bronze A	C67500	Oxygen-Free Copper with Silver (OFS)	C10500
Manganese Bronze B	C67000	Oxygen-Free Copper with Silver (OFS)	C10700
Medium Leaded Brass, 62%	C35000	Oxygen-Free Electronic Copper (OFE)	C10100
Medium Leaded Brass, 64-½%	C34000	Oxygen-Free Extra Low Phosphorus Copper FXLP)	C10300
Misc. Copper Zinc Alloy	C66410	Oxygen-Free Low Phosphorus Copper (OFLP)	C10800
Muntz Metal, 60%	C28000	Phosphor Bronze	C50100
Naval Brass, 63-½%	C46200	Phosphor Bronze	C50200
Naval Brass, Arsenical	C46500	Phosphor Bronze	C50700
Naval Brass, High Leaded	C48500	Phosphor Bronze	C51100
Naval Brass, Medium Leaded	C48200	Phosphor Bronze	C51800
Naval Brass, Uninhibited	C46400	Phosphor Bronze B-1	C53400
Naval Brass, Welding and Brazing Rod	C47000	Phosphor Bronze B-2	C54400
Nickel Aluminum Bronze Welding Filler Metal	C63280	Phosphor Bronze, 1.25% E	C50500
Nickel Brass Welding and Brazing Filler Metal	C77300	Phosphor Bronze, 10% D	C52400
Nickel Silver	C73500	Phosphor Bronze, 5% A	C51000
Nickel Silver	C74000	Phosphor Bronze, 8%	C52100
Nickel Silver	C76200	Phosphorus Deoxidized Copper, Arsenical (DPA)	C14200
Nickel Silver	C76400	Phosphorus Deoxidized Copper, Tellurium Bearing	C14500

COMMON NAMES OF COPPER & COPPER ALLOYS WITH UNS No. (Continued)

Common Name	UNS	Common Name	UNS
Phosphorus Deoxidized, High Residual Phosphorus Copper (DHP)	C12200	Tin Brass	C41500
Phosphorus Deoxidized, Low Residual Phosphorus Copper (DLP)	C12000	Tin Brass	C42200
Silicon Brass	C69710	Tin Brass	C42500
Silicon Bronze	C64700	Tin Brass	C43000
Silicon Bronze	C65600	Tin Brass	C43400
Silicon Bronze	C66100	Tough Pitch Copper with Silver (STP)	C11300
Silicon Red Brass	C69400	Tough Pitch Copper with Silver (STP)	C11400
Silicon Red Brass	C69430	Tough Pitch Copper with Silver (STP)	C11500
Silver-Bearing, High Residual Phosphorus Copper (DHP)	C12300	Tough Pitch Copper with Silver (STP)	C11600
Silver-Bearing, Low Residual Phosphorus Copper (DLP)	C12100	Yellow Brass, 63%	C27400
Sulfur-Bearing Copper	C14700	Yellow Brass, 65%	C27000
Tin Brass	C40500	Yellow Brass, 66%	C26800
Tin Brass	C41100	Zirconium Copper	C15000
Tin Brass	C41300	Zirconium Copper	C15100

BRITISH BSI SPECIFICATIONS - COPPER & COPPER ALLOYS

BS	Title
61	Threads for Light Gauge Copper Tubes and Fittings
125	Hard-Drawn Copper and Copper-Cadmium Conductors for Overhead Power Transmission Purposes
159	High-Voltage Busbars and Busbar Connections
181	Copper-Cadmium Jointing Sleeves for Telegraph and Telephone Purposes
1184	Copper and Copper Alloy Traps
1306	Copper and Copper Alloy Pressure Piping Systems
1400	Copper Alloy Ingots and Copper Alloy and High Conductivity Copper Castings
1431	Wrought Copper and Wrought Zinc Rainwater Goods
1432	Copper for Electrical Purposes : High Conductivity Copper Rectangular Conductors with Drawn or Rolled Edges

BRITISH BSI SPECIFICATIONS - COPPER & COPPER ALLOYS (Continued)

BS	Title
1433	Copper for Electrical Purposes. Rod and Bar
1434	Copper for Electrical Purposes: Copper Sections in Bars, Blanks and Segments for Commutators
1977	High Conductivity Copper Tubes for Electrical Purposes
2755	Copper and Copper- Cadmium Stranded Conductors for Overhead Electric Traction Systems
2767	Manually Operated Copper Alloy Valves for Radiators
2870	Rolled Copper and Copper Alloys. Sheet, Strip and Foil
2871 Part 1	Copper and Copper Alloys. Tubes Part 1: Copper Tubes for Water, Gas and Sanitation
2871 Part 2	Copper and Copper Alloys. Tubes Part 2: Tubes for General Purposes
2871 Part 3	Copper and Copper Alloys. Tubes Part 3: Tubes for Heat Exchangers
2872	Copper and Copper Alloy Forging Stock and Forgings
2873	Copper and Copper Alloys. Wire
2874	Copper and Copper Alloy Rods and Sections (Other Than Forging Stock)
2875	Copper and Copper Alloys. Plate
2901 Part 3	Filler Rods and Wires for Gas-Shielded Arc Welding Part 3: Specification for Copper and Copper Alloys
3071	Nickel-Copper Alloy Castings
3839	Oxygen-Free High- Conductivity Copper for Electronic Tubes and Semi-Conductor Devices
4608	Copper for Electrical Purposes. Rolled Sheet, Strip and Foil
5561	Spools for Copper Wire
5624	The Lining of Vessels and Equipment for Chemical Processes: Copper and Copper Alloys
5899	Method for Hydrogen Embrittlement Test for Copper
5909	Scale Adhesion Test for Oxygen-Free Copper
6017	Copper Refinery Shapes
6926	Copper for Electrical Purposes: High Conductivity Copper Wire Rod
6931	Glossary of Terms for Copper and Copper Alloys
7428	Estimating the Average Grain Size of Copper and Copper Alloys
CP143 Part 12	Sheet Roof and Wall Coverings Part 12: Copper. Metric Units
MA 60	Summary and Application of Cooper and Copper Alloy Tubes for Ships' Pipework Systems

BRITISH BSI SPECIFICATIONS - COPPER & COPPER ALLOYS (Continued)	
BS	**Title**
MA 75 Part 1	Bulkhead Pieces and Tank Pads Part 1: Bulkhead Pieces of Fabricated Steel or Copper Alloy and Steel Pad Pieces
B 23	Copper-Aluminium-Nickel-Iron Alloy Rods, Sections, Forging Stock And Forgings
B 24	Copper-Tin-Phosphorous Alloy Rods and Sections
B 25	Copper-Nickel-Silicon Alloy Rods, Sections, Forging Stock and Forgings (Heat Treated)
B 26	Copper-Nickel-Silicon Alloy Rods and Sections (Cold Worked and Heat Treated)
B 27	Copper-Zinc-Aluminium- Nickel-Silicon Alloy Tube
B 28	Copper-Beryllium Alloy Strip, Foil and Parts (Solution Treated (W) and Precipitated)
B 29	Copper-Beryllium Alloy Strip, Foil and Parts (Solution Treated, Cold Rolled: Quarter Hard (W(1/4H)) and Precipitated)
B 30	Copper-Beryllium Alloy Strip, Foil and Parts (Solution Treated, Cold Rolled: Half Hard (W(½H)) and Precipitated)
B 31	Copper-Beryllium Alloy Strip, Foil and Parts (Solution Treated, Cold Rolled: Full Hard (W(H)) and Precipitated)
B 32	Copper-Beryllium Alloy Rods, Sections and Parts (Solution Treated (W) and Precipitated)
B 33	Copper-Beryllium Alloy Wire and Springs (Solution Treated (W) and Precipitated)
B 100	Inspection, Testing and Acceptance of Wrought Copper Alloys
2B 8	Specification For Aircraft Material. Phosphor Bronze Casting for Bearings.
HC 502	Copper Base Aluminium- Nickel-Iron Alloy Castings (620 MPa) (Al 9.5,Ni 5, Fe 5)
HR 205	Nickel-Copper Alloy Sheet and Strip: Annealed (Nickel Base, Cu 31)

GERMAN DIN SPECIFICATIONS - COPPER & COPPER ALLOYS	
DIN	**Title**
1705	Copper-Tin and Copper- Tin-Zinc Casting Alloys; (Cast Tin Bronze and Gunmetal); Castings
1705 Suppl. 1	Copper-Tin and Copper- Tin-Zinc Casting Alloys; (Cast Tin Bronze and Gunmetal); Castings; Reference Data on Mechanical and Physical Properties
1708	Copper; Cathodes and Refinery Shapes
1714	Copper-Aluminium Casting Alloys; (Cast Aluminium Bronze); Castings
1714 Suppl. 1	Copper-Aluminium Casting Alloys; (Cast Aluminium Bronze); Castings; Reference Data on Mechanical and Physical Properties
1716	Copper-Lead-Tin Casting Alloys; (Cast Tin-Lead Bronze); Castings

GERMAN DIN SPECIFICATIONS - COPPER & COPPER ALLOYS (Continued)

DIN	Title
1716 Suppl. 1	Copper-Lead-Tin Casting Alloys; (Cast Tin-Lead Bronze); Castings; Reference Data on Mechanical and Physical Properties
1718	Copper Alloys; Definitions
1733 Pt 1	Filler Metals for Welding Copper and Copper Alloys; Composition, Application and Technical Delivery Conditions
1751	Sheet and Sheet Cut to Length of Copper and Wrought Copper Alloys; Cold Rolled; Dimensions
1754 Pt 1	Copper Tubes; Seamless Drawn; Dimension Ranges and Coordination of Tolerances
1754 Pt 2	Copper Tubes; Seamless Drawn; Preferred Dimensions for General Purposes
1754 Pt 3	Copper Tubes; Seamless Drawn; Preferred Dimensions for Pipelines
1755 Pt 1	Wrought Copper Alloy Tubes; Seamless Drawn; Dimension Ranges and Coordination of Tolerances
1755 Pt 2	Copper Wrought Alloy Tubes; Seamless Drawn; Preferred Dimensions for General Purposes
1755 Pt 3	Copper Wrought Alloy Tubes; Seamless Drawn; Preferred Dimensions for Pipelines
1756	Round Rod of Copper and Wrought Copper Alloys; Drawn; Dimensions
1757	Wire of Copper and Wrought Copper Alloys, Drawn; Dimensions
1759	Rectangular Bars of Copper and Wrought Copper Alloys; Drawn, with Sharp Edges; Dimensions, Permissible Variations, Static Values
1761	Square Rod of Copper and Copper Wrought Alloys; Drawn, with Sharp Edges; Dimensions
1763	Hexagon Rod of Copper and Wrought Copper Alloys; Drawn, with Sharp Edges; Dimensions
1777	Wrought Copper Alloy Strip for Springs; Technical Delivery Conditions
1782	Round Rod of Copper and Wrought Copper Alloys; Extruded; Dimensions
1785	Wrought Copper and Copper Alloy Tubes for Condensers and Heat Exchangers
1786	Seamless Drawn Copper Tubes for Piping Systems
1787	Copper; Half-Finished Products
1791	Strip and Strip Cut to Length of Copper and Wrought Copper Alloys; Cold Rolled; Dimensions
1850 Pt 1	Bushes for Plain Bearings; Made from Copper Alloys, Solid
3352 Pt 11	Flanged Copper Alloy Gate Valves
3352 Pt 12	Socket End Copper Alloy Gate Valves
8513 Pt 1	Brazing and Braze Weld Filler Metals; Copper Base Brazing Alloys; Composition, Use, Technical Conditions of Delivery
8552 Pt 3	Edge Preparation for Welding; Groove Forms on Copper and Copper Alloys; Gas Welding and Gas-Shielded Arc Welding
17650	Copper Sheet and Strip for Use in Building Construction; Technical Delivery Conditions

GERMAN DIN SPECIFICATIONS - COPPER & COPPER ALLOYS (Continued)	
DIN	**Title**
17652	Copper Drawing Stock
17655	Unalloyed and Low Alloy Copper Materials for Casting; Castings
17656	Copper Casting Alloys; Ingot Metals; Composition
17657	Copper Master Alloys; Composition Copper-Nickel Casting Alloys; Castings
17660	Wrought Copper Alloys; Copper-Zinc Alloys; (Brass); (Special Brass); Composition
17662	Wrought Copper Alloys; Copper-Tin Alloys; (Tin Bronze); Composition
17663	Wrought Copper Alloys; Copper-Nickel-Zinc Alloys; (Nickel Silver); Composition
17664	Wrought Copper Alloys; Copper-Nickel Alloys; Composition
17665	Wrought Copper Alloys; Copper-Aluminium Alloys; (Aluminium Bronze); Composition
17666	Low Alloy Wrought Copper Alloys;
17670 Pt 1	Wrought Copper and Copper Alloy Plate, Sheet and Strip; Properties
17670 Pt 2	Plate, Sheet and Strip of Copper and Wrought Copper Alloys; Technical Conditions of Delivery
17671 Pt 1	Wrought Copper and Copper Alloy Tubes; Properties
17671 Pt 2	Tubes of Copper and Wrought Copper Alloys; Technical Conditions of Delivery
17672 Pt 1	Wrought Copper and Copper Alloy Rod and Bar; Properties
17672 Pt 2	Bars of Copper and Wrought Copper Alloys; Technical Conditions of Delivery
17673 Pt 1	Wrought Copper and Copper Alloy Drop Forgings; Properties
17673 Pt 2	Drop Forgings of Copper and Wrought Copper Alloys; Technical Conditions of Delivery
17673 Pt 3	Drop Forgings of Copper and Wrought Copper Alloys; Design Principles
17673 Pt 4	Drop Forgings of Copper and Wrought Copper Alloys; Permissible Variations
17674 Pt 1	Wrought Copper and Copper Alloy Extruded Sections; Properties
17674 Pt 2	Copper and Wrought Copper Alloy Extruded Sections; Technical Conditions of Delivery
17674 Pt 3	Copper and Wrought Copper Alloy Extruded Sections; Design
17674 Pt 4	Copper and Copper Wrought Alloy Extruded Sections; Extruded, Permissible Variations
17674 Pt 5	Copper and Wrought Copper Alloy Extruded Sections; Drawn; Permissible Variations
17675 Pt 1	Plates of Copper and Wrought Copper Alloys for Condensers and Heat Exchangers; Strength Properties
17675 Pt 2	Plates of Copper and Wrought Copper Alloys for Condensers and Heat Exchangers; Technical Conditions of Delivery

GERMAN DIN SPECIFICATIONS - COPPER & COPPER ALLOYS (Continued)

DIN	Title
17675 Pt 3	Plates of Copper and Wrought Copper Alloys for Condensers and Heat Exchangers; Dimensions
17677 Pt 1	Wrought Copper and Copper Alloy Wire; Properties
17677 Pt 2	Wires of Copper and Wrought Copper Alloys; Technical Conditions of Delivery
17678 Pt 1	Wrought Copper and Copper Alloy Hand Forgings; Properties
17678 Pt 2	Hand Forgings of Copper and Wrought Copper Alloys; Technical Conditions of Delivery
17678 Pt 3	Hand Forgings of Copper and Wrought Copper Alloys; Basis of Design
17678 Pt 4	Hand Forgings of Copper and Wrought Copper Alloys; Permissible Variations
17679	Wrought Copper and Copper Alloy Tubes with Rolled Fins for Use in Heat Exchangers
17682	Round Spring Wires Made of Wrought Copper Alloys; Strength Properties; Technical Conditions of Delivery
40500 Pt 4	Copper for Electrical Engineering; Wires of Copper and Copper-Silver Alloy; Technical Conditions of Delivery
40500 Pt 5	Copper for Electrical Purposes; Tinned Wire; Technical Delivery Conditions
46211	Stamped Cable Sockets for Copper Conductors
46415	Copper and Copper Alloy Band; Cold Rolled with Rounded (Lightly Rolled) Edges; Dimensions
46416 Pt 4	Winding Wires; Round Copper Wires, Insulated; Enameled, Heat Resistant with a Temperature Index of 155, Type W 155, Technical Conditions of Delivery
46431	Round Copper Wires for Electrical Purposes; Precision Drawn; Dimensions
48203 Pt 1	Copper Wires and Copper Stranded Conductors; Technical Delivery Conditions
48203 Pt 2	Wrought Copper Alloy (Bz) Wires and Conductors; Technical Delivery Conditions
48203 Pt 7	Copper Covered Steel Wires and Copper Covered Steel Stranded Conductors; Technical Delivery Conditions
50916 Pt 2	Testing of Copper Alloys; Stress Corrosion Cracking Test Using Ammonia; Testing of Components
86086	Copper-Nickel Alloy Welding Pipe Fittings; Technical Delivery Conditions
LN 29557 Pt 2	Aerospace; Laminated Shim of Wrought Copper Alloy; Dimensions, Masses

JAPANESE JIS SPECIFICATIONS - COPPER AND COPPER ALLOYS	
Wrought Alloys	
JIS	**Title**
H 3100	Copper and Copper Alloy Sheets, Plates and Strips
H 3110	Phosphor Bronze and Nickel Silver Sheets, Plates and Strips
H 3130	Copper Beryllium Alloy, Phosphor Bronze and Nickel Silver Sheets, Plates and Strips for Springs
H 3140	Copper Bus Bars
H 3250	Copper and Copper Alloy Rods and Bars
H 3260	Copper and Copper Alloy Wires
H 3270	Copper Beryllium Alloy, Phosphor Bronze and Nickel Silver Rods, Bars and Wires
H 3300	Copper and Copper Alloy Seamless Pipes and Tubes
H 3320	Copper and Copper Alloy Welded Pipes and Tubes
H 3401	Pipe Fittings of Copper and Copper Alloy
H 3510	Oxygen Free Copper Sheet, Plate, Strip, Seamless Pipe and Tube, Rod, Bar and Wire for Electron Devices
Castings	
H 5100	Copper Castings
H 5101	Brass Castings
H 5102	High Strength Brass Castings
H 5111	Bronze Castings
H 5112	Silicon Bronze Castings
H 5113	Phosphor Bronze Castings
H 5114	Aluminum Bronze Castings
H 5115	Leaded Tin Bronze Castings

CHEMICAL COMPOSITION OF COPPER & COPPER ALLOYS	
UNS	**Chemical Composition**
C10100	Cu 99.99 min Ag 0.0025 max As 0.0005 max Bi 0.0001 max Cd 0.0001 max Fe 0.0010 max Hg 0.0001 max Mn 0.0005 max Ni 0.0010 max O_2 0.0005 max P 0.0003 max Pb 0.0005 max S 0.0018 max Sb 0.0004 max Se 0.0010 max Sn 0.0002 max Te 0.0002 max Zn 0.0001 max Other total As, Bi, Mn, Sb, Se, Sn, and Te 0.0040 max
C10200	Cu (incl Ag) 99.95 min
C10300	Cu (incl Ag) 99.95 min P 0.001-0.005
C10400	Cu (incl Ag) 99.95 min Ag 0.027 min
C10500	Cu (incl Ag) 99.95 min Ag 0.034 min
C10700	Cu (incl Ag) 99.95 min Ag 0.085 min
C10800	Cu (incl Ag+P) 99.95 min P 0.005-0.012
C11000	Cu (incl Ag) 99.90 min
C11010	Cu (incl Ag) 99.90 min Other Unspecified oxygen and trace elements
C11020	Cu (incl Ag) 99.90 min Other Unspecified oxygen and trace elements
C11030	Cu (incl Ag) 99.90 min Other Unspecified oxygen and trace elements
C11040	Cu 99.90 min Ag 0.0025 max As 0.0005 max Bi 0.00010 max Fe 0.0010 max Ni 0.0010 max O_2 0.010-0.065 Pb 0.0005 max S 0.0015 max Sb 0.0004 max Se 0.0002 max Sn 0.0005 max Te 0.0002 max Other Bi+Se+Te 0.0003 max total, Total all (except Cu+O_2) 0.0065 max
C11100	Cu (incl Ag) 99.90 min Note: Small amounts of Cd or other elements may be added to improve the resistance to softening at elevated temperatures
C11300	Cu (incl Ag) 99.90 min Ag 0.027 min Note: Oxygen and trace elements may vary depending on the process
C11400	Cu (incl Ag) 99.90 min Ag 0.034 min Note: Oxygen and trace elements may vary depending on the process
C11500	Cu (incl Ag) 99.90 min Ag 0.054 min Note: Oxygen and trace elements may vary depending on the process
C11600	Cu (incl Ag) 99.90 min Ag 0.085 min Note: Oxygen and trace elements may vary depending on the process
C12000	Cu (incl Ag) 99.90 min P 0.004-0.012
C12100	Cu (incl Ag) 99.90 min Ag 0.014 min P 0.005-0.012
C12200	Cu (incl Ag) 99.90 min P 0.015-0.040
C12300	Cu (incl Ag) 99.90 min P 0.015-0.040
C12900	Cu (incl Ag) 99.88 min Ag 0.054 min As 0.012 max Bi 0.003 max Ni 0.050 max Pb 0.004 max Sb 0.003 max Te 0.025 max Other Se included in Te
C14180	Cu (incl Ag) 99.90 min Al 0.01 max P 0.075 max Pb 0.02 max
C14181	Cu (incl Ag) 99.90 min C 0.005 max Cd 0.002 max P 0.002 max Pb 0.002 max Zn 0.002

CHEMICAL COMPOSITION OF COPPER & COPPER ALLOYS (Continued)

UNS	Chemical Composition
C14200	Cu (incl Ag) 99.4 min As 0.15-0.50 P 0.015-0.040
C14420	Cu (incl Ag+Sn+Te) 99.90 min Sn 0.05-0.15 Te 0.02-0.05
C14500	Cu (incl Ag+Te) 99.90 min P 0.004-0.012 Te 0.40-0.70 Note: Other deoxidizers may be used instead of P
C14510	Cu (incl Ag+Te) 99.90 min P 0.010-0.030 Pb 0.05 max Te 0.30-0.70
C14520	Cu (incl Ag+Te) 99.40 min P 0.004-0.020 Te 0.40-0.7
C14700	Cu (incl Ag+P+S) 99.90 min S 0.20-0.50 P 0.002-0.005 Note: For oxygen-free and deoxidized grades: O, P, B, Li, etc., limits as agreed upon
C15000	Cu (incl Ag) 99.80 min Zr 0.10-0.20 Other Cu+sum of named elements 99.9 min
C15100	Cu (incl Ag) 99.82 min Al 0.005 max Mn 0.005 max Zr 0.05-0.15
C16200	Cu (incl Ag) rem Cd 0.7-1.2 Fe 0.02 max Other Cu+sum of named elements 99.5 min
C16500	Cu (incl Ag) rem Cd 0.6-1.0 Fe 0.02 max Sn 0.50-0.7 Other Cu+sum of named elements 99.5 min
C17000	Cu (incl Ag) rem Al 0.20 max Be 1.60-1.79 Si 0.20 max; Ni+Co 0.20 min Other Ni+Fe+Co 0.6 max; Cu+sum of named elements 99.5 min
C17200	Cu (incl Ag) rem Al 0.20 max Be 1.80-2.00 Si 0.20 max; Ni+Co 0.20 min Other Ni+Fe+Co 0.6 max; Cu+sum of named elements 99.5 min
C17300	Cu (incl Ag) rem Al 0.20 max Be 1.80-2.00 Pb 0.20-0.6 Si 0.20 max; Ni+Co 0.20 min Other Ne+Fe+Co 0.6 max; Cu+sum of named elements 99.5 min
C17410	Cu (incl Ag) rem Al 0.20 max Be 0.15-0.50 Co 0.35-0.6 Fe 0.20 max Si 0.20 max Other Cu+sum of named elements 99.5 min
C17500	Cu (incl Ag) rem Be 0.40-0.7 Co 2.4-2.7 Fe 0.10 max Other Cu+sum of named elements 99.5 min
C17510	Cu (incl Ag) rem Al 0.20 max Be 0.20-0.6 Co 0.30 max Fe 0.10 max Ni 1.4-2.2 Other Cu+sum of named elements 99.5 min
C18200	Cu (incl Ag) rem Cr 0.6-1.2 Fe 0.10 max Pb 0.05 max Si 0.10 max Other Cu+sum of named elements 99.5 min
C18700	Cu (incl Ag) rem Pb 0.8-1.5 Other Cu+sum of named elements 99.5 min
C18980	Cu (incl Ag) 98.0 min Mn 0.50 max P 0.15 max Pb 0.02 max Si 0.50 max Sn 1.0 max Other Cu+sum of named elements 99.5 min
C19200	Cu 98.7 min Fe 0.8-1.2 P 0.01-0.04 Zn 0.20 max Other Cu+sum of named elements 99.8 min
C19400	Cu 97.0 min Fe 2.1-2.6 P 0.015-0.15 Pb 0.03 max Zn 0.05-0.20 Other Cu+sum of named elements 99.8 min
C19500	Cu 96.0 min Al 0.02 max Co 0.30-1.3 Fe 1.0-2.0 P 0.01-0.35 Pb 0.02 max Sn 0.10-1.0 Zn 0.20 max Other Cu+sum of named elements 99.8 min
C21000	Cu 94.0-96.0 Fe 0.05 max Pb 0.03 max Zn rem Other Cu+sum of named elements 99.8 min
C22000	Cu 89.0-91.0 Fe 0.05 max Pb 0.05 max Zn rem Other Cu+sum of named elements 99.8 min
C22600	Cu 86.0-89.0 Fe 0.05 max Pb 0.05 max Zn rem Other Cu+sum of named elements 99.8 min
C23000	Cu 84.0-86.0 Fe 0.05 max Pb 0.05 max Zn rem Other Cu+sum of named elements 99.8 min

CHEMICAL COMPOSITION OF COPPER & COPPER ALLOYS (Continued)	
UNS	**Chemical Composition**
C24000	Cu 78.5-81.5 Fe 0.05 max Pb 0.05 max Zn rem Other Cu+sum of named elements 99.8 min
C26000	Cu 68.5-71.5 Fe 0.05 max Pb 0.07 max Zn rem Other Cu+sum of named elements 99.7 min
C26800	Cu 64.0-68.5 Fe 0.05 max Pb 0.15 max Zn rem Other Cu+sum of named elements 99.7 min
C27000	Cu 63.0-68.5 Fe 0.07 max Pb 0.10 max Zn rem Other Cu+sum of named elements 99.7 min
C27200	Cu 62.0-65.0 Fe 0.07 max Pb 0.07 max Zn rem Other Cu+sum of named elements 99.7 min
C27400	Cu 61.0-64.0 Fe 0.05 max Pb 0.10 max Zn rem Other Cu+sum of named elements 99.7 min
C28000	Cu 59.0-63.0 Fe 0.07 max Pb 0.30 max Zn rem Other Cu+sum of named elements 99.7 min
C31400	Cu 87.5-90.5 Fe 0.10 max Ni 0.7 max Pb 1.3-2.5 Zn rem Other Cu+sum of named elements 99.6 min
C31600	Cu 87.5-90.5 Fe 0.10 max Ni 0.7-1.2 P 0.04-0.10 Pb 1.3-2.5 Zn rem Other Cu+sum of named elements 99.6 min
C32000	Cu 83.5-86.5 Fe 0.10 max Ni 0.25 max Pb 1.5-2.2 Zn rem Other Cu+sum of named elements 99.6 min
C33000	Cu 65.0-68.0 Fe 0.07 max Pb 0.25-0.7 Zn rem Other Cu+sum of named elements 99.6 min
C33200	Cu 65.0-68.0 Fe 0.07 max Pb 1.5-2.5 Zn rem Other Cu+sum of named elements 99.6 min
C33500	Cu 62.0-65.0 Fe 0.15 max Pb 0.25-0.7 Zn rem Other Cu+sum of named elements 99.6 min Note: Fe 0.010 max for flat products
C34000	Cu 62.0-65.0 Fe 0.15 max Pb 0.8-1.5 Zn rem Other Cu+sum of named elements 99.6 min Note: Fe 0.010 max for flat products
C34200	Cu 62.0-65.0 Fe 0.15 max Pb 1.5-2.5 Zn rem Other Cu+sum of named elements 99.6 min Note: Fe 0.010 max for flat products
C34500	Cu 62.0-65.0 Fe 0.15 max Pb 1.5-2.5 Zn rem Other Cu+sum of named elements 99.6 min
C35000	Cu 60.0-63.0 Fe 0.15 max Pb 0.8-2.0 Zn rem Other Cu+sum of named elements 99.6 min Note: For rod, Cu 61.0 min; for flat products, Fe 0.10 max
C35300	Cu 60.0-63.0 Fe 0.15 max Pb 1.5-2.5 Zn rem Other Cu+sum of named elements 99.5 min Note: For rod, Cu 61.0 min; for flat products, Fe 0.10 max
C35600	Cu 60.0-63.0 Fe 0.15 max Pb 2.0-3.0 Zn rem Other Cu+sum of named elements 99.5 min Note: For flat products, Fe 0.10 max
C36000	Cu 60.0-63.0 Fe 0.35 max Pb 2.5-3.7 Zn rem Other Cu+sum of named elements 99.5 min
C36500	Cu 58.0-61.0 Fe 0.15 max Pb 0.25-0.7 Sn 0.25 max Zn rem Other Cu+sum of named elements 99.6 min
C37000	Cu 59.0-62.0 Fe 0.15 max Pb 0.8-1.5 Zn rem Other Cu+sum of named elements 99.6 min
C37700	Cu 58.0-61.0 Fe 0.30 max Pb 1.5-2.5 Zn rem Other Cu+sum of named elements 99.5 min
C38000	Cu 55.0-60.0 Al 0.50 max Fe 0.35 max Pb 1.5-2.5 Sn 0.30 max Zn rem Other Cu+sum of named elements 99.5 min
C38500	Cu 55.0-59.0 Fe 0.35 max Pb 2.5-3.5 Zn rem Other Cu+sum of named elements 99.5 min
C40500	Cu 94.0-96.0 Fe 0.05 Pb 0.05 max Sn 0.7-1.3 Zn rem Other Cu+sum of named elements 99.7 min
C41100	Cu 89.0-92.0 Fe 0.05 max Pb 0.10 max Sn 0.30-0.7 Zn rem Other Cu+sum of named elements 99.7 min

CHEMICAL COMPOSITION OF COPPER & COPPER ALLOYS (Continued)

UNS	Chemical Composition
C41300	Cu 89.0-93.0 Fe 0.05 max Pb 0.10 max Sn 0.7-1.3 Zn rem Other Cu+sum of named elements 99.7 min
C41500	Cu 89.0-93.0 Fe 0.05 max Pb 0.10 max Sn 1.5-2.2 Zn rem Other Cu+sum of named elements 99.7 min
C42200	Cu 86.0-89.0 Fe 0.05 max P 0.35 max Pb 0.05 max Sn 0.8-1.4 Zn rem Other Cu+sum of named elements 99.7 min
C42500	Cu 87.0-90.0 Fe 0.05 max P 0.35 max Pb 0.05 max Sn 1.5-3.0 Zn rem Other Cu+sum of named elements 99.7 min
C43000	Cu 84.0-87.0 Fe 0.05 max Pb 0.10 max Sn 1.7-2.7 Zn rem Other Cu+sum of named elements 99.7 min
C43400	Cu 84.0-87.0 Fe 0.05 max Pb 0.05 max Sn 0.40-1.0 Zn rem Other Cu+sum of named elements 99.7 min
C44300	Cu 70.0-73.0 As 0.02-0.06 Fe 0.06 max Pb 0.07 max Sn 0.8-1.2 Zn rem Other For tubular products, Sn 0.9 min Other Cu+sum of named elements 99.6 min
C44400	Cu 70.0-73.0 Fe 0.06 max Pb 0.07 max Sb 0.02-0.10 Sn 0.8-1.2 Zn rem Other For tubular products, Sn 0.9 min Other Cu+sum of named elements 99.6 min
C44500	Cu 70.0-73.0 Fe 0.06 max P 0.02-0.10 Pb 0.07 max Sn 0.8-1.2 Zn rem Other For tubular products, Sn 0.9 min Other Cu+sum of named elements 99.6 min
C46200	Cu 62.0-65.0 Fe 0.10 max Pb 0.20 max Sn 0.50-1.0 Zn rem Other Cu+sum of named elements 99.6 min
C46400	Cu 59.0-62.0 Fe 0.10 max Pb 0.20 max Sn 0.50-1.0 Zn rem Other Cu+sum of named elements 99.6 min
C46500	Cu 59.0-62.0 As 0.02-0.06 Fe 0.10 max Pb 0.20 max Sn 0.50-1.0 Zn rem Other Cu+sum of named elements 99.6 min
C47000	Cu 57.0-61.0 Al 0.01 max Pb 0.05 max Sn 0.25-1.0 Zn rem Other Cu+sum of named elements 99.6 min
C47940	Cu 63.0-66.0 Fe 0.10-1.0 Ni (incl Co) 0.10-0.50 Pb 1.0-2.0 Zn rem
C48200	Cu 59.0-62.0 Fe 0.10 max Pb 0.40-1.0 Sn 0.50-1.0 Zn rem Other Cu+sum of named elements 99.6 min
C48500	Cu 59.0-62.0 Fe 0.10 max Pb 1.3-2.2 Sn 0.50-1.0 Zn rem Other Cu+sum of named elements 99.6 min
C50100	Cu rem Fe 0.05 max P 0.01-0.05 Pb 0.05 max Sn 0.50-0.8 Other Cu+sum of named elements 99.5 min
C50200	Cu rem Fe 0.10 max P 0.04 max Pb 0.05 max Sn 1.0-1.5 Other Cu+sum of named elements 99.5 min
C50500	Cu rem Fe 0.10 max P 0.03-0.35 Pb 0.05 max Sn 1.0-1.7 Zn 0.30 max Other Cu+sum of named elements 99.5 min
C50700	Cu rem Fe 0.10 max P 0.30 max Pb 0.05 max Sn 1.5-2.0 Other Cu+sum of named elements 99.5 min
C51000	Cu rem Fe 0.10 max P 0.03-0.35 Pb 0.05 max Sn 4.2-5.8 Zn 0.30 max Other Cu+sum of named elements 99.5 min
C51100	Cu rem Fe 0.10 max P 0.03-0.35 Pb 0.05 max Sn 3.5-4.9 Zn 0.30 max Other Cu+sum of named elements 99.5 min
C51800	Cu rem Al 0.01 max P 0.10-0.35 Pb 0.02 max Sn 4.0-6.0 Other Cu+sum of named elements 99.5 min
C52100	Cu rem Fe 0.10 max P 0.03-0.35 Pb 0.05 max Sn 7.0-9.0 Zn 0.20 max Other Cu+sum of named elements 99.5 min

CHEMICAL COMPOSITION OF COPPER & COPPER ALLOYS (Continued)

UNS	Chemical Composition
C52400	Cu rem Fe 0.10 max P 0.03-0.35 Pb 0.05 max Sn 9.0-11.0 Zn 0.20 max Other Cu+sum of named elements 99.5 min
C53400	Cu rem Fe 0.10 max P 0.03-0.35 Pb 0.8-1.2 Sn 3.5-5.8 Zn 0.30 max Other Cu+sum of named elements 99.5 min
C54400	Cu rem Fe 0.10 max P 0.01-0.50 Pb 3.5-4.5 Sn 3.5-4.5 Zn 1.5-4.5 Other Cu+sum of named elements 99.5 min
C55180	Cu rem P 4.8-5.2 Other Cu+P 99.85 min
C55181	Cu rem P 7.0-7.5 Other Cu+P 99.85 min
C55280	Cu rem Ag 1.8-2.2 P 6.8-7.2 Other Cu + all named elements 99.85 min
C55281	Cu rem Ag 4.8-5.2 P 5.8-6.2 Other Cu + all named elements 99.85 min
C55282	Cu rem Ag 4.8-5.2 P 6.5-7.0 Other Cu + all named elements 99.85 min
C55283	Cu rem Ag 5.8-6.2 P 7.0-7.5 Other Cu + all named elements 99.85 min
C55284	Cu rem Ag 14.5-15.5 P 4.8-5.2 Other Cu + all named elements 99.85 min
C60800	Cu (incl Ag) rem Al 5.0-6.5 As 0.2-0.35 Fe 0.10 max Pb 0.10 max Other Cu+sum of named elements 99.5 min
C61000	Cu (incl Ag) rem Al 6.0-8.5 Fe 0.50 max Pb 0.02 max Si 0.10 max Zn 0.20 max Other Cu+sum of named elements 99.5 min
C61300	Cu (incl Ag) rem Al 6.0-7.5 Fe 2.0-3.0 Mn 0.20 max Ni (incl Co) 0.15 max P 0.015 max Pb 0.01 max Si 0.10 max Sn 0.20-0.50 Zn 0.10 max Other Cu+sum of named elements 99.8 min Note: When the product is for subsequent welding applications and is so specified, the Cr, Cd, Zr, and Zn limits shall be 0.05 max each
C61400	Cu (incl Ag) rem Al 6.0-8.0 Fe 1.5-3.5 Mn 1.0 max P 0.015 max Pb 0.01 max Zn 0.20 max Other Cu+sum of named elements 99.5 min
C61800	Cu (incl Ag) rem Al 8.5-11.0 Fe 0.50-1.5 Pb 0.02 max Si 0.10 max Zn 0.02 max Other Cu+sum of named elements 99.5 min
C61900	Cu (incl Ag) rem Al 8.5-10.0 Fe 3.0-4.5 Pb 0.02 max Sn 0.06 max Zn 0.08 max Other Cu+sum of named elements 99.5 min
C62300	Cu (incl Ag) rem Al 8.5-10.0 Fe 2.0-4.0 Mn 0.50 max Ni (incl Co) 1.0 max Si 0.25 max Sn 0.6 max Other Cu+sum of named elements 99.5 min
C62400	Cu (incl Ag) rem Al 10.0-11.5 Fe 2.0-4.5 Mn 0.30 max Si 0.25 max Sn 0.20 max Other Cu+sum of named elements 99.5 min
C63000	Cu (incl Ag) rem Al 9.0-11.0 Fe 2.0-4.0 Mn 1.5 max Ni (incl Co) 4.0-5.5 Si 0.25 max Sn 0.20 max Zn 0.30 max Other u+sum of named elements 99.5 min
C63200	Cu (incl Ag) rem Al 8.7-9.5 Fe 3.5-4.3 Mn 1.2-2.0 Ni (incl Co) 4.0-4.8 Pb 0.02 max Si 0.10 max Other Cu+sum of named elements 99.5 min Note: Fe content shall not exceed Ni content
C63280	Cu (incl Ag) rem Al 8.5-9.5 Fe 3.0-5.0 Mn 0.6-3.5 Ni 4.0-5.5 Pb 0.02 max Si 0.10 max Other Cu+sum of named elements 99.5 min Note: Fe content shall not exceed Ni content

CHEMICAL COMPOSITION OF COPPER & COPPER ALLOYS (Continued)

UNS	Chemical Composition
C63380	Cu (incl Ag) rem Al 7.0-8.5 Fe 2.0-4.0 Mn 11.0-14.0 Ni 1.5-3.0 Pb 0.02 max Si 0.10 max Zn 0.15 max Other Cu+sum of named elements 99.5 min Note: Fe content shall not exceed Ni content
C63800	Cu (incl Ag) rem Al 2.5-3.1 Co 0.25-0.55 Fe 0.20 max Mn 0.10 max Ni 0.20 max Pb 0.05 max Si 1.5-2.1 Zn 0.8 max Other Cu+sum of named elements 99.5 min Note: Ni does not include Co
C64200	Cu (incl Ag) rem Al 6.3-7.6 As 0.15 max Fe 0.30 max Mn 0.10 max Ni (incl Co) 0.25 max Pb 0.05 max Si 1.5-2.2 Sn 0.20 max Zn 0.50 max Other Cu+sum of named elements 99.5 min
C64210	Cu (incl Ag) rem Al 6.3-7.0 As 0.15 max Fe 0.30 max Mn 0.10 max Ni (incl Co) 0.25 max Pb 0.05 max Si 1.5-2.0 Sn 0.20 max Zn 0.50 max Other Cu+sum of named elements 99.5 min
C64700	Cu (incl Ag) rem Fe 0.10 max Ni (incl Co) 1.6-2.2 Pb 0.10 max Si 0.40-0.8 Zn 0.50 max Other Cu+sum of named elements 99.5 min
C65100	Cu (incl Ag) rem Fe 0.8 max Mn 0.7 max Pb 0.05 max Si 0.8-2.0 Zn 1.5 max Other Cu+sum of named elements 99.5 min
C65400	Cu (incl Ag) Cr 0.01-0.12 rem Pb 0.05 max Si 2.7-3.4 Sn 1.2-1.9 Zn 0.50 max Other Cu+sum of named elements 99.5 min
C65500	Cu (incl Ag) rem Fe 0.8 max Mn 0.50-1.3 Ni (incl Co) 0.6 max Pb 0.05 max Si 2.8-3.8 Zn 1.5 max Other Cu+sum of named elements 99.5 min
C65600	Cu (incl Ag) rem Al 0.01 max Fe 0.50 max Mn 1.5 max Pb 0.02 max Si 2.8-4.0 Sn 1.5 max Zn 1.5 max Other Cu+sum of named elements 99.5 min
C66100	Cu (incl Ag) rem Fe 0.25 max Mn 1.5 max Pb 0.20-0.8 Si 2.8-3.5 Zn 1.5 max Other Cu+sum of named elements 99.5 min
C66400	Cu (incl Ag) rem Co 0.30-0.7 Fe 1.3-1.7 Mn 0.05 max Ni 0.05 max P 0.02 max Pb 0.015 max Si 0.05 max Sn 0.05 max Zn 11.0-12.0 Other Fe+Co 1.8-2.3 Other Cu+sum of named elements 99.5 min
C66410	Cu (incl Ag) rem Al 0.05 max As 0.05 max Fe 1.8-2.3 Mn 0.05 max Ni 0.05 max P 0.02 max Pb 0.015 max Si 0.05 max Sn 0.05 max Zn 11.0-12.0 Other Cu+sum of named elements 99.5 min
C66700	Cu (incl Ag) 68.5-71.5 Fe 0.10 max Mn 0.8-1.5 Pb 0.07 max Zn rem Other Cu+sum of named elements 99.5 min
C67000	Cu (incl Ag) 63.0-68.0 Al 3.0-6.0 Fe 2.0-4.0 Mn 2.5-5.0 Pb 0.20 max Sn 0.50 max Zn rem Other Cu+sum of named elements 99.5 min
C67300	Cu (incl Ag) 58.0-63.0 Al 0.25 max Fe 0.50 max Mn 2.0-3.5 Ni (incl Co) 0.25 max Pb 0.40-3.0 Si 0.50-1.5 Sn 0.30 max Zn rem Other Cu+sum of named elements 99.5 min
C67400	Cu (incl Ag) 57.0-60.0 Al 0.50-2.0 Fe 0.35 max Mn 2.0-3.5 Ni (incl Co) 0.25 max Pb 0.50 max Si 0.50-1.5 Sn 0.30 max Zn rem Other Cu+sum of named elements 99.5 min
C67500	Cu (incl Ag) 57.0-60.0 Al 0.25 max Fe 0.8-2.0 Mn 0.05-0.50 Pb 0.20 max Sn 0.50-1.5 Zn rem Other Cu+sum of named elements 99.5 min
C68000	Cu (incl Ag) 56.0-60.0 Al 0.01 max Fe 0.25-1.25 Mn 0.01-0.50 Ni (incl Co) 0.20-0.8 Pb 0.05 max Si 0.04-0.15 Sn 0.75-1.10 Zn rem Other Cu+sum of named elements 99.5 min

CHEMICAL COMPOSITION OF COPPER & COPPER ALLOYS (Continued)

UNS	Chemical Composition
C68100	Cu (incl Ag) 56.0-60.0 Al 0.01 max Fe 0.25-1.25 Mn 0.01-0.50 Pb 0.05 max Si 0.04-0.15 Sn 0.75-1.10 Zn rem Other Cu+sum of named elements 99.5 min
C68700	Cu (incl Ag) 76.0-79.0 Al 1.8-2.5 As 0.02-0.06 Fe 0.06 max Pb 0.07 max Zn rem Other Cu+sum of named elements 99.5 min
C68800	Cu (incl Ag) rem Al 3.0-3.8 Co 0.25-0.55 Fe 0.20 max Pb 0.05 max Zn 21.3-24.1 Other Cu+sum of named elements 99.5 min
C69100	Cu (incl Ag) 81.0-84.0 Al 0.7-1.2 Fe 0.25 max Mn 0.10 max Ni (incl Co) 0.8-1.4 Pb 0.05 max Si 0.8-1.3 Sn 0.10 max Zn rem Other Cu+sum of named elements 99.5 min
C69400	Cu (incl Ag) 80.0-83.0 Fe 0.20 max Pb 0.30 max Si 3.5-4.5 Zn rem Other Cu+sum of named elements 99.5 min
C69430	Cu (incl Ag) 80.0-83.0 As 0.03-0.06 Fe 0.20 max Pb 0.30 max Si 3.5-4.5 Zn rem Other Cu+sum of named elements 99.5 min
C69700	Cu (incl Ag) 75.0-80.0 Fe 0.20 max Mn 0.40 max Pb 0.50-1.5 Si 2.5-3.5 Zn rem Other Cu+sum of named elements 99.5 min
C69710	Cu (incl Ag) 75.0-80.0 As 0.03-0.06 Fe 0.20 max Mn 0.40 max Pb 0.50-1.5 Si 2.5-3.5 Zn Rem Other Cu+sum of named elements 99.5 min
C70200	Cu (incl Ag) rem Fe 0.10 max Mn 0.40 max Ni (incl Co) 2.0-3.0 Pb 0.05 max Other Cu+sum of named elements 99.5 min
C70400	Cu (incl Ag) rem Fe 1.3-1.7 Mn 0.30-0.8 Ni (incl Co) 4.8-6.2 Pb 0.05 max Zn 1.0 max Other Cu+sum of named elements 99.5 min
C70600	Cu (incl Ag) rem Fe 1.0-1.8 Mn 1.0 max Ni (incl Co) 9.0-11.0 Pb 0.05 max Zn 1.0 max Other Cu+sum of named elements 99.5 min Note: When the product is for subsequent welding applications, C shall be 0.05 max, Zn 0.50 max, P 0.02 max, Pb 0.02 max, and S 0.02 max
C70690	Cu (incl Ag) rem Al 0.002 max As 0.001 max Bi 0.001 max C 0.03 max Co 0.02 max Fe 0.005 max Hg 0.0005 max Mn 0.001 max Ni (incl Co) 9.0-11.0 P 0.001 max Pb 0.001 max S 0.003 max Sb 0.001 max Si 0.02 max Sn 0.001 max Ti 0.001 max Zn 0.001 max Other Cu+sum of named elements 99.5 min
C71000	Cu (incl Ag) rem Fe 1.0 max Mn 1.0 max Ni (incl Co) 19.0-23.0 Pb 0.05 max Zn 1.0 max Other Cu+sum of named elements 99.5 min
C71500	Cu (incl Ag) rem Fe 0.40-1.0 Mn 1.0 max Ni (incl Co) 29.0-33.0 Pb 0.05 max Zn 1.0 max Other Cu+sum of named elements 99.5 min Note: When the product is for subsequent welding applications, C shall be 0.05 max, Pb 0.02 max, P 0.02 max, S 0.02 max, and Zn 0.50 max
C71580	Cu (Incl Ag) rem Al 0.05 max C 0.07 max Fe 0.50 max Mn 0.30 max Ni (incl Co) 29.0-33.0 P 0.03 max Pb 0.05 max S 0.024 max Si 0.15 max Zn 0.05 max Other Cu+sum of named elements 99.5 min
C71581	Cu (incl Ag) rem Fe 0.40-0.7 Mn 1.0 max Ni (incl Co) 29.0-32.0 P 0.02 max Pb 0.02 max S 0.01 max Si 0.25 max Ti 0.20-0.50 Other Cu+sum of named elements 99.5 min
C71590	Cu (incl Ag) rem Al 0.002 max As 0.001 max Bi 0.001 max C 0.02 max Co 0.05 max Fe 0.15 max Hg 0.0005 max Mg 0.10 max Mn 0.50 max Ni 29.0-31.0 P 0.001 max Pb 0.001 max S 0.003 max Sb 0.001 max Si 0.015 max Sn 0.001 max Ti 0.001 max Zn 0.001 max
C71640	Cu (incl Ag) rem C 0.06 max Fe 1.7-2.3 Mn 1.5-2.5 Ni (incl Co) 29.0-32.0 Pb 0.01 max S 0.03 max Other Cu+sum of named elements 99.5 min

CHEMICAL COMPOSITION OF COPPER & COPPER ALLOYS (Continued)

UNS	Chemical Composition
C72200	Cu (incl Ag) rem C 0.03 max Cr 0.30-0.7 Fe 0.5-1.0 Mn 1.0 max Ni (incl Co) 15.0-18.0 Pb 0.05 max Si 0.03 max Ti 0.03 max Zn 1.0 max Other Cu+sum of named elements 99.8 min Note: When product is for subsequent welding applications, C shall be 0.05 max, Pb 0.02 max, P 0.02 max, S 0.02 max, and Zn 0.50 max
C72500	Cu (incl Ag) rem Fe 0.6 max Mn 0.20 max Ni (incl Co) 8.50-10.5 Pb 0.05 max Sn 1.8-2.8 Zn 0.50 max Other Cu+sum of named elements 99.8 min
C72700	Cu (incl Ag) rem Fe 0.50 max Mg 0.15 max Mn 0.05-0.30 Nb 0.10 max Ni (incl Co) 8.5-9.5 Pb 0.02 max Sn 5.5-6.5 Zn 0.50 max Other Cu+sum of named elements 99.7 min
C72900	Cu (incl Ag) rem Fe 0.50 max Mg 0.15 max Mn 0.30 max Nb 0.10 max Ni (incl Co) 14.5-15.5 Pb 0.02 Sn 7.5-8.5 Zn 0.50 max Other Cu+sum of named elements 99.7 min
C73500	Cu (incl Ag) 70.5-73.5 Fe 0.25 max Mn 0.50 max Ni (incl Co) 16.5-19.5 Pb 0.10 max Zn rem Other Cu+sum of named elements 99.5 min
C74000	Cu (incl Ag) 69.0-73.5 Fe 0.25 max Mn 0.50 max Ni (incl Co) 9.0-11.0 Pb 0.10 max Zn rem Other Cu+sum of named elements 99.5 min
C74500	Cu (incl Ag) 63.5-66.5 Fe 0.25 max Mn 0.50 max Ni (incl Co) 9.0-11.0 Pb 0.10 max Zn rem Other Cu+sum of named elements 99.5 min
C75200	Cu (incl Ag) 63.5-66.5 Fe 0.25 max Mn 0.50 max Ni (incl Co) 16.5-19.5 Pb 0.05 max Zn rem Other Cu+sum of named elements 99.5 min
C75700	Cu (incl Ag) 63.5-66.5 Fe 0.25 max Mn 0.50 max Ni (incl Co) 11.0-13.0 Pb 0.05 max Zn rem Other Cu+sum of named elements 99.5 min
C76200	Cu (incl Ag) 57.0-61.0 Fe 0.25 max Mn 0.50 max Ni (incl Co) 11.0-13.5 Pb 0.10 max Zn rem Other Cu+sum of named elements 99.5 min
C76400	Cu (incl Ag) 58.5-61.5 Fe 0.25 max Mn 0.50 max Ni (incl Co) 16.5-19.5 Pb 0.05 max Zn rem Other Cu+sum of named elements 99.5 min
C77000	Cu (incl Ag) 53.5-56.5 Fe 0.25 max Mn 0.50 max Ni (incl Co) 16.5-19.5 Pb 0.05 max Zn rem Other Cu+sum of named elements 99.5 min
C77300	Cu (incl Ag) 46.0-50.0 Al 0.01 max Ni (incl Co) 9.0-11.0 P 0.25 max Pb 0.05 max Si 0.04-0.25 Zn rem Other Cu+sum of named elements 99.5 min
C77400	Cu (incl Ag) 43.0-47.0 Ni (incl Co) 9.0-11.0 Pb 0.20 max Zn rem Other Cu+sum of named elements 99.5 min
C79200	Cu (incl Ag) 59.5-66.5 Fe 0.25 max Mn 0.50 max Ni (incl Co) 11.0-13.0 Pb 0.8-1.4 Zn rem Other Cu+sum of named elements 99.5 min
C81400	Cu 98.5 min Be 0.02-0.10 Cr 0.6-1.0 Other Cu+sum of named elements 99.5 min
C82000	Cu 95.0 min Al 0.10 max Be 0.45-0.8 Co (incl Ni) 2.4-2.7 Cr 0.10 max Fe 0.10 max Ni 0.20 max Pb 0.02 max Si 0.15 max Sn 0.10 max Zn 0.10 max Other Cu+sum of named elements 99.5 min
C82200	Cu 96.5 min Be 0.35-0.8 Ni 1.0-2.0 Other Cu+sum of named elements 99.5 min
C82400	Cu 96.4 min Al 0.15 max Be 1.65-1.75 Co 0.20-0.40 Cr 0.10 max Fe 0.20 max Ni 0.10 max Pb 0.02 max Sn 0.10 max Zn 0.10 max Other Cu+sum of named elements 99.5 min
C82500	Cu 95.5 min Al 0.15 max Be 1.90-2.15 Co (incl Ni) 0.35-0.7 Cr 0.10 max Fe 0.25 max Ni 0.20 max Pb 0.02 max Si 0.20-0.35 Sn 0.10 max Zn 0.10 max Other Cu+sum of named elements 99.5 min

CHEMICAL COMPOSITION OF COPPER & COPPER ALLOYS (Continued)

UNS	Chemical Composition
C82510	Cu 95.5 min Al 0.15 max Be 1.90-2.15 Co 1.0-2.0 Cr 0.10 max Fe 0.25 max Ni 0.20 max Pb 0.02 max Si 0.20-0.35 Sn 0.10 max Zn 0.10 max Other Cu+sum of named elements 99.5 min
C82600	Cu 95.2 min Al 0.15 max Be 2.25-2.45 Co 0.35-0.7 Cr 0.10 max Fe 0.25 max Ni 0.20 max Pb 0.02 max Si 0.20-0.35 Sn 0.10 max Zn 0.10 max Other Cu+sum of named elements 99.5 min
C82800	Cu 94.8 min Al 0.15 max Be 2.50-2.75 Co (incl Ni) 0.35-0.7 Cr 0.10 max Fe 0.25 max Ni 0.20 max Pb 0.02 max Si 0.20-0.35 Sn 0.10 max Zn 0.10 max Other Cu+sum of named elements 99.5 min
C83400	Cu (incl Ni) 88.0-92.0 Pb 0.50 max Sn 0.20 max Zn 8.0-12.0 Other Cu+sum of named elements 99.3 min
C83450	Cu 87.0-89.0 Al 0.005 max Fe 0.30 max Ni (incl Co) 0.8-2.0 P 0.03 max Pb 1.5-3.0 S 0.08 max Sb 0.25 max Si 0.005 max Sn 2.0-3.5 Zn 5.5-7.5 Other Cu+sum of named elements 99.3 min Note: For continuous castings, P 1.5 max
C83600	Cu (incl Ni) 84.0-86.0 Al 0.005 max Fe 0.30 max Ni (incl Co) 1.0 max P 0.05 max Pb 4.0-6.0 S 0.08 max Sb 0.25 max Si 0.005 max Sn 4.0-6.0 Zn 4.0-6.0 Other Cu+sum of named elements 99.3 min Note: For continuous casting, P 1.5 max
C83800	Cu (incl Ni) 82.0-83.8 Al 0.005 max Fe 0.30 max Ni (incl Co) 1.0 max P 0.03 max Pb 5.0-7.0 S 0.08 max Sb 0.25 max Si 0.005 max Sn 3.3-4.2 Zn 5.0-8.0 Other Cu+sum of named elements 99.3 min Note: For continuous casting, P 1.5 max
C84200	Cu (incl Ni) 78.0-82.0 Al 0.005 max Fe 0.40 max Ni (incl Co) 0.8 max P 0.05 max Pb 2.0-3.0 S 0.08 max Sb 0.25 max Si 0.005 max Sn 4.0-6.0 Zn 10.0-16.0 Other Cu+sum of named elements 99.2 min Note: For continuous casting, P 1.5 max
C84400	Cu (incl Ni) 78.0-82.0 Al 0.005 max Fe 0.40 max Ni (incl Co) 1.0 max P 0.2 max Pb 6.0-8.0 S 0.08 max Sb 0.25 max Si 0.005 max Sn 2.3-3.5 Zn 7.0-10.0 Other Cu+sum of named elements 99.2 min; Ingot 123 Note: For continuous casting, P 1.5 max
C84800	Cu (incl Ni) 75.0-77.0 Al 0.005 max Fe 0.40 max Ni (incl Co) 1.0 max P 0.02 max Pb 5.5-7.0 S 0.08 max Sb 0.25 max Si 0.005 max Sn 2.0-3.0 Zn 13.0-17.0 Other Cu+sum of named elements 99.2 min Note: For continuous casting, P 1.5 max
C85200	Cu (incl Ni) 70.0-74.0 Al 0.005 max Fe 0.6 max Ni (incl Co) 1.0 max P 0.02 max Pb 1.5-3.8 S 0.05 max Sb 0.20 max Si 0.05 max Sn 0.7-2.0 Zn 20.0-27.0 Other Cu+sum of named elements 99.1 min
C85400	Cu (incl Ni) 65.0-70.0 Al 0.35 max Fe 0.7 max Ni (incl Co) 1.0 max Pb 1.5-3.8 Si 0.05 max Sn 0.50-1.5 Zn 24.0-32.0 Other Cu+sum of named elements 98.9 min
C85700	Cu (incl Ni) 58.0-64.0 Al 0.8 max Fe 0.7 max Ni (incl Co) 1.0 max Pb 0.8-1.5 Si 0.05 max Sn 0.50-1.5 Zn 32.0-40.0 Other Cu+sum of named elements 98.7 min
C85800	Cu (incl Ni) 57.0 min Al 0.50 max As 0.05 max Fe 0.50 max Mn 0.25 max Ni (incl Co) 0.50 max P 0.01 max Pb 1.5 max S 0.05 max Sb 0.05 max Si 0.25 max Sn 1.5 max Zn 31.0-41.0 Other Cu+sum of named elements 98.7 min

CHEMICAL COMPOSITION OF COPPER & COPPER ALLOYS (Continued)	
UNS	**Chemical Composition**
C86200	Cu (incl Ni) 60.0-66.0 Al 3.0-4.9 Fe 2.0-4.0 Mn 2.5-5.0 Ni (incl Co) 1.0 max Pb 0.20 max Sn 0.20 max Zn 22.0-28.0 Other Cu+sum of named elements 99.0 min
C86300	Cu (incl Ni) 60.0-66.0 Al 5.0-7.5 Fe 2.0-4.0 Mn 2.5-5.0 Ni (incl Co) 1.0 max Pb 0.20 max Sn 0.20 max Zn 22.0-28.0 Other Cu+sum of named elements 99.0 min
C86400	Cu (incl Ni) 56.0-62.0 Al 0.50-1.5 Fe 0.40-2.0 Mn 0.10-1.0 Ni (incl Co) 1.0 max Pb 0.50-1.5 Sn 0.50-1.5 Zn 34.0-42.0 Other Cu+sum of named elements 99.0 min
C86500	Cu (incl Ni) 55.0-60.0 Al 0.50-1.5 Fe 0.40-2.0 Mn 1.0-1.5 Ni (incl Co) 1.0 max Pb 0.40 max Sn 1.0 max Zn 36.0-42.0 Other Cu+sum of named elements 99.0 min
C86700	Cu (incl Ni) 53.0-60.0 Al 1.0-3.0 Fe 1.0-3.0 Mn 1.0-3.5 Ni (incl Co) 1.0 max Pb 0.5-1.5 max Sn 1.5 Zn 30.0-38.0 Other Cu+sum of named elements 99.0 min
C87300	Cu 94.0 min Fe 0.20 max Mn 0.8-1.5 Pb 0.20 max Si 3.5-4.5 Zn 0.25 max Other Cu+sum of named elements 99.5 min
C87400	Cu 79.0 min Al 0.8 max Pb 1.0 max Si 2.5-4.0 Zn 12.0-16.0 Other Cu+sum of named elements 99.2 min
C87500	Cu 79.0 min Al 0.50 max Pb 0.50 max Si 3.0-5.0 Zn 12.0-16.0 Other Cu+sum of named elements 99.5 min
C87600	Cu 88.0 min Fe 0.20 max Mn 0.25 max Pb 0.50 max Si 3.5-4.5 Zn 4.0-7.0 Other Cu+sum of named elements 99.5 min
C87610	Cu 90.0 min Fe 0.20 max Mn 0.25 max Pb 0.20 max Si 3.0-5.0 Zn 3.0-5.0 Other Cu+sum of named elements 99.5 min
C87800	Cu 80.0 min Al 0.15 max As 0.05 max Fe 0.15 max Mg 0.01 max Mn 0.15 max Ni (incl Co) 0.20 max P 0.01 max Pb 0.15 max S 0.05 max Sb 0.05 max Si 3.8-4.2 Sn 0.25 max Zn 12.0-16.0 Other Cu+sum of named elements 99.5 min
C90300	Cu (incl Ni) 86.0-89.0 Al 0.005 max Fe 0.20 max Ni (incl Co) 1.0 max P 0.05 max Pb 0.30 max S 0.05 max Sb 0.20 max Si 0.005 Sn 7.5-9.0 Zn 3.0-5.0 Other Cu+sum of named elements 99.4 min Note: For continuous casting, P 1.5 max
C90500	Cu (incl Ni) 86.0-89.0 Al 0.005 max Fe 0.20 max Ni (incl Co) 1.0 max P 0.05 max Pb 0.30 max S 0.05 max Sb 0.20 max Si 0.005 max Sn 9.0-11.0 Zn 1.0-3.0 Other Cu+sum of named elements 99.7 min Note: For continuous casting, P 1.5 max
C90700	Cu (incl Ni) 88.0-90.0 Al 0.005 max Fe 0.15 max Ni (incl Co) 0.50 max P 0.30 max Pb 0.50 max S 0.05 max Sb 0.20 max Si 0.005 max Sn 10.0-12.0 Zn 0.50 max Other Cu+sum of named elements 99.4 min Note: For continuous casting, P 1.5 max
C90800	Cu (incl Ni) 85.0-89.0 Al 0.005 max Fe 0.15 max Ni (incl Co) 0.50 max P 0.30 max Pb 0.25 max S 0.05 max Sb 0.20 max Si 0.005 max Sn 11.0-13.0 Zn 0.25 max Other Cu+sum of named elements 99.4 Note: For continuous casting, P 1.5 max
C91000	Cu (incl Ni) 84.0-86.0 Al 0.005 max Fe 0.10 max Ni (incl Co) 0.8 max P 0.05 max Pb 0.20 max S 0.5 max Sb 0.20 max Si 0.005 max Sn 14.0-16.0 Zn 1.5 max Other Cu+sum of named elements 99.4 min Note: For continuous casting, P 1.5 max

CHEMICAL COMPOSITION OF COPPER & COPPER ALLOYS (Continued)	
UNS	**Chemical Composition**
C91100	Cu (incl Ni) 82.0-85.0 Al 0.005 max Fe 0.25 max Ni (incl Co) 0.50 max P 1.0 max Pb 0.25 max S 0.05 max P 0.20 max Si 0.005 max Sn 15.0-17.0 Zn 0.25 max Other Cu+sum of named elements 99.4 min Note For continuous casting, P 1.5 max
C91300	Cu (incl Ni) 79.0-82.0 Al 0.005 max Fe 0.25 max Ni (incl Co) 0.50 max P 1.0 max Pb 0.25 max S 0.05 max Sb 0.20 max Si 0.005 max Sn 18.0-20.0 Zn 0.25 max Other Cu+sum of named elements 99.4 min Note: For continuous casting, P 1.5 max
C91600	Cu (incl Ni) 86.0-89.0 Al 0.005 max Fe 0.20 max Ni (incl Co) 1.2-2.0 P 0.30 max Pb 0.25 max S 0.05 max Sb 0.20 max Si 0.005 max Sn 9.7-10.8 Zn 0.25 max Other Cu+sum of named elements 99.4 min Note: For continuous casting, P 1.5 max
C91700	Cu (incl Ni) 84.0-87.0 Al 0.005 max Fe 0.20 max Ni (incl Co) 1.2-2.0 P 0.30 max Pb 0.25 max S 0.05 max Sb 0.20 max Si 0.005 max Sn 11.3-12.5 Zn 0.25 max Other Cu+sum of named elements 99.4 min Note: For continuous casting, P 1.5 max
C92200	Cu (incl Ni) 86.0-90.0 Al 0.005 max Fe 0.25 max Ni (incl Co) 1.0 max P 0.05 max Pb 1.0-2.0 S 0.05 max Sb 0.25 max Si 0.005 max Sn 5.5-6.5 Zn 3.0-5.0 Other Cu+sum of named elements 99.3 min Note: For continuous casting, P 1.5 max
C92300	Cu (incl Ni) 85.0-89.0 Al 0.005 max Fe 0.25 max Ni (incl Co) 1.0 max P 0.05 max Pb 0.30-1.0 S 0.05 max Sb 0.25 max Si 0.005 max Sn 7.5-9.0 Zn 2.5-5.0 Other Cu+sum of named elements 99.3 min Note: For continuous casting, P 1.5 max
C92500	Cu 85.0-88.0 Al 0.005 max Fe 0.30 max Ni (incl Co) 0.8-1.5 P 0.30 max S 0.05 max Pb 1.0-1.5 Sb 0.25 max Si 0.005 max Sn 10.0-12.0 Zn 0.50 max Other Cu+sum of named elements 99.3 min Note: For continuous casting, P 1.5 max
C92600	Cu (incl Ni) 86.0-88.5 Al 0.005 max Fe 0.20 max Ni (incl Co) 0.7 max P 0.03 max Pb 0.8-1.5 S 0.05 max Sb 0.25 max Si 0.005 max Sn 9.3-10.5 Zn 1.3-2.5 Other Cu+sum of named elements 99.3 min Note: For continuous casting, P 1.5 max
C92700	Cu (incl Ni) rem Al 0.005 max Fe 0.20 max Ni (incl Co) 1.0 max P 0.25 max Pb 1.0-2.5 S 0.05 max Sb 0.25 max Si 0.005 max Sn 9.0-11.0 Zn 0.7 max Other Cu+sum of named elements 99.3 min Note: For continuous casting, P 1.5 max
C92800	Cu (incl Ni) 78.0-82.0 Al 0.005 max Fe 0.20 max Ni (incl Co) 0.8 max P 0.05 max Pb 4.0-6.0 S 0.05 max Sb 0.25 max Si 0.005 max Sn 15.0-17.0 Zn 0.8 max Other Cu+sum of named elements 99.3 min Note: For continuous casting, P 1.5 max
C92900	Cu 82.0-86.0 Al 0.005 max Fe 0.20 max Ni (incl Co) 2.8-4.0 P 0.50 max Pb 2.0-3.2 S 0.05 max Sb 0.25 max Si 0.005 max Sn 9.0-11.0 Zn 0.25 max Other Cu+sum of named elements 99.3 min Note: For continuous casting, P 1.5 max
C93200	Cu (incl Ni) 81.0-85.0 Al 0.005 max Fe 0.20 max Ni (incl Co) 1.0 max P 0.15 max Pb 6.0-8.0 S 0.08 max Sb 0.35 max Si 0.005 max Sn 6.3-7.5 Zn 1.0-4.0 Other Cu+sum of named elements 99.2 min Note: For continuous casting, P 1.5 max
C93400	Cu (incl Ni) 82.0-85.0 Al 0.005 max Fe 0.20 max Ni (incl Co) 1.0 max P 0.50 max Pb 7.0-9.0 S 0.08 max Sb 0.50 max Si 0.005 max Sn 7.0-9.0 Zn 0.8 max Other Cu+sum of named elements 99.2 min Note: For continuous casting, P 1.5 max
C93500	Cu (incl Ni) 83.0-86.0 Al 0.005 max Fe 0.20 max Ni (incl Co) 1.0 max P 0.05 max Pb 8.0-10.0 S 0.08 max Sb 0.30 max Si 0.005 max Sn 4.3-6.0 Zn 2.0 max Other Cu+sum of named elements 99.4 min Note: For continuous casting, P 1.5 max

CHEMICAL COMPOSITION OF COPPER & COPPER ALLOYS (Continued)	
UNS	**Chemical Composition**
C93600	Cu (incl Ni) 79.0-83.0 min Al 0.005 max Fe 0.20 max Ni (incl Co) 1.0 max P 0.15 max Pb 11.0-13.0 S 0.08 max Sb 0.55 max Si 0.005 max Sn 6.0-8.0 Zn 1.0 max Other Cu+sum of named elements 99.3 min Note: For continuous casting, P 1.5 max
C93700	Cu (incl Ni) 78.0-82.0 Al 0.005 max Fe 0.7 max Ni (incl Co) 0.50 max P 0.10 max Pb 8.0-11.0 S 0.08 max Sb 0.50 max Si 0.005 max Sn 9.0-11.0 Zn 0.8 max Si 0.005 max Sn 9.0-11.0 Zn 0.8 max Other Cu may include Ni; Fe 0.35 max for steel-backed bearings; Cu+sum of named elements 99.0 min Note: For continuous casting, P 1.5 max
C93720	Cu (incl Ni) 83.0 min Fe 0.7 max Ni (incl Co) 0.50 max P 0.10 max Pb 7.0-9.0 Sb 0.50 max Sn 3.5-4.5 Zn 4.0 max Other Cu+sum of named elements 99.0 min Note: For continuous casting, P 1.5 max
C93800	Cu (incl Ni) 75.0-79.0 Al 0.005 max Fe 0.15 max Ni (incl Co) 1.0 max P 0.05 max Pb 13.0-16.0 S 0.08 max Sb 0.8 max Si 0.005 max Sn 6.3-7.5 Zn 0.8 max Other Cu+sum of named elements 98.9 min Note: For continuous casting, P 1.5 max
C93900	Cu (incl Ni) 76.5-79.5 Al 0.005 max Fe 0.40 max Ni (incl Co) 0.8 max P 1.5 max Pb 14.0-18.0 S 0.08 max Sb 0.50 max Pb 14.0-18.0 S 0.08 max Sb 0.50 max Si 0.005 max Sn 5.0-7.0 Zn 1.5 max Other Cu+sum of named elements 98.9 min
C94000	Cu (incl Ni) 69.0-72.0 Al 0.005 max Fe 0.25 max Ni (incl Co) 0.50-1.0 P 0.05 max Pb 14.0-16.0 S 0.08 max Sb 0.50 max Si 0.005 max Sn 12.0-14.0 Zn 0.50 max Other Cu+sum of named elements 98.7 min Note: For continuous casting, P 1.5 max and S 0.25 max
C94100	Cu (incl Ni) 72.0-79.0 Al 0.005 max Fe 0.25 max Ni (incl Co) 1.0 max P 0.05 max Pb 18.0-22.0 S 0.08 max Sb 0.8 max Si 0.005 max Sn 4.5-6.5 Zn 1.0 max Other Cu+sum of named elements 98.7 min Note: For continuous casting, P 1.5 max and S 0.25 max
C94300	Cu (incl Ni) 67.0-72.0 Al 0.005 max Fe 0.15 max Ni (incl Co) 1.0 max P 0.08 max Pb 22.0-25.0 S 0.08 max Sb 0.8 max Si 0.005 max Sn 4.5-6.0 Zn 0.8 max Other Cu+sum of named elements 99.0 Note: For continuous casting, P 1.5 max and S 0.25 max
C94320	Cu (incl Ni) rem Fe 0.35 max Pb 24.0-32.0 Sn 4.0-7.0 Other Cu+sum of named elements 99.0 min
C94330	Cu (incl Ni) 68.5-75.5 Fe 0.7 max Ni (incl Co) 0.50 max P 0.10 max Pb 21.0-25.0 Sb 0.50 max Sn 3.0-4.0 Zn 3.0 max Other Cu+sum of named elements 99.0 min Note: For continuous casting, P 1.5 max
C94400	Cu (incl Ni) rem Al 0.005 max Fe 0.15 max Ni (incl Co) 1.0 max P 0.50 max Pb 9.0-12.0 S 0.08 max Sb 0.08 max Si 0.005 max Sn 7.0-9.0 Zn 0.8 max Other Cu+sum of named elements 99.0 min Note: For continuous casting, P 1.5 max
C94500	Cu (incl Ni) rem Al 0.005 max Fe 0.15 max Ni (incl Co) 1.0 max P 0.05 max Pb 16.0-22.0 S 0.08 max Sb 0.08 max Si 0.005 max Sn 6.0-8.0 Zn 1.2 max Other Cu+sum of named elements 99.0 min Note: For continuous casting, P 1.5 max
C94700	Cu 85.0-89.0 Al 0.005 max Fe 0.25 max Mn 0.20 max Ni 4.5-6.0 P 0.05 max Pb 0.10 max S 0.05 max Sb 0.15 max Si 0.005 max Sn 4.5-6.0 Zn 1.0-2.5 Other Cu+sum of named elements 99.3 min
C94800	Cu 84.0-89.0 Al 0.005 max Fe 0.25 max Mn 0.20 max Ni 4.5-6.0 P 0.05 max Pb 0.30-1.0 S 0.05 max Sb 0.15 max Si 0.005 max Sn 4.5-6.0 Zn 1.0-2.5 Other Cu+sum of named elements 99.3 min

CHEMICAL COMPOSITION OF COPPER & COPPER ALLOYS (Continued)	
UNS	**Chemical Composition**
C94900	Cu 79.0-81.0 Al 0.005 max Fe 0.30 max Mn 0.10 max Ni 4.0-6.0 P 0.05 max Pb 4.0-6.0 S 0.08 max Sb 0.25 max Si 0.005 max Sn 4.0-6.0 Zn 4.0-6.0 Other Cu+sum of named elements 99.2 min
C95200	Cu 86.0 min Al 8.5-9.5 Fe 2.5-4.0 Other Cu+sum of named elements 99.0 min
C95300	Cu 86.0 min Al 9.0-11.0 Fe 0.8-1.5 Other Cu+sum of named elements 99.0 min
C95400	Cu 83.0 min Al 10.0-11.5 Fe 3.0-5.0 Mn 0.50 max Ni (incl Co) 1.5 max Other Cu+sum of named elements 99.5 min
C95410	Cu 83.0 min Al 10.0-11.5 Fe 3.0-5.0 Mn 0.50 max Ni (incl Co) 1.5-2.5 Other Cu+sum of named elements 99.5 min
C95420	Cu 83.5 min Al 10.5-12.0 Fe 3.0-4.3 Mn 0.50 max Ni (incl Co) 0.50 max Other Cu+sum of named elements 99.5 min
C95500	Cu 78.0 min Al 10.0-11.5 Fe 3.0-5.0 Mn 3.5 max Ni (incl Co) 3.0-5.5 Other Cu+sum of named elements 99.5 min
C95510	Cu 78.0 min Al 9.7-10.9 Fe 2.0-3.5 Mn 1.5 max Ni (incl Co) 4.5-5.5 Sn 0.20 max Zn 0.30 max Other Cu+sum of named elements 99.8 min Note: Cu included Ag and Ni included Co; Cu + all named elements 99.8 min
C95520	Cu 74.5 min Al 10.5-11.5 Co 0.20 max Cr 0.05 max Fe 4.0-5.5 Mn 1.5 max Ni (incl Co) 4.2-6.0 Pb 0.03 max Si 0.15 max Sn 0.25 max Zn 0.30 max Other Cu+sum of named elements 99.5 min
C95600	Cu 88.0 min Al 6.0-8.0 Ni (incl Co) 0.25 max Si 1.8-3.3 Other Cu+sum of named elements 99.0 min
C95700	Cu 71.0 min Al 7.0-8.5 Fe 2.0-4.0 Mn 11.0-14.0 Ni (incl Co) 1.5-3.0 Pb 0.03 max Si 0.10 max Other Cu+sum of named elements 99.5 min
C95800	Cu 79.0 min Al 8.5-9.5 Fe 3.5-4.5 Mn 0.8-1.5 Ni (incl Co) 4.0-5.0 Pb 0.03 max Si 0.10 max Other Fe less than Ni, total named elements 99.5 min Note: Fe content shall not exceed Ni content
C95900	Cu rem Fe Al 12.0-13.5 3.0-5.0 Mn 1.5 max Ni (incl Co) 0.50 max Other Cu+sum of named elements 99.5 min
C96200	Cu rem C 0.10 max Fe 1.0-1.8 Mn 1.5 max Nb 0.50-1.0 Ni (incl Co) 9.0-11.0 P 0.02 max Pb 0.01 max S 0.02 max Si 0.50 max Other Cu+sum of named elements 99.5 min
C96400	Cu rem C 0.15 max Fe 0.25-1.5 Mn 1.5 max Nb 0.50-1.5 Ni (incl Co) 28.0-32.0 P 0.02 max Pb 0.03 max S 0.02 max Si 0.50 max Other Cu+sum of named elements 99.5 min Note: For welding grades, Pb 0.01 max
C96700	Cu rem Be 1.1-1.2 Fe 0.7-1.0 Mn 0.7 max Ni (incl Co) 29.0-33.0 Pb 0.01 max Si 0.15 max Ti 0.010-0.20 Zr 0.10-0.20 Other Cu+sum of named elements 99.5 min
C96800	Cu rem Al 0.10 max B 0.001 max Bi 0.001 max Fe 0.50 max Mg 0.005-0.15 Mn 0.05-0.30 Nb 0.10-0.30 Ni (incl Co) 9.5-10.5 P 0.005 max Pb 0.005 max S 0.0025 max Sb 0.02 max Si 0.05 max Sn 7.5-8.5 Ti 0.01 max Zn 1.0 max Other Cu+sum of named elements 99.5 min
C97300	Cu 53.0-58.0 Al 0.005 max Fe 1.5 max Mn 0.50 max Ni (incl Co) 11.0-14.0 P 0.05 max Pb 8.0-11.0 S 0.08 max Sb 0.35 max Si 0.15 max Sn 1.5-3.0 Zn 17.0-25.0 Other Cu+sum of named elements 99.0 min

CHEMICAL COMPOSITION OF COPPER & COPPER ALLOYS (Continued)

UNS	Chemical Composition
C97600	Cu 63.0-67.0 Al 0.005 max Fe 1.5 max Mn 1.0 max Ni (incl Co) 19.0-21.5 P 0.05 max Pb 3.0-5.0 S 0.08 max Sb 0.25 max Si 0.15 max Sn 3.5-4.5 Zn 3.0-9.0 Other Cu+sum of named elements 99.7 min
C97800	Cu 64.0-67.0 Al 0.005 max Fe 1.5 max Mn 1.0 max Ni (incl Co) 24.0-27.0 P 0.05 max Pb 1.0-2.5 S 0.08 max Sb 0.20 max Si 0.15 max Sn 4.0-5.5 Zn 1.0-4.0 Other Cu+sum of named elements 99.6 min
C98200	Cu 73.0-79.0 Fe 0.7 max Ni 0.50 max P 0.10 max Pb 21.0-27.0 Sb 0.50 max Sn 0.6-2.0 Zn 0.50 max Other Cu+sum of named elements 99.5 min
C98400	Cu rem Ag 1.5 max Fe 0.7 max Ni 0.50 max P 0.10 max Pb 26.0-33.0 Sb 0.50 max Sn 0.50 max Zn 0.50 max Other Cu+sum of named elements 99.5 min
C98820	Cu rem Fe 0.35 max Pb 40.0-44.0 Sn 1.0-5.0
C98840	Cu rem Fe 0.35 max Pb 44.0-58.0 Sn 1.0-5.0 Other Cu+sum of named elements 99.5 min
C99400	Cu rem Al 0.50-2.0 Fe 1.0-3.0 Mn 0.50 max Ni 1.0-3.5 Pb 0.25 max Si 0.50-2.0 Zn 0.50-5.0 Other Cu+sum of named elements 99.7 min
C99500	Cu rem Al 0.50-2.0 Fe 3.0-5.0 Mn 0.50 max Ni 3.5-5.5 Pb 0.25 max Si 0.50-2.0 Zn 0.50-2.0 Other Cu+sum of named elements 99.7 min
C99700	Cu 54.0 min Al 0.50-3.0 Fe 1.0 max Mn 11.0-15.0 Nb 4.0-6.0 Ni 4.0-6.0 Pb 2.0 max Sn 1.0 max Zn 19.0-25.0 Other Cu+sum of named elements 99.7 min
C99750	Cu 55.0-61.0 Al 0.25-3.0 Fe 1.0 max Mn 17.0-23.0 Ni 5.0 max Pb 0.50-2.5 Sn 0.50-2.5 Zn 17.0-23.0 Other Cu+sum of named elements 99.7 min

CHEMICAL COMPOSITION OF COPPER & COPPER ALLOY - WELD FILLER METALS

AWS A5.6

Common Name	Classification	Chemical Composition
Copper	ECu	Cu+Ag rem Mn 0.1 Fe 0.2 Si 0.1 Al 0.1 Total other elements 0.50
Copper-Silicon (Silicon Bronze)	ECuSi	Cu+Ag rem Sn 1.5 Mn 1.5 Fe 0.50 Si 2.4-4.0 Al 0.01 Pb 0.02 Total other elements 0.50
Copper-Tin (Phosphor Bronze)	ECuSn-A	Cu+Ag rem Sn 4.0-6.0 Fe 0.25 P 0.05-0.35 Al 0.01 Pb 0.02 Total other elements 0.50
Copper-Tin (Phosphor Bronze)	ECuSn-C	Cu+Ag rem Sn 7.0-9.0 Fe 0.25 P 0.05-0.35 Al 0.01 Pb 0.02 Total other elements 0.50
Copper-Nickel	ECuNi[b]	Cu+Ag rem Mn 1.00-2.50 Fe 0.40-0.75 Si 0.50 Ni 29.0-33.0 P 0.020 Pb 0.02 Ti 0.50 Total other elements 0.50

CHEMICAL COMPOSITION OF COPPER & COPPER ALLOY - WELD FILLER METALS (Continued)		
AWS A5.6 (Continued)		
Common Name	**Classification**	**Chemical Composition**
Copper-Aluminum (Aluminum Bronze)	ECuAl-A2	Cu+Ag rem Fe 0.5-5.0 Si 1.0 Al 7.0-9.0 Pb 0.02 Total other elements 0.60
Copper-Aluminum (Aluminum Bronze)	ECuAl-B	Cu+Ag rem Fe 2.5-5.0 Si 1.0 Al 8.0-10.0 Pb 0.02 Total other elements 0.60
Copper-Aluminum (Aluminum Bronze)	ECuNiAl	Cu+Ag rem Mn 0.5-3.5 Fe 3.0-6.0 Si 1.0 Ni 4.0-6.0 Al 6.5-8.5 Pb 0.02 Total other elements 0.60
Copper-Aluminum (Aluminum Bronze)	ECuMnNiAl	Cu+Ag rem Mn 11.0-13.0 Fe 2.0-6.0 Si 1.5 Ni 1.0-2.5 Al 5.5-7.5 Pb 0.02 Total other elements 0.60
AWS A5.7		
Copper	ERCu	Cu+Ag 98.0 min Sn 1.0 Mn 0.5 Si 0.50 P 0.15 Al 0.01 Pb 0.02 Total other elements 0.50
Copper-Silicon (Silicon Bronze)	ERCuSi-A	Cu+Ag 94.0 min Zn 1.5 Sn 1.5 Mn 1.5 Fe 0.5 Si 2.8-4.0 Al 0.01 Pb 0.02 Total other elements 0.50
Copper-Tin (Phosphor Bronze)	ERCuSn-A	Cu+Ag 93.5 min Sn 4.0-6.0 P 0.10-0.35 Al 0.01 Pb 0.02 Total other elements 0.50
Copper-Nickel	ERCuNi[a]	Cu+Ag rem Mn 1.00 Fe 0.40-0.75 Si 0.25 Ni+Co 29.0-32.0 P 0.02 Pb 0.02 Ti 0.20-0.50 Total other elements 0.50
Copper-Aluminum (Aluminum Bronze)	ERCuAl-A1	Cu+Ag rem Zn 0.10 Mn 0.50 Si 0.10 Al 6.0-9.0 Pb 0.02 Total other elements 0.50
Copper-Aluminum (Aluminum Bronze)	ERCuAl-A2	Cu+Ag rem Zn 0.02 Fe 1.5 Si 0.10 Al 9.0-11.0 Pb 0.02 Total other elements 0.50
Copper-Aluminum (Aluminum Bronze)	ERCuAl-A3	Cu+Ag rem Zn 0.10 Fe 3.0-5.0 Si 0.10 Al 10.0-11.0 Pb 0.02 Total other elements 0.50
Copper-Aluminum (Aluminum Bronze)	ERCuNiA1	Cu+Ag rem Zn 0.10 Mn 0.60-3.50 Fe 3.0-5.0 Si 0.10 Ni+Co 4.00-5.50 Al 8.50-9.50 Pb 0.02 Total other elements 0.50
Copper-Aluminum (Aluminum Bronze)	ERCuMnNiA1	Cu+Ag rem Zn 0.15 Mn 11.0-14.0 Fe 2.0-4.0 Si 0.10 Ni+Co 1.5-3.0 Al 7.0-8.5 Pb 0.02 Total other elements 0.50

CHEMICAL COMPOSITION OF COPPER & COPPER ALLOY - WELD FILLER METALS (Continued)		
AWS A5.27		
Common Name	**Classification**	**Chemical Composition**
Copper	ERCu	Cu+Ag 98.0 Sn 1.0 Mn 0.5 Si 0.50 P 0.15 Al 0.01 Pb 0.02 Total other elements 0.50
Copper-Silicon (Silicon Bronze)	ERCuSi-A	Cu+Ag 94.0 Zn 1.5 Sn 1.5 Mn 1.5 Fe 0.5 Si 2.8-4.0 Al 0.01 Pb 002 Total other elements 0.50
Copper-Nickel	ERCuNi[a]	Cu+Ag rem Mn 1.00 Fe 0.40-0.70 Si 0.15 Ni+Co 29.0-32.0 P 0.02 Pb 0.02 Ti 0.20-0.50 Total other elements 0.50
Naval Brass	RBCuZn-A	Cu+Ag 57.0-61.0 Zn rem Sn 0.25-1.00 Al 0.01 Pb 0.05 Total other elements 0.50
Low-Fuming Brass (Ni)	RCuZn-B	Cu+Ag 56.0-60.0 Zn rem Sn 0.80-1.10 Mn 0.01-0.50 Fe 0.25-1.20 Si 0.04-0.15 Ni+Co 0.20-0.80 Al 0.01 Pb 0.05 Total other elements 0.50
Low-Fuming Brass	RCuZn-C	Cu+Ag 56.0-60.0 Zn rem Sn 0.80-1.10 Mn 0.01-0.50 Fe 0.25-1.20 Si 0.04-0.15 Al 0.01 Pb 0.05 Total other elements 0.50
Nickel Brass	RBCuZn-D	Cu+Ag 46.0-50.0 Zn rem Si 0.04-0.25 Ni+Co 9.0-11.0 P 0.25 Al 0.01 Pb 0.05 Total other elements 0.50

a. Sulphur is restricted to 0.01% maximum.
b. Sulphur is restricted to 0.015% maximum.
Single values shown are maximum percentages except where otherwise noted.

MECHANICAL PROPERTIES OF COPPER ALLOYS - PLATE & SHEET FOR PRESSURE VESSELS, CONDENSERS & HEAT EXCHANGERS					
ASTM B 171					
Copper Alloy UNS No.	**Thickness, in.**	**T.S. ksi**	**Y.S.[a] ksi**	**Y.S.[b] ksi**	**% El**
C36500, C36600, C36700, C36800	2 and under	50	20	20	35
	over 2 to 3.5, incl	45	15	15	35
	over 3.5 to 5, incl	40	12	12	35
C44300, C44400, and C44500	4 and under	45	15	15	35
C46400, C46500, C46600, C46700	3 and under	50	20	20	35
	over 3 to 5, incl	50	18	18	35

MECHANICAL PROPERTIES OF COPPER ALLOYS -
PLATE & SHEET FOR PRESSURE VESSELS, CONDENSERS & HEAT EXCHANGERS (Continued)

ASTM B 171 Copper Alloy UNS No.	Thickness, in.	T.S. ksi	Y.S.[a] ksi	Y.S.[b] ksi	% El
C61300	2 and under	75	37	36	30
	over 2 to 3, incl	70	30	28	35
	over 3, to 5, incl	65	28	26	35
C61400	2 and under	70	30	28	35
	over 2 to 5, incl	65	28	26	35
C63000, C63200	2 and under	90	36	34	10
	over 2 to 3.5, incl	85	33	31	10
	over 3.5 to 5, incl	80	30	28	10
C70600	2.5 and under	40	15	15	30
	over 2.5 to 5, incl	40	15	15	30
C71500	2.5 and under	50	20	20	30
	over 2.5 to 5, incl	45	18	18	30
C72200	2.5 and under	42	16	16	35

a. Yield Strength determined as the stress producing in elongation of 0.5% under load, that is, 0.01 in. in a gage length of 2 in.
b. Yield strength at 0.2% offset, minimum.
T.S. - Tensile Strength; Y.S. - Yield Strength.
All values are minimum, unless otherwise noted.

MECHANICAL PROPERTIES OF COPPER ALLOYS -
U-BEND SEAMLESS COPPER & COPPER ALLOY HEAT EXCHANGER & CONDENSER TUBES

ASTM B 395	Temper Designation		T.S.	Y.S.	% El
Copper or Copper Alloy UNS No.	**Standard**	**Former**	**ksi, min**	**ksi, min**	**min**
C10200, C10300, C10800, C12000, C12200, C14200	H55	light drawn	36	30	---
C19200	H55	light drawn	40	35	---
C19200	O61	annealed	38	12	---
C23000	O61	annealed	40	12	---
C44300, C44400, C44500	O61	annealed	45	15	---
C60800	O61	annealed	50	19	---
C68700	O61	annealed	50	18	---
C70400	O61	annealed	38	12	---
C70400	H55	light drawn	40	30	---
C70600	O61	annealed	40	15	---
C70600	H55	light drawn	45	35	---
C71000	O61	annealed	45	16	---
C71500:	O61	annealed	52	18	---
For wall thicknesses up to 0.048 in., incl	HR50	drawn, stress-relieved	72	50	12
For wall thicknesses over 0.048 in.	HR50	drawn, stress-relieved	72	50	15
C72200	O61	annealed	45	16	---
C72200	H55	light drawn	50	45	---

T.S. - Tensile Strength; Y.S. - Yield Strength.
All values are minimum, unless otherwise noted.

MECHANICAL PROPERTIES OF GEAR BRONZE ALLOY CASTINGS[a]

ASTM B 427		Tensile Strength		Yield Strength			Hardness
Copper Alloy UNS No.	Cast Type	ksi	MPa	ksi	MPa	% El	HB[b]
C90700, C90800, C91700	Static or Centrifugally Chill Cast	50	345	28	193	12	95
C91600	Static or Centrifugally Chill Cast	45	310	25	172	10	85
C90700, C90800, C91600, C91700	Sand Cast	35	241	17	117	10	65
C92900	Sand or Chill Cast	45	310	25	172	8	75

a. The properties of a separate cast test specimen shall meet the minimum values listed. b. Brinell hardness test, 500 kg load on bar or casting, min. All values are minimum. Metric values listed are for information purposes only and are not mandatory by ASTM B 427.

MECHANICAL PROPERTIES OF COPPER ALLOY SAND CASTINGS FOR GENERAL APPLICATIONS[a]

ASTM B 584	Tensile Strength		Yield Strength[b]		% Elongation
Copper Alloy UNS No.	ksi, min	MPa, min	ksi, min	MPa, min	min
C83450	30	207	14	97	25
C83600	30	207	14	97	20
C83800	30	207	13	90	20
C84400	29	200	13	90	18
C84800	28	193	12	83	16
C85200	35	241	12	83	25
C85400	30	207	11	76	20
C85700	40	276	14	97	15
C86200	90	621	45	310	18
C86300	110	758	60	414	12
C86400	60	414	20	138	15
C86500	65	448	25	172	20
C86700	80	552	32	221	15
C87300	45	310	18	124	20

MECHANICAL PROPERTIES OF COPPER ALLOY SAND CASTINGS FOR GENERAL APPLICATIONS[a] (Continued)

ASTM B 584	Tensile Strength		Yield Strength[b]		% Elongation
Copper Alloy UNS No.	ksi, min	MPa, min	ksi, min	MPa, min	min
C87400	50	345	21	145	18
C87500	60	414	24	165	16
C87600	60	414	30	207	16
C87610	45	310	18	124	20
C90300	40	276	18	124	20
C90500	40	276	18	124	20
C92200	34	234	16	110	22
C92300	36	248	16	110	18
C92600	40	276	18	124	20
C93200	30	207	14	97	15
C93500	28	193	12	83	15
C93700	30	207	12	83	15
C93800	26	179	14	97	12
C94300	24	165	---	---	10
C94700	45	310	20	138	25
C94700(HT)	75	517	50	345	5
C94800	40	276	20	138	20
C94900	38	262	15	103	15
C96800	125	862	100	689	3
C96800 (HT)	135	931	120	821	---
C97300	30	207	15	103	8
C97600	40	276	17	117	10
C97800	50	345	22	152	10

a. Metric values listed are for information purposes only and are not mandatory by ASTM B 584.

b. Yield strength is determined as the stress producing an elongation under load of 0.5%, that is, 0.01 in. (0.254 mm) in gage length of 2 in. or 50 mm.

All values are minimum.

ASME P-No. - BASE METAL COPPER & COPPER ALLOYS						
ASME Spec.	**Condition**	**Size(s) or Thickness, in.**	**TS ksi[a]**	**UNS No.**	**Nominal Composition**	**Product Form**
P No. 31						
SB-42	061	All	30	C10200	(99.95Cu+Ag)	Seamless Pipe
	61	All	30	C12000	(99.9Cu+Ag)	Seamless Pipe
	61	All	30	C12200	(99.9Cu+Ag)	Seamless Pipe
	H80	⅛-2, incl.	45	C10200	(99.95Cu+Ag)	Seamless Pipe
	H55	2 ½-12	36	---	---	Seamless Pipe
	H80	⅛-2, incl.	45	C12000	(99.9Cu+Ag)	Seamless Pipe
	H55	2 ½-12	36	---	---	Seamless Pipe
	H80	⅛-2, incl.	45	C12200	(99.9Cu+Ag)	Seamless Pipe
	H55	2 ½ -12	36	---	---	Seamless Pipe
SB-75	Annealed	All	30	C10200	(99.95Cu+Ag)	Seamless Tube
	Annealed	All	30	C10300	(99.95Cu+P)	Seamless Tube
	Annealed	All	30	C12000	(99.9Cu+Ag)	Seamless Tube
	Annealed	All	30	C12200	(99.9Cu+Ag)	Seamless Tube
	Annealed	All	30	C14200	(99.4Cu+Ag)	Seamless Tube
	Light Drawn	All	36	C10200	(99.95Cu+Ag)	Seamless Tube
	Light Drawn	All	36	C10300	(99.95Cu+P)	Seamless Tube
	Light Drawn	All	36	C12200	(99.9Cu+Ag)	Seamless Tube
	Light Drawn	All	36	C14200	(99.4Cu+Ag)	Seamless Tube
	Hard Drawn	All	45	C10200	(99.95Cu+Ag)	Seamless Tube
	Hard Drawn	All	45	C10300	(99.95Cu+P)	Seamless Tube
	Hard Drawn	All	45	C12000	(99.9Cu+Ag)	Seamless Tube
	Hard Drawn	All	45	C12200	(99.9Cu+Ag)	Seamless Tube
	Hard Drawn	All	45	C14200	(99.4Cu+Ag)	Seamless Tube
SB-111	Annealed	---	38	C19200	(99.7Cu+Fe)	Seamless Tube
	Light Drawn	---	36	C10200	(99.95Cu+Ag)	Seamless Tube

ASME P-No. - BASE METAL COPPER & COPPER ALLOY (Continued)S

ASME Spec.	Condition	Size(s) or Thickness, in.	TS ksi[a]	UNS No.	Nominal Composition	Product Form
P No. 31 (Continued)						
SB-111	Light Drawn	---	36	C12000	(99.9Cu+Ag)	Seamless Tube
(con't)	Light Drawn	---	36	C12200	(99.9Cu+Ag)	Seamless Tube
	Light Drawn	---	36	C14200	(99.4Cu+Ag)	Seamless Tube
	Hard Drawn	---	45	C10200	(99.95Cu+Ag)	Seamless Tube
	Hard Drawn	---	45	C12000	(99.9Cu+Ag)	Seamless Tube
	Hard Drawn	---	45	C12200	(99.9Cu+Ag)	Seamless Tube
	Hard Drawn	---	45	C14200	(99.4Cu+Ag)	Seamless Tube
SB-133	060	All	30	C10200	(99.95Cu+Ag)	Rod and Bar
	---	---	---	C11000	(99.90Cu+Ag)	Rod and Bar
	---	---	---	C12000	(99.9Cu+Ag)	Rod and Bar
	---	---	---	C12200	(99.9Cu+Ag)	Rod and Bar
	---	---	---	C12500	(99.88Cu+Ag)	Rod and Bar
	---	---	---	C14200	(99.4Cu+Ag)	Rod and Bar
SB-152	Hot Rolled Tempers	---	30	C10200	(99.95Cu+Ag)	Sheet, Strip, Plate, Bar
	Hot Rolled and Annealed	---	30	C10400	(99.95Cu+Ag)	Sheet, Strip, Plate, Bar
	Hot Rolled and Annealed	---	30	C10500	(99.95Cu+Ag)	Sheet, Strip, Plate, Bar
	Hot Rolled and Annealed	---	30	C10700	(99.95Cu+Ag)	Sheet, Strip, Plate, Bar
	Hot Rolled and Annealed	---	30	C12200	(99.9Cu+Ag)	Sheet, Strip, Plate, Bar
	Hot Rolled and Annealed	---	30	C12300	(99.90Cu+Ag)	Sheet, Strip, Plate, Bar
	Hot Rolled and Annealed	---	---	C12500	(99.88Cu+Ag)	Sheet, Strip, Plate, Bar
	Hot Rolled and Annealed	---	---	C14200	(99.4Cu+Ag)	Sheet, Strip, Plate, Bar
SB-359	Annealed	---	30	C10200	(99.95Cu+Ag)	Seamless Tube
	Annealed	---	30	C12000	(99.9Cu+Ag)	Seamless Tube
	Annealed	---	30	C12200	(99.9Cu+Ag)	Seamless Tube
	Annealed	---	30	C14200	(99.4Cu+Ag)	Seamless Tube

ASME P-No. - BASE METAL COPPER & COPPER ALLOY (Continued)S						
ASME Spec.	**Condition**	**Size(s) or Thickness, in.**	**TS ksi[a]**	**UNS No.**	**Nominal Composition**	**Product Form**
P No. 31 (Continued)						
SB-359	Annealed	---	38	C19200	(99.7Cu+Fe)	Seamless Tube
(con't)	Light Drawn	---	36	C10200	(99.95Cu+Ag)	Seamless Tube
	Light Drawn	---	36	C12000	(99.9Cu+Ag)	Seamless Tube
	Light Drawn	---	36	C12200	(99.9Cu+Ag)	Seamless Tube
	Light Drawn	---	36	C14200	(99.4Cu+Ag)	Seamless Tube
SB-395	Annealed	---	38	C19200	(99.7Cu+Fe)	Seamless Tube
	Light Drawn	---	36	C10200	(99.95Cu+Ag)	Seamless Tube
	Light Drawn	---	36	C12000	(99.9Cu+Ag)	Seamless Tube
	Light Drawn	---	36	C12200	(99.9Cu+Ag)	Seamless Tube
	Light Drawn	---	36	C14200	(99.4Cu+Ag)	Seamless Tube
SB-543	As-Welded from Annealed Strip	---	32	C12200	(99.9Cu+Ag)	Welded Tube
	Light Drawn	---	36	C12200	(99.9Cu+Ag)	Welded Tube
	Hard Drawn	---	45	C12200	(99.9Cu+Ag)	Welded Tube
	As-Welded from Annealed Strip	---	45	C19400	(97.5Cu+Fe+Zn)	Welded Tube
	As-Welded and Annealed	---	45	C19400	(97.5Cu+Fe+Zn)	Welded Tube
	As-Welded from Cold Rolled Strip	---	50	C19400	(97.5Cu+Fe+Zn)	Welded Tube
P No.32						
SB-43	Annealed	All	40	C23000	(15Zn)	Seamless Pipe
SB-111	Annealed	---	40	C23000	(15Zn)	Seamless Pipe
	Annealed	---	50	C28000	(40Zn)	Seamless Tube
	Annealed	---	45	C44300	(28Zn-1Sn-0.06As)	Seamless Tube
	Annealed	---	45	C44400	(28Zn-1Sn-0.06Sb)	Seamless Tube
	Annealed	---	45	C44500	(28Zn-1Sn-0.06P)	Seamless Tube
	Annealed	---	50	C68700	(20Zn-2Al)	Seamless Tube
SB-135	Annealed	---	40	C23000	(15Zn)	Seamless Tube

ASME P-No. - BASE METAL COPPER & COPPER ALLOY (Continued)S

ASME Spec.	Condition	Size(s) or Thickness, in.	TS ksi[a]	UNS No.	Nominal Composition	Product Form
P No. 32 (Continued)						
SB-171	Annealed	≤ 2	50	C36500	(40.5Zn)	Plate
	Annealed	≥ 2-3.5	45	C36500	(40.5Zn)	Plate
	Annealed	≥ 3.5-5	40	C36500	(40.5Zn)	Plate
	Annealed	≤ 2	50	C36600	(40.5Zn-0.06As)	Plate
	Annealed	≥ 2-3.5	45	C36600	(40.5Zn-0.06As)	Plate
	Annealed	≥ 3.5-5	40	C36600	(40.5Zn-0.06As)	Plate
	Annealed	≤ 2	50	C36700	(40.5Zn-0.06Sb)	Plate
	Annealed	≥ 2-3.5	45	C36700	(40.5Zn-0.06Sb)	Plate
	Annealed	≥ 3.5-5	40	C36700	(40.5Zn-0.06Sb)	Plate
	Annealed	≤ 2	50	C36800	(40.5Zn-0.06P)	Plate
	Annealed	≥ 2-3.5	45	C36800	(40.5Zn-0.06P)	Plate
	Annealed	≥ 3.5-5	40	C36800	(40.5Zn-0.06P)	Plate
	Annealed	≤ 4	45	C44300	(28Zn-1Sn-0.06As)	Plate
	Annealed	≤ 4	45	C44400	(28Zn-1Sn-0.06Sb)	Plate
	Annealed	≤ 4	45	C44500	(28Zn-1Sn-0.06P)	Plate
	Annealed	< 5	50	C46400	(39.5Zn)	Plate
	Annealed	≤ 5	50	C46500	(39.5Zn-0.06As)	Plate
	Annealed	≤ 5	50	C46600	(39.5Zn-0.06Sb)	Plate
	Annealed	≤ 5	50	C46700	(39.5Zn-0.06P)	Plate
SB-359	Annealed	---	40	C23000	(15Zn)	Seamless Tube
	Annealed	---	45	C44300	(28Zn-1Sn-0.06As)	Seamless Tube
	Annealed	---	45	C44400	(28Zn-1Sn-0.06Sb)	Seamless Tube
	Annealed	---	45	C44500	(28Zn-1Sn-0.06P)	Seamless Tube
	Annealed	---	50	C68700	(20Zn-2Al)	Seamless Tube
SB-395	Annealed	---	40	C23000	(15Zn)	Seamless Tube

ASME P-No. - BASE METAL COPPER & COPPER ALLOY (Continued)S						
ASME Spec.	**Condition**	**Size(s) or Thickness, in.**	**TS ksi[a]**	**UNS No.**	**Nominal Composition**	**Product Form**
P No. 32 (Continued)						
SB-395	Annealed	---	45	C44300	(28Zn-1Sn-0.06As)	Seamless Tube
(con't)	Annealed	---	45	C44400	(28Zn-1Sn-0.06Sb)	Seamless Tube
	Annealed	---	45	C44500	(28Zn-1Sn-0.06P)	Seamless Tube
	Annealed	---	50	C68700	(20Zn-2Al)	Seamless Tube
SB-543	Welded and Annealed	---	40	C23000	(15Zn)	Welded Tube
	Annealed	≤ 4	45	C44300	(28Zn-1Sn-0.06As)	Plate
	As-Welded from Annealed Strip	---	42	C23000	(15Zn)	Welded Tube
	Welded and Annealed	---	45	C44300	(28Zn-1Sn-0.06As)	Welded Tube
	Welded and Annealed	---	45	C44400	(28Zn-1Sn-0.06Sb)	Welded Tube
	Welded and Annealed	---	45	C44500	(28Zn-1Sn-0.06P)	Welded Tube
	Welded and Annealed	---	50	C68700	(20Zn-2Al)	Welded Tube
P No. 33						
SB-96	Annealed	---	50	C65500	(3.3Si)	Plate and Sheet
SB-98	Soft	---	40	C65100	(1.6Si)	Rod, Bar, Shape
	Half Hard	< 2 diam.	55	C65100	(1.6Si)	Rod, Bar, Shape
	Bolt Temper	< ½	85	C65100	(1.6Si)	Rod, Bar, Shape
	Bolt Temper	½-1	75	C65100	(1.6Si)	Rod, Bar, Shape
	Bolt Temper	> 1-1 ½	75	C65100	(1.6Si)	Rod, Bar, Shape
	Soft	---	52	C65500	(3.3Si)	Rod, Bar, Shape
	Quarter Hard	---	55	C65500	(3.3Si)	Rod, Bar, Shape
	Half Hard	< 2 diam.	70	C65500	(3.3Si)	Rod, Bar, Shape
	Soft	---	52	C66100	(3.2Si)	Rod, Bar, Shape
	Quarter Hard	---	55	C66100	(3.2Si)	Rod, Bar, Shape
	Half Hard	< 2 diam.	70	C66100	(3.2Si)	Rod, Bar, Shape
SB-315	Annealed	---	50	C65500	(3.3Si)	Seamless Pipe and Tube

ASME P-No. - BASE METAL COPPER & COPPER ALLOY (Continued)S						
ASME Spec.	**Condition**	**Size(s) or Thickness, in.**	**TS ksi[a]**	**UNS No.**	**Nominal Composition**	**Product Form**
P No. 34						
SB-111	Annealed	---	38	C70400	(5.5Ni)	Seamless Tube
	Light Drawn	---	40	C70400	(5.5Ni)	Seamless Tube
	Annealed	---	40	C70600	(10Ni)	Seamless Tube
	Light Drawn	---	45	C70600	(10Ni)	Seamless Tube
	Annealed	---	45	C71000	(20Ni)	Seamless Tube
	Annealed	---	52	C71500	(30Ni)	Seamless Tube
	Drawn and Stress Relvd.	---	72	C71500	(30Ni)	Seamless Tube
	061 Annealed	---	63	C71640	(30 ½Ni-2Fe-2Mn)	Seamless Tube
	HR50 Drawn and Stress Relived.	---	81	C71640	(30 ½Ni-2Fe-2Mn)	Seamless Tube
	061 Annealed	---	45	C72200	(16 ½Ni-¾Fe-½Cr)	Seamless Tube
	H55 Light Drawn	---	50	C72200	(16 ½Ni-¾Fe-½Cr)	Seamless Tube
SB151	Soft	All	38	C70600	(10Ni)	Rod and Bar
SB-171	Annealed	≤ 2 ½	40	C70600	(10Ni)	Plate
	Annealed	≤ 2 ½	50	C71500	(30Ni)	Plate
	Annealed	Over 2 ½-5, incl.	45	C71500	(30Ni)	Plate
SB-359	Annealed	---	38	C70400	(5.5Ni)	Seamless Tube
	Annealed	---	40	C70600	(10Ni)	Seamless Tube
	Annealed	---	45	C71000	(20Ni)	Seamless Tube
	Annealed	---	52	C71500	(30Ni)	Seamless Tube
SB-369	As Cast	---	45	C96200	(10Ni-1.4Fe-0.75Nb)	Casting
SB-395	Annealed	---	40	C70600	(10Ni)	Seamless Tube
	Annealed	---	45	C71000	(20Ni)	Seamless Tube
	Annealed	---	52	C71500	(30Ni)	Seamless Tube
	Drawn and Stress Relvd.	---	72	C71500	(30Ni)	Seamless Tube
SB-466	Annealed	---	38	C70600	(10Ni)	Seamless Pipe and Tube

ASME P-No. - BASE METAL COPPER & COPPER ALLOY (Continued)S

ASME Spec.	Condition	Size(s) or Thickness, in.	TS ksi[a]	UNS No.	Nominal Composition	Product Form
P No. 34 (Continued)						
SB-466	Annealed	---	45	C71000	(20Ni)	Seamless Pipe and Tube
(con't)	Annealed	---	50	C71500	(30Ni)	Seamless Pipe and Tube
SB-467	As-Welded and Annealed	< 4 ½ O.D.	40	C70600	(10Ni)	Welded Pipe
	As-Welded and Annealed	> 4 ½ O.D.	38	C70600	(10Ni)	Welded Pipe
	As-Welded from Annealed Strip	< 4 ½ O.D.	45	C70600	(10Ni)	Welded Pipe
	As-Welded from Cold Rolled Strip	< 4 ½ O.D.	54	C70600	(10Ni)	Welded Pipe
	As-Welded and Annealed	< 4 ½ O.D.	50	C71500	(30Ni)	Welded Pipe
	As-Welded and Annealed	> 4 ½ O.D.	45	C71500	(30Ni)	Welded Pipe
SB-543	As-Welded and Annealed	---	38	C70400	(5.5Ni)	Welded Tube
	As-Welded and Annealed	---	40	C70600	(10Ni)	Welded Tube
	As-Welded from Annealed Strip	---	45	C70600	(10Ni)	Welded Tube
	As-Welded and Annealed	---	52	C71500	(30Ni)	Welded Tube
	Annealed	---	63	C71640	(30 ½Ni-2Fe-2Mn)	Welded Tube
	Lt. Cold Work	---	75	C71640	(30 ½Ni-2Fe-2Mn)	Welded Tube
P No. 35						
SB-148	As Cast	---	65	C95200	(9Al)	Casting
SB-111	Annealed	---	50	C60800	(5.8Al)	Seamless Tube
	As Cast	---	75	C95400	(11Al)	Casting
SB-150	Annealed	≤ ½	80	C61400	(7Al-2.5Fe)	Round
	Annealed	≥ ½-1	75	C61400	(7Al-2.5Fe)	Round
	Annealed	Over 1-3, incl.	70	C61400	(7Al-2.5Fe)	Round
	Annealed	All	75	C62300	(9Al-3Fe)	Rod, Bar, Shape
	Std. Strength	½-1, incl.	100	C63000	(10Al-5Ni-3Fe)	Rod, Bar
	Std. Strength	Over 1-2, incl.	90	C63000	(10Al-5Ni-3Fe)	Rod, Bar
	Std. Strength	Over 2-4, incl.	85	C63000	(10Al-5Ni-3Fe)	Rod, Bar

ASME P-No. - BASE METAL COPPER & COPPER ALLOY (Continued)S						
ASME Spec.	**Condition**	**Size(s) or Thickness, in.**	**TS ksi[a]**	**UNS No.**	**Nominal Composition**	**Product Form**
P No. 35 (Continued)						
SB-150	---	All	85	C63000	(10Al-5Ni-3Fe)	Shape
(con't)	Annealed	½ and under	90	C64200	(7Al-2Si)	Round
	Annealed	Over ½-1, incl.	85	C64200	(7Al-2Si)	Round
	Annealed	Over 1-2, incl.	80	C64200	(7Al-2Si)	Round
	Annealed	Over 2-3, incl.	75	C64200	(7Al-2Si)	Round
SB-169	Annealed	½ and under	72	C61400	(7Al-2.5Fe)	Plate, Sheet, Strip, Bar
	Annealed	Over ½-2, include.	70	C61400	(7Al-2.5Fe)	Plate, Sheet, Strip, Bar
	Annealed	Over 2-5, incl.	65	C61400	(7Al-2.5Fe)	Plate, Sheet, Strip, Bar
SB-171	Annealed	≤ 2	70	C61400	(7Al-2.5Fe)	Plate
	Annealed	Over 2-5, incl.	65	C61400	(7Al-2.5Fe)	Plate
	Annealed	≤ 2	90	C63000	(10Al-5Ni-3Fe)	Plate
	Annealed	Over 2-3.5, incl.	85	C63000	(10Al-5Ni-3Fe)	Plate
	Annealed	Over 3.5-5, incl.	80	C63000	(10Al-5Ni-3Fe)	Plate
SB-271	As Cast	---	65	C95200	(9Al)	Centrifugal Casting
	As Cast	---	75	C95400	(11Al)	Centrifugal Casting
SB-359	Annealed	---	50	C60800	(5.8Al)	Finned Tube
SB-395	Annealed	---	50	C60800	(5.8Al)	Seamless Tube
SB-505	As Cast	---	68	C95200	(9Al)	Continuous Casting

a. Minimum specified tensile strength.

ASME F No. - FILLER METAL COPPER & COPPER ALLOYS		
F No.	**ASME Spec.**	**AWS Classification**
31	SFA-5.6	ECu
31	SFA-5.7	ER Cu
31	SFA-5.27	ER Cu
32	SFA-5.6	ECuSi
32	SFA-5.7	ER CuSi-A
32	SFA-5.27	ER CuSi-A
33	SFA-5.6	ECuSn-A, ECuSn-C
33	SFA-5.7	ER CuSn-A
34	SFA-5.6	ECuNi
34	SFA-5.7	ER CuNi
34	SFA-5.30	IN 67
35	SFA-5.27	RB CuZn-A, RB CuZn-B, RB CuZn-C, RB CuZn-D
36	SFA-5.6	ECuAl-A2, ECuAl-B
	SFA-5.7	ER CuAl-A1, ER CuAl-A2, ER CuAl-A3
37	SFA-5.6	ECuNiAl, ECuMnNiAl
	SFA-5.7	ER CuNiAl, ER CuMnNiAl

INTERNATIONAL CROSS REFERENCES - COPPER & COPPER ALLOYS[a]				
USA UNS	**BRITAIN BS**	**GERMANY DIN**	**JAPAN JIS**	**INTERNATIONAL ISO**
C10200	C 103	OF-Cu	C 1020 P, C 1020 R, C 1020 B, C 1020 T	Cu-OF
C11000	C 101, C 102	E-Cu57	C 1100 P, C 1100 R, C 1100 PP, C1100 B, C 1100 T	Cu-ETP
C12000	C 106	SW-Cu	C 1201 P, C 1201 R, C 1201 B, C 1201 T	Cu-DLP
C12200	C 106	SF-Cu	C 1220 P, C 1220 R, C 1221 P, C 1221 R, C 1221 PP, C 1220 B, C 1220 T	Cu-DHP
C17000	---	CuBe1.7	C 1700 P, C 1700 R	CuBe1.7

INTERNATIONAL CROSS REFERENCES - COPPER & COPPER ALLOYS[a] (Continued)

USA UNS	BRITAIN BS	GERMANY DIN	JAPAN JIS	INTERNATIONAL ISO
C17200	CB 101	CuBe2	C 1720 P, C 1720 R, C 1720 B, C 1720 W	CuBe2
C21000	---	CuZn5	C 2100 P, C 2100 R, C 2100 W	CuZn5
C22000	CZ 101	CuZn10	C 2200 P, C 2200 R, C 2200 W, C 2200 T	CuZn10
C23000	CZ 102	CuZn15	C 2300 P, C 2300 R, C 2300 W, C 2300 T	CuZn15
C24000	CZ 103	CuZn20	C 2400 P, C 2400 R, C 2400 W	CuZn20
C26000	CZ 106	CuZn30	C 2600 P, C 2600 R, C 2600 B, C 2600 W, C 2600 T	CuZn30
C26800	CZ 107	CuZn36	C 2680 P, C 2680 R	CuZn35
C27000	---	CuZn36	C 2700 T	CuZn35
C27200	CZ 108	CuZn37	C 2720 P, C 2720 R	CuZn37
---	CZ 109	CuZn40	C 2800 B	CuZn40
C28000	CZ 119	CuZn40	C 2801 P, C 2801 R, C 2800 T	CuZn40
---	CZ 114	CuZn40Al1	C 6782 B, C 6783 B	CuZn39AlFeMn
C35600	CZ 119	CuZn36Pb1.5	C 3560 P, C 3560 R, C 3561 P, C 3561 R	CuZn36Pb3
---	CZ 120	CuZn38Pb1.5	C 3713 P, C 3713 R	CuZn38Pb2
---	CZ 123	CuZn39Pb0.5	C 3710 P, C 3710 R, C 3712 B	CuZn39Pb1
C36000	CZ 124	CuZn36Pb3	C 3601 B, C 3602 B	CuZn36Pb3
C37700	CZ 122	CuZn39Pb2	C 3771 B	CuZn39Pb2
C42500	---	---	C 4250P, C 4250 R	---
C44300	CZ 111	CuZn28Sn1	C 4430 P, C 4430 R, C 4430 T	CuZn28Sn1
C46200	CZ 122	CuZn38Sn1	C 4622 B	CuZn38Sn1
C46400	---	CuZn38Sn1	C 4640 P, C4641 B	CuZn38Sn1
C51000	PB 102	CuSn4	C 5111 P, C 5111 R, C 5102 P, C 5102 R	CuSn4, CuSn5
---	PB 103	CuSn6	C 5191 P, 5191 R	CuSn6
C52100	PB 104	CuSn8	C 5212 P, 5212 R	CuSn8
C54400	---	---	C 5441 B	CuSn4Pb4Zn3
C61400	CA 106	---	C 6140 P	CuAl8Fe3

INTERNATIONAL CROSS REFERENCES - COPPER & COPPER ALLOYS[a] (Continued)				
USA UNS	BRITAIN BS	GERMANY DIN	JAPAN JIS	INTERNATIONAL ISO
---	CA 103	CuAl8Fe3	C 6161 P	CuAl10Fe3
---	CA 105	CuAl10Ni5Fe4	C 6280 P	CuAl10Ni5Fe4
C61900	CA 104	CuAl10Fe3Mn2	C 6191 B	CuAl10Fe3
C62400	---	CuAl10Fe3Mn2	C 6241 B	CuAl10Fe3
C63000	CA 104	CuAl10Ni5Fe4	C 6301 P	CuAl10Ni5Fe4
C68700	CZ 110	CuZn20Al2	C 6870 T, C 6871 T, C 6872 T	CuZn20Al2
C70600	CN 102	CuNi10Fe1Mn	C 7060 PC 7060 T	CuNi10Fe1Mn
C71000	---	---	C 7100 T	---
C71500	CN 107	CuNi30Mn1Fe	C 7150 P, C 7150 T	CuNi30Mn1Fe
C71640	CN 108	---	C 7164 T	CuNi30Fe2Mn2
C73500	---	---	C 7351 P, C 7351 R	---
C74500	NS 103	CuNi12Zn24	C 7451 P, C 7451 R, C 7451 W	CuNi10Zn27
C75200	NS 106	CuNi18Zn20	C 7521 P, C 7521 R, C 7521 B, C 7521 W	CuNi18Zn20
---	NS 105	CuNi12Zn24	C 7541 P, C 7541 R, C 7541 B, C 7541 W	CuNi15Zn21
C77000	NS 107	CuNi18Zn27	C 7701 P, C 7701 R, C 7701 B, C 7701 W	CuNi18Zn27
C83600	LG 2	CuSn5ZnPb	BC 6, BC 6 C	CuPb5Sn5Zn5
C85400	SCB 3	CuZn33Pb	YBsC2	CuZn33Pb2
C85700	DCB 3	CuZn37Pb	YBsC 3	CuZn40Pb
C86200	HTB 3	CuZn25Al5	HBsC 3, HBsC 3 C	CuZn26Al4Fe3Mn3
C86300	HTB 3	CuZn25Al5	HBsC 4, HBsC 4 C	CuZn25Al6FE3Mn3
C86400	HTB 1	CuZn34Al1	HBsC 2, HBsC 2 C	CuZn35AlFeMn
C86500	HTB 1	CuZn35Al1	HBSC 1	CuZn35AlFeMn
C86500	---	---	HBsC 1 C	CuZn35AlFeMn
C84400	LG 1	---	BC 1, BC 1 C	---
C87400	---	---	SzBC 1	---
C87500	---	CuZn15Si4	SzBC 2	---

INTERNATIONAL CROSS REFERENCES - COPPER & COPPER ALLOYS[a] (Continued)

USA UNS	BRITAIN BS	GERMANY DIN	JAPAN JIS	INTERNATIONAL ISO
C90300	---	---	BC 2, BC 2 C	---
C90500	G 1	CuSn10Zn	BC 3, BC 3 C	CuSn10Zn2
C90700	PB 2	CuSn10	PBC 2, PBC 2 C	CuSn10
C91000	PB 2	CuSn12	PBC 3 B, PBC 3 C	CuSn12
C92200	LG 4	---	BC 7, BC 7 C	---
C93700	LB 2	CuPb10Sn	LBC 3, LBC 3 C	CuPb10Sn10
C93800	LB 1	CuPb15Sn	LBC 4, LCB 4 C	CuPb15Sn8
C95200	AB 1	CuAl10Fe	AlBC 1	CuAl9
C95400	---	CuAl9Ni	AlBC 2, AlBC 2 C	CuAl10Fe3
C95800	AB 2	CuAl10Ni	AlBC 3	CuAl10Fe5Ni5
C95700	CMA 2	---	AlBC 4	---
---	LB 5	CuPb20Sn	LBC 5	CuPb20Sn5

a. It is not practical to directly correlate the various metal designations from country to country, let alone comparing several countries and their metal designations; from the view that chemical composition may be similar, but not identical, and that manufacturing technologies may differ greatly. Consequently, the cross references made in this table are, at best, only listed as a guide to assist in finding comparable metal designations, rather than equivalent metal designations.
UNS - Unified Numbering System; BS - British Standards; DIN - Deutsches Institut für Normung; JIS - Japanese Industrial Standards; ISO - International Organization for Standardization.

Appendix

1

HARDNESS CONVERSION TABLES

APPROXIMATE HARDNESS CONVERSION NUMBERS FOR NICKEL & HIGH-NICKEL ALLOYS									
Vickers[a]	Brinell[b]	Rockwell Hardness Number[c]							
HV	HB	HRA	HRB	HRC	HRD	HRE	HRF	HRG	HRK
513	479	75.5	---	50.0	63.0	---	---	---	---
481	450	74.5	---	48.0	61.5	---	---	---	---
452	425	73.5	---	46.0	60.0	---	---	---	---
427	403	72.5	---	44.0	58.5	---	---	---	---
404	382	71.5	---	42.0	57.0	---	---	---	---
382	363	70.5	---	40.0	55.5	---	---	---	---
362	346	69.5	---	38.0	54.0	---	---	---	---
344	329	68.5	---	36.0	52.5	---	---	---	---
326	313	67.5	---	34.0	50.5	---	---	---	---
309	298	66.5	106	32.0	49.5	---	116.5	94.0	---
285	275	64.5	104	28.5	46.5	---	115.5	91.0	---
266	258	63.0	102	25.5	44.5	---	114.5	87.5	---
248	241	61.5	100	22.5	42.0	---	113.0	84.5	---
234	228	60.5	98	20.0	40.0	---	112.0	81.5	---
220	215	59.0	96	17.0	38.0	---	111.0	78.5	100.0
209	204	57.5	94	14.5	36.0	---	110.0	75.5	98.0
198	194	56.5	92	12.0	34.0	---	108.5	72.0	96.5
188	184	55.0	90	9.0	32.0	108.5	107.5	69.0	94.5
179	176	53.5	88	6.5	30.0	107.0	106.5	65.5	93.0
171	168	52.5	86	4.0	28.0	106.0	105.0	62.5	91.0
164	161	51.5	84	2.0	26.5	104.5	104.0	59.5	89.0
157	155	50.0	82	---	24.5	103.0	103.0	56.5	87.5
151	149	49.0	80	---	22.5	102.0	101.5	53.0	85.5
145	144	47.5	78	---	21.0	100.5	100.5	50.0	83.5
140	139	46.5	76	---	19.0	99.5	99.5	47.0	82.0
135	134	45.5	74	---	17.5	98.0	98.5	43.5	80.0

APPROXIMATE HARDNESS CONVERSION NUMBERS FOR NICKEL & HIGH-NICKEL ALLOYS (Continued)

Vickers[a]	Brinell[b]	Rockwell Hardness Number[c]							
HV	HB	HRA	HRB	HRC	HRD	HRE	HRF	HRG	HRK
130	129	44.0	72	---	16.0	97.0	97.0	40.5	78.0
126	125	43.0	70	---	14.5	95.5	96.0	37.5	76.5
122	121	42.0	68	---	13.0	94.5	95.0	34.5	74.5
119	118	41.0	66	---	11.5	93.0	93.5	31.0	72.5
115	114	40.0	64	---	10.0	91.5	92.5	---	71.0
112	111	39.0	62	---	8.0	90.5	91.5	---	69.0
108	108	---	60	---	---	89.0	90.0	---	67.5
106	106	---	58	---	---	88.0	89.0	---	65.5
103	103	---	56	---	---	86.5	88.0	---	63.5
100	100	---	54	---	---	85.5	87.0	---	62.0
98	98	---	52	---	---	84.0	85.5	---	60.0
95	95	---	50	---	---	83.0	84.5	---	58.0
93	93	---	48	---	---	81.5	83.5	---	56.5
91	91	---	46	---	---	80.5	82.0	---	54.5
89	89	---	44	---	---	79.0	81.0	---	52.5
87	87	---	42	---	---	78.0	80.0	---	51.0
85	85	---	40	---	---	76.5	79.0	---	49.0
83	83	---	38	---	---	75.0	77.5	---	47.0
81	81	---	36	---	---	74.0	76.5	---	45.5
79	79	---	34	---	---	72.5	75.5	---	43.5
78	78	---	32	---	---	71.5	74.0	---	42.0
77	77	---	30	---	---	70.0	73.0	---	40.0

a. Vickers Hardness Number, Vickers indenter, 1.5, 10, 30-kgf load.

b. Brinell Hardness Number, 10 mm ball, 3000 kgf load.

APPROXIMATE HARDNESS CONVERSION NUMBERS FOR NICKEL & HIGH-NICKEL ALLOYS (Continued)

c. A Scale - 60-kgf load, diamond penetrator; B Scale - 100-kgf load, 1/16 in. (1.588 mm) ball; C Scale - 150-kgf load, diamond penetrator; D Scale - 100-kgf, diamond penetrator; E Scale - 100-kgf load, 1/8 in. (3.175 mm) ball; F Scale - 60-kgf load, 1/16 in. (1.588 mm) ball; G Scale - 150-kgf load, 1/16 in. (1.588 mm) ball; K Scale - 150-kgf load, 1/8 in. (3.175 mm) ball.

Note that in Table 5 of ASTM Test Method E 10, the use of a 3000-kgf load is recommended (but not mandatory) for material in the hardness range 96 to 600 HV, and a 1500-kgf load is recommended (but not mandatory) for material in the hardness range 48 to 300 HV. These recommendations are designed to limit impression diameters to the range 2.50 to 6.0 mm. The Brinell hardness numbers in this conversion table are based on tests using a 3000-kgf load. When the 1500-kgf load is used for the softer nickel and high-nickel alloys, these conversion relationships do not apply.

APPROXIMATE HARDNESS CONVERSION NUMBERS FOR NICKEL & HIGH-NICKEL ALLOYS

Vickers[a]	Brinell[b]	Rockwell Superficial Hardness[c]					
HV	HB	HR15-N	HR30-N	HR45-N	HR15-T	HR30-T	HR45-T
513	479	85.5	68.0	54.5	---	---	---
481	450	84.5	66.5	52.5	---	---	---
452	425	83.5	64.5	50.0	---	---	---
427	403	82.5	63.0	47.5	---	---	---
404	382	81.5	61.0	45.5	---	---	---
382	363	80.5	59.5	43.0	---	---	---
362	346	79.5	58.0	41.0	---	---	---
344	329	78.5	56.0	38.5	---	---	---
326	313	77.5	54.5	36.0	---	---	---
309	298	76.5	52.5	34.0	94.5	85.5	77.0
285	275	75.0	49.5	30.0	94.0	84.5	75.0
266	258	73.5	47.0	26.5	93.0	83.0	73.0
248	241	72.0	44.5	23.0	92.5	81.5	71.0
234	228	70.5	42.0	20.0	92.0	80.5	69.0
220	215	69.0	39.5	17.0	91.0	79.0	67.0
209	204	68.0	37.5	14.0	90.5	77.5	65.0

APPROXIMATE HARDNESS CONVERSION NUMBERS FOR NICKEL & HIGH-NICKEL ALLOYS (Continued)							
Vickers[a]	Brinell[b]	Rockwell Superficial Hardness[c]					
HV	HB	HR15-N	HR30-N	HR45-N	HR15-T	HR30-T	HR45-T
198	194	66.5	35.5	11.0	89.5	76.0	63.0
188	184	65.0	32.5	7.5	89.0	75.0	61.0
179	176	64.0	30.5	5.0	88.0	73.5	59.5
171	168	62.5	28.5	2.0	87.5	72.0	57.5
164	161	61.5	26.5	-0.5	87.0	70.5	55.5
157	155	---	---	---	86.0	69.5	53.5
151	149	---	---	---	85.5	68.0	51.5
145	144	---	---	---	84.5	66.5	49.5
140	139	---	---	---	84.0	65.5	47.5
135	134	---	---	---	83.0	64.0	45.5
130	129	---	---	---	82.5	62.5	43.5
126	125	---	---	---	82.0	61.0	41.5
122	121	---	---	---	81.0	60.0	39.5
119	118	---	---	---	80.5	58.5	37.5
115	114	---	---	---	79.5	57.0	35.5
112	111	---	---	---	79.0	56.0	33.5
108	108	---	---	---	78.5	54.5	31.5
106	106	---	---	---	77.5	53.0	29.5
103	103	---	---	---	77.0	51.5	27.5
100	100	---	---	---	76.0	50.5	25.5
98	98	---	---	---	75.5	49.0	23.5
95	95	---	---	---	74.5	47.5	21.5
93	93	---	---	---	74.0	46.5	19.5
91	91	---	---	---	73.5	45.0	17.0
89	89	---	---	---	72.5	43.5	14.5
87	87	---	---	---	72.0	42.0	12.5

APPROXIMATE HARDNESS CONVERSION NUMBERS FOR NICKEL & HIGH-NICKEL ALLOYS (Continued)

Vickers[a]	Brinell[b]	Rockwell Superficial Hardness[c]					
HV	HB	HR15-N	HR30-N	HR45-N	HR15-T	HR30-T	HR45-T
85	85	---	---	---	71.0	41.0	10.0
83	83	---	---	---	70.5	39.5	7.5
81	81	---	---	---	70.0	38.0	5.5
79	79	---	---	---	69.0	36.5	3.0
78	78	---	---	---	68.5	35.5	1.0
77	77	---	---	---	67.5	34.0	-1.5

a. Vickers Hardness Number, Vickers indenter, 1.5, 10, 30-kgf load.

b. Brinell Hardness Number, 10 mm ball, 3000 kgf load. Note that in Table 5 of ASTM Test Method E 10, the use of a 3000-kgf load is recommended (but not mandatory) for material in the hardness range 96 to 600 HV, and a 1500-kgf load is recommended (but not mandatory) for material in the hardness range 48 to 300 HV. These recommendations are designed to limit impression diameters to the range 2.50 to 6.0 mm. The Brinell hardness numbers in this conversion table are based on tests using a 3000-kgf load. When the 1500-kgf load is used for the softer nickel and high-nickel alloys, these conversion relationships do not apply.

c. 15-N Scale - 15-kgf load, superficial diamond penetrator; 30-N Scale - 30-kgf load, superficial diamond penetrator; 45-N Scale - 45-kgf load, superficial diamond penetrator; 15-T Scale - 15-kgf load, 1⁄16 in. (1.588 mm) ball; 30-T Scale - 30-kgf load, 1⁄16 in. (1.588 mm) ball; 45-T Scale - 45-kgf load, 1⁄16 in. (1.588 mm) ball.

APPROXIMATE HARDNESS CONVERSION NUMBERS FOR CARTRIDGE BRASS (70% COPPER - 30% ZINC ALLOY)

Vickers	Brinell	Rockwell Superficial Hardness Number[a]				
HV	HB	HRB	HRF	HR15-T	HR30-T	HR45-T
196	169	93.5	110.0	90.0	77.5	66.0
194	167	---	109.5	---	---	65.5
192	166	93.0	---	---	77.0	65.0
190	164	92.5	109.0	---	76.5	64.5
188	162	92.0	---	89.5	---	64.0
186	161	91.5	108.5	---	76.0	63.5
184	159	91.0	---	---	75.5	63.0
182	157	90.5	108.0	89.0	---	62.5

APPROXIMATE HARDNESS CONVERSION NUMBERS FOR CARTRIDGE BRASS (70% COPPER - 30% ZINC ALLOY) (Continued)

Vickers	Brinell	Rockwell Superficial Hardness Number[a]				
HV	HB	HRB	HRF	HR15-T	HR30-T	HR45-T
180	156	90.0	107.5	---	75.0	62.0
178	154	89.0	---	---	74.5	61.5
176	152	88.5	107.0	---	---	61.0
174	150	88.0	---	88.5	74.0	60.5
172	149	87.5	106.5	---	73.5	60.0
170	147	87.0	---	---	---	59.5
168	146	86.0	106.0	88.0	73.0	59.0
166	144	85.5	---	---	72.5	58.5
164	142	85.0	105.5	---	72.0	58.0
162	141	84.0	105.0	87.5	---	57.5
160	139	83.5	---	---	71.5	56.5
158	138	83.0	104.5	---	71.0	56.0
156	136	82.0	104.0	87.0	70.5	55.5
154	135	81.5	103.5	---	70.0	54.5
152	133	80.5	103.0	---	---	54.0
150	131	80.0	---	86.5	69.5	53.5
148	129	79.0	102.5	---	69.0	53.0
146	128	78.0	102.0	---	68.5	52.5
144	126	77.5	101.5	86.0	68.0	51.5
142	124	77.0	101.0	---	67.5	51.0
140	122	76.0	100.5	85.5	67.0	50.0
138	121	75.0	100.0	---	66.5	49.0
136	120	74.5	99.5	85.0	66.0	48.0
134	118	73.5	99.0	---	65.5	47.5
132	116	73.0	98.5	84.5	65.0	46.5
130	114	72.0	98.0	84.0	64.5	45.5

APPROXIMATE HARDNESS CONVERSION NUMBERS FOR CARTRIDGE BRASS (70% COPPER - 30% ZINC ALLOY) (Continued)

Vickers	Brinell	Rockwell Superficial Hardness Number[a]				
HV	HB	HRB	HRF	HR15-T	HR30-T	HR45-T
128	113	71.0	97.5	---	63.5	45.0
126	112	70.0	97.0	83.5	63.0	44.0
124	110	69.0	96.5	---	62.5	43.0
122	108	68.0	96.0	83.0	62.0	42.0
120	106	67.0	95.5	---	61.0	41.0
118	105	66.0	95.0	82.5	60.5	40.0
116	103	65.0	94.5	82.0	60.0	39.0
114	101	64.0	94.0	81.5	59.5	38.0
112	99	63.0	93.0	81.0	58.5	37.0
110	97	62.0	92.6	80.5	58.0	35.5
108	95	61.0	92.0	---	57.0	34.5
106	94	59.5	91.2	80.0	56.0	33.0
104	92	58.0	90.5	79.5	55.0	32.0
102	90	57.0	89.8	79.0	54.5	30.5
100	88	56.0	89.0	78.5	53.5	29.5
98	86	54.0	88.0	78.0	52.5	28.0
96	85	53.0	87.2	77.5	51.5	26.5
94	83	51.0	86.3	77.0	50.5	24.5
92	82	49.5	85.4	76.5	49.0	23.0
90	80	47.5	84.4	75.5	48.0	21.0
88	79	46.0	83.5	75.0	47.0	19.0
86	77	44.0	82.3	74.5	45.5	17.0
84	76	42.0	81.2	73.5	44.0	14.5
82	74	40.0	80.0	73.0	43.0	12.5
80	72	37.5	78.6	72.0	41.0	10.0
78	70	35.0	77.4	71.5	39.5	7.5

APPROXIMATE HARDNESS CONVERSION NUMBERS FOR CARTRIDGE BRASS (70% COPPER - 30% ZINC ALLOY) (Continued)

Vickers	Brinell	Rockwell Superficial Hardness Number[a]				
HV	HB	HRB	HRF	HR15-T	HR30-T	HR45-T
76	68	32.5	76.0	70.5	38.0	4.5
74	66	30.0	74.8	70.0	36.0	1.0
72	64	27.5	73.2	69.0	34.0	---
70	63	24.5	71.8	68.0	32.0	---
68	62	21.5	70.0	67.0	30.0	---
66	61	18.5	68.5	66.0	28.0	---
64	59	15.5	66.8	65.0	25.5	---
62	57	12.5	65.0	63.5	23.0	---
60	55	10.0	62.5	62.5	---	---
58	53	---	61.0	61.0	18.0	---
56	52	---	58.8	60.0	15.0	---
54	50	---	56.5	58.5	12.0	---
52	48	---	53.5	57.0	---	---
50	47	---	50.5	55.5	---	---
49	46	---	49.0	54.5	---	---
48	45	---	47.0	53.5	---	---
47	44	---	45.0	---	---	---
46	43	---	43.0	---	---	---
45	42	---	40.0	---	---	---

a. B Scale - 100-kgf load, 1/16 in. (1.588 mm) ball; F Scale - 60-kgf load, 1/16 in. (1.588 mm) ball; 15-T Scale - 15-kgf load, 1/16 in. (1.588 mm) ball; 30-T Scale - 30-kgf load, 1/16 in. (1.588 mm) ball; 45-T Scale - 45-kgf load, 1/16 in. (1.588 mm) ball.

APPROXIMATE HARDNESS CONVERSION NUMBERS FOR COPPER, No. 102 to 142 Inclusive								
Vickers		Knoop		Rockwell Hardness Superficial Number				
HV 1-kgf	HV 100-gf	HK 1-kgf	HK 500-gf	HR15-T[a]	HR15-T[b]	HR15-T[c]	HR30-T[c]	HR45-T[c]
130	127.0	138.7	133.8	---	85.0	---	69.5	49.0
128	125.2	126.8	132.1	83.0	84.5	87.0	68.5	48.0
126	123.6	134.9	130.4	---	84.0	---	67.5	46.5
124	121.9	133.0	128.7	82.5	83.5	86.0	66.5	45.0
122	121.1	131.0	127.0	---	83.0	85.5	66.0	44.0
120	118.5	129.0	125.2	82.0	82.5	---	65.0	42.5
118	116.8	127.1	123.5	81.5	---	85.0	64.0	41.0
116	115.0	125.1	121.7	---	82.0	---	63.0	40.0
114	113.5	123.2	119.9	81.0	81.5	84.5	62.0	38.5
112	111.8	121.4	118.1	80.5	81.0	---	61.0	37.0
110	109.9	119.5	116.3	80.0	---	84.0	60.0	36.0
108	108.3	117.5	114.5	---	80.5	83.5	59.0	34.5
106	106.6	115.6	112.6	79.5	80.0	---	58.0	33.0
104	104.9	113.5	110.1	79.0	79.5	83.0	57.0	32.0
102	103.2	111.5	108.0	78.5	79.0	82.5	56.0	30.0
100	101.5	109.4	106.0	78.0	78.0	82.0	55.0	28.5
98	99.8	107.3	104.0	77.5	77.5	81.0	53.5	26.5
96	98.0	105.3	102.1	77.0	77.0	80.5	52.0	25.5
94	96.4	103.2	100.0	76.5	76.5	80.0	51.0	23.0
92	94.7	101.0	98.0	76.0	75.5	79.0	49.0	21.0
90	93.0	98.9	96.0	75.5	75.0	78.0	47.5	19.0
88	91.2	96.9	94.0	75.0	74.5	77.0	46.0	16.5
86	89.7	95.5	92.0	74.5	73.5	76.0	44.0	14.0
84	87.9	92.3	90.0	74.0	73.0	75.0	43.0	12.0
82	86.1	90.1	87.9	73.5	72.0	74.5	41.0	9.5
80	84.5	87.9	86.0	72.5	71.0	73.5	39.5	7.0

APPROXIMATE HARDNESS CONVERSION NUMBERS FOR COPPER, No. 102 to 142 Inclusive (Continued)								
Vickers		Knoop		Rockwell Hardness Superficial Number				
HV 1-kgf	HV 100-gf	HK 1-kgf	HK 500-gf	HR15-T[a]	HR15-T[b]	HR15-T[c]	HR30-T[c]	HR45-T[c]
78	82.8	85.7	84.0	72.0	70.0	72.5	36.0	2.0
76	81.0	83.5	81.9	71.5	69.5	71.5	36.0	2.0
74	79.2	81.1	79.9	71.0	68.5	70.0	34.0	---
72	77.6	78.9	78.7	70.0	67.5	69.0	32.0	---
70	75.8	76.8	76.6	69.5	66.5	67.5	30.0	---
68	74.3	74.1	74.4	69.0	65.5	66.0	28.0	---
66	72.6	71.9	71.9	68.0	64.5	64.5	25.5	---
64	70.9	69.5	70.0	67.5	63.5	63.5	23.5	---
62	69.1	67.0	67.9	66.5	62.0	61.0	21.0	---
60	67.5	64.6	65.9	66.0	61.0	59.0	18.0	---
58	65.8	62.0	63.8	65.0	60.0	57.0	15.5	---
56	64.0	59.8	61.8	64.5	58.5	55.0	13.0	---
54	62.3	57.4	59.5	63.5	57.5	53.0	10.0	---
52	60.7	55.0	57.2	63.0	56.0	51.5	7.5	---
50	58.9	52.8	55.0	62.0	55.0	49.5	4.5	---
48	57.3	50.3	52.7	61.0	53.5	47.5	1.5	---
46	55.8	48.0	50.2	60.5	52.0	45.0	---	---
44	53.9	45.9	47.8	59.5	51.0	43.0	---	---
42	52.2	43.7	45.2	58.5	49.5	41.0	---	---
40	51.3	40.2	42.8	57.5	48.0	38.5	---	---

a. Strip 0.010 in. (0.25 mm).

b. Strip 0.020 in. (0.51 mm).

c. Strip 0.040 in. (1 mm) and greater. Rockwell superficial hardness: 15-T Scale - 15-kgf load, 1⁄16 in. (1.588 mm) ball; 30-T Scale - 30-kgf load, 1⁄16 in. (1.588 mm) ball; 45-T Scale - 45-kgf load, 1⁄16 in. (1.588 mm) ball.

APPROXIMATE HARDNESS CONVERSION NUMBERS FOR COPPER, No. 102 to 142 Inclusive							
Vickers		Knoop		Brinell		Rockwell[c]	
HV 1-kgf	HV 100-gf	HK 1-kgf	HK 500-gf	HB 500-kgf[a]	HB 20-kgf[b]	HRB	HRF
130	127.0	138.7	133.8	---	119.0	67.0	99.0
128	125.2	126.8	132.1	---	117.5	66.0	98.0
126	123.6	134.9	130.4	120.0	115.0	65.0	97.0
124	121.9	133.0	128.7	117.5	113.0	64.0	96.0
122	121.1	131.0	127.0	115.0	111.0	62.5	95.5
120	118.5	129.0	125.2	112.0	109.0	61.0	95.0
118	116.8	127.1	123.5	110.0	107.5	59.5	94.0
116	115.0	125.1	121.7	107.0	105.5	58.5	93.0
114	113.5	123.2	119.9	105.0	103.5	57.0	92.5
112	111.8	121.4	118.1	102.0	102.0	55.0	91.5
110	109.9	119.5	116.3	99.5	100.0	53.5	91.0
108	108.3	117.5	114.5	97.0	98.0	52.0	90.5
106	106.6	115.6	112.6	94.5	96.0	50.0	89.5
104	104.9	113.5	110.1	92.0	94.0	48.0	88.5
102	103.2	111.5	108.0	89.5	92.0	46.5	87.5
100	101.5	109.4	106.0	87.0	90.0	44.5	87.0
98	99.8	107.3	104.0	84.5	88.0	42.0	85.5
96	98.0	105.3	102.1	82.0	86.5	40.0	84.5
94	96.4	103.2	100.0	79.5	85.0	38.0	83.0
92	94.7	101.0	98.0	77.0	83.0	35.5	82.0
90	93.0	98.9	96.0	74.5	81.0	33.0	81.0
88	91.2	96.9	94.0	---	79.0	30.5	79.5
86	89.7	95.5	92.0	---	77.0	28.0	78.0
84	87.9	92.3	90.0	---	75.0	25.5	76.5
82	86.1	90.1	87.9	---	73.0	23.0	74.5
80	84.5	87.9	86.0	---	71.5	20.0	73.0

APPROXIMATE HARDNESS CONVERSION NUMBERS FOR COPPER, No. 102 to 142 Inclusive (Continued)

Vickers		Knoop		Brinell		Rockwell[c]	
HV 1-kgf	HV 100-gf	HK 1-kgf	HK 500-gf	HB 500-kgf[a]	HB 20-kgf[b]	HRB	HRF
78	82.8	85.7	84.0	---	69.5	17.0	71.0
76	81.0	83.5	81.9	---	67.5	14.5	69.0
74	79.2	81.1	79.9	---	66.0	11.5	67.5
72	77.6	78.9	78.7	---	64.0	8.5	66.0
70	75.8	76.8	76.6	---	62.0	5.0	64.0
68	74.3	74.1	74.4	---	60.5	2.0	62.0
66	72.6	71.9	71.9	---	58.5	---	60.0
64	70.9	69.5	70.0	---	57.0	---	58.0
62	69.1	67.0	67.9	---	55.0	---	56.0
60	67.5	64.6	65.9	---	53.0	---	54.0
58	65.8	62.0	63.8	---	51.5	---	51.5
56	64.0	59.8	61.8	---	49.5	---	49.0
54	62.3	57.4	59.5	---	48.0	---	47.0
52	60.7	55.0	57.2	---	46.5	---	44.0
50	58.9	52.8	55.0	---	44.5	---	41.5
48	57.3	50.3	52.7	---	42.0	---	39.0
46	55.8	48.0	50.2	---	41.0	---	36.0
44	53.9	45.9	47.8	---	---	---	33.5
42	52.2	43.7	45.2	---	---	---	30.5
40	51.3	40.2	42.8	---	---	---	28.0

a. Strip 0.080 in. (2.03 mm), 10 mm diameter ball.
b. Strip 0.040 in. (1 mm), 2 mm ball.
c. Strip 0.040 in. (1 mm) and greater. Rockwell hardness: HRB - 100-kgf load, 1⁄16 in. (1.588 mm) ball; HRF - 60-kgf load, 1⁄16 in. (1.588 mm) ball.

APPROXIMATE HARDNESS CONVERSION NUMBERS FOR WROUGHT ALUMINUM PRODUCTS

Brinell	Vickers	Rockwell[a]			Rockwell Superficial[b]		
HB 500-kgf	HV 15-kgf	HRB	HRE	HRH	HR15-T	HR30-T	HR15-H
160	189	91	---	---	89	77	95
155	183	90	---	---	89	76	95
150	177	89	---	---	89	75	94
145	171	87	---	---	88	74	94
140	165	86	---	---	88	73	94
135	159	84	---	---	87	71	93
130	153	81	---	---	87	70	93
125	147	79	---	---	86	68	92
120	141	76	101	---	86	67	92
115	135	72	100	---	86	65	91
110	129	69	99	---	85	63	91
105	123	65	98	---	84	61	91
100	117	60	---	---	83	59	90
95	111	56	96	---	82	57	90
90	105	51	94	108	81	54	89
85	98	46	91	107	80	52	89
80	92	40	88	106	78	50	88
75	86	34	84	104	76	47	87
70	80	28	80	102	74	44	86
65	74	---	75	100	72	---	85
60	68	---	70	97	70	---	83
55	62	---	65	94	67	---	82
50	56	---	59	91	64	---	80
45	50	---	53	87	62	---	79
40	44	---	46	83	59	---	77

APPROXIMATE HARDNESS CONVERSION NUMBERS FOR WROUGHT ALUMINUM PRODUCTS (Continued)

a. Rockwell hardness: B Scale - 100-kgf load, 1/16 in. (1.588 mm) ball; E Scale - 100-kgf load, 1/8 in. (3.175 mm) ball; H Scale - 60-kgf 1/8 in. (3.175 mm) ball.

b. Rockwell superficial hardness: 15-T Scale - 15-kgf load, 1/16 in. (1.588 mm) ball; 30-T Scale - 30-kgf load, 1/16 in. (1.588 mm) ball; 15-H Scale - 15-kgf load, 1/8 in. (3.175 mm) ball.

Appendix

2

METRIC CONVERSIONS

METRIC CONVERSION FACTORS

To Convert From	To	Multiply By	To Convert From	To	Multiply By
Angle			**Mass per unit length**		
degree	rad	1.745 329 E -02	lb/ft	kg/m	1.488 164 E + 00
Area			lb/ft	kg/m	1.785 797 E + 01
in.^{2}	mm^2	6.451 600 E + 02	**Mass per unit time**		
in.^{2}	cm^2	6.451 600 E + 00	lb/h	kg/s	1.259 979 E - 04
in.^{2}	m^2	6.451 600 E - 04	lb/min	kg/s	7.559 873 E - 03
ft^2	m^2	9.290 304 E - 02	lb/s	kg/s	4.535 924 E - 01
Bending moment or torque			**Mass per unit volume (includes density)**		
lbf - in.	N - m	1.129 848 E - 01	g/cm^3	kg/m^3	1.000 000 E + 03
lbf - ft	N - m	1.355 818 E + 00	lb/ft^3	g/cm^3	1.601 846 E - 02
kgf - m	N - m	9.806 650 E + 00	lb/ft^3	kg/m^3	1.601 846 E + 01
ozf - in.	N-m	7.061 552 E - 03	lb/in.^{3}	g/cm^3	2.767 990 E + 01
Bending moment or torque per unit length			lb/in.^{3}	kg/m^3	2.767 990 E + 04
lbf - in./in.	N - m/m	4.448 222 E + 00	**Power**		
lbf - ft/in.	N - m/m	5.337 866 E + 01	Btu/s	kW	1.055 056 E + 00
Corrosion rate			Btu/min	kW	1.758 426 E - 02
mils/y	mm/yr	2.540 000 E - 02	Btu/h	W	2.928 751 E - 01
mils/y	µ/yr	2.540 000 E + 01	erg/s	W	1.000 000 E - 07
Current density			ft - lbf/s	W	1.355 818 E + 00
A/in.^{2}	A/cm^2	1.550 003 E - 01	ft - lbf/min	W	2.259 697 E - 02
A/in.^{2}	A/mm^2	1.550 003 E - 03	ft - lbf/h	W	3.766 161 E - 04
A/ft^2	A/m^2	1.076 400 E + 01	hp (550 ft - lbf/s)	kW	7.456 999 E - 01
Electricity and magnetism			hp (electric)	kW	7.460 000 E - 01
gauss	T	1.000 000 E - 04			

METRIC CONVERSION FACTORS (Continued)

To Convert From	To	Multiply By	To Convert From	To	Multiply By
Electricity and magnetism (Continued)			**Power density**		
maxwell	μWb	1.000 000 E - 02	W/in.^{2}	W/m^2	1.550 003 E + 03
mho	S	1.000 000 E + 00	**Pressure (fluid)**		
Oersted	A/m	7.957 700 E + 01	atm (standard)	Pa	1.013 250 E + 05
Ω - cm	Ω - m	1.000 000 E - 02	bar	Pa	1.000 000 E + 05
Ω circular - mil/ft	μΩ - m	1.662 426 E - 03	in. Hg (32°F)	Pa	3.386 380 E + 03
Energy (impact other)			in. Hg (60°F)	Pa	3.376 850 E + 03
ft - lbf	J	1.355 818 E + 00	lbf/in.^{2} (psi)	Pa	6.894 757 E + 03
Btu (thermochemical)	J	1.054 350 E + 03	torr (mm Hg, 0°C)	Pa	1.333 220 E + 02
cal (thermochemical)	J	4.184 000 E + 00	**Specific heat**		
kW - h	J	3.600 000 E + 06	Btu/lb - °F	J/kg - K	4.186 800 E + 03
W - h	J	3.600 000 E + 03	cal/g - °C	J/kg - K	4.186 800 E + 03
Flow rate			**Stress (force per unit area)**		
ft^3/h	L/min	4.719 475 E - 01	tonf/in.^{2} (tsi)	MPa	1.378 951 E + 01
ft^3/min	L/min	2.831 000 E + 01	kgf/mm^2	MPa	9.806 650 E + 00
gal/h	L/min	6.309 020 E - 02	ksi	MPa	6.894 757 E + 00
gal/min	L/min	3.785 412 E + 00	lbf/in.^{2} (psi)	MPa	6.894 757 E - 03
Force			MN/m^2	MPa	1.000 000 E + 00
lbf	N	4.448 222 E + 00	**Temperature**		
kip (1000 lbf)	N	4.448 222 E + 03	°F	°C	5/9 (°F - 32)
tonf	kN	8.896 443 E + 00	R	K	5/9
kgf	N	9.806 650 E + 00	**Temperature interval**		
			°F	°C	5/9
Force per unit length			**Thermal conductivity**		
lbf/ft	N/m	1.459 390 E + 01	Btu - in./s - ft^2 - °F	W/m - K	5.192 204 E + 02
lbf/in.	N/m	1.751 268 E + 02	Btu/ft - h - °F	W/m - K	1.730 735 E + 00

METRIC CONVERSION FACTORS (Continued)

To Convert From	To	Multiply By
Fracture toughness		
ksi √in.	MPa √m	1.098 800 E + 00
Heat content		
Btu/lb	kJ/kg	2.326 000 E + 00
cal/g	kJ/kg	4.186 800 E + 00
Heat input		
J/in.	J/m	3.937 008 E + 01
kJ/in.	kJ/m	3.937 008 E + 01
Length		
A	nm	1.000 000 E - 01
µin.	µm	2.540 000 E - 02
mil	µm	2.540 000 E + 01
in.	mm	2.540 000 E + 01
in.	cm	2.540 000 E + 00
ft	m	3.048 000 E - 01
yd	m	9.144 000 E -01
mile	km	1.609 300 E + 00
Mass		
oz	kg	2.834 952 E - 02
lb	kg	4.535 924 E - 01
ton (short 2000 lb)	kg	9.071 847 E + 02
ton (short 2000 lb)	kg x 10^3	9.071 847 E - 01
ton (long 2240 lb)	kg	1.016 047 E + 03
kg x 10^3 = 1 metric ton		

To Convert From	To	Multiply By
Thermal conductivity (Continued)		
Btu - in./h . ft^2 - °F	W/m - K	1.442 279 E - 01
cal/cm - s - °C	W/m - K	4.184 000 E + 02
Thermal expansion		
in./in. - °C	m/m - K	1.000 000 E + 00
in./in. - °F	m/m - K	1.800 000 E + 00
Velocity		
ft/h	m/s	8.466 667 E - 05
ft/min	m/s	5.080 000 E - 03
ft/s	m/s	3.048 000 E - 01
in./s	m/s	2.540 000 E - 02
km/h	m/s	2.777 778 E - 01
mph	km/h	1.609 344 E + 00
Velocity of rotation		
rev/min (rpm)	rad/s	1.047 164 E - 01
rev/s	rad/s	6.283 185 E + 00
Viscosity		
poise	Pa - s	1.000 000 E - 01
stokes	m^2/S	1.000 000 E - 04
ft^2/s	m^2/s	9.290 304 E - 02
$in.^2/s$	mm^2/s	6.451 600 E + 02
Volume		
$in.^3$	m^3	1.638 706 E - 05
ft^3	m^3	2.831 685 E - 02
fluid oz	m^3	2.957 353 E - 05
gal (U.S. liquid)	m^3	3.785 412 E - 03

METRIC CONVERSION FACTORS (Continued)

To Convert From	To	Multiply By	To Convert From	To	Multiply By
Mass per unit area			**Volume per unit time**		
oz/in.^{2}	kg/m^2	4.395 000 E + 01	ft^3/min	m^3/S	4.719 474 E - 04
oz/ft^2	kg/m^2	3.051 517 E - 01	ft^3/S	m^3/s	2.831 685 E - 02
oz/yd^2	kg/m^2	3.390 575 E - 02	in.^{3}/min	m^3/S	2.731 177 E - 07
lb/ft^2	kg/m^2	4.882 428 E + 00	**Wavelength**		
			A	nm	1.000 000 E - 01

THE GREEK ALPHABET

Α, α - Alpha	Ι, ι - Iota	Ρ, ρ - Rho
Β, β - Beta	Κ, κ - Kappa	Σ, σ - Sigma
Γ, γ - Gamma	Λ, λ - Lambda	Τ, τ - Tau
Δ, δ - Delta	Μ, μ - Mu	Υ, υ - Upsilon
Ε, ε - Epsilon	Ν, ν - Nu	Φ, ϕ - Phi
Ζ, ζ - Zeta	Ξ, ξ - Xi	Χ, χ - Chi
Η, η - Eta	Ο, ο - Omicron	Ψ, ψ - Psi
Θ, θ - Theta	Π, π - Pi	Ω, ω - Omega

SI PREFIXES

Prefix	Symbol	Exponential Expression	Multiplication Factor
exa	E	10^{18}	1 000 000 000 000 000 000
peta	P	10^{15}	1 000 000 000 000 000
tera	T	10^{12}	1 000 000 000 000
giga	G	10^{9}	1 000 000 000
mega	M	10^{6}	1 000 000
kilo	k	10^{3}	1 000
hecto	h	10^{2}	100
deka	da	10^{1}	10
Base Unit	---	10^{0}	1
deci	d	10^{-1}	0.1
centi	c	10^{-2}	0.01
milli	m	10^{-3}	0.001
micro		10^{-6}	0.000 001
nano	n	10^{-9}	0.000 000 001
pico	p	10^{-12}	0.000 000 000 001
femto	f	10^{-15}	0.000 000 000 000 001
atto	a	10^{-18}	0.000 000 000 000 000 001

Appendix 3

IMPERIAL UNITS

DECIMAL EQUIVALENT OF FRACTIONS		
Fraction (in.)	**Decimal (in.)**	**Millimeter (mm)**
1/64	0.015 625	0.396 875
1/32	0.031 250	0.793 750
3/64	0.046 875	1.190 625
1/16	0.062 500	1.587 500
5/64	0.078 125	1.984 375
3/32	0.093 750	2.381 250
7/64	0.109 375	2.778 125
1/8	0.125 000	3.175 000
9/64	0.140 625	3.571 875
5/32	0.156 250	3.968 750
11/64	0.171 875	4.365 625
3/16	0.187 500	4.762 500
13/64	0.203 125	5.159 375
7/32	0.218 750	5.556 250
15/64	0.234 375	5.953 125
1/4	0.250 000	6.350 000
17/64	0.265 625	6.746 875
9/32	0.281 250	7.143 750
19/64	0.296 875	7.540 625
15/16	0.312 500	7.937 500
21/64	0.328 125	8.334 375
11/32	0.343 750	8.731 250
23/64	0.359 375	9.128 125
3/8	0.375 000	9.525 000
25/64	0.390 625	9.921 875
13/32	0.406 250	10.318 750
27/64	0.421 875	10.715 625

DECIMAL EQUIVALENT OF FRACTIONS (Continued)

Fraction (in.)	Decimal (in.)	Millimeter (mm)
7/16	0.437 500	11.112 500
29/64	0.453 125	11.509 375
15/32	0.468 750	11.906 250
31/64	0.484 375	12.303 125
1/2	0.500 000	12.700 000
33/64	0.515 625	13.096 875
17/32	0.531 250	13.493 750
35/64	0.546 875	13.890 625
9/16	0.562 500	14.287 500
37/64	0.578 125	14.684 375
19/32	0.593 750	15.081 250
39/64	0.609 375	15.478 125
5/8	0.625 000	15.875 000
41/64	0.640 625	16.271 875
21/32	0.656 250	16.668 750
43/64	0.671 875	17.065 625
11/16	0.687 500	17.462 500
45/64	0.703 125	17.859 375
23/32	0.718 750	18.256 250
47/64	0.734 375	18.653 125
3/4	0.750 000	19.050 000
49/64	0.765 625	19.446 875
25/32	0.781 250	19.843 750
51/64	0.796 875	20.240 625
13/16	0.812 500	20.637 500
53/64	0.828 125	21.034 375
27/32	0.843 750	21.431 250

DECIMAL EQUIVALENT OF FRACTIONS (Continued)

Fraction (in.)	Decimal (in.)	Millimeter (mm)
55/64	0.859 375	21.828 125
7/8	0.875 000	22.225 000
57/64	0.890 625	22.621 875
29/32	0.906 250	23.018 750
59/64	0.921 875	23.415 625
15/16	0.937 500	23.812 500
61/64	0.953 125	24.209 375
31/12	0.968 750	24.606 250
63/64	0.984 375	25.003 125
1	1.000 000	25.400 000

Appendix

4

PIPE DIMENSIONS

DIMENSIONS OF WELDED AND SEAMLESS PIPE

Nominal Pipe Size	Outside Diameter	Nominal Wall Thickness (in) For						
		Schedule 5S	Schedule 10S	Schedule 10	Schedule 20	Schedule 30	Schedule Standard	Schedule 40
⅛	0.405	---	0.049	---	---	---	0.068	0.068
¼	0.540	---	0.065	---	---	---	0.088	0.088
⅜	0.675	---	0.065	---	---	---	0.091	0.091
½	0.840	0.065	0.083	---	---	---	0.109	0.109
¾	1.050	0.065	0.083	---	---	---	0.113	0.113
1	1.315	0.065	0.109	---	---	---	0.133	0.133
1 ¼	1.660	0.065	0.109	---	---	---	0.140	0.140
1 ½	1.900	0.065	0.109	---	---	---	0.145	0.145
2	2.375	0.065	0.109	---	---	---	0.154	0.154
2 ½	2.875	0.083	0.120	---	---	---	0.203	0.203
3	3.5	0.083	0.120	---	---	---	0.216	0.216
3 ½	4.0	0.083	0.120	---	---	---	0.226	0.226
4	4.5	0.083	0.120	---	---	---	0.237	0.237
5	5.563	0.109	0.134	---	---	---	0.258	0.258
6	6.625	0.109	0.134	---	---	---	0.280	0.280
8	8.625	0.109	0.148	---	0.250	0.277	0.322	0.322
10	10.75	0.134	0.165	---	0.250	0.307	0.365	0.365
12	12.75	0.156	0.180	---	0.250	0.330	0.375	0.406
14 O.D.	14.0	0.156	0.188	0.250	0.312	0.375	0.375	0.438
16 O.D.	16.0	0.165	0.188	0.250	0.312	0.375	0.375	0.500
18 O.D.	18.0	0.165	0.188	0.250	0.312	0.438	0.375	0.562
20 O.D.	20.0	0.188	0.218	0.250	0.375	0.500	0.375	0.594
22 O.D.	22.0	0.188	0.218	0.250	0.375	0.500	0.375	---
24 O.D.	24.0	0.218	0.250	0.250	0.375	0.562	0.375	0.688
26 O.D.	26.0	---	---	0.312	0.500	---	0.375	---

DIMENSIONS OF WELDED AND SEAMLESS PIPE (Continued)

Nominal Pipe Size	Outside Diameter	Nominal Wall Thickness (in) For						
		Schedule 5S	Schedule 10S	Schedule 10	Schedule 20	Schedule 30	Schedule Standard	Schedule 40
28 O.D.	28.0	---	---	0.312	0.500	0.625	0.375	---
30 O.D.	30.0	0.250	0.312	0.312	0.500	0.625	0.375	---
32 O.D.	32.0	---	---	0.312	0.500	0.625	0.375	0.688
34 O.D.	34.0	---	---	0.312	0.500	0.625	0.375	0.688
36 O.D.	36.0	---	---	0.312	0.500	0.625	0.375	0.750
42 O.D.	42.0	---	---	---	---	---	0.375	---

See next page for heavier wall thicknesses; all units are inches.

DIMENSIONS OF WELDED AND SEAMLESS PIPE

Nominal Pipe Size	Outside Diameter	Nominal Wall Thickness (in) For							
		Schedule 60	Extra Strong	Schedule 80	Schedule 100	Schedule 120	Schedule 140	Schedule 160	XX Strong
⅛	0.405	---	0.095	0.095	---	---	---	---	---
¼	0.540	---	0.119	0.119	---	---	---	---	---
⅜	0.675	---	0.126	0.126	---	---	---	---	---
½	0.840	---	0.147	0.147	---	---	---	0.188	0.294
¾	1.050	---	0.154	0.154	---	---	---	0.219	0.308
1	1.315	---	0.179	0.179	---	---	---	0.250	0.358
1 ¼	1.660	---	0.191	0.191	---	---	---	0.250	0.382
1 ½	1.900	---	0.200	0.200	---	---	---	0.281	0.400
2	2.375	---	0.218	0.218	---	---	---	0.344	0.436
2 ½	2.875	---	0.276	0.276	---	---	---	0.375	0.552
3	3.5	---	0.300	0.300	---	---	---	0.438	0.600
3 ½	4.0	---	0.318	0.318	---	---	---	---	---

DIMENSIONS OF WELDED AND SEAMLESS PIPE (Continued)

Nominal Pipe Size	Outside Diameter	Nominal Wall Thickness (in) For							
		Schedule 60	Extra Strong	Schedule 80	Schedule 100	Schedule 120	Schedule 140	Schedule 160	XX Strong
4	4.5	---	0.337	0.337	---	0.438	---	0.531	0.674
5	5.563	---	0.375	0.375	---	0.500	---	0.625	0.750
6	6.625	---	0.432	0.432	---	0.562	---	0.719	0.864
8	8.625	0.406	0.500	0.500	0.594	0.719	0.812	0.906	0.875
10	10.75	0.500	0.500	0.594	0.719	0.844	1.000	1.125	1.000
12	12.75	0.562	0.500	0.688	0.844	1.000	1.125	1.312	1.000
14 O.D.	14.0	0.594	0.500	0.750	0.938	1.094	1.250	1.406	---
16 O.D.	16.0	0.656	0.500	0.844	1.031	1.219	1.438	1.594	---
18 O.D.	18.0	0.750	0.500	0.938	1.156	1.375	1.562	1.781	---
20 O.D.	20.0	0.812	0.500	1.031	1.281	1.500	1.750	1.969	---
22 O.D.	22.0	0.875	0.500	1.125	1.375	1.625	1.875	2.125	---
24 O.D.	24.0	0.969	0.500	1.218	1.531	1.812	2.062	2.344	---
26 O.D.	26.0	---	0.500	---	---	---	---	---	---
28 O.D.	28.0	---	0.500	---	---	---	---	---	---
30 O.D.	30.0	---	0.500	---	---	---	---	---	---
32 O.D.	32.0	---	0.500	---	---	---	---	---	---
34 O.D.	34.0	---	0.500	---	---	---	---	---	---
36 O.D.	36.0	---	0.500	---	---	---	---	---	---
42 O.D.	42.0	---	0.500	---	---	---	---	---	---

All units are inches.

Appendix

5

PERIODIC TABLE

Periodic Table of the Elements

← Metals → ← Nonmetals →

Transition Elements: groups IIIb–VIII, Ib

Key to chart: Atomic Number → **50**; Symbol → Sn; Atomic Weight → 118.69; −18−18−4 ← Electron Configuration; +2 +4 ← Oxidation States

Iª	IIª	IIIᵇ	IVᵇ	Vᵇ	VIᵇ	VIIᵇ	VIII			Iᵇ	IIᵇ	IIIª	IVª	Vª	VIª	VIIª	0	Orbit
1 H 1.0079 1 +1 −1																	**2** He 4.00260 2 0	K
3 Li 6.939 2−1 +1	**4** Be 9.0122 2−2 +2											**5** B 10.81 2−3 +3	**6** C 12.011 2−4 +2 +4 −4	**7** N 14.0067 2−5 +1 +2 +3 +4 +5 −1 −2 −3	**8** O 15.9994 2−6 −2	**9** F 18.998403 2−7 −1	**10** Ne 10.17_9 2−8 0	K−L
11 Na 22.9898 2−8−1 +1	**12** Mg 24.312 2−8−2 +2											**13** Al 26.98154 2−8−3 +3	**14** Si 28.08 2−8−4 +2 +4 −4	**15** P 30.97376 2−8−5 +3 +5 −3	**16** S 32.06 2−8−6 +4 +6 −2	**17** Cl 35.453 2−8−7 +1 +5 +7 −1	**18** Ar 39.948 2−8−8 0	K−L−M
19 K 39.09 −8−8−1 +1	**20** Ca 40.08 −8−8−2 +2	**21** Sc 44.9559 −8−9−2 +3	**22** Ti 47.9 −8−10−2 +2 +3 +4	**23** V 50.941 −8−11−2 +2 +3 +4 +5	**24** Cr 51.996 −8−13−1 +2 +3 +6	**25** Mn 54.9380 −8−13−2 +2 +3 +4 +7	**26** Fe 55.847 −8−14−2 +2 +3	**27** Co 58.9332 −8−15−2 +2 +3	**28** Ni 58.71 −8−16−2 +2 +3	**29** Cu 63.54 −8−18−1 +1 +2	**30** Zn 65.38 −8−18−2 +2	**31** Ga 69.72 −8−18−3 +3	**32** Ge 72.59 −8−18−4 +2 +4	**33** As 74.9216 −8−18−5 +3 +5 −3	**34** Se 78.96 −8−18−6 +4 +6 −2	**35** Br 79.904 −8−18−7 +1 +5 −1	**36** Kr 83.80 −8−18−8 0	−L−M−N
37 Rb 85.467 −18−8−1 +1	**38** Sr 87.62 −18−8−2 +2	**39** Y 88.9059 −18−9−2 +3	**40** Zr 91.22 −18−10−2 +4	**41** Nb 92.9064 −18−12−1 +3 +5	**42** Mo 95.94 −18−13−1 +6	**43** Tc 98.9062 −18−13−2 +4 +6 +7	**44** Ru 101.07 −18−15−1 +3	**45** Rh 102.905 −18−16−1 +3	**46** Pd 106.4 −18−18−0 +2 +4	**47** Ag 107.868 −18−18−1 +1	**48** Cd 112.40 −18−18−2 +2	**49** In 114.82 −18−18−3 +3	**50** Sn 118.69 −18−18−4 +2 +4	**51** Sb 121.75 −18−18−5 +3 +5 −1	**52** Te 127.60 −18−18−6 +4 +6 −2	**53** I 126.9045 −18−18−7 +1 +5 +7 −3	**54** Xe 131.30 −18−18−8 0	−M−N−O
55 Cs 132.9054 −18−8−1 +1	**56** Ba 137.3 −18−8−2 +2	**57*** La 138.9055 −18−9−2 +3	**72** Hf 178.49 −32−10−2 +4	**73** Ta 180.948 −32−11−2 +5	**74** W 183.85 −32−12−2 +6	**75** Re 186.207 −32−13−2 +4 +6 +7	**76** Os 190.2 −32−14−2 +3 +4	**77** Ir 192.9 −32−15−2 +3 +4	**78** Pt 195.09 −32−16−2 +2 +4	**79** Au 196.9665 −32−18−1 +1 +3	**80** Hg 200.59 −32−18−2 +1 +2	**81** Tl 204.37 −32−19−3 +1 +3	**82** Pb 207.19 −32−18−4 +2 +4	**83** Bi 208.980 −32−18−5 +3 +5	**84** Po (209) −32−18−6 +2 +4	**85** At (210) −32−18−7	**86** Rn (222) −32−18−8 0	−N−O−P
87 Fr (223) −18−8−1 +1	**88** Ra 226.0254 −18−8−2 +2	**89**** Ac (227) −18−9−2 +3	**104** Rf (261) −32−10−2 +4	**105** Ha (262) −32−11−2	**106** (263) −32−12−2													−O−P−Q

Appendix 6

INTERNATIONAL STANDARDS ORGANIZATIONS, TECHNICAL ASSOCIATIONS & SOCIETIES

International Standards Organizations

AENOR	Asoiación Española de Normalización y Cetificación (Spain) tel +34 1 310 48 51, fax +34 1 310 49 76
AFNOR	Association Française de Normalisation (France) tel +33 1 42 91 55 55, fax +33 1 42 91 56 56
ANSI	American National Standards Institute (USA) tel +212 642 4900, fax +212 302 1286
BSI	British Standards Institution (England) tel +44 181 996 70 00, fax +44 181 996 70 01
CSA	Canadian Standards Association (Canada) tel +416 747 4044, fax +416 747 2475
CSCE	Canadian Society for Chemical Engineers (613) 526-4652
DIN	Deutches Institut für Normung e.V. (Germany) tel +49 30 26 01 2260, fax +49 30 2601 1231
DS	Dansk Standard (Denmark) tel +45 39 77 01 01, fax +45 39 77 02 02
ELOT	Hellenic Organization for Standardization (Greece) tel +30 1 201 50 25, fax +30 1 202 07 76
IBN/BIN	Institut Belge de Normalisation/Belgisch Instituut voor Normalisatie (Belgium) tel +32 2 738 01 11, fax +32 2 733 42 64
IPQ	Instituto Português da Qualidade (Potugal) tel +351 1 294 81 00, fax +351 1 294 81 01
ISO	International Organization for Standardization (Switzerland) tel +41 22 749 01 11, fax +41 22 733 34 30
ITM	Inspection du Travail et des Mines (Luxembourg) tel +352 478 61 54, fax +352 49 14 47

International Standards Organizations (Continued)

JSA	Japanese Standards Association (Japan) tel +03 3583 8074, fax + 033582 2390
NNI	Nederlands NormalisatieiInstituut (Netherlands) tel +31 15 69 03 90, fax +31 15 69 01 90
NSAI	National Standards Authority of Ireland tel +353 1 837 01 01, fax +353 1 836 98 21
NFS	Norges Standardiseringsforbund (Norway) tel +47 22 46 60 94, fax +47 22 46 44 57
ON	Österreichisches Normungsindtitut (Austria) tel +43 1 213 00, fax +43 1 213 00 650
SA	Standards Australia tel +08 373 1540, fax +08 373 1051
SCC	Standards Council of Canada tel +800 267 8220, fax +613 995 4564
SIS	Standardiseringskommissione n i Sverige (Sweden) tel +46 8 613 52 00, fax +46 8 411 70 35
SFS	Suomen Standardisoimisliitto r.y. (Findland) tel +358 0 149 93 31, fax +358 0 146 49 25
SNV	Schweizerische Normen-Vereinigung (Switzerland) tel +41 1 254 54 54, fax +41 1 254 54 74
STRI	Technological Institute of Iceland tel +354 587 70 02, fax +354 587 74 09
UNI	Ente Nazionale Italiano di Unificazione (Italy) tel +39 2 70 02 41, fax +39 2 70 10 61 06

Technical Associations & Societies[a]

AA	The Aluminum Association (202) 862-5100
AEE	The Association of Energy Engineers (404) 447-5083
AFS	American Foundrymen's Society (312) 824-0181
AISI	Association of Iron and Steel Engineers (412) 281-6323
AIChE	American Institute of Chemical Engineers (212) 705-7338
AMEC	Advanced Materials Engineering Centre (902) 425-4500
ASEE	American Society for Engineering Education (202) 331-3500
ASM	ASM International - The Materials Information Society (800) 336-5152 or (216) 338-5151
ASME	American Society of Mechanical Engineers (212) 705-7722
ASNT	American Society for Nondestructive Testing (614) 274-6003
ASQC	American Society for Quality Control (414) 272-8575
ASTM	American Society for Testing and Materials (215) 299-5400
AWS	American Welding Society (305) 443-9353 or (800) 443-9353
CAIMF	Canadian Advanced Industrial Materials Forum (416) 798-8055

Technical Associations & Societies[a] (Continued)

CASI	Canadian Aeronautic & Space Institute (613) 234-0191
CCA	Canadian Construction Association (613) 236-9455
CCPE	Canadian Council of Professional Engineers (613) 232-2474
CCS	Canadian Ceramics Society (416) 491-2886
CDA	Copper Development Association (212) 251-7200
CEN	European Committee for Standardization +32 2 550 08 11
CIE	Canadian Institute of Energy (403) 262-6969
CIM	Canadian Institute for Mining and Metallurgy (514) 939-2710
CMA	Canadian Manufacturing Association (416) 363-7261
CNS	Canadian Nuclear Society (416) 977-6152
CPI	Canadian Plastics Institute (416) 441-3222
CPIC	Canadian Professional Information Centre (905) 624-1058
CSEE	Canadian Society of Electronic Engineers (514) 651-6710
CSME	Canadian Society of Mechanical Engineers (514) 842-8121
CSNDT	Canadian Society for Nondestructive Testing (416) 676-0785
El	Engineering Information Inc. (212) 705-7600
FED	Federal & Military Standards (215) 697-2000
IEEE	Institute of Electrical & Electronic Engineers (212) 705-7900
IES	Institute of Environmental Sciences (312) 255-1561
IIE	Institute of Industrial Engineers (404) 449-0460
IMMS	International Material Management Society (705) 525-4667
ISA	Instrument Society of America (919) 549-8411
ISA	Instrument Society of America (919) 549-8411
ISS	Iron and Steel Society (412) 776-1535
ITI	International Technology Institute (412) 795-5300
ITRI	International Tin Research Institute (614) 424-6200
MSS	Manufactures Standardization Society of Valves & Fittings Industry (703) 281-6613
MTS	Marine Technology Society (202) 775-5966
NACE	National Association of Corrosion Engineers (713) 492-0535
NAPE	National Association of Power Engineers (212) 298-0600
NAPEGG	Association of Professional Engineers, Geologists and Geophysicists of theNorthwest Territories (403) 920-4055
NiDI	Nickel Development Institute (416) 591-7999
PIA	Plastics Institute of America (201) 420-5553
RIA	Robotic Industries Association (313) 994-6088
SAE	Society of Automotive Engineers (412) 776-4841
SAME	Society of American Military Engineers (703) 549-3800
SAMPE	Society for the Advancement of Materials and Processing Engineering (818) 331-0616
SCC	Standards Council of Canada (800) 267-8220
SCTE	Society of Carbide & Tool Engineers (216) 338 5151

Technical Associations & Societies[a] (Continued)

SDCE	Society of Die Casting Engineers (312) 452-0700
SME	Society of Manufacturing Engineers (313) 271-1500
SPE	Society of Petroleum Engineers (214) 669-3377
SSIUS	Specialty Steel Industry of the United States (202) 342-8630
SSPC	Steel Structures Painting Council (412) 268-3327
STC	Society for Technical Communications (202) 737-0035
STLE	Society of Tribologists and Lubrication Engineers (312) 825-5536
TDA	Titanium Development Association (303) 443-7515
TMS	The Minerals, Metals, and Materials Society (412) 776-9000
WIC	Welding Institute of Canada (905) 257-9881

a. Telephone numbers only.

Appendix 7

TRADENAMES & TRADEMARKS

TRADEMARK	COMPANY
ALCOA	Aluminum Company of America
AL 6XN, AL 6X	Allegheny Ludlum Corp
ALUMEL	Hoskins Manufacturing Co.
ARMCO 20-45-5	Armco Steel Corp.
CARPENTER, 20Cb3, 20Mo6	Carpenter Technology Corp.
PYROMET	Carpenter Technology Corp.
CHROMEL	Hoskin Manufacturing Co.
COLMONOY	Wall Colmonoy Corp.
CREUSOT UR SB 8	Creusot-Loire
DURATHERM	Vacuumschmelze GmbH
EATONITE	Eaton Corp.
ELGILOY	Elgiloy Limited Partnership
EVERDUR	Anaconda Co.
HASTELLOY, HAYNES, 242, 214, 230	Haynes International Inc.
230W, 556, B-3, C-22, G-30, G-50	Haynes International Inc.
H-9M, HR-120, HR-160	Haynes International Inc.
HAVAR	Hamilton Technology Inc.
IN-100, IN-102, RENE 41	Cannon-Muskegon Corp.
INCO, INCOLOY, INCONEL, NIMONIC	Inco Alloys International Inc.
MAR-M Alloy	Martin-Marietta Corp.
NICHROME	Harrison Alloys
RA330, RA330-04, RA-333	Rolled Alloys Inc.
SANICRO 28	Sandvik Steel Corp.
SM-2060	Svenska Metallverken A.B.
STELLITE, TRIBALOY	Stoody Deloro Stellite Inc.
TPM	American Manufacture
UDIMET	Specialty Metals inc.
VITALIUM	Howmet Corp.
WASPALOY	United Technologies

INDEX